Plant Cell Wall Proteins and Development

Plant Cell Wall Proteins and Development

Editors

Elisabeth Jamet
Christophe Dunand

MDPI • Basel • Beijing • Wuhan • Barcelona • Belgrade • Manchester • Tokyo • Cluj • Tianjin

Editors
Elisabeth Jamet
CNRS (Centre National de la recherche Scientifique)
France

Christophe Dunand
Université de Toulouse
France

Editorial Office
MDPI
St. Alban-Anlage 66
4052 Basel, Switzerland

This is a reprint of articles from the Special Issue published online in the open access journal *International Journal of Molecular Sciences* (ISSN 1422-0067) (available at: https://www.mdpi.com/journal/ijms/special_issues/plant_cell_wall).

For citation purposes, cite each article independently as indicated on the article page online and as indicated below:

LastName, A.A.; LastName, B.B.; LastName, C.C. Article Title. *Journal Name* **Year**, *Volume Number*, Page Range.

ISBN 978-3-0365-7590-2 (Hbk)
ISBN 978-3-0365-7591-9 (PDF)

Contents

About the Editors

Elisabeth Jamet

Elisabeth Jamet gained a Ph.D. in 1985 at Strasbourg University (France) on so-called plant-pathogenesis-related proteins. She was the first to be able to purify these proteins which were considered at that time as degradation products due to their great stability. She was hired as a permanent researcher at the National Center for Scientific Research (CNRS) in 1986 at the Institute of Plant Molecular Biology (IBMP) in Strasbourg. She then developed a research project dealing with plant cell dedifferentiation through protoplasts. She focused on the role of the cell wall in protoplast recovery, a critical step to acquire the capability to divide, form a callus, and regenerate a plant. In 2001, as a research director at CNRS, she moved to the Plant Science Research Laboratory (LRSV) close to Toulouse (France) where she developed her own team focusing on the role of cell wall proteins in plant development. She has greatly contributed to the establishment of diverse strategies to characterize the plant cell wall proteome, including experimental approaches and bioinformatics treatment of mass spectrometry data. In parallel, she has contributed to the unraveling of the function of several families of cell wall proteins, notably the PAC (proline-rich, arabinogalactan proteins, conserved cysteines) domain proteins. She is presently the co-leader of the Dynamics and Evolution of Plant Cell Walls team at LRSV. She has co-authored 100 articles in international journals, and she is an editor of *Frontiers in Plant Science*, the *International Journal of Molecular Science*, and *Cells*. She has been/is a member of different scientific evaluation committees at national and international levels. She was the head of LRSV between 2011 and 2015, and she presently exercises responsibilities at Toulouse University.

Christophe Dunand

Christophe Dunand gained a Ph.D. in 1997 at Grenoble University (France) in plant biochemistry. He has demonstrated the binding and biological effects of oligosaccharides on the enzymatic activity of plant protoplast extracts. After gaining his Ph.D., he completed three post-doctoral internships. First, he completed 1 year at the University J. Fourier (Grenoble) on Arabidopsis floral development and then 18 months at the University of Connecticut on the study of cell walls. In 2000, he joined the Plant Physiology Laboratory of Geneva University as an assistant professor. Until 2008, he strongly developed bioinformatics and molecular biology approaches dedicated to plant peroxidase genes. He was hired as a full professor at the University of Toulouse in 2008 at the Plant Science Research Laboratory (LRSV) close to Toulouse (France). He then developed his own team to continue studies on plant peroxidase functions and evolution. His research interests mainly focus on the model plant Arabidopsis, and his research has taken a more integrative direction with the use of less typical models and approaches such as hepatic or aquatic plants. He is presently the co-leader of the Dynamics and Evolution of Plant Cell Walls team at LRSV. He has co-authored 90 articles in international journals, and he is an editor of *Frontiers in Plant Science*, the *International Journal of Molecular Science*, and *Cells*.

Preface to "Plant Cell Wall Proteins and Development"

This book focuses on plant cell wall proteins and their roles during development and in response to abiotic stresses. Plant cell walls are mainly composed of polymers, such as polysaccharides or lignins. Although proteins are minor components, they play crucial roles in signaling and in the dynamics of the cell wall architecture, enabling growth, differentiation, and the adaptation of plants to their changing environment. This collection of research reports and reviews published in a Special Issue of *IJMS* includes a broad range of topics and experimental approaches which illustrate recent advances in cell wall biology.

Elisabeth Jamet and Christophe Dunand
Editors

International Journal of
Molecular Sciences

MDPI

Editorial

Plant Cell Wall Proteins and Development

Elisabeth Jamet * and Christophe Dunand *

Laboratoire de Recherche en Sciences Végétales, Université de Toulouse, CNRS, UPS, 31320 Auzeville Tolosane, France

* Correspondence: jamet@lrsv.ups-tlse.fr (E.J.); dunand@lrsv.ups-tlse.fr (C.D.)

Received: 9 April 2020; Accepted: 13 April 2020; Published: 15 April 2020

Plant cell walls surround cells and provide both external protection and a means of cell-to-cell communication. They mainly comprise polymers like polysaccharides (cellulose, hemicelluloses, and pectins) and lignin in lignified secondary walls and a small amount of cell wall proteins (CWPs) [1,2]. CWPs are major players of cell wall remodeling and signaling. Cell wall proteomics, as well as numerous genetic or biochemical studies, have revealed the high diversity of CWPs, among which proteins acting on polysaccharides, proteases, oxido-reductases, lipid-related proteins, and structural proteins ([3–7]). CWPs may have enzymatic activities such as cutting/ligating polymers or processing/degrading proteins [8]. They may also contribute to the supra-molecular assembly of cell walls via protein/protein or protein/polysaccharide interactions [9–11]. Thanks to these biochemical activities, they contribute to the dynamics and functionality of cell walls. Even though much research has already been pursued to shed light on the many roles of CWPs, many functions still remain to be discovered, especially for proteins identified in cell wall proteomes with yet unknown function.

This Special Issue "Plant Cell Wall Proteins and Development" has welcomed a selection of articles in the field of cell wall biology, which were focused on cell wall proteins and their roles during development. Eight experimental articles, nine up-to-date review articles, as well as a concept article, have been published. We wish to thank all the authors for their great contribution to this unique collection of articles as well as the International Journal of Molecular Science supporting team.

The content of this Special Issue embraces several topics, all of them stressing the roles of cell wall proteins: cell wall proteomics studies on monocot species [7,12]; the role of cell wall proteins during plant development [13–15] or in response to environmental stresses [16–19]; overviews on several cell wall protein families either from green microalgae [20] or from plants, i.e., fasciclin arabinogalactan proteins (FLAs) [21,22], membrane-bound class III peroxidases (Class III Prxs) [23], pectin methylesterases inhibitors [24], DUF642 (Domain of Unknown Function 642) proteins [25], and Proline-rich, Arabinogalactan proteins, conserved Cysteines (PAC) domain-proteins [26]; and the role of fasciclin arabinogalactan proteins (FLAs) in Ca^{2+} signaling during plant morphogenesis [27,28].

For two decades, cell wall proteomics has become a powerful experimental approach and has revealed the diversity of the cell wall protein families. *Arabidopsis thaliana* has been the most studied plant species, and almost half of its expected cell wall proteome has been described so far (see *WallProtDB*, http://www.polebio.lrsv.ups-tlse.fr/WallProtDB/). The monocotyledon species have been studied more recently thanks to the sequencing of additional genomes like those of *Oryza sativa* [29], *Brachypodium distachyon* [30], and *Triticum aestivum* [31] as well as the availability of transcriptomics data as for *Saccharum* spp [32]. Calderan-Rodrigues et al. [7] provide a comparison of monocotyledon and dicotyledon cell wall proteomes and have discussed the specificities of the former. Such specificities were related to the differences between the composition and structure of monocotyledon and dicotyledon cell walls [1,33]. Also, Cherkaoui et al. [12] report on the comparison between cell wall proteomes of the endosperm, and the outer layers of the wheat grain. They reveal a strong metabolic activity in the cell wall during endosperm differentiation, whereas the accumulation of proteins was more important at an earlier stage of development in the outer layers.

As mentioned above, the cell wall composition and structure varys during development, and these changes can allow further differentiation processes. Betekhtin et al. [13] provide a fine mapping of cell wall epitopes in zygotic embryos of *B. distachyon* at a mature stage of development, including antibodies recognizing extensins and arabinogalactan proteins (AGPs), which are structural proteins involved in the cell wall architecture and proteins assumed to be involved in signaling, respectively. The plasma membrane is the interface between the cytoplasm and the cell wall. Its composition can vary locally in the domains characterized by particular lipid compositions. Kubátová et al. [15] show that two plasma membrane domains with a distinct lipid composition are located close to the Ortmannian ring, a cell wall domain-specific to trichomes. These plasma membrane domains are generated thanks to exocysts complex containing EXO70 subunits recognizing the target membrane. Cell-to-cell communication can be ensured through plasmodesmata [34]. Han et al. [14] provide a review article focusing on the cytoskeleton and on plasmodesmata-associated cell wall proteins like callose synthase and callose hydrolase, which are involved in the regulation of plasmodesmata closure.

Environmental cues induce modifications of the cell wall. In particular, nutrient availability can regulate cell wall composition. The absorption of nutrients by roots occurs through the apoplastic pathway. This pathway is blocked by the deposition of lignin and later of suberin at the level of the Casparian strips around endodermis cells in differentiated roots. In their review article, Ogden et al. [19] focus on the changes observed in the modulation of the suberization of the root endodermal walls in response to nutrient availability, showing that the plasticity of suberin accumulation is an adaptative response. They also show that the availability of nitrate or phosphorus modulates the development of lateral roots and/or of root hairs and has a direct effect on the transcription of genes encoding proteins involved in the biosynthesis of cell wall components or regulating the oxidative status in the cell wall. Wu et al. [18] focus on a few cell wall proteins playing critical roles during phosphorus deficiency such as expansins, Pro-rich proteins, oxidoreductases, and purple acid phosphatases. Abiotic stresses like flooding or temperature can also induce changes in the cell wall. Song et al. [17] show that xyloglucan endotransglycosylases/hydrolases (XTHs), which are hemicelluloses remodeling enzymes *in muro*, play roles in the regulation of stress responses to flooding. Indeed, the overexpression of the *A. thaliana AtXTH31* gene in soybean plants leads to increasing of resistance to flooding. Pinski et al. [16] observe changes in the accumulation of extensin and AGP epitopes in *B. distachyon* leaves exposed to cold and hot temperature stresses.

Cell wall proteins are mostly encoded by multigene families, which can comprise a large number of members like class III Prxs [35] or pectin methyl esterase inhibitors [36] (73 and 71 members in *A. thaliana*, respectively). Each member has its own regulatory pathway during development or upon stress, and even if the proteins of a give family share the same functional domains, subtle differences can confer different biological activities. As an example, AtPrx36 plays a particular role in mucilage release because of the timely regulation of expression of its gene during seed development, and of its anchoring in a cell wall microdomain [37]. Most cell wall protein families are conserved in the green lineage. This is illustrated in four articles of this Special Issue. Guerriero et al. [20] describe a family of green microalgal cellulases. Seifert et al. [21] show the conservation of the fasciclin 1 domain (FAS1) in all the kingdoms of life, suggesting a role in the mechanisms mediating interactions between the cells and their environment. He et al. [14] describe the evolution of FLAs which are possibly involved in signaling. Nguyen-Kim et al. [26] explore the PAC domain-proteins family possibly forming non-covalent networks with polysaccharides and *O*-glycoproteins.

Since cell wall proteins families contain many members, it is interesting to consider each of them independently to fully uncover their roles in cell wall biology. Three review articles present such overviews. Lüthje and Martinez-Cortes [23] describe the sub-family of membrane-bound class III Prxs which are located at the plasma membrane or in the tonoplast and are assumed to play roles in membrane protection or repair. Wormit and Usadel [24] give an overview of the roles of pectin methylesterase inhibitors (PMEIs). These proteins participate in the regulation of the degree of methylesterification of the pectic homogalacturonans, which in turn contributes to cell adhesion, cell wall porosity, and plasticity. Finally, Cruz-Valderrama et al. [25] propose a role for the DUF642 protein family in development and in response to environmental stresses by modulating directly, or indirectly, the degree of methylation of homogalacturonans. These proteins were first described as abundant proteins in cell wall proteomes [38] and were until recently considered proteins with unknown function.

This Special issue was also open to new concepts. Two articles by Lamport et al. [27,28] propose new roles for the arabinogalactan protein (AGP) family in root and shoot morphogenesis, as well as in phyllotaxis patterning. Such molecules are actually proteoglycans with a proportion of glycans of up to 90% [39], which are assumed to play roles in signaling. However, the molecular mechanisms underlying this function were not deciphered until recently when its role as an extracellular calcium capacitor was proposed [40].

Altogether, we believe that this Special Issue will provide a collection of articles allowing both experts and newcomers in the field to get a valuable update on plant cell wall biology. A combination of research articles, reviews, and concept articles allows a survey of several topics of interest today regarding the many roles of cell wall proteins.

References

1. Carpita, N.C.; Gibeaut, D.M. Structural models of primary cell walls in flowering plants, consistency of molecular structure with the physical properties of the walls during growth. *Plant J.* **1993**, *3*, 1–30. [CrossRef] [PubMed]
2. Zhong, R.; Ye, Z.-H. Secondary cell walls: biosynthesis, patterned deposition and transcriptional regulation. *Plant Cell. Physiol.* **2015**, *56*, 195–214. [CrossRef] [PubMed]
3. Albenne, C.; Canut, H.; Jamet, E. Plant cell wall proteomics: the leadership of *Arabidopsis thaliana*. *Front. Plant Sci.* **2013**, *4*, 111. [CrossRef] [PubMed]
4. Jamet, E.; Albenne, C.; Boudart, G.; Irshad, M.; Canut, H.; Pont-Lezica, R. Recent advances in plant cell wall proteomics. *Proteomics* **2008**, *8*, 893–908. [CrossRef] [PubMed]
5. Jamet, E.; Canut, H.; Boudart, G.; Pont-Lezica, R. Cell wall proteins: a new insight through proteomics. *Trends Plant Sci.* **2006**, *11*, 33–39. [CrossRef] [PubMed]
6. Guerra-Guimarães, L.; Pinheiro, C.; Chaves, I.; Barros, D.; Ricardo, C. Protein dynamics in the plant extracellular space. *Proteomes* **2016**, *4*, 22. [CrossRef]
7. Calderan-Rodrigues, M.J.; Guimarães Fonseca, J.; de Moraes, F.E.; Setem, L.V.; Carmanhanis Begossi, A.; Labate, C.A. Plant cell wall proteomics: A focus on monocot species, *Brachypodium distachyon*, *Saccharum*spp. and *Oryza sativa*. *Int. J. Mol. Sci.* **2019**, *20*, 1975. [CrossRef]
8. Franková, L.; Fry, S.C. Biochemistry and physiological roles of enzymes that 'cut and paste' plant cell-wall polysaccharides. *J. Exp. Bot.* **2013**, *64*, 3519–3550. [CrossRef]
9. Hijazi, M.; Roujol, D.; Nguyen-Kim, H.; del Rocio Cisneros Castillo, L.; Saland, E.; Jamet, E.; Albenne, C. Arabinogalactan protein 31 (AGP31), a putative network-forming protein in *Arabidopsis thaliana* cell walls? *Ann. Bot.* **2014**, *114*, 1087–1097. [CrossRef]
10. Tan, L.; Eberhard, S.; Pattathil, S.; Warder, C.; Glushka, J.; Yuan, C.; Hao, Z.; Zhu, X.; Avci, U.; Miller, J.S.; et al. An Arabidopsis cell wall proteoglycan consists of pectin and arabinoxylan covalently linked to an arabinogalactan protein. *Plant Cell* **2013**, *25*, 270–287. [CrossRef]
11. Cannon, M.C.; Terneus, K.; Hall, Q.; Tan, L.; Wang, Y.; Wegenhart, B.L.; Chen, L.; Lamport, D.T.; Chen, Y.; Kieliszewski, M.J. Self-assembly of the plant cell wall requires an extensin scaffold. *Proc. Natl. Acad. Sci. USA* **2008**, *105*, 2226–2231. [CrossRef] [PubMed]
12. Cherkaoui, M.; Lollier, V.; Geairon, A.; Bouder, A.; Larré, C.; Rogniaux, H.; Jamet, E.; Guillon, F.; Francin-Allami, M. Cell wall proteome of wheat grain endosperm and outer layers at two key stages of early development. *Int. J. Mol. Sci.* **2020**, *21*, 239.
13. Betekhtin, A.; Milewska-Hendel, A.; Lusinska, J.; Chajec, L.; Kurczynska, E.; Hasterok, R. Organ and tissue-specific localisation of selected cell wall epitopes in the zygotic embryo of *Brachypodium distachyon*. *Int. J. Mol. Sci.* **2018**, *19*, 725. [CrossRef] [PubMed]

14. Han, X.; Huang, L.-J.; Feng, D.; Jiang, W.; Miu, W.; Li, N. Plasmodesmata-related structural and functional proteins: The long sought-after secrets of a cytoplasmic channel in plant cell walls. *Int. J. Mol. Sci.* **2019**, *20*, 2946. [CrossRef]
15. Kubátová, Z.; Pejchar, P.; Potocký, M.; Sekereš, J.; Žárský, V.; Kulich, I. Arabidopsis trichome contains two plasma membrane domains with different lipid compositions which attract distinct EXO70 subunits . *Int. J. Mol. Sci.* **2019**, *20*, 3803. [CrossRef]
16. Pinski, A.; Betekhtin, A.; Sala, K.; Godel-Jedrychowska, K.; Kurczynska, E.; Hasterok, R. Hydroxyproline-rich glycoproteins as markers of temperature stress in the leaves of *Brachypodium distachyon*. *Int. J. Mol. Sci.* **2019**, *20*, 2571. [CrossRef]
17. Song, L.; Valliyodan, B.; Prince, S.; Wan, J.; Nguyen, H. Characterization of the *XTH* gene family: New insight to the roles in soybean flooding tolerance. *Int. J. Mol. Sci.* **2018**, *19*, 2705. [CrossRef]
18. Wu, W.; Zhu, S.; Chen, Q.; Lin, Y.; Tian, J.; Liang, C. Cell wall proteins play critical roles in plant adaptation to phosphorus deficiency. *Int. J. Mol. Sci.* **2019**, *20*, 5259. [CrossRef]
19. Ogden, M.; Hoefgen, R.; Roessner, U.; Persson, S.; Abbas, G. Feeding the walls: How does nutrient availability regulate cell wall composition? *Int. J. Mol. Sci.* **2018**, *19*, 2691. [CrossRef]
20. Guerriero, G.; Sergeant, K.; Legay, S.; Hansman, J.-F.; Cauchie, H.-M.; Ahmad, I.; Siddiqui, K. Novel insights from comparatvie *in silico* analysis of green microalgal cellulases. *Int. J. Mol. Sci.* **2018**, *19*, 1782. [CrossRef]
21. Seifert, G. Fascinating fasciclins: A surprisingly widespread family of proteins that mediate interactions between the cell exterior and the cell surface. *Int. J. Mol. Sci.* **2018**, *19*, 1628. [CrossRef] [PubMed]
22. He, J.; Zhao, H.; Cheng, Z.; Ke, Y.; Liu, J.; Ma, H. Evolution analysis of the fasciclin-like arabinogalactan proteins in plants shows variable fasciclin-AGP domain constitution. *Int. J. Mol. Sci.* **2019**, *20*, 1945. [CrossRef] [PubMed]
23. Lüthje, S.; Martinez-Cortes, T. Membrane-bound class III peroxidases: Unexpected enzymes with exciting functions. *Int. J. Mol. Sci.* **2018**, *19*, 2876. [CrossRef] [PubMed]
24. Wormit, A.; Usadel, B. The multifaceted role of pectin methylesterase inhibitors (PMEI). *Int. J. Mol. Sci.* **2018**, *19*, 2878. [CrossRef] [PubMed]
25. Cruz-Valderrama, J.; Gómez-Maqueo, X.; Salazar-Iribe, A.; Zúñiga-Sánchez, E.; Hernández-Barrera, A.; Quezada-Rodríguez, E.; Gamboa-deBuen, A. Overview of the role of cell wall DUF642 proteins in plant development. *Int. J. Mol. Sci.* **2019**, *20*, 3333. [CrossRef] [PubMed]
26. Nguyen-Kim, H.; San Clemente, H.; Laimer, J.; Lackner, P.; Gadermaier, G.; Dunand, C.; Jamet, E. The cell wall PAC (Proline-rich, Arabinogalactan Proteins, conserved Cysteines) domain-proteins are conserved in the green lineage. *Int. J. Mol. Sci.* **2020**, *21*, 2488. [CrossRef]
27. Lamport, D.; Tan, L.; Held, M.; Kieliszewski, M. Phyllotaxis turns over a new leaf—A new hypothesis. *Int. J. Mol. Sci.* **2020**, *21*, 1145. [CrossRef]
28. Lamport, D.; Tan, L.; Held, M.; Kieliszewski, M. The role of primary cell wall in plant morphogenesis. *Int. J. Mol. Sci.* **2018**, *19*, 2674. [CrossRef]
29. Yu, J.; Hu, S.; Wang, J.; Wong, G.; Li, S.; Liu, B.; Deng, Y.; Dai, L.; Zhou, Y.; Zhang, X.; et al. A draft sequence of the rice genome (*Oryza sativa* L. ssp. indica). *Science* **2002**, *296*, 79–92. [CrossRef]
30. International Brachypodium initiative. Genome sequencing and analysis of the model grass *Brachypodium distachyon*. *Nature* **2010**, *463*, 763–768. [CrossRef]
31. International Wheat Genome Sequencing Consortium (IWGSC). Shifting the limits in wheat research and breeding using a fully annotated reference genome. *Science* **2018**, *361*, eaar7191. [CrossRef] [PubMed]
32. Vettore, A.; da Silva, F.; Kemper, E.; Souza, G.; da Silva, A.; Ferro, M.; Henrique-Silva, F.; Giglioti, E.; Lemos, M.; Coutinho, L.; et al. Analysis and functional annotation of an expressed sequence tag collection for tropical crop sugarcane. *Genome Res.* **2003**, *13*, 2725–2735. [CrossRef] [PubMed]
33. Vogel, J. Unique aspects of the grass cell wall. *Curr. Opin. Plant Biol.* **2008**, *11*, 301–307. [CrossRef] [PubMed]
34. Sager, R.; JY, L. Plasmodesmata at a glance. *J. Cell Sci.* **2018**, *131*, jcs209346. [CrossRef]
35. Tognolli, M.; Penel, C.; Greppin, H.; Simon, P. Analysis and expression of the class III peroxidase large gene family in *Arabidopsis thaliana*. *Gene* **2002**, *288*, 129–138. [CrossRef]
36. Wang, M.; Yuan, D.; Gao, W.; Li, Y.; Tan, J.; Zhang, X. A comparative genome analysis of PME and PMEI families reveals the evolution of pectin metabolism in plant cell walls. *PLoS ONE* **2013**, *8*, e72082. [CrossRef]

37. Francoz, E.; Ranocha, P.; Le Ru, A.; Martinez, Y.; Fourquaux, I.; Jauneau, A.; Dunand, C.; Burlat, V. Pectin demethylesterification generates platforms that anchor peroxidases to remodel plant cell wall domains. *Dev. Cell* **2019**, *48*, 261–276. [CrossRef]
38. Vázquez-Lobo, A.; Roujol, D.; Zuñiga-Sánchez, E.; Albenne, C.; Piñero, D.; Gamboa de Buen, A.; Jamet, E. The highly conserved spermatophyte cell wall DUF642 protein family: phylogeny and first evidence of interaction with cell wall polysaccharides *in vitro*. *Mol. Phylogenet. Evol.* **2012**, *63*, 510–520. [CrossRef]
39. Seifert, G.J.; Roberts, K. The biology of arabinogalactan proteins. *Annu. Rev. Plant Biol.* **2007**, *58*, 137–161. [CrossRef]
40. Lamport, D.T.; Várnai, P. Periplasmic arabinogalactan glycoproteins act as a calcium capacitor that regulates plant growth and development. *New Phytol.* **2013**, *197*, 58–64. [CrossRef]

International Journal of
Molecular Sciences

MDPI

Review

Plant Cell Wall Proteomics: A Focus on Monocot Species, *Brachypodium distachyon*, *Saccharum* spp. and *Oryza sativa*

Maria Juliana Calderan-Rodrigues *, Juliana Guimarães Fonseca, Fabrício Edgar de Moraes, Laís Vaz Setem, Amanda Carmanhanis Begossi and Carlos Alberto Labate

Department of Genetics, Max Feffer Laboratory of Plant Genetics, "Luiz de Queiroz" College of Agriculture, University of São Paulo, CP 83, 13400-970 Piracicaba, SP, Brazil; j.g.fonseca@usp.br (J.G.F.); fabricioedgar.m@gmail.com (F.E.d.M.); lala_arnz@hotmail.com (L.V.S.); amanda.lpp@hotmail.com (A.C.B.); calabate@usp.br (C.A.L.)

* Correspondence: mariajuliana@usp.br; Tel.: +55-019-3429-43-17

Received: 24 January 2019; Accepted: 7 March 2019; Published: 23 April 2019

Abstract: Plant cell walls mostly comprise polysaccharides and proteins. The composition of monocots' primary cell walls differs from that of dicots walls with respect to the type of hemicelluloses, the reduction of pectin abundance and the presence of aromatic molecules. Cell wall proteins (CWPs) differ among plant species, and their distribution within functional classes varies according to cell types, organs, developmental stages and/or environmental conditions. In this review, we go deeper into the findings of cell wall proteomics in monocot species and make a comparative analysis of the CWPs identified, considering their predicted functions, the organs analyzed, the plant developmental stage and their possible use as targets for biofuel production. *Arabidopsis thaliana* CWPs were considered as a reference to allow comparisons among different monocots, i.e., *Brachypodium distachyon*, *Saccharum* spp. and *Oryza sativa*. Altogether, 1159 CWPs have been acknowledged, and specificities and similarities are discussed. In particular, a search for *A. thaliana* homologs of CWPs identified so far in monocots allows the definition of monocot CWPs characteristics. Finally, the analysis of monocot CWPs appears to be a powerful tool for identifying candidate proteins of interest for tailoring cell walls to increase biomass yield of transformation for second-generation biofuels production.

Keywords: plant cell wall; proteome; monocot; stiff brome; rice; sugarcane; *Brachypodium distachyon*; *Saccharum* spp.; *Oryza sativa*

---

1. Introduction

The plant cell wall confines the cell volume and serves as protection against stresses, being responsible for the plant shape, enabling trees to be several meters high. In addition to these functions, the cell wall is the most external part of the cell, and as such, interacts with the apoplast, which is also essential in virtually all cell processes, including division, expansion, differentiation [1], growth and signaling [2]. This versatility implies that the mechanisms involved in a great deal of the cell wall functions have not been completely depicted.

The plant cell wall is a dynamic structure that undergoes changes during development [3]. The cell wall is mainly composed of wall polysaccharides, such as cellulose, hemicellulose, pectin, and proteins [4]. In some cell types, lignin and other compounds might be found, as well. Carbohydrates account for around 90% of the cell wall mass, and proteins around 10% for dicots [5] and 1% for monocots [6]. Cell wall proteins (CWPs) are understood to be proteins directed towards the secretory pathway, such as structural proteins linked to the wall and those secreted into the apoplast and extracellularly [7].

Some characteristics are commonly found in classical CWPs: an N-terminal cleavage signal-peptide, responsible for guiding these proteins to the endoplasmic reticulum (ER) [8]; lack of the ER-retention C-terminal motif KDEL/HDEL, which avoids the secretion of proteins [9,10]; and the absence of a transmembrane domain that retains proteins in the plasma membrane [11]. The secretory pathway directs proteins from ER to the Golgi apparatus, where they are packed into vesicles and guided to the cell wall and the extracellular matrix.

CWPs encompass hundreds of molecules presenting distinctive roles and are related to the modification of the cell wall components, signaling, interaction with the apoplast and the plasma membrane, and with virtually all the cell wall-related processes [12,13]. CWPs are usually divided into categories with respect to their functionality, such as proteins acting on carbohydrates, oxido-reductases, proteins related to lipid metabolism, and proteases, which represent the most abundant classes [7].

Besides being divided regarding their functional roles, CWPs can also be classified according to their type of link to the cell wall, such as structural proteins, which are strongly bound to the wall components, and loosely and weakly wall-bound proteins [13]. Loosely bound proteins are not linked to the cell wall polysaccharides and are able to move unconnected in the intercellular space. Weakly bound proteins interact with the matrix by Van der Waals, hydrogen bonds or ionic forces, and are related to remodeling, signaling, interactions with the plasma membrane and defense. Strongly bound proteins can be connected by covalent bonds [14].

To increase the comprehension of cell wall-related processes, several reports have characterized plant CWPs. To accomplish this task, specific methods of CWP extraction and data analysis had to be established, since these proteins are usually lost or in a low abundance in total proteins extraction surveys [15]. The constitution and structure of the CWPs collection are diverse due to the developmental stage, organ, tissue and species of the analyzed material; thus, it is possible to make a comparison among these different treatments and relate them physiologically [7]. Initial studies have been made on *Arabidopsis thaliana* [11,16–22]. Since then, CWPs from several other species have been identified by mass spectrometry. In *A. thaliana*, about one half of the predicted existing CWPs were identified, being the most studied plant at present, with 935 proteins already identified [11,16–22]. In addition to providing an overview of proteins present in a biological context, the identification of CWPs in *A. thaliana* enabled the selection of candidates to have their structure and function further investigated.

Over the course of time, several other studies have been performed with model plants and crops: Brassica [23,24], *Zea mays* [25], *Cicer arietinum* [26], *Vitis vinifera* [27], *Oriza sativa* (rice) [28–31], *Nicotiana tabacum* [32], *Brachypodium distachyon* (stiff brome) [33–36], *Saccharum* spp. (sugarcane) [37–40], among others. Due to this functional diversity, CWPs data present differences between monocots and dicots related to the cell wall structure and composition [41], probably due to both metabolisms' specificities. As other C3-dicots, *A. thaliana* has type I walls [42], whereas grasses, such as *Saccharum* spp., *B. distachyon*, *Sorghum bicolor* and *Setaria viridis*, present C4 metabolism and type II cell walls. *O. sativa* is a C3-monocot that also possesses type II walls [42]. Characteristic differences can be predicted when comparing CWPs collection from dicots and monocots, as their cell wall structures and components are diverse. For instance, the cell wall proteome of grasses generally consists of a higher proportion of oxido-reductases, and it has been suggested that this is due to the presence of aromatic compounds in the type II primary cell walls [33]. In summary, type II walls are characterized by a low proportion of pectins and xyloglucans, and high content of glucuronoarabinoxylans and mixed linked β-D-glucan, resulting in absent woody stems and branches [42,43], in comparison to dicots. Type II cell walls also contain ferulic and coumaric acids and more complex arabinoxylans in secondary walls. Moreover, the arabinosyl side chains inside arabinoxylans may be cross-linked with lignin through feruloyl esters [6,42]. One hypothesis for the selective advantage for the appearance of type II wall plants is that they are not as high and as reinforced as, for example, tall trees, so they focus their energy on rapid growth and more efficient reproductive strategies directed to occupy habitats that are not very suitable for trees [43].

Many of the monocots have C4 metabolism, presenting a higher photosynthetic efficiency, making them commercially valuable plants. As an example, *Saccharum* spp. has been used from more than three decades ago up to the present day as a raw material for the production of food, energy and co-products in Brazil [44]. Moreover, the study of CWPs in crops presents extra challenges compared to model plants, as less genetic and molecular data is usually available.

There is a need for more information about cell wall structure, components and their roles. The identification of CWPs allows new targets for further investigation and elucidation of pathways in order to better understand cell wall functions. Knowing that more than 10% of the plant's genome is related to the cell wall biogenesis [45], and as there are fewer studies on CWPs in monocots than dicots, it is important to gather and analyze CWP data from monocots, which would provide new insights about their unique metabolism and the specificities of their cell walls.

In this review, we compiled the information on monocots cell wall proteomics till date, except for studies using special treatments or stress-related data. Thus, *B. distachyon*, *Saccharum* spp. and *O. sativa* were chosen for this review. For each species, we searched for the *A. thaliana* sequences with the highest identities (BLASTp) in order to enable a comparison among all of them. Similar and identical *A. thaliana* sequences were found in monocots, and similarity in functional classes could also be established, as well as some specificities for each species. In this review, when it is mentioned that one *A. thaliana* CWPs was identified in the repertoire of the monocots, it means that the monocot CWP sequence matched that *A. thaliana* protein after BLASTp analysis.

2. Methods of Monocots CWPs Extraction and Analysis

Different methods of extraction of CWPs have been developed over the years. One of the biggest challenges is to isolate the cell wall with minimum contamination by intracellular or membrane proteins. Thus, the subcellular fractionation before proteomic analysis can be a useful strategy for acquiring a representative extract of CWPs and reducing contamination by proteins from other organelles. Protocols involving tissue grinding and centrifugation to generate a density gradient enable the separation of fractions that are highly enriched with specific cell compartments. However, such methods will also lead to plasma membrane disruption, which may result in subsequent contamination of cell wall subfraction. On the other hand, methodologies that do not break the cell structure can be less efficient, as the extraction buffers need to come into contact with the CWPs in order to extract them [17]. Thereby, methods of CWPs extraction can be divided into destructive and non-destructive techniques.

Regarding CWP extraction, the situation is not just one-method-fits-all, but rather the species, the organ and the targeted subset of proteins have to be considered in order to choose the most adequate protocol(s). The destructive techniques used for monocot CWP extraction utilize tissue grinding followed by a growing gradient of sucrose with a low ionic strength buffer to allow the sedimentation and isolation of the cell wall while preserving the ionic bonds. This gradient is able to eliminate organelles and other molecules less dense than the cell wall polysaccharides. Then, the last step is dedicated to washing away intracellular proteins that have remained trapped in the cell wall matrix through extensive washing on a polymer net. After cell wall isolation, the destructive protein extraction usually relies on salts such as calcium and lithium chloride to successfully liberate the wall-bound proteins [46]. Calcium chloride ($CaCl_2$) has the ability to collect CWPs because acidic and neutral carbohydrates strongly chelate calcium, and thus proteins weakly bound to the walls' polysaccharides can be solubilized by $CaCl_2$ through a competition mechanism [47]. In addition, lithium chloride (LiCl) is used to extract mostly hydroxyproline-rich glycoproteins [17]. The non-destructive CWPs extraction technique used for the monocots here revised is based on vacuum infiltration of the plant samples with solutions containing the same salts used for the destructive method mentioned above, followed by centrifugation of these samples to release the extracted CWPs [17].

Regarding mass spectrometry (MS), different strategies have been used for the study of monocot CWPs. These include previous separation by 1D-polyacrylamide gel electrophoresis (1D-PAGE) prior to tryptic digestion or shotgun analysis by liquid chromatography-tandem mass spectrometry

(LC-MS/MS). Following the identification of proteins through MS and bioinformatics, data analysis is essential for subcellular and functional predictions. Table 1 summarizes the plant samples, isolation of CWPs and MS techniques used in monocot CWP studies.

B. distachyon is a model species for temperate grasses with a fully sequenced genome, in addition to being easily grown and closely related to biomass crops. In this species, nine different plant samples have been analyzed, including young and mature leaves, apical and basal internodes, seeds and seedlings at different developmental stages [33–36]. These studies could point to specificities in different organs and developmental stages analyzed, especially in glycoside hydrolases and oxido-reductases. Fewer proteins were extracted from mature organs in comparison to young ones, which was attributed to a higher level of cell wall polymers cross-linking [33]. *Saccharum* spp. is a crop of special economic interest, as it is one of the major sources of sugar and bioethanol. For this crop, ten different plant samples were surveyed, including suspension cells culture, young and mature leaves, and basal and apical internodes at different developmental stages [37–40,48]. *O. sativa* was one of the first plant species with a sequenced genome and it is an economically relevant crop worldwide. *O. sativa* CWP surveys used five plant samples, such as suspension cells culture, culture media, roots and leaves [28–31]. In the next sections, when a CWP is mentioned as identified or not in monocots or *A. thaliana*, it means that this protein was or was not identified by MS in the mentioned CWP studies, respectively. When discussing the gene sequences corresponding to these CWPs or functional studies on them, this information is specified in the text and referenced properly.

3. Functional Class Distribution in Monocots

The number of non-redundant proteins was collected from each species and sorted into functional classes [28–31,33–40,48]. Although the results are not equally comparable, since different organs and several methods of CWP extraction and MS analysis were used, we were able to retrieve information regarding the functional classes and unique proteins in each species. Currently, the highest number of identified CWPs in monocots is from *B. distachyon*, comprising 594 proteins. *Saccharum* spp. is the second, with 283, and then *O. sativa*, with 270 identified CWPs. Altogether, 1159 CWPS proteins were identified, corresponding to 466 *A. thaliana* sequences. These proteins were divided into nine functional classes, according to Jamet et al. [7]: proteins acting on carbohydrates (PACs), oxidoreductases (ORs), proteases (Ps), proteins related to lipid metabolism (LMs), proteins possibly involved in signaling (Ss), proteins with predicted interaction domains (IDs), miscellaneous proteins (Ms), proteins of unknown function (UFs) and structural proteins (SPs). The functional classification of the CWPs from the monocots can be seen in Figure 1.

Table 1. Summary of the plant samples, extraction and mass spectrometry methods used in monocot CWP studies.

Species	Plant Sample	Extraction Method	Pipeline of MS Analyses	Total of CWPs Identified
B. distachyon	young and mature leaves, apical and basal internodes [33]	destructive technique followed by salt-based extraction	separation through 1D-PAGE and LC-MS/MS	594
	young seedlings [35]	destructive technique followed by SDS and phenol-based extraction, and enrichment by ConA affinity chromatography	2D-LC-MS/MS	
	seeds [36]	destructive technique followed by salt-based extraction	separation through 1D-PAGE and LC-MS/MS	
	seeds (9, 13 and 19 days after flowering (DAF)) [34]	destructive technique followed by salt-based extraction	separation through 1D-PAGE and LC-MS/MS	
Saccharum spp.	7 days cells suspension cultures [37]	destructive technique followed by salt-based extraction	2D-LC-MS/MS	283
	2 month-old internodes [38]	destructive and non-destructive techniques followed by salt-based extractions	2D-LC-MS/MS	
	4 month-old young and mature leaves and apical and basal internodes [39]	destructive technique followed by salt-based extraction	2D-LC-MS/MS	
	7 month-old young and mature leaves [40,48]	destructive and non-destructive techniques followed by salt-based extractions	2D-LC-MS/MS	
	7 month-old apical and basal internodes [40,48]	non-destructive technique followed by salt-based extraction	2D-LC-MS/MS	
O. sativa	5 days cultures medium of cell suspension cultures [28]	non-destructive technique followed by TCAAEB-based extraction	separation through 2D-PAGE and LC-MS/MS	270
	2–3 weeks cell suspension cultures [29]	non-destructive technique followed by salt-based extraction	separation through 1D-PAGE and 2D-LC-MS/MS	
	3 week-old 4th leaf [28]	non-destructive technique followed by Tween-20, CTAB and TCAAEB-based extraction	separation through 2D-PAGE and LC-MS/MS	
	2–3 weeks culture medium of cell suspension cultures [30]	non-destructive technique followed by salt-based extraction	separation through 1D-PAGE and 2D-LC-MS/MS	
	roots [31]	non-destructive technique followed by salt-based extraction	separation through 2D-PAGE and MALDI-TOF/TOF MS	

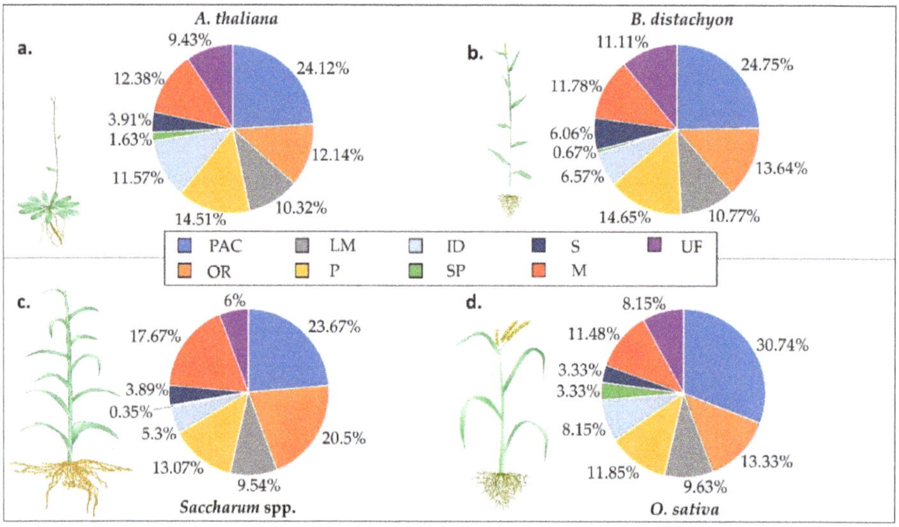

Figure 1. Percentage of CWP functional classes in the species *A. thaliana* (**a**), *B. distachyon* (**b**), *Saccharum* spp. (**c**) and *O. sativa* (**d**). Plant drawings are not to scale. Functional class abbreviations: proteins acting on carbohydrates (PACs), oxidoreductases (ORs), proteases (Ps), proteins related to lipid metabolism (LMs), proteins possibly involved in signaling (Ss), proteins with predicted interaction domains (IDs), miscellaneous proteins (Ms), proteins of unknown function (UFs) and structural proteins (SPs).

To allow some comparison among the different CWPs on monocots, we performed BLASTP with *A. thaliana* protein sequences against *B. distachyon*, *O. sativa*, and *Saccharum* spp. sequences (using default parameters) and selected the first-ranked *A. thaliana* sequence. These monocot species were selected since they have a comparable number of CWPs already identified by mass spectrometry with no particular treatment. WallProtDB [48] was the database used to retrieve these data. Only proteins predicted to be secreted were considered.

Taking the mentioned data together [28–31,33–40,48], as can be seen in Figure 1, the proportion of PACs inside the CWPs were similar among the three species, and ranged from around 20 (*Saccharum* spp.) to 30% (*O. sativa*), which is comparable to *A. thaliana* (~24%). For *O. sativa* and *B. distachyon*, the percentages of ORs were similar to *A. thaliana*—about 12 to 13%. Conversely, *Saccharum* spp. presented the highest proportion of ORs—20%. The LM percentages in all three species varied from around 9 to 11%, which was also similar to that of *A. thaliana*. Except for *B. distachyon*, the percentages of Ps in monocots were slightly lower than in *A. thaliana*. IDs were much lower in all the monocots in comparison with *A. thaliana*. In grasses, the percentage of SP was much lower than in *A. thaliana*. The percentages of Ms and UFs were higher and lower in *Saccharum* spp. in comparison with the other species, respectively.

3.1. Proteins Acting on Carbohydrates

Regarding PACs, from the 297 identified in cell wall proteomes, 111 are non-redundant *A. thaliana* sequences identified through BLASTp. Of these, 79 were also identified in *A. thaliana* CWP studies. Among the 297, several members of Glycosyl Hydrolase (GH) families 1 and 17 were identified. Other families, such as GH5, 13, 16, 32, 35 and 38, were also present. GHs are proteins involved in cell wall carbohydrates remodeling and can be regulated during development. These families are used in enzymatic cocktails for biomass degradation in second-generation ethanol production [49]. Literature data suggests that grasses present fewer GH1, GH16, GH28 and GH35 members than dicots and more

GH5, GH13, GH18 and GH51 members [33,50], which is consistent with our comparative analysis on monocots *vs. A. thaliana*, in general.

The GH1 AT1G61820 is a β-glucosidase named BGLU46, identified in *A. thaliana* and *B. distachyon* CWP studies. It has been suggested that coniferin is the substrate for BGLU46 and BGLU45, and that monolignol glucosides are a source of storage monolignols instead of direct precursors of lignin in Angiosperms. These storage monolignols could be metabolized under stress conditions, which would lead to lignin synthesis de novo [51].

The proportion of GH3 was slightly higher in monocots. Some *Saccharum* spp. GH3 have been found to be possible β-xylosidases, which catalyze the hydrolysis of xylose from xylo-oligosaccharides. In barley, a type II-wall C3, a binding site for (1 → 3, 1 → 4)-β-D-glucans was identified [52].

GH5 are more numerous in monocots (Os05g0244500, Os10g0370500, Bradi2g31690, Bradi5g13550 and Bradi5g13560), and this data may correlate with the fact that (1,3) (1,4)-β-D-glucans are their putative substrates [53]. *O. sativa* studies identified several GH13 (Os08g0473600, Os09g0457800, Os08g0473900, Os09g0457400, Os06g0713800, Os02g0765600 and Os09g0457600), all α-amylases that are capable of hydrolizing 1,4-α-glucosidic linkages [49], which is consistent with *O. sativa* endogenous metabolism. Starch is a plant carbohydrate often linked to storage organs, and its breakdown is mediated by amylases. In grains, there is an extracellular matrix enriched in starch, which is degraded by secreted enzymes, the alpha-amylases [4].

Saccharum spp. GH17, such as GH1, are probably β-glucosidases [49]. The proportion of GH17 was also higher in monocots (see Supplementary Table S1 for accession numbers). The presence of GH17 in monocots is not surprising, since, as mentioned, type II cell walls present mixed (1,3)(1,4)-β-D-glucans as the principal hemicellulose [54], which is the substrate for GH17 that displays glucan-1,3-β-glucosidase activity. These enzymes are used in enzymatic cocktails for biomass deconstruction and are considered one of the most efficient enzymes in breaking glycosidic bonds in hemicelluloses [55].

GH18 substrates are not currently known, but they could be xylanase inhibitors that promote cell wall extension or chitinases involved in cell signaling or pathogen response [56–58]. The proteins Bradi3g26840, Bradi3g26850, Os10g0416100 and SCQGRT3044B10 are all GH18 that did not match any *A. thaliana* sequence record. Interestingly, these *B. distachyon* GH18 (Bradi3g26840 and Bradi3g26850) were phylogenetically grouped into a separate clade in the GH18 family, and indeed, *A. thaliana* did not present any GH inside this clade [50]. Thus, because of their high specificity, the GH18 enzymes are excellent candidates for further investigation in order to unravel the structure of monocot cell walls. Additionally, as grasses are used as raw materials for biofuels production, the identification of GH functions could be valuable in solving bottlenecks related to biomass deconstruction.

*At*CWIN1 (AT3G13790) is a cell wall invertase from the GH32 family, and was identified in all CWP studies, except in *O. sativa*. *At*CWIN1 regulates carbon partitioning by cleaving apoplastic sucrose and helping in the process of carbon import into the cell, which is the role of membrane sugar transporters [59].

GH families 16 and 35 have more members in the *A. thaliana* proteome, and a possible explanation is that their substrates are xyloglucans and galactans, the last related to pectins. Galactan is involved in xyloglucan structure and mediates the interaction between xyloglucan and cellulose, in *A. thaliana*. Interestingly, the *At*BGAL1 (AT3G13750) from GH35 family, identified in *A. thaliana* and *B. distachyon*, act with BGAL3, identified in *A.thaliana* only, during cell elongation [60]. Whether this enzyme was poorly identified in monocots because of the low content of xyloglucans needs further investigation. The principal xyloglucan β-galactosidase in *A. thaliana*, *At*BGAL10, was also identified in all cell wall proteomes, with the exception of *O. sativa*.

More members of GH51 were found in *B. distachyon* (Bradi4g26270, Bradi1g63990, Bradi4g43710 and Bradi1g57017) than in *A. thaliana*. These GH51 are probably α-L-arabinofuranosidases [49], a type of enzyme used for monocots biomass deconstruction, as they are rich in arabinoxylans. α-L-arabinofuranosidases act on hydrolysis of α-L-arabinofuranoside in α-L-arabinosides, together with hemicellulases, resulting in hemicellulose hydrolysis [49]. It is important to mention that more studies related to substrate specificities and protein structure are necessary to establish the function of these proteins.

In addition to cellulose and hemicellulose, pectin is one of the main constituents of the primary cell wall. Wall porosity, charge density and microfibril spacing are some of the functional roles of pectin [61]. Consistently with the fact that type II-wall plants have lower pectin content, Pectin Methyl Esterases (PMEs) and Pectate Lyase-like proteins were more represented in *A. thaliana*. After the transport from the Golgi apparatus to the cell wall, pectin is partially deesterified by PMEs, exposing a carboxyl group on galacturonosyl residues and allowing the pectin to be stiffened by ionic crossbonding with calcium ions [62]. The degree of methylation impacts on the wall stiffening and access to enzymes [63].

3.2. Oxidoreductases

ORs mostly comprise several class III peroxidases (Prxs), multicopper oxidases, plastocyanins, berberine-bridge enzymes (BBEs) and blue copper-binding proteins. Prxs, part of large multigenic families, can either oxidize phenolic compounds, and consume hydrogen peroxide or generate reactive oxygen species [64]. They have been involved in several functional roles, such as cell elongation, lignin metabolism, stress responses and germination (reviewed by [64]). As Prxs are versatile proteins, they can both promote cell wall expansion or the crosslinking of its components, favoring cell wall strengthening [65]; it is difficult to establish a correlation between their higher or lower proportion and their metabolic function. Among the ORs, monocots show a slightly higher percentage of Prxs, which may be related to the fact that Poaceae presents additional groups of paralogous Prxs genes in comparison to *A. thaliana* and other dicots [66]. Sugarcane commercial varieties are highly polyploid and aneuploid plants, usually resulting from the interspecific hybridization of *Saccharum officinarum* and *S. spontaneum* [67]. The higher proportion of ORs in *Saccharum* spp., mostly Prxs, compared to the other monocots may be due the high level of ploidy of this crop, but this observation is speculative. Different Prxs were identified when using destructive and non-destructive CWPs extraction in young *Saccharum* spp. culms [38], which was suggested to be due to a differential level of pectin-binding capacity, as Prx with a Ca^{+2}-pectate binding domain would be more difficult to extract using the infiltration technique. Interestingly, *At*Prx34 (AT3G49120) and 36 (AT3G50990) were only identified in the *A. thaliana* cell wall proteome, which is consistent with the fact that the first has a putative binding site to the calcium-mediated conformation of a pectin structure [68], and the second is a promoter of pectin solubilization [69], and thus all these Prxs that have some level of relation with pectin are expected to be less numerous in type II walls.

Associated with lignin biosynthesis, *At*Prx52 (AT5G05340) [70] and the monocot homologs (Os01g0205900, Bradi3g09120, SCQSST3114C09) were present in all analyzed cell wall proteomes. Similarly, the proteins related to lignification *At*Prx72 [71] (Os01g0327100, Bradi2g40590 and SCEPRZ1011A06) and *At*Prx64 [72] (Bradi2g37060 and SCJFLR1035D05) were identified in the monocots. *At*Prx16 (Os04g0656800, Bradi1g33740 and SCJFLR1035D02) and *At*Prx53 (Os10g0109300 and Bradi1g68900) were both identified among *A. thaliana* and the monocots' CWPs. The first is related to germination [73] and the second to cell elongation inhibition and cell wall strengthening [74].

*At*Prx17 was also linked to lignin content when induced by the transcription factor AGAMOUS-LIKE15, which controls the lignification of tissues and changes the cell wall properties [75]. This could point to a conserved pathway in plants, as *At*Prx17 was identified in *A. thaliana* and in monocots´ CWP studies. Also identified in the four cell wall proteomes, *At*Prx39 was linked to higher production of reactive oxygen species that led to cold tolerance [76], evidencing the multiple roles of peroxidases in plant development. Because of the numerous functions of Prxs, more targeted studies are needed to determine the reason of their high variety in the plant cell wall, both in dicots and monocots.

BBEs catalyze the formation of berberine bridges, but in plants their function is vastly unexplored. In *A. thaliana*, it has previously been shown that some BBEs can be identified as monolignol oxidoreductases, and are related to lignin formation [77]. A much higher proportion of BBEs is found in *A. thaliana*. Phylogenetic analysis shows that not only does *A. thaliana* have more BBE members, but these enzymes also present several types of active sites, and few of type IV, which is exactly the one that is found most in grasses. In the course of evolution, in addition to expanding the amount of BBEs, it seems that the number of BBEs with active site type IV decreased and type I increased. Four BBEs only identified in *A. thaliana*—AT1G01980, AT4G20840, AT1G11770 and AT4G20830, the last with a type I active site—are able to inactivate oligogalacturonides, which is suggested to strengthen the immune response to fungal polygalacturonases [78], pointing to an evolved mechanism. Previously, it has been shown that BBEs in monocots seem to lack the catalytic and substrate coordination motifs linked to monolignol oxidoreductase activity, which was linked to lignin formation [77]. There are few functional studies in plant BBEs with type IV active sites, but fungal BBEs with this site have been related to oligosaccharide oxidation and plant immune response [79].

3.3. Proteins Related to Lipid Metabolism

Under this category, we highlight the Lipid Transfer Proteins (LTPs), Glycerophosphodiester Phosphodiesterases (GDPD)/GPDP-Like (GDPDL) and lipases GDSL. LTPs and lipases are proportionally more and less numerous in the monocot cell wall proteome, respectively [28–31,33–40,48]. LTPs are encoded by large multigenic families, which are considered to be essential to land colonization by plants, and are among the most abundant secreted proteins, but their exact in vivo role is still unclear. It has been suggested that LTPs mediate the transference and adhesion of molecules required for the composition of lipid barriers that are water-resistant, such as cutin, suberin and wax (reviewed by Edqvist et al. [80]). In the leaves of C4-metabolism plants, such as *Saccharum* spp. and *B. distachyon*, suberin surrounds the plasma membrane of bundle sheath cells, inhibiting CO_2 diffusion [4], which could be a possible explanation for increased LTPs in these species. Accordingly, in *Saccharum* spp., some LTPs were only identified in leaves. LTPs were also associated with lipid deposition for cell expansion, as their transcripts were differentially expressed in maize elongating internodes in comparison to non-elongating ones [81]. Curiously, homologs of LTP12 were identified in *B. distachyon* (Bradi4g25750) and *O. sativa* (Os12g0115100), but not in *A. thaliana* (AT3G51590), where it was thought to be pollen-specific [82]. Monocot homologs of AT5G01870 were unique to the suspension cell culture or young plants, which could be an indication that it is related to growing tissues. In *B. distachyon* and *Saccharum* spp., AT5G01870 homologs seem to be organ-specific, and are found only in leaves.

AtLTP3 was identified in all four cell wall proteomes analyzed. This protein negatively regulates plant defense mechanisms through the regulation of the antagonism between abscisic and salicylic acids, as it is induced by the former. LTP3 is proposed to be a disease-related marker, and it is also thought that LTP3 and 4 show some level of redundancy in plant immunity [83], but curiously, LTP4 was not identified in the monocots CWPs, and whether this overlay occurs in grasses could be an interesting research topic. Identified in all four species, the AtGDPDL3 (AT4G26690) is linked with lipid rafts in root-hair tip growth, suggesting that root hairs could be used as a model to study lipid rafts in plant development [84].

3.4. Proteases

Essentially, Ps break peptide bonds and control several relevant plant processes, such as protein transport, activity and half-lives [85], being generally divided into aspartyl (Asp), serine, cysteine, metallo and threonine proteases. The proportions of Asp Ps seem to be slightly higher in monocots. Phylogenetic analyses indicate that S8, C1A and A1 plant proteases functions were established even before the evolutive divergence of monocots and eudicots. This is corroborated by the conservation patterns of intron/exon arrangements and phylogeny analysis from monocots and dicots [85]. Accordingly, most of the proteases identified in the CWP studies from the monocots showed the highest identities with accessions also identified in the *A. thaliana* cell wall proteome (see Supplementary Tables S1–S4). However, only targeted functional analysis would reveal whether Ps display specific activities in monocots.

The senescence-associated subtilisin (AT3G14067) is a serine protease identified in all four species cell wall proteomes (Bradi2g51440/Bradi3g57140/Bradi3g57130, SCJFRT2057F03/SCRFHR1007E04, Os02g0779200), which confirms the fact that orthologs were identified in monocots crops [86]. Subtilisin was associated with the regulation of abscisic acid (ABA) signaling and drought tolerance [87], probably through a conserved mechanism between dicots and monocots, given the relevant role it plays.

The cysteine protease papain-like AtSAG2 (also named AALP) (AT5G60360) has been associated with senescence and necrotic cell death [88], as its expression increases along with leaf development. Reasonably, AtSAG2 was mostly identified in mature organs in *A. thaliana* and in the monocot cell wall proteomes, pointing to its use as a senescence marker in monocots.

3.5. Proteins with Interacting Domains

IDs encompass Pectin Methyl Esterase Inhibitors (PMEIs), proteins with leucine-rich repeat (LRR) and Lysm domains, protease inhibitors such as cystatins, Bowman-Birk inhibitors, lectins and jacalins. PMEIs were proportionally less numerous in monocots, as might be expected for pectin-poor type II walls. Interestingly, Hocq et al. [89] showed that *At*PMEI9 is a strong inhibitor of *At*PME3, which were both identified in *A. thaliana* and *B. distachyon* cell wall proteomes. It has been suggested that PMEIs diverge with respect to their ability to bind PME at different pHs, which results in varied modulation of the pectin structure. Additionally, they act in pairs formed by the enzyme plus its inhibitor, an interaction mode more tightly controlled by the structural determinants from the inhibitor than the enzyme [89].

O. sativa showed a higher proportion of proteins with Lysm domains, which was associated with both plant immunity and symbiotic interactions in this species [90]. All Bowman-Birk serine protease inhibitors (BBIs) were only identified in members of Poaceae and Fabaceae families [91], and thus they have only been found in *Saccharum* spp. (SCJFLR1013A04 and SCRUFL3062D08), *B. distachyon* (Bradi2g24810, Bradi1g03510, Bradi2g01920 and Bradi2g24820) and *O. sativa* (Os01g0132000) cell wall proteomes. Protease inhibitors regulate protease activities, and BBIs, in particular, display an essential role in defense mechanisms directed towards protection against pathogens and pests [91]. Another type of protease inhibitor, *At*Cys6 homologs were identified in *B. distachyon* (Bradi2g52670) and *Saccharum* spp. (SCEPLR1051C09), which are associated with abiotic stresses and nucleic acid degradation [92].

Overall, the LRR-containing domain is conserved throughout evolution in the plants, displaying activity in the innate immune system through the sensing of pathogen-associated molecular patterns [93]. *At*PGIP1 (an LRR-domain protein) and its monocots corresponding accessions were identified in cell wall proteomes. *At*PGIP1 was associated with reduced damage caused by infection of a root nematode by inducing plant camalexin and indole-glucosinolate pathways [94]. The levels of the transcripts of another protein classified as ID, *At*Cys-5, are increased upon nematode infection [95] and ABA [96]. This cystatin was also identified in all four species' CWP surveys. Perhaps both proteins could be part of the conserved defense mechanisms against nematodes in dicots and grasses.

3.6. Proteins Possibly Related to Signaling

This class of CWPs is composed by fasciclin-like arabinogalactans (FLAs), leucine-rich repeat receptor protein kinases (LRR-RKs) (actually trans-membrane proteins), and COBRA-like proteins (COBLs), among others. The proportion of proteins from the S class is similar in all four species, being higher in *B. distachyon* (see Figure 1). In proportion, FLAs seem to be more numerous in the cell wall proteomes of monocots. However, phylogenetic analyses showed that *A. thaliana* has 21 and *O. sativa* 15 FLA genes with conserved functions [97]. FLAs are related to cell-to-cell adhesion, mechanical strength for secondary cell walls and cellulose biosynthesis [98], in addition to elasticity [99]. Identified in all four species, FLA1 is supposed to act on the lateral root and shoot formation in tissue culture [100], and two *B. distachyon* FLAs (Bradi2g00220 and Bradi4g33490) were only identified in internodes, indicating that they could display organ-specific activities. In contrast, COBLs were only identified in *A. thaliana* CWPs, with the exception of one COBL in *O. sativa* (Os10g0497700). These are glycosylphosphatidylinositol-anchored specific plant proteins, and are associated with cell expansion and cellulose level of crystallinity, predominantly in elongating tissues [101]. More members of COBL are found in dicots, as several duplications occurred after the separation of dicots and monocots during evolution [102]. Recently, it has been demonstrated that a *Sorghum bicolor* COBL protein is linked to cellulose biosynthesis in the secondary wall, affecting plant mechanical strength [103], and this provided evidence of a cellulose-related role for this protein family in grasses. Overall, the LRR domain-containing proteins are conserved in both dicots and monocots [104].

3.7. Miscellaneous

Proteins with diverse functions are grouped under this category. Invariably, in several studies on CWPs, these consist of dirigent proteins, germins, thaumatins, gibberellic acid-stimulated proteins, purple acid phosphatases, phosphate-induced (phi) proteins, aldose epimerases, carbonic anhydrases, metallophosphoesterases, ribonucleases, pathogenesis-related proteins, low-molecular-weight cysteine-rich proteins and strictosidine synthases. Identified in all four species, *At*PAP10 has been proven to be transcriptionally regulated by MYB-CC factors, which control plant responses to inorganic phosphate starvation [105]. A germin protein (AT1G72610, and the corresponding Os08g0460000 and Bradi3g37680) seems to be leaf-specific, as it was mostly identified in this organ. Another protein from the same family (AT1G18970 and the corresponding Os03g0804500) was found to be auxin-responsive [106].

Dirigent proteins have more members in *B. distachyon* and *Saccharum* spp. CWP surveys. The family of dirigent proteins is possibly linked to lignin polymerization [107], and its higher content may be associated with the presence of aromatic molecules such as ferulic acids in some type II primary walls [4].

Nucleoside phosphatases were only identified in *Saccharum* spp. One of them, SCCCRZ1C01H06, is an apyrase associated with calcium signaling and has been suggested to be a messenger for sucrose accumulation [108], which is consistent with this species' high sugar content.

3.8. Structural Proteins

The reduced number of SPs in CWP studies on monocots is probably due their low levels in type II walls [6]. They are usually covalently linked to the wall, and thus present extra difficulty with respect to extraction. Extensins (EXTs), Proline- and Glycine-Rich Proteins are the most represented structural protein families in *A. thaliana* CWPs. In monocots, 10 leucine-rich repeat extensins (LRXs) have been identified altogether. EXTs are basic cell wall glycoproteins, rich in hydroxyproline residues with alternating hydrophilic and hydrophobic motifs [109], and have been associated with cell wall strengthening after different stresses and with pectin to create a coacervate that may serve as a template for cell wall deposition [110], reinforcing their role in type I cell expansion. Intriguingly, they are one of the most relevant families of CWPs in C3. In addition, specifically in *B. distachyon*, EXTs are essential to embryo regeneration and germination [111]. Three (Bradi2g05080, Bradi3g03370 and Bradi2g42477) of the four LRXs found in *B. distachyon* CWP studies [33–36] were also identified in the bioinformatic work of Liu et al. [112]. In a previous analysis, eight LRX genes were identified in *O. sativa* genome and were thought to form two distinct clades for vegetative and reproductive organs, which could reflect adaptations to different cell wall types. In addition to their role in cell expansion, it is suggested that LRXs might act on cell differentiation [113].

3.9. Proteins of Unkown Function

UFs mostly consist of several proteins with Domains of Unknown Function (DUFs). DUF642, for example, is present in the three monocot species plus *A. thaliana* (AT3G08030, Bradi1g04670, Os03g0807700, SCCCCL4009G04, AT5G11420, Os01g0611000, SCCCLB1001G04, AT5G25460 and AT4G32460). According to this categorization, it would only be speculative to propose for them a functional role, but this family is thought to be conserved, and has been proposed to be a new family of carbohydrate-binding proteins [114]. Some of these proteins have shown their ability to bind cellulose [114], have been shown to interact with PME [115], with auxin flux and hypocotyl elongation [116] and germination [117].

Several CWPs classified as UFs possess the BURP domain. Identified in *O. sativa* and *B. distachyon* (Os01g0733500 and Bradi2g49000), AT5G25610 has high sequence identity with *Gh*RDL1. *Gh*RDL1 interacts with a cotton α-expansin, and together they promote plant growth when overexpressed simultaneously [118]. The great number of UFs identified in the four species, and the fact that several UFs are suggested to have roles linked to the cell wall polysaccharides, reinforce the need for more functional studies on them to bring more information regarding the biology of monocot cell walls.

4. Applicative Aspects of Research on CWPs

Glycoside hydrolases are good candidates to be used in order to increase plant biomass or decrease recalcitrance destined to bioenergy production. Considering the enzymatic cocktails made from enzymes of microorganisms and the data on CWPs identified in monocots, the focus may be on GH families 1, 3, 17, 27, 35 and 51 [49]. The expression of a bacterial GH5 in *A. thaliana* led to a less recalcitrant wall without harming plant growth. It is suggested that the β-1,4 linkages of cellulose could be cut in the wall in an appropriate time and result in beneficial effects only [119]. As grasses present more GH5 and are the material used for biofuel production, they could be enzymes to watch with respect to their manipulation in the course of plant development.

As Prxs might be involved with lignification, they need to be mentioned when discussing ways to improve biomass and to facilitate its conversion. In this sense, *OsPrx38* overexpression in *A. thaliana* increased biomass and seed yield under arsenic stress [120], which could be an alternative pathway to be engineered.

The overexpression of a plantacyanin (AT2G02850, an OR), identified in the cell wall proteomes of all species analyzed here, led to decreased plant biomass and seed yield in *A. thaliana*. As this gene is regulated through a microRNA (miR408) whose overexpression results in augmented biomass, this could be further studied [121].

In a previous work, Endo et al. [122] fused the promoter of an LTP gene (AT3G18280), identified in *A. thaliana* and monocot CWP studies, named *At*TED4, with several other genes. *At*TED4 was able to regulate the transition from the immature to the mature developmental stage, as expected for an early xylem-specific promoter, which could be used to target specialized genes in biomass engineering.

5. Conclusions

Although a comparison among a variety of studies, and modes of extraction, analysis and mass spectrometry techniques is still far from a complete understanding, this review compiled the most extensive studies on monocot CWPs, gathering their common characteristics and comparing them to *A. thaliana* data. As not all studies have used quantitative analysis, the abundance of these proteins should be considered in a more detailed way in the future before drawing conclusions. The different GH families and the low number of both pectin-related CWPs and SPs were related to the specific type II-wall characteristics, such as the presence of mixed β-D-glucans, and lower contents of pectin and structural proteins. Furthermore, specificities were indicated, such as the monocot proteins from the GH18 family, and some questions remain unanswered, such as the roles of the Prx, Bowman-Birk inhibitor and dirigent proteins in type II cell wall processes.

Despite the particularities displayed by type II cell walls, the nucleotide sequences of genes encoding some of the proteins that are lacking in grasses are conserved among dicots and monocots [123]. This observation points to the fact that grasses have the genetic capacity to produce xyloglucan, but it remains inactive [43]. Furthermore, we need to increase our knowledge of CWPs from both dicots and monocots to provide novel insights into the specialization of type II cell walls and their adaptive advantages. This would aid in genetically tailoring plants to improve efficiency and biomass for the production of commercially important products such as second-generation biofuels. In addition to the gains brought by the first surveys on the CWPs of monocots, data mining and integration using several -omics and protein–protein interaction studies are needed to establish the roles of CWPs.

Supplementary Materials: Supplementary materials can be found at http://www.mdpi.com/1422-0067/20/8/1975/s1. Table S1. List of CWPs identified in *A. thaliana* (retrieved from WALLPROTDB [48]) and the accessions from *B. distachyon*, *Saccharum* spp. and *O. sativa* identified through BLASTp analysis; Table S2. List of CWPs identified in *B. distachyon* (retrieved from WALLPROTDB [48]) and the *A. thaliana* accessions identified through BLASTp analysis; Table S3. List of CWPs identified in *Saccharum* spp. (retrieved from WALLPROTDB [48]) and the *A. thaliana* accessions identified through BLASTp analysis; Table S4. List of CWPs identified in *O. sativa* (retrieved from WALLPROTDB [48]) and the *A. thaliana* accessions identified through BLASTp analysis. The published and unpublished proteomics data were retrieved from WALLPROTDB [48]. Phytozome nomenclature [124] was used for *O. sativa*.

Funding: Financial support was provided by a BIOEN/FAPESP PRONEX project (Process number: 2008/56100-5) as well as by INCT—Instituto Nacional de Ciência e Tecnologia do Bioetanol, CNPq (Process number: 142784/2007-9) and FAPESP (Process number: 2007/59327-8).

Conflicts of Interest: The authors declare no conflict of interest.

Abbreviations

1D-PAGE	one-dimensional polyacrylamide gel electrophoresis
2D	two-dimensional
2D-PAGE	two-dimensional polyacrylamide gel electrophoresis
ABA	abscisic acid
Asp	aspartyl
BBE	berberine-bridge enzyme
BBI	bowman-birk serine protease inhibitor
CaCl2	calcium chloride
CWP	cell wall protein
COBL	COBRA-like protein
DAF	days after flowering
DUF	protein with domains of unknown function
ER	endoplasmic reticulum
EXT	extensin
FLA	fasciclin-like arabinogalactan
GDPD	glycerophosphodiester phosphodiesterase
GDPDL	glycerophosphodiester phosphodiesterase-like
GH	glycosyl hydrolase
ID	protein with predicted interaction domains
LiCl	lithium chloride
LC-MS/MS	liquid chromatography-tandem mass spectrometry
LM	protein related to lipid metabolism
LRR	leucine-rich repeat
LRR-RK	leucine-rich repeat receptor kinase
LRX	leucine-rich repeat extensin
LTP	lipid transfer protein
M	miscellaneous protein
MS	mass spectrometry
OR	oxidoreductase
P	protease
PAC	proteins acting on carbohydrate
PME	pectin methyl esterase
PMEI	pectin methyl esterase inhibitor
Prx	class III peroxidase
S	protein possibly involved in signaling
SP	structural protein
UF	proteins of unknown function

References

1. Roberts, K. The plant extracellular matrix: In a new expansive mood. *Curr. Opin. Cell Biol.* **1994**, *6*, 688–694. [CrossRef]
2. Brownlee, C. Role of the extracellular matrix in cell-cell signalling: Paracrine paradigms. *Curr. Opin. Plant Biol.* **2002**, *5*, 396–401. [CrossRef]
3. Franková, L.; Fry, S.C. Biochemistry and physiological roles of enzymes that 'cut and paste' plant cell-wall polysaccharides. *J. Exp. Bot.* **2013**, *64*, 3519–3550. [CrossRef] [PubMed]
4. Buchanan, B.B.; Gruissem, W.; Jones, R.L. *Biochemistry & Molecular Biology of Plants*, 2nd ed.; John Wiley & Sons: Chichester, UK, 2015; pp. 1–1221. ISBN 9780470714225.
5. Cassab, G.I.; Varner, J.E. Cell Wall Proteins. *Annu. Rev. Plant Physiol. Plant Mol. Biol.* **1988**, *39*, 321–353. [CrossRef]
6. Vogel, J. Unique aspects of the grass cell wall. *Curr. Opin. Plant Biol.* **2008**, *11*, 301–307. [CrossRef] [PubMed]
7. Jamet, E.; Albenne, C.; Boudart, G.; Irshad, M.; Canut, H.; Pont-Lezica, R. Recent advances in plant cell wall proteomics. *Proteomics* **2008**, *8*, 893–908. [CrossRef]
8. Rapoport, T.A. Transport of proteins across the endoplasmic reticulum membrane. *Science* **1992**, *258*, 931–936. [CrossRef]
9. Vitale, A.; Denecke, J. The endoplasmic reticulum—Gateway of the secretory pathway. *Plant Cell* **1999**, *11*, 615–628. [CrossRef]
10. Crofts, A.; Leborgne-Castel, N.; Hillmer, S.; Robinson, D.G.; Phillipson, B.; Carlsson, L.E.; Ashford, D.A.; Denecke, J. Saturation of the endoplasmic reticulum retention machinery reveals anterograde bulk flow. *Plant Cell* **1999**, *11*, 2233–2248. [CrossRef] [PubMed]
11. Chivasa, S.; Ndimba, B.K.; Simon, W.J.; Robertson, D.; Yu, X.L.; Knox, J.P.; Bolwell, P.; Slabas, A.R. Proteomic analysis of the *Arabidopsis thaliana* cell wall. *Electrophoresis* **2002**, *23*, 1754–1765. [CrossRef]
12. Carpita, N.; Tierney, M.; Campbell, M. Molecular biology of the plant cell wall: Searching for the genes that define structure, architecture and dynamics. *Plant Mol. Biol.* **2001**, *47*, 1–5. [CrossRef]
13. Jamet, E.; Canut, H.; Boudart, G.; Pont-Lezica, R.F. Cell wall proteins: A new insight through proteomics. *Trends Plant Sci.* **2006**, *11*, 33–39. [CrossRef]
14. Tan, L.; Eberhard, S.; Pattathil, S.; Warder, C.; Glushka, J.; Yuan, C.; Hao, Z.; Zhu, X.; Avci, U.; Miller, J.S.; et al. An *Arabidopsis* cell wall proteoglycan consists of pectin and arabinoxylan covalently linked to an arabinogalactan protein. *Plant Cell* **2013**, *25*, 270–287. [CrossRef] [PubMed]
15. Lee, S.-J.; Saravanan, R.S.; Damasceno, C.M.B.; Yamane, H.; Kim, B.-D.; Rose, J.K.C. Digging deeper into the plant cell wall proteome. *Plant Physiol. Biochem.* **2004**, *42*, 979–988. [CrossRef]
16. Borderies, G.; Jamet, E.; Lafitte, C.; Rossignol, M.; Jauneau, A.; Boudart, G.; Monsarrat, B.; Esquerré-Tugayé, M.-T.; Boudet, A.; Pont-Lezica, R. Proteomics of loosely bound cell wall proteins of *Arabidopsis thaliana* cell suspension cultures: A critical analysis. *Electrophoresis* **2003**, *24*, 3421–3432. [CrossRef] [PubMed]
17. Boudart, G.; Jamet, E.; Rossignol, M.; Lafitte, C.; Borderies, G.; Jauneau, A.; Esquerré-Tugayé, M.-T.; Pont-Lezica, R. Cell wall proteins in apoplastic fluids of *Arabidopsis thaliana* rosettes: Identification by mass spectrometry and bioinformatics. *Proteomics* **2005**, *5*, 212–221. [CrossRef] [PubMed]
18. Bayer, E.M.; Bottrill, A.R.; Walshaw, J.; Vigouroux, M.; Naldrett, M.J.; Thomas, C.L.; Maule, A.J. *Arabidopsis* cell wall proteome defined using multidimensional protein identification technology. *Proteomics* **2006**, *6*, 301–311. [CrossRef]
19. Irshad, M.; Canut, H.; Borderies, G.; Pont-Lezica, R.; Jamet, E. A new picture of cell wall protein dynamics in elongating cells of *Arabidopsis thaliana*: Confirmed actors and newcomers. *BMC Plant Biol.* **2008**, *8*, 94. [CrossRef]
20. Nguyen-Kim, H.; San Clamente, H.; Balliau, T.; Zivy, M.; Dunand, C.; Albenne, C.; Jamet, E. *Arabidopsis thaliana* root cell wall proteomics: Increasing the proteome coverage using a combinatorial peptide ligand library and description of unexpected Hyp in peroxidase amino acid sequences. *Proteomics* **2016**, *16*, 491–503. [CrossRef]
21. Hervé, V.; Duruflé, H.; San Clemente, H.; Albenne, C.; Balliau, T.; Zivy, M.; Dunand, C.; Jamet, E. An enlarged cell wall proteome of *Arabidopsis thaliana* rosettes. *Proteomics* **2016**, *16*, 3183–3187. [CrossRef] [PubMed]
22. Duruflé, H.; San Clemente, H.; Balliau, T.; Zivy, M.; Dunand, C.; Jamet, E. Proline hydroxylation in cell wall proteins: Is it yet possible to define rules? *Front. Plant Sci.* **2017**, *8*, 1802. [CrossRef]

23. Kehr, J.; Buhtz, A.; Giavalisco, P. Analysis of xylem sap proteins from Brassica napus. *BMC Plant Biol.* **2005**, *5*, 11. [CrossRef] [PubMed]
24. Ligat, L.; Lauber, E.; Albenne, C.; San Clemente, H.; Valot, B.; Zivy, M.; Pont-Lezica, R.; Arlat, M.; Jamet, E. Analysis of the xylem sap proteome of *Brassica oleracea* reveals a high content in secreted proteins. *Proteomics* **2011**, *11*, 1798–1813. [CrossRef] [PubMed]
25. Zhu, J.; Alvarez, S.; Marsh, E.L.; LeNoble, M.E.; Cho, I.-J.; Sivaguru, M.; Chen, S.; Nguyen, H.T.; Wu, Y.; Schachtman, D.P.; et al. Cell wall proteome in the maize primary root elongation zone. II. Region-specific changes in water soluble and lightly ionically bound proteins under water deficit. *Plant Physiol.* **2007**, *145*, 1533–1548. [CrossRef] [PubMed]
26. Bhushan, D.; Pandey, A.; Chattopadhyay, A.; Choudhary, M.K.; Chakraborty, S.; Datta, A.; Chakraborty, N. Extracellular matrix proteome of chickpea (*Cicer arietinum* L.) illustrates pathway abundance, novel protein functions and evolutionary perspect. *J. Proteome Res.* **2006**, *5*, 1711–1720. [CrossRef]
27. Negri, A.S.; Prinsi, B.; Scienza, A.; Morgutti, S.; Cocucci, M.; Espen, L. Analysis of grape berry cell wall proteome: A comparative evaluation of extraction methods. *J. Plant Physiol.* **2008**, *165*, 1379–1389. [CrossRef]
28. Jung, Y.-H.; Jeong, S.-H.; Kim, S.H.; Singh, R.; Lee, J.; Cho, Y.-S.; Agrawal, G.K.; Rakwal, R.; Jwa, N.-S. Systematic secretome analyses of rice leaf and seed callus suspension-cultured cells: Workflow development and establishment of high-density two-dimensional gel reference maps. *J. Proteome Res.* **2008**, *7*, 5187–5210. [CrossRef] [PubMed]
29. Chen, X.Y.; Kim, S.T.; Cho, W.K.; Rim, Y.; Kim, S.; Kim, S.W.; Kang, K.Y.; Park, Z.Y.; Kim, J.Y. Proteomics of weakly bound cell wall proteins in rice calli. *J. Plant Physiol.* **2009**, *166*, 675–685. [CrossRef] [PubMed]
30. Cho, W.K.; Chen, X.Y.; Chu, H.; Rim, Y.; Kim, S.; Kim, S.T.; Kim, S.-W.; Park, Z.-Y.; Kim, J.-Y. Proteomic analysis of the secretome of rice calli. *Physiol. Plant.* **2009**, *135*, 331–341. [CrossRef]
31. Zhou, L.; Bokhari, S.A.; Dong, C.-J.; Liu, J.-Y. Comparative proteomics analysis of the root apoplasts of rice seedlings in response to hydrogen peroxide. *PLoS ONE* **2011**, *6*, e16723. [CrossRef]
32. Millar, D.J.; Whitelegge, J.P.; Bindschedler, L.V.; Rayon, C.; Boudet, A.-M.; Rossignol, M.; Borderies, G.; Bolwell, G.P. The cell wall and secretory proteome of a tobacco cell line synthesising secondary wall. *Proteomics* **2009**, *9*, 2355–2372. [CrossRef] [PubMed]
33. Douché, T.; San Clemente, H.; Burlat, V.; Roujol, D.; Valot, B.; Zivy, M.; Pont-Lezica, R.; Jamet, E. *Brachypodium distachyon* as a model plant toward improved biofuel crops: Search for secreted proteins involved in biogenesis and disassembly of cell wall polymers. *Proteomics* **2013**, *13*, 2438–2454. [CrossRef] [PubMed]
34. Francin-Allami, M.; Lollier, V.; Pavlovic, M.; San Clemente, H.; Rogniaux, H.; Jamet, E.; Guillon, F.; Larré, C. Understanding the remodelling of cell walls during *Brachypodium distachyon* grain development through a sub-cellular quantitative proteomic approach. *Proteomes* **2016**, *4*, 21. [CrossRef] [PubMed]
35. Zhang, M.; Chen, G.-X.; Lv, D.-W.; Li, X.-H.; Yan, Y.-M. N-Linked Glycoproteome profiling of seedling leaf in *Brachypodium distachyon* L. *J. Proteome Res.* **2015**, *14*, 1727–1738. [CrossRef]
36. Francin-Allami, M.; Merah, K.; Albenne, C.; Rogniaux, H.; Pavlovic, M.; Lollier, V.; Sibout, R.; Guillon, F.; Jamet, E.; Larré, C. Cell wall proteomic of *Brachypodium distachyon* grains: A focus on cell wall remodeling proteins. *Proteomics* **2015**, *15*, 2296–2306. [CrossRef]
37. Calderan-Rodrigues, M.J.; Jamet, E.; Bonassi, M.B.C.R.; Guidetti-Gonzalez, S.; Begossi, A.C.; Setem, L.V.; Franceschini, L.M.; Fonseca, J.G.; Labate, C.A. Cell wall proteomics of sugarcane cell suspension cultures. *Proteomics* **2014**, *14*, 738–749. [CrossRef] [PubMed]
38. Calderan-Rodrigues, M.J.; Jamet, E.; Douché, T.; Bonassi, M.B.R.; Cataldi, T.R.; Fonseca, J.G.; San Clemente, H.; Pont-Lezica, R.; Labate, C.A. Cell wall proteome of sugarcane stems: Comparison of a destructive and a non-destructive extraction method showed differences in glycoside hydrolases and peroxidases. *BMC Plant Biol.* **2016**, *16*, 14. [CrossRef] [PubMed]
39. Fonseca, J.G.; Calderan-Rodrigues, M.J.; de Moraes, F.E.; Cataldi, T.R.; Jamet, E.; Labate, C.A. Cell wall proteome of sugarcane young and mature leaves and stems. *Proteomics* **2018**, *18*, 9853. [CrossRef] [PubMed]
40. Labate, C.A.; Calderan-Rodrigues, M.J.; Cataldi, T.R. Cell Wall Proteome of Seven Month-Old Sugarcane Leaves and Internodes. 2019. Available online: http://proteomecentral.proteomexchange.org/cgi/GetDataset?ID=PXD012761 (accessed on 11 February 2019).
41. Albenne, C.; Canut, H.; Hoffmann, L.; Jamet, E. Plant cell wall Proteins: A large body of data, but what about runaways? *Proteomes* **2014**, *2*, 224–242. [CrossRef]

42. Carpita, N.C.; Gibeaut, D.M. Structural models of primary cell walls in flowering plants: Consistency of molecular structure with the physical properties of the walls during growth. *Plant J.* **1993**, *3*, 1–30. [CrossRef]
43. Sarkar, P.; Bosneaga, E.; Auer, M. Plant cell walls throughout evolution: Towards a molecular understanding of their design principles. *J. Exp. Bot.* **2009**, *60*, 3615–3635. [CrossRef]
44. Nogueira, L.A.H.; Leal, M.R.L.V.; Fernandes, E.; Chum, H.L.; Diaz-Chavez, R.; Endres, J.; Mahakhant, A.; Otto, M.; Seebaluck, V.; van der Wielen, L. Sustainable development and innovation. In *SCOPE Bioenergy & Sustainability: Bridging the Gaps*, 1st ed.; Souza, G.M., Victoria, R.I., Joly, C.A., Verdade, L.M., Eds.; Scientific Committee on Problems of the Environment (SCOPE): Paris, France, 2015; Volume 1, pp. 184–217. ISBN 978-2-9545557-0-6.
45. Yong, W.; Link, B.; O'Malley, R.; Tewari, J.; Hunter, C.T.; Lu, C.A.; Li, X.; Bleecker, A.B.; Koch, K.E.; McCann, M.C.; et al. Genomics of plant cell wall biogenesis. *Planta* **2005**, *221*, 747–751. [CrossRef] [PubMed]
46. Feiz, L.; Irshad, M.; Pont-Lezica, R.F.; Canut, H.; Jamet, E. Evaluation of cell wall preparations for proteomics: A new procedure for purifying cell walls from Arabidopsis hypocotyls. *Plant Methods* **2006**, *2*, 10. [CrossRef]
47. Jamet, E.; Boudart, G.; Borderies, G.; Charmont, S.; Lafitte, C.; Rossignol, M.; Canut, H.; Pont-Lezica, R. Isolation of plant cell wall proteins. In *Methods in Molecular Biology, Applications and Protocols in Expression Proteomics*, 1st ed.; Posch, A., Ed.; Humana Press Inc.: Totowa, NJ, USA, 2007; Volume 2, pp. 187–201.
48. WallProtDB. Available online: http://www.polebio.lrsv.ups-tlse.fr/WallProtDB/ (accessed on 29 October 2018).
49. Calderan-Rodrigues, M.J.; Fonseca, J.G.; Clemente, H.S.; Labate, C.A.; Jamet, E. Glycoside hydrolases in plant cell wall proteomes: Predicting functions that could be relevant for improving biomass transformation processes. In *Advances in Biofuels and Bioenergy*, 1st ed.; Nageswara-Rao, M., Ed.; IntechOpen: London, UK, 2018; Volume 1, pp. 165–182.
50. Tyler, L.; Bragg, J.N.; Wu, J.; Yang, X.; Tuskan, G.A.; Vogel, J.P. Annotation and comparative analysis of the glycoside hydrolase genes in *Brachypodium distachyon*. *BMC Genom.* **2010**, *11*, 600. [CrossRef]
51. Chapelle, A.; Morreel, K.; Vanholme, R.; Le-Bris, P.; Morin, H.; Lapierre, C.; Boerjan, W.; Jouanin, L.; Demont-Caulet, N. Impact of absence of stem-specific β-glucosidases on lignin and monolignols. *Plant Physiol.* **2012**, *160*, 1204–1217. [CrossRef] [PubMed]
52. Varghese, J.N.; Hrmova, M.; Fincher, G.B. Three-dimensional structure of a barley beta-D-glucan exohydrolase, a family 3 glycosyl hydrolase. *Structure* **1999**, *7*, 179–190. [CrossRef]
53. Minic, Z. Physiological roles of plant glycoside hydrolases. *Planta* **2008**, *227*, 723–740. [CrossRef] [PubMed]
54. Minic, Z.; Jouanin, L. Plant glycoside hydrolases involved in cell wall polysaccharide degradation. *Plant Physiol. Biochem.* **2006**, *44*, 435–449. [CrossRef]
55. Jordan, D.B.; Wagschal, K. Properties and applications of microbial β-D-xylosidases featuring the catalytically efficient enzyme from *Selenomonas ruminantium*. *Appl. Microbiol. Biotechnol.* **2010**, *86*, 1647–1658. [CrossRef] [PubMed]
56. Gebruers, K.; Brijs, K.; Courtin, C.; Fierens, K.; Goesaert, H.; Rabijns, A.; Raedschelders, G.; Robben, J.; Sansen, S.; Sørensen, J.F.; et al. Properties of TAXI-type endoxylanase inhibitors. *Biochim. Biophys. Acta Proteins Proteomics* **2004**, *1696*, 213–221. [CrossRef]
57. Brunner, F.; Stintzi, A.; Fritig, B.; Legrand, M. Substrate specificities of tobacco chitinases. *Plant J.* **1998**, *14*, 225–234. [CrossRef]
58. Sasaki, C.; Vårum, K.M.; Itoh, Y.; Tamoi, M.; Fukamizo, T. Rice chitinases: Sugar recognition specificities of the individual subsites. *Glycobiology* **2006**, *16*, 1242–1250. [CrossRef]
59. Veillet, F.; Gaillard, C.; Coutos-Thévenot, P.; La Camera, S. Targeting the *At*CWIN1 gene to explore the role of invertases in sucrose transport in roots and during *Botrytis cinerea* infection. *Front. Plant Sci.* **2016**, *7*, 1899. [CrossRef] [PubMed]
60. Moneo-Sánchez, M.; Alonso-Chico, A.; Knox, J.P.; Dopico, B.; Labrador, E.; Martín, I. β-(1,4)-Galactan remodelling in Arabidopsis cell walls affects the xyloglucan structure during elongation. *Planta* **2018**, 1–12. [CrossRef] [PubMed]
61. Wieczorek, K.; Elashry, A.; Quentin, M.; Grundler, F.M.; Favery, B.; Seifert, J.G.; Bohlmann, H. A distinct role of pectate lyases in the formation of feeding structures induced by cyst and root-knot nematodes. *Mol. Plant-Microbe Interact.* **2014**, *27*, 901–912. [CrossRef]
62. Guan, Y.; Nothnagel, E.A. Binding of arabinogalactan proteins by Yariv phenylglycoside triggers wound-like responses in Arabidopsis cell cultures. *Plant Physiol.* **2004**, *135*, 1346–1366. [CrossRef]

63. Atmodjo, M.A.; Hao, Z.; Mohnen, D. Evolving views of pectin biosynthesis. *Annu. Rev. Plant Biol.* **2013**, *64*, 747–779. [CrossRef]
64. Shigeto, J.; Tsutsumi, Y. Diverse functions and reactions of class III peroxidases. *New Phytol.* **2016**, *209*, 1395–1402. [CrossRef] [PubMed]
65. Passardi, F.; Penel, C.; Dunand, C. Performing the paradoxical: How plant peroxidases modify the cell wall. *Trends Plant Sci.* **2004**, *9*, 534–540. [CrossRef]
66. Welinder, K.G.; Justesen, A.F.; Kjærsgård, I.V.; Jensen, R.B.; Rasmussen, S.K.; Jespersen, H.M.; Duroux, L. Structural diversity and transcription of class III peroxidases from *Arabidopsis thaliana*. *Eur. J. Biochem.* **2002**, *269*, 6063–6081. [CrossRef] [PubMed]
67. Costa, E.A.; Anoni, C.O.; Mancini, M.C.; Santos, F.R.C.; Marconi, T.G.; Gazaffi, R.; Pastina, M.M.; Perecin, D.; Mollinari, M.; Xavier, M.A.; et al. QTL mapping including codominant SNP markers with ploidy level information in a sugarcane progeny. *Euphytica* **2016**, *211*, 1–16. [CrossRef]
68. Shah, K.; Penel, C.; Gagnon, J.; Dunand, C. Purification and identification of a Ca^{2+}-pectate binding peroxidase from *Arabidopsis* leaves. *Phytochemistry* **2004**, *65*, 307–312. [CrossRef]
69. Kunieda, T.; Shimada, T.; Kondo, M.; Nishimura, M.; Nishitani, K.; Hara-Nishimura, I. Spatiotemporal secretion of PEROXIDASE36 is required for seed coat mucilage extrusion in *Arabidopsis*. *Plant Cell* **2013**, *25*, 1355–1367. [CrossRef]
70. Fernández-Pérez, F.; Pomar, F.; Pedreño, M.A.; Novo-Uzal, E. The suppression of *AtPrx52* affects fibers but not xylem lignification in *Arabidopsis* by altering the proportion of syringyl units. *Physiol. Plant.* **2015**, *154*, 395–406. [CrossRef]
71. Herrero, J.; Fernández-Pérez, F.; Yebra, T.; Novo-Uzal, E.; Pomar, F.; Pedreño, M.Á.; Cuello, J.; Guéra, A.; Esteban-Carrasco, A.; Zapata, J.M. Bioinformatic and functional characterization of the basic peroxidase 72 from *Arabidopsis thaliana* involved in lignin biosynthesis. *Planta* **2013**, *237*, 1599–1612. [CrossRef] [PubMed]
72. Lee, Y.; Rubio, M.C.; Alassimone, J.; Geldner, N. A mechanism for localized lignin deposition in the endodermis. *Cell* **2013**, *153*, 402–412. [CrossRef] [PubMed]
73. Linkies, A.; Schuster-Sherpa, U.; Tintelnot, S.; Leubner-Metzger, G.; Müller, K. Peroxidases identified in a subtractive cDNA library approach show tissue-specific transcript abundance and enzyme activity during seed germination of *Lepidium sativum*. *J. Exp. Bot.* **2010**, *61*, 491–502. [CrossRef] [PubMed]
74. Jin, J.; Hewezi, T.; Baum, T.J. Arabidopsis peroxidase AtPRX53 influences cell elongation and susceptibility to *Heterodera schachtii*. *Plant Signal. Behav.* **2011**, *6*, 1778–1786. [CrossRef] [PubMed]
75. Cosio, C.; Ranocha, P.; Francoz, E.; Burlat, V.; Zheng, Y.; Perry, S.E.; Ripoll, J.-J.; Yanofsky, M.; Dunand, C. The class III peroxidase PRX17 is a direct target of the MADS-box transcription factor AGAMOUS-LIKE15 (AGL15) and participates in lignified tissue formation. *New Phytol.* **2017**, *213*, 250–263. [CrossRef] [PubMed]
76. Kim, B.H.; Kim, S.Y.; Nam, K.H. Genes encoding plant-specific class III peroxidases are responsible for increased cold tolerance of the *brassinosteroid-insensitive* 1 mutant. *Mol. Cells* **2012**, *34*, 539–548. [CrossRef]
77. Daniel, B.; Pavkov-Keller, T.; Steiner, B.; Dordic, A.; Gutmann, A.; Nidetzky, B.; Sensen, C.W.; van der Graaff, E.; Wallner, S.; Gruber, K.; et al. Oxidation of monolignols by members of the berberine bridge enzyme family suggests a role in plant cell wall metabolism. *J. Biol. Chem.* **2015**, *290*, 18770–18781. [CrossRef]
78. Benedetti, M.; Verrascina, I.; Pontiggia, D.; Locci, F.; Mattei, B.; De Lorenzo, G.; Cervone, F. Four Arabidopsis berberine bridge enzyme-like proteins are specific oxidases that inactivate the elicitor-active oligogalacturonides. *Plant J.* **2018**, *94*, 260–273. [CrossRef] [PubMed]
79. Daniel, B.; Konrad, B.; Toplak, M.; Lahham, M.; Messenlehner, J.; Winkler, A.; Macheroux, P. The family of berberine bridge enzyme-like enzymes: A treasure-trove of oxidative reactions. *Arch. Biochem. Biophys.* **2017**, *632*, 88–103. [CrossRef]
80. Edqvist, J.; Blomqvist, K.; Nieuwland, J.; Salminen, T.A. Plant lipid transfer proteins: Are we finally closing in on the roles of these enigmatic proteins? *J. Lipid Res.* **2018**, *59*, 1374–1382. [CrossRef]
81. Bosch, M.; Mayer, C.D.; Cookson, A.; Donnison, I.S. Identification of genes involved in cell wall biogenesis in grasses by differential gene expression profiling of elongating and non-elongating maize internodes. *J. Exp. Bot.* **2011**, *62*, 3545–3561. [CrossRef]
82. Ariizumi, T.; Amagai, M.; Shibata, D.; Hatakeyama, K.; Watanabe, M.; Toriyama, K. Comparative study of promoter activity of three anther-specific genes encoding lipid transfer protein, xyloglucan endotransglucosylase/hydrolase and polygalacturonase in transgenic *Arabidopsis thaliana*. *Plant Cell Rep.* **2002**, *21*, 90–96. [CrossRef]

83. Gao, S.; Guo, W.; Feng, W.; Liu, L.; Song, X.; Chen, J.; Hou, W.; Zhu, H.; Tang, S.; Hu, J. *LTP3* contributes to disease susceptibility in *Arabidopsis* by enhancing abscisic acid (ABA) biosynthesis. *Mol. Plant Pathol.* **2016**, *17*, 412–426. [CrossRef]
84. Jones, M.A.; Raymond, M.J.; Smirnoff, N. Analysis of the root-hair morphogenesis transcriptome reveals the molecular identity of six genes with roles in root-hair development in Arabidopsis. *Plant J.* **2006**, *45*, 83–100. [CrossRef] [PubMed]
85. Beers, E.P.; Jones, A.M.; Dickerman, A.W. The S8 serine, C1A cysteine and A1 aspartic protease families in Arabidopsis. *Phytochemistry* **2004**, *65*, 43–58. [CrossRef] [PubMed]
86. Martinez, D.E.; Borniego, M.L.; Battchikova, N.; Aro, E.-M.; Tyystjärvi, E.; Guiamét, J.J. SASP, a senescence-associated subtilisin protease, is involved in reproductive development and determination of silique number in *Arabidopsis*. *J. Exp. Bot.* **2015**, *66*, 161–174. [CrossRef]
87. Wang, Q.; Guo, Q.; Guo, Y.; Yang, J.; Wang, M.; Duan, X.; Niu, J.; Liu, S.; Zhang, J.; Lu, Y.; et al. Arabidopsis subtilase *SASP* is involved in the regulation of ABA signaling and drought tolerance by interacting with OPEN STOMATA 1. *J. Exp. Bot.* **2018**, *69*, 4403–4417. [CrossRef]
88. Pathirana, R.; West, P.; Hedderley, D.; Eason, J. Cell death patterns in Arabidopsis cells subjected to four physiological stressors indicate multiple signalling pathways and cell cycle phase specificity. *Protoplasma* **2017**, *254*, 635–647. [CrossRef]
89. Hocq, L.; Sénéchal, F.; Lefebvre, V.; Lehner, A.; Domon, J.-M.; Mollet, J.-C.; Dehors, J.; Pageau, J.; Marcelo, P.; Guérineau, F.; et al. Combined experimental and computational approaches reveal distinct pH dependence of pectin methylesterase inhibitors. *Plant Physiol.* **2017**, *173*, 1075–1093. [CrossRef]
90. Carotenuto, G.; Chabaud, M.; Miyata, K.; Capozzi, M.; Takeda, N.; Kaku, H.; Shibuya, N.; Nakagawa, T.; Barker, D.G.; Genre, A. The rice LysM receptor-like kinase *Os*CERK1 is required for the perception of short-chain chitin oligomers in arbuscular mycorrhizal signaling. *New Phytol.* **2017**, *214*, 1440–1446. [CrossRef]
91. Qi, R.-F.; Song, Z.-W.; Chi, C.-W. Structural features and molecular evolution of Bowman-Birk protease inhibitors and their potential application. *Acta Biochim. Biophys. Sin.* **2005**, *37*, 283–292. [CrossRef]
92. Guo, K.; Bu, Y.; Takano, T.; Liu, S.; Zhang, X. *Arabidopsis* cysteine proteinase inhibitor *At*CYSb interacts with a Ca^{2+}-dependent nuclease, AtCaN2. *FEBS Lett.* **2013**, *587*, 3417–3421. [CrossRef]
93. Helft, L.; Reddy, V.; Chen, X.; Koller, T.; Federici, L.; Fernández-Recio, J.; Gupta, R.; Bent, A. LRR conservation mapping to predict functional sites within protein leucine-rich repeat domains. *PLoS ONE* **2011**, *6*, e21614. [CrossRef]
94. Shah, S.J.; Anjam, M.S.; Mendy, B.; Anwer, M.A.; Habash, S.S.; Lozano-Torres, J.L.; Grundler, F.M.W.; Siddique, S. Damage-associated responses of the host contribute to defence against cyst nematodes but not root-knot nematodes. *J. Exp. Bot.* **2017**, *68*, 5949–5960. [CrossRef]
95. Labudda, M.; Różańska, E.; Szewińska, J.; Sobczak, M.; Dzik, J.M. Protease activity and phytocystatin expression in *Arabidopsis thaliana* upon *Heterodera schachtii* infection. *Plant Physiol. Biochem.* **2016**, *109*, 416–429. [CrossRef]
96. Song, C.; Kim, T.; Chung, W.S.; Lim, C.O. The *Arabidopsis* phytoscystatin AtCYS5 enhances seed germination and seedling growth under heat stress conditions. *Mol. Cells* **2017**, *40*, 577–586. [CrossRef]
97. Johnson, K.L.; Jones, B.J.; Bacic, A.; Schultz, C.J. The fasciclin-like arabinogalactan proteins of Arabidopsis. A multigene family of putative cell adhesion molecules. *Plant Physiol.* **2003**, *133*, 1911–1925. [CrossRef]
98. Kotake, T.; Hirata, N.; Kitazawa, K.; Soga, K.; Tsumuraya, Y. Arabinogalactan-proteins in the evolution of gravity resistance in land plants. *Soc. Biol. Sci. Space* **2009**, *23*, 143–149. [CrossRef]
99. MacMillan, C.P.; Mansfield, S.D.; Stachurski, Z.H.; Evans, R.; Southerton, S.G. Fasciclin-like arabinogalactan proteins: Specialization for stem biomechanics and cell wall architecture in Arabidopsis and *Eucalyptus*. *Plant J.* **2010**, *62*, 689–703. [CrossRef]
100. Johnson, K.L.; Kibble, N.A.; Bacic, A.; Schultz, C.J. A Fasciclin-Like Arabinogalactan-Protein (FLA) Mutant of *Arabidopsis thaliana, fla1*, Shows Defects in Shoot Regeneration. *PLoS ONE* **2011**, *6*, e25154. [CrossRef]
101. Schindelman, G.; Morikami, A.; Jung, J.; Baskin, T.I.; Carpita, N.C.; Derbyshire, P.; McCann, M.C.; Benfey, P.N. COBRA encodes a putative GPI-anchored protein, which is polarly localized and necessary for oriented cell expansion in *Arabidopsis*. *Genes Dev.* **2001**, *15*, 1115–1127. [CrossRef]

102. Niu, E.; Shang, X.; Cheng, C.; Bao, Z.; Zeng, Y.; Cai, C.; Du, X.; Guo, W. Comprehensive analysis of the *COBRA-Like (COBL)* gene family in *Gossypium* identifies two COBLs potentially associated with fiber quality. *PLoS ONE* **2015**, *10*, e0145725. [CrossRef] [PubMed]
103. Li, P.; Liu, Y.; Tan, W.; Chen, J.; Zhu, M.; Lv, Y.; Liu, Y.; Yu, S.; Zhang, W.; Cai, H. *Brittle Culm 1* encodes a COBRA-like protein involved in secondary cell wall cellulose biosynthesis in Sorghum. *Plant Cell Physiol.* **2018**. [CrossRef]
104. Holton, N.; Nekrasov, V.; Ronald, P.C.; Zipfel, C. The phylogenetically-related pattern recognition receptors EFR and XA21 recruit similar immune signaling components in monocots and dicots. *PLoS Pathog.* **2015**, *11*, e1004602. [CrossRef] [PubMed]
105. Sun, L.; Song, L.; Zhang, Y.; Zheng, Z.; Liu, D. Arabidopsis PHL2 and PHL1 act redundantly as the key components of the central regulatory system controlling transcriptional responses to phosphate starvation. *Plant Physiol.* **2016**, *170*, 499–514. [CrossRef]
106. Yin, K.; Han, X.; Xu, Z.; Xue, H. *Arabidopsis* GLP4 is localized to the Golgi and binds auxin *in vitro*. *Acta Biochim. Biophys. Sin.* **2009**, *41*, 478–487. [CrossRef] [PubMed]
107. Pickel, B.; Pfannstiel, J.; Steudle, A.; Lehmann, A.; Gerken, U.; Pleiss, J.; Schaller, A. A model of dirigent proteins derived from structural and functional similarities with allene oxide cyclase and lipocalins. *FEBS J.* **2012**, *279*, 1980–1993. [CrossRef] [PubMed]
108. Papini-Terzi, F.S.; Rocha, F.R.; Vêncio, R.Z.N.; Felix, J.M.; Branco, D.S.; Waclawovsky, A.J.; Del Bem, L.E.V.; Lembke, C.G.; Costa, M.D.L.; Nishiyama, M., Jr.; et al. Sugarcane genes associated with sucrose content. *BMC Genomics* **2009**, *10*, 1–21. [CrossRef]
109. Lamport, D.T.A.; Kieliszewski, M.J.; Chen, Y.; Cannon, M.C. Role of the extensin superfamily in primary cell wall architecture. *Plant Physiol.* **2011**, *156*, 11–19. [CrossRef] [PubMed]
110. Cannon, M.C.; Terneus, K.; Hall, Q.; Tan, L.; Wang, W.; Wegenhart, B.L.; Chen, L.; Lamport, D.T.A.; Chen, Y.; Kieliszewski, M.J. Self-assembly of the plant cell wall requires an extensin scaffold. *Proc. Natl. Acad. Sci. USA* **2008**, *105*, 2226–2231. [CrossRef]
111. Betekhtin, A.; Milewska-Hendel, A.; Lusinska, J.; Chajec, L.; Kurczynska, E.; Hasterok, R. Organ and Tissue-Specific Localisation of selected cell wall epitopes in the zygotic embryo of *Brachypodium distachyon*. *Int. J. Mol. Sci.* **2018**, *19*, 725. [CrossRef]
112. Liu, X.; Wolfe, R.; Welch, L.R.; Domozych, D.S.; Popper, Z.A.; Showalter, A.M. Bioinformatic identification and analysis of extensins in the plant kingdom. *PLoS ONE* **2016**, *11*, e0150177. [CrossRef] [PubMed]
113. Baumberger, N.; Doesseger, B.; Guyot, R.; Diet, A.; Parsons, R.L.; Clark, M.A.; Simmons, M.P.; Bedinger, P.; Goff, S.A.; Ringli, C.; et al. Whole-genome comparison of leucine-rich repeat extensins in Arabidopsis and rice. A conserved family of cell wall proteins form a vegetative and a reproductive clade. *Plant Physiol.* **2003**, *131*, 1313–1326. [CrossRef]
114. Vázquez-Lobo, A.; Roujol, D.; Zuñiga-Sánchez, E.; Albenne, C.; Piñero, D.; Gamboa de Buen, A.; Jamet, E. The highly conserved spermatophyte cell wall DUF642 protein family: Phylogeny and first evidence of interaction with cell wall polysaccharides *in vitro*. *Mol. Phylogenet. Evol.* **2012**, *63*, 510–520. [CrossRef]
115. Zúñiga-Sanchez, E.; Soriano, D.; Martínez-Barajas, E.; Orozco-Segovia, A.; Gamboa-deBuen, A. *BIIDXI*, the *At4g32460* DUF642 gene, is involved in pectin methyl esterase regulation during *Arabidopsis thaliana* seed germination and plant development. *BMC Plant Biol.* **2014**, *14*, 338. [CrossRef]
116. Salazar-Iribe, A.; Cruz-Valderrama, J.E.; Jímenez-Durán, K.; Gómez-Maqueo, X.; Gamboa-deBuen, A. BIIDXI, a DUF642 cell wall protein, is involved in hypocotyl growth via auxin efflux. *J. Plant Physiol.* **2018**, *231*, 105–109. [CrossRef]
117. Garza-Caligaris, L.E.; Avendaño-Vázquez, A.O.; Alvarado-López, S.; Zúñiga-Sánchez, E.; Orozco-Segovia, A.; Pérez-Ruíz, R.V.; Gamboa-DeBuen, A. *At3g08030* transcript: A molecular marker of seed ageing. *Ann. Bot.* **2012**, *110*, 1253–1260. [CrossRef]
118. Xu, B.; Gou, J.-Y.; Li, F.-G.; Shangguan, X.X.; Zhao, B.; Yang, C.-Q.; Wang, L.-J.; Yuan, S.; Liu, C.-J.; Chen, X.-Y. A cotton BURP domain protein interacts with α-expansin and their co-expression promotes plant growth and fruit production. *Mol. Plant* **2013**, *6*, 945–958. [CrossRef]
119. Donohoe, B.S.; Wei, H.; Mittal, A.; Shollenberger, T.; Lunin, V.V.; Himmel, M.E.; Brunecky, R. Towards an understanding of enhanced biomass digestibility by *In Planta* expression of a family 5 glycoside hydrolase. *Sci. Rep.* **2017**, *7*, 4389. [CrossRef]

120. Kidwai, M.; Dhar, Y.V.; Gautam, N.; Tiwari, M.; Ahmad, I.Z.; Asif, M.H.; Chakrabarty, D. *Oryza sativa* class III peroxidase (*OsPRX38*) overexpression in *Arabidopsis thaliana* reduces arsenic accumulation due to apoplastic lignification. *J. Hazard. Mater* **2019**, *362*, 383–393. [CrossRef]
121. Song, Z.; Zhang, L.; Wang, Y.; Li, H.; Li, S.; Zhao, H.; Zhang, H. Constitutive expression of miR408 improves biomass and seed yield in Arabidopsis. *Front. Plant Sci.* **2018**, *8*, 2114. [CrossRef]
122. Endo, S.; Iwamoto, K.; Fukuda, H. Overexpression and cosuppression of xylem-related genes in an early xylem differentiation stage-specific manner by the AtTED4 promoter. *Plant Biotechnol. J.* **2018**, *16*, 451–458. [CrossRef]
123. Popper, Z. Evolution and diversity of green plant cell walls. *Curr. Opin. Plant Biol.* **2008**, *11*, 286–292. [CrossRef]
124. Phytozome. Available online: https://phytozome.jgi.doe.gov/pz/portal.html (accessed on 23 January 2019).

International Journal of
Molecular Sciences

MDPI

Article

Cell Wall Proteome of Wheat Grain Endosperm and Outer Layers at Two Key Stages of Early Development

Mehdi Cherkaoui [1], Virginie Lollier [1], Audrey Geairon [1], Axelle Bouder [1], Colette Larré [1], Hélène Rogniaux [1], Elisabeth Jamet [2], Fabienne Guillon [1] and Mathilde Francin-Allami [1,*]

1 INRAE, UR BIA, F-44316 Nantes, France; mehdicherkaoui04@gmail.com (M.C.); virginie.lollier@inra.fr (V.L.); audrey.geairon@inra.fr (A.G.); axelle.bouder@inra.fr (A.B.); colette.larre@inra.fr (C.L.); helene.rogniaux@inra.fr (H.R.); fabienne.guillon@inra.fr (F.G.)

2 Laboratoire de Recherche en Sciences Végétales, Université de Toulouse, CNRS, UPS, 31326 Castanet Tolosan, France; jamet@lrsv.ups-tlse.fr

* Correspondence: mathilde.francin-allami@inra.fr

Received: 15 November 2019; Accepted: 23 December 2019; Published: 29 December 2019

Abstract: The cell wall is an important compartment in grain cells that fulfills both structural and functional roles. It has a dynamic structure that is constantly modified during development and in response to biotic and abiotic stresses. Non-structural cell wall proteins (CWPs) are key players in the remodeling of the cell wall during events that punctuate the plant life. Here, a subcellular and quantitative proteomic approach was carried out to identify CWPs possibly involved in changes in cell wall metabolism at two key stages of wheat grain development: the end of the cellularization step and the beginning of storage accumulation. Endosperm and outer layers of wheat grain were analyzed separately as they have different origins (maternal and seed) and functions in grains. Altogether, 734 proteins with predicted signal peptides were identified (CWPs). Functional annotation of CWPs pointed out a large number of proteins potentially involved in cell wall polysaccharide remodeling. In the grain outer layers, numerous proteins involved in cutin formation or lignin polymerization were found, while an unexpected abundance of proteins annotated as plant invertase/pectin methyl esterase inhibitors were identified in the endosperm. In addition, numerous CWPs were accumulating in the endosperm at the grain filling stage, thus revealing strong metabolic activities in the cell wall during endosperm cell differentiation, while protein accumulation was more intense at the earlier stage of development in outer layers. Altogether, our work gives important information on cell wall metabolism during early grain development in both parts of the grain, namely the endosperm and outer layers. The wheat cell wall proteome is the largest cell wall proteome of a monocot species found so far.

Keywords: cell wall; grain; remodeling; polysaccharide; proteome; wheat

1. Introduction

The wheat grain is a caryopsis composed of an embryo separated from the starchy endosperm and aleurone cells by the scutellum. The progeny is surrounded by the maternal tissues, which are composed of several outer layers [1]. They consist of the pericarp, the testa, and the nucellar epidermis [2]. The pericarp includes epidermal cells, parenchyma cells, and cross and tube cells. Cytological and physiological features of wheat endosperm and pericarp were characterized all along the grain development [3–5].

At the early developmental stage, the outer layers constitute most of the grain volume [3]. By around 150 growing degree days (GDD), the layers of the outer integument are the first to degenerate, whereas endosperm cells continue to divide. The beginning of grain filling occurs around 220 GDD. The aleurone layer becomes recognizable. Endosperm cells stop dividing and the grains reach their maximum fresh mass at around 440 GDD. Concomitantly with the fusion of the pericarp with the maternal epidermis, the expanding endosperm compresses the outer layers. At grain maturity, except for the aleurone cells and embryo, all tissues degenerate, and only cell walls remain.

Cell walls play key roles during grain development. They have various compositions depending on the cell type and developmental stages. Due to their importance in nutrient transport, as well as cell size, shape, and stiffness, cell walls are essential for the determination of the final grain size and the nutritional quality of cereal grain [6–8]. They account for 3% of the endosperm, and around 70% of the outer layers in the mature wheat grain [9]. The cell walls are the primary contributors to cereal dietary fiber with important implications for human nutrition and disease prevention [10]. Cell walls also have important effects on cereal processes, such as milling, brewing, and breadmaking quality [11]. They are mostly composed of polysaccharides, together with a smaller amount of proteins, lipids, and phenolic compounds that participate in the cell wall metabolism.

Cell walls of developing endosperm have recently been described in detail [12–17]. Despite their important roles, both in grain physiology and from a nutritional point of view, cell walls of the outer layers of cereal grain have been poorly studied so far. Callose was the first polysaccharide to be detected during the development of wheat grain [18]. Arabinoxylans (AXs) and mixed β-glucans (MLGs) are the main polysaccharides of wheat grain, and of grasses more broadly. The remaining polysaccharides consist of cellulose, mannans, xyloglucans, and pectins [14,15,18]. MLGs are deposited early during the cellularization step and throughout the grain development. AXs are detected at the beginning of the differentiation in the endosperm cells. Spatial and temporal variations of the AX structure were observed in wheat grain. This structural heterogeneity is mainly explained by the variation in the substitution degree of AXs by arabinose, but also by ferulic acid, and to a lesser extent, p-coumaric acid, which binds to the arabinose residues [12,19]. Although less abundant than in dicot plants, cellulose was observed not only in the outer layers, but also in endosperm cells in a proportion that was not expected [20]. Recently, pectins were detected in the endosperm and the outer layers of wheat grain [14]. An impressive deposition of homogalacturonans (HGs) in the testa was reported at 250 GDD, with a decrease in their methylation degree during grain development. Rhamnogalacturonans of type I (RGI) were mainly detected in the endosperm and mostly in aleurone cells. Spatial and temporal variations of the RGI structure have been observed [14].

In addition to variability of the polysaccharide distribution and structures, the cell walls of the outer layers contain significant amounts of hydrophobic and polyphenolic polymers that compose the lignin and/or cutins. Lignification occurs earlier than previously reported in the outer layers and long before the grain reaches its final size [2]. Several cuticles are formed early in the development and persist at grain maturity. They are considered to be physically associated with the cell walls and have some overlapping functions. These cuticles are mainly composed of cutins and soluble waxes. Polysaccharides are also embedded into the cuticular layers [21]. Recently, xyloglucans were detected in the cuticle above the testa in the grain of the wild grass *Brachypodium distachyon* [22].

As described above, the composition and structure of cell wall polysaccharides vary during grain development and according to grain tissue to answer to the physiological needs of plants. Cell wall polysaccharide modifications imply numerous modifying enzymes, among which, glycosyl hydrolases (GHs) act on polysaccharides and oxidoreductases. They are mainly located in the cell wall and participate in the cell elongation, cuticle deposition, and cell wall structural variation that occurs during grain development. They may be transported by vesicles from the cytoplasm to the extracellular compartment [23]. The identification and functional understanding of these protein actors are essential to gain access to cell wall dynamics. To date, only a few experiments have been reported on grass cell wall proteomes: *Brachypodium distachyon* [24–26], *Saccharum* spp (sugarcane) [27–29], *Oryza sativa* (rice) [30–33], and *Triticum aestivum* (wheat) [34]. Transcriptomic data on the developing wheat grain allowed for the identification of GH genes [15], and recently, we reported the first data on the cell wall proteome of endosperm and the outer layers of wheat grain at a given developmental stage [34].

More than 600 cell wall proteins (CWPs) were identified, half of which were present in both tissues, suggesting common and specific tissue remodeling activities.

In this study, we investigated the cell wall proteome targeting both the endosperm and outer layers of the wheat grain at two key developmental stages in order to report on the cell wall dynamics during grain development. We identified proteins annotated as CWPs that are potentially involved in cell wall polysaccharide remodeling and cell wall assembly. A quantitative analysis allowed for describing the cell wall proteome content during the grain developmental stage for two tissues: the endosperm and the outer layers.

2. Results

2.1. Developing Grain Stages

The aim of this work was to identify CWPs in order to better understand the cell wall dynamics during the development of wheat grains. For this, the experiments were conducted from grains harvested at two developmental stages. The first one corresponded to an early stage of development (150 GDD, Figure 1), where the cellularization step was almost achieved but endosperm cells continued to divide. Aleurone and transfer cells were not yet differentiated. The outer layers derived from maternal tissues were already well differentiated. The outer layers were composed of the pericarp, the testa, and the nucellus epidermis (Figure 1).

The second developmental stage corresponded to the beginning of the accumulation of storage products. At this stage, around 250 GDD, endosperm was well differentiated. Aleurone and transfer cells were easily detected. In addition, the outer tissues progressively began to degenerate (Figure 1).

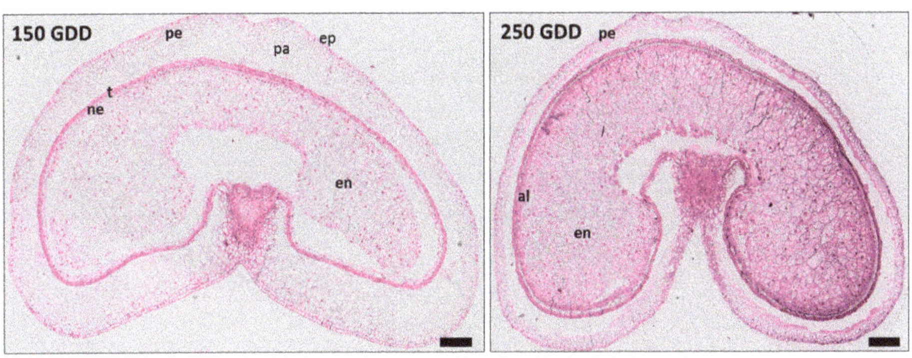

Figure 1. Tissue differentiation and development of wheat grain. Wheat grain cross-sections at (**left**) 150 and (**right**) 250 growing degree days (GDD) using bright-field micrographs of toluidine blue-stained sections. pe: pericarp, ep: external pericarp, pa: pericarp parenchyma, ne: nucellus epidermis, t: testa, al: aleurone layer, en: endosperm. Scale bars: 250 μm.

2.2. Protein Extraction and Proteomic Analysis

For our analysis, the endosperm and outer layers were manually separated from grains at 150 and 250 GDD. Then proteins were extracted from three independent biological replicates per condition. Our procedure led to an enrichment of CWPs in the extracts. The electrophoretic profiles of replicates were similar and large number of bands was observed between 10 and 120 kDa (Figure S1). For each replicate, around 0.5 mg and 0.8 mg of proteins were obtained from 500 mg of fresh mass of endosperm and outer layers materials, respectively. Fifty micrograms of proteins were separated using 1DE (one dimension electrophoresis) and each lane was cut into five slices prior to being trypsin-digested and analyzed using LC-MS/MS. Only proteins identified as having at least two specific peptides were conserved. Interrogation of the UniProt databank restricted to *T. aestivum* led to the identification of 2454 proteins (Table S2). The principal component analysis (PCA) indicated a good clustering of biological replicates and a significant discrimination between the four experimental conditions (Figure 2). The first two components explained 59.5% of the total variance. The different replicates clustered quite separately according to the tissue along the first principal component, and the development stage along the second principal component.

The predicted subcellular localization and functional annotation were achieved using the *ProtAnnDB* pipeline. A dataset of 734 total CWPs was selected, as described in Material and Methods (Table S2), which represents about 30% of all the 2454 identified proteins. In the outer layers, the CWP enrichment was 39%, while in the endosperm it was 22%. CWPs identified from the developing grains of *B. distachyon* represented around 40% of the total number of identified proteins, i.e., in the same range as what was obtained for the outer layers of the wheat grain [25]. Given that the complete genome annotation of wheat has currently not been achieved, a particularly large number of proteins were annotated as "uncharacterized protein". By searching predicted functional domains using the *ProtAnnDB* pipeline, the annotation of our data was refined, leading to the functional annotation of a significant number of additional proteins, thus drastically decreasing the amount of "uncharacterized proteins".

In addition, a statistical analysis of the quantitative data obtained for CWPs was performed in order to consider the variation of CWP abundance between the two developmental stages and the two parts of the grain. The relative quantification of proteins was performed from the signal intensity of their respective peptides in mass spectroscopy (MS), using a label-free and non-targeted approach. Quantification was achieved from a minimum of two peptides per protein. A total of 243 proteins was shown to present variation in their abundance according to parts of the grain and/or developmental stages (Table S3).

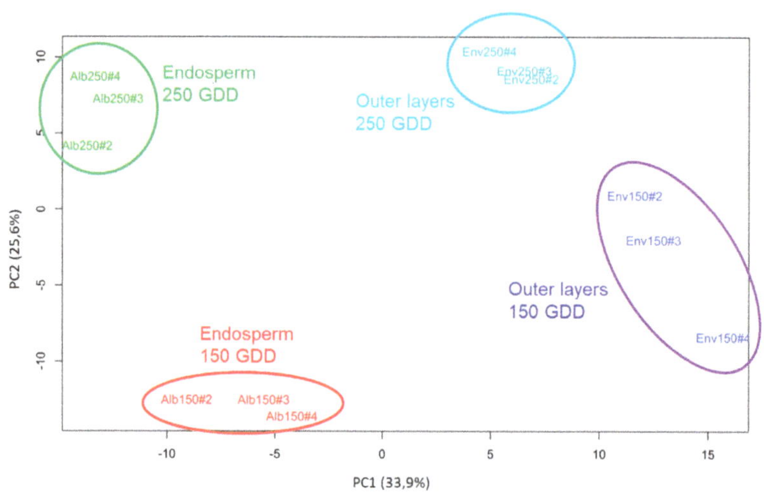

Figure 2. Principal component analysis (PCA) performed using the relative amounts of cell wall proteins (CWPs) for both parts of the wheat grain and at the two developmental stages (150 and 250 GDD). Individual maps of the biological replicates (red: endosperm samples at 150 GDD, green: endosperm samples at 250 GDD, dark blue: outer layers samples at 150 GDD, light blue: endosperm samples at 250 GDD) along principal components 1 and 2.

2.3. Validation of Cell Wall Localization for Three of the Identified Proteins

In order to validate our proteomic analysis and bioinformatics treatment, we decided to check the cell wall localization for three randomly selected CWPs. These proteins were respectively annotated as a pectin methylesterase (PME), a PME inhibitor (PMEI), and a glycoside hydrolase belonging to family 152 (GH152). We therefore generated C-terminal green fluorescent protein (GFP) fusions under the control of the cauliflower mosaic virus (CaMV) 35S constitutive promoter using an appropriate binary vector. The sequences used for the constructs encoded the full length CWPs, including their signal peptides. The constructs were transiently expressed in *N. tabacum* leaves after *A. tumefaciens* infiltration. The subcellular localization of the fusion proteins was analyzed using confocal laser scanning microscopy. The fluorescence of the protein fusions PME::GFP, GH152::GFP, and PMEI::GFP was detected at the periphery of the cells; this was consistent with their localization in the cell wall. A plasmolysis of the infiltrated leaf pieces was performed to separate the plasma membranes from the cell walls. The results indicated that the three fusion proteins were present in the cell wall and in the apoplastic compartment, as observed with the apoplastic marker Sec-mRFP (Figure 3).

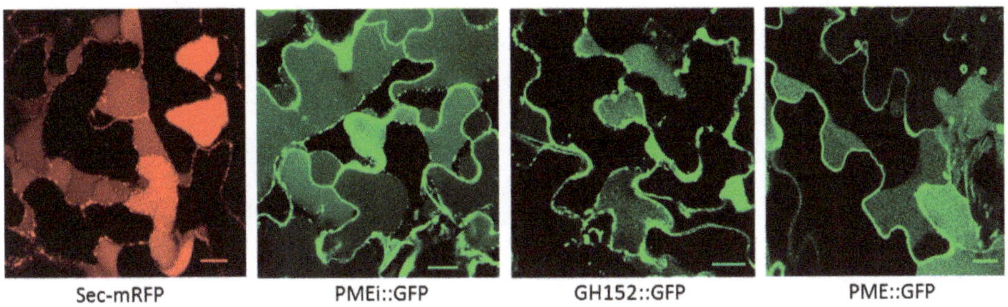

Figure 3. Subcellular localization of a PMEI (A0A1D6BK49), a GH152 (A0A1D6DD47), and a PME (A0A1D5V0T8) fused to a green fluorescent protein (GFP) in plasmolyzed *N. tabacum* leaf epidermal cells. Confocal images of plasmolyzed cells expressing the apoplastic marker Sec-mRFP, PMEI::GFP, GH152::GFP, and PME::GFP in leaf epidermal cells 2–3 days after agroinfiltration. Scale bars: 20 μm.

2.4. Distribution of the CWPs in Functional Classes

The CWPs are distributed into eight of the nine functional classes, as defined by the *ProtAnnDB* tool [35] (Figure 4A). No protein was annotated as a structural protein that has been currently observed in cell wall proteomes of monocots [36]. The most effective functional class was that of "proteins acting on cell wall polysaccharides" (PACs) with around a quarter of the CWPs. Inside this functional class, we mainly found GHs, which was consistent with previous cell wall proteome studies [25,26,28,37] (Table S2). PACs also contained expansins, carbohydrate esterases, and to a lesser extent, pectate lyases, PNGases (*N*-glycosidases), and COBRA-like proteins. Altogether, 136 GHs were identified. They belong to 21 different GH families (http://www.cazy.org/). They are thought to play various functions in cell expansion, signaling, plant defense, and the hydrolysis of glycans from glycoproteins and polysaccharides [38]. GHs were fairly distributed between the developmental stages and grain parts. Among them, the best-represented families were GH3 and GH17, as previously found in *B. distachyon* grains [24]. The classes of "proteases" and "proteins with interacting domains" were also well represented. The functional class of "signaling proteins" accounted for 4% of the total CWPs, as currently found in other cell wall proteome of monocots (3.4% for overall cell wall proteomes of monocots [39]; 6% if only the cell wall proteome of *B. distachyon* grains is considered [25]). Among the proteins of yet unknown function, DUF642 proteins were found as in all the previously studied cell wall proteomes (see *WallProtDB*).

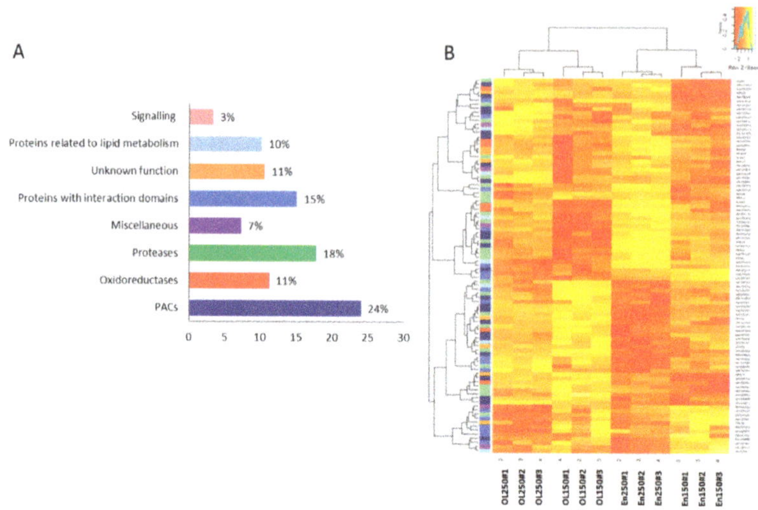

Figure 4. Distribution of CWPs and variation of their abundance according to the different conditions. (**A**) Distribution of CWPs into functional classes according to their predicted functions. PACs: proteins acting on cell wall polysaccharides. (**B**) Heatmap of the relative amounts of CWPs identified in the four conditions (in the endosperm (EN), in the outer layers (OL), and at 150 and 250 GDD).

2.5. *Distribution of CWPs and Their Abundance Variation According to Tissue*

Among the CWPs, almost two times more proteins were identified in the outer layers compared to the endosperm (Figure 5). More than 300 CWPs were shared between the two parts of the grain.

Among the CWPs identified in the endosperm, few of them were specific to this tissue. Several of them were annotated as proteins with leucine-rich repeats (LRR proteins), class III peroxidases, multicopper oxidases, or plant invertases/PME inhibitors, and others corresponded to the COBRA-like, DUF296, DUF239, and antimicrobial peptide (AMP) protein families. No GH family was found exclusively in the endosperm. In contrast, the outer layers accounted for numerous specific CWPs (Figure 5, Table S2) with proteins belonging to families of pectate lyases, acetyl xylan esterases (AXEs), GH10, GH13, GH32, laccases, and some DUF proteins (DUF642, 1929, 568, and 760).

Although numerous protein families were identified, both in the endosperm and outer layers, some of them differed by the number of their members between these two tissues. Some protein families were best represented in one of these two grain parts. This was the case for the fasciclin-like arabinogalactan proteins (FLAs), GDSL lipases/esterases, GH31, GH35, class III peroxidases, purple acid phosphatases (PAPs), and pectin acetylesterases (PAEs), which were more represented in the outer layers. Due to the relatively low number of CWPs identified in the endosperm compared to the outer layers, it was more difficult to reveal the protein families best represented in this tissue. However, our results clearly indicated that proteins with a protease-associated (PA) domain, α-amylase inhibitors, GH28, GH29, and plant invertases/PME inhibitors were well represented in the endosperm (Table S2).

Considering the developmental stages, only a third of the CWPs were shared by the endosperm and outer layers (Figure 5). This was partly due to the big difference in the numbers of CWPs identified in the endosperm and outer layers. The classes of "oxidoreductases," "proteins related to the lipid metabolism," and "proteins possibly involved in signaling proteins" were more numerous in the outer layers compared to the endosperm (Figure S2). Inversely, the class of "proteins with interacting domains" was best represented in the endosperm, with a significant proportion of CWPs annotated as plant invertase/PME inhibitors (Figure S2, Table S2).

From a statistical point of view, variations of the CWPs' abundance were observed between the endosperm and outer layers at a given developmental stage (150 GDD or 250 GDD) (Table S3, Figure S2). Among the CWPs shared by the endosperm and outer layers at 150 GDD, 60 were accumulated at a higher level in the outer layers, and only 28 in the endosperm. At 250 GDD, the numbers of CWPs accumulated at a high level in the outer layers and endosperm were closer (57 CWPs in outer layers versus 47 in endosperm), revealing a strong cell wall dynamic in both parts of the grain. More dirigent proteins, oxidoreductases, and thaumatins were accumulated at a higher level in the outer layers than in the endosperm, especially at 150 GDD. Some GHs (GH1, GH17, and GH79) were also accumulated in the outer layers, whatever the developmental stage. In the endosperm, plant invertases/PME inhibitors were found to have greater levels of accumulation compared to the outer layers, as well as some protease inhibitors and GH3.

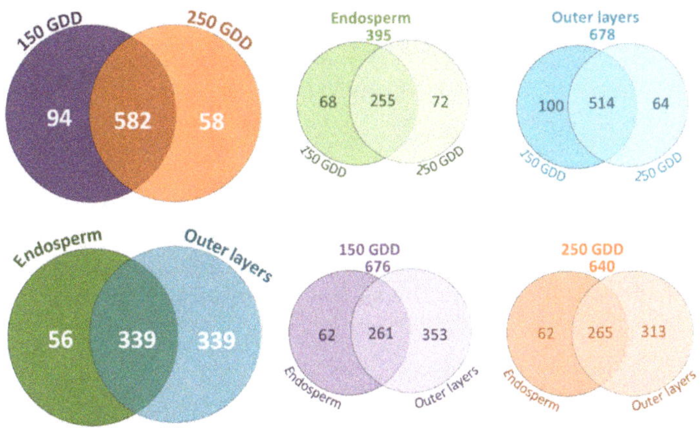

Figure 5. Venn diagrams showing the distribution of CWPs according to the wheat developmental stages and grain parts.

2.6. Distribution of CWPs and Their Abundance Variation According to the Developmental Stage

Few differences were noticed between the total number of CWPs identified at 150 GDD and 250 GDD (676 vs. 640 CWPs) (Figure 5, Table S2). A large majority of them were shared by the two stages. Only a few differences were observed for each protein family between the two developmental stages. However, a slightly higher number of expansins, lipid transfert proteins (LTPs), invertase/PME inhibitors, class III peroxidases, and plastocyanins was noticed at 150 GDD compared to 250 GDD. On the other side, arabinofuranosidases (GH51 and GH146), proteins with predicted protease-associated (PA) or cupin domains, and some proteases (subtilases) were slightly more represented at 250 GDD compared to 150 GDD. Whereas the GH10, DUF760, and DUF296 families were only identified at 150 GDD, lectins and COBRA proteins were only found at 250 GDD (Table S2).

Among the CWPs identified at 150 GDD, around one hundred were specific to this stage, with more proteins identified specifically in the outer layers compared to those identified in the endosperm (Figure 5, Table S2). While some invertase/PME inhibitors were specific to the endosperm at 150 GDD, several AXEs, class III peroxidases, and DUF538 proteins were specifically found in the outer layers at this earlier developmental stage. At 250 GDD, we identified 58 CWPs specific to this stage, with twice as many identified proteins in the outer layers compared to the endosperm, two arabinofuranosidases and two subtilases among them (Table S2).

Considering the endosperm, more than twice as many CWPs were significantly accumulated at higher level at 250 GDD compared to 150 GDD (52 vs. 21). The opposite was observed in the outer layers (Table S3, Figure S3). This could reflect an intensive cell wall metabolism in the outer layers at 150 GDD, whereas the cell wall metabolism in the endosperm seemed to be more dynamic at 250 GDD.

In this way, numerous proteases, but also FLAs, leucine rich repeat-containing proteins (LRR-proteins), and GH9 were accumulated at 250 GDD in the endosperm. Inversely, in the outer layers, proteins annotated as AXEs, GDSL lipases/esterases, or proteases were particularly abundant at 150 GDD (Table S3).

3. Discussion

In this work, we investigated the cell wall proteome of two distinct parts of the wheat grain. The outer layers and endosperm are composed of distinct cell layers that have different functions in grain. Thus, the cell wall structures are very different and we were expecting to obtain distinct CWP profiles between these two parts of grain. Two early developmental stages, 150 GDD (end of cellularization step) and 250 GDD (differentiation step and beginning of storage accumulation) were considered since clear differences at the histological and cell levels (Figure 1) and in the cell wall composition have been described between these two stages [12,14,17]. Although numerous cell wall events occur after 250 GDD, the accumulation of storage proteins, such as glutenins and gliadins in wheat endosperm, makes wall preparation and proteomic analysis more difficult. Indeed, due to the amount of storage proteins being particularly high from 300 GDD, the detection of CWPs, which are by weight minor cell wall components, would be very hard.

Although the outer layers can be easily detached from the endosperm of early harvested wheat via manual dissection [40], we cannot totally exclude protein cross-contaminations between these two tissues. However, the PCA indicated a strong discrimination between the endosperm and outer layers samples collected at 150 or 250 GDD, as well as a low variability between replicates (Figure 2). Because most of the polysaccharides were deposited both in the endosperm and outer layers, and since several cell wall remodeling events occurred in both grain tissues, numerous CWPs were expected to be identified in both of them. Forty-six percent of the identified CWPs were indeed found in both parts of grain.

To analyze the cell wall proteome, we performed cell wall enrichment and the proteins were extracted with $CaCl_2$ and LiCl buffers using a protocol successfully used on different plant species and tissues [25,26,28,29,37,41,42]. In this work, we identified a total of 2454 proteins, 734 of them being predicted as secreted proteins present in the cell wall or the apoplast, and called CWPs. CWPs were more numerous in the outer layers compared to the endosperm. The outer layers contained several cell layers, whereas endosperm tissues only contained starchy endosperm and aleurone cells. This could explain why more CWPs were identified in the outer layers, reflecting the presence of more different cell types. In addition, the outer layers contained more cell walls than the endosperm, leading to a higher protein amount. Also, we cannot exclude a better extractability of proteins from the outer layers than from endosperm. Conversely, almost the same number of CWPs were identified at both developmental stages of grains, with around 90% being shared proteins. These results, reinforced by the PCA data, suggest that in our experiments, the tissue effect was stronger than the developmental stage effect. However, many cell wall metabolism events differed from 150 to 250 GDD, such as the aleurone cell differentiation or the start of the pericarp degeneration, which may involve different CWPs [3,5]. Additionally, these distinct metabolic events were certainly reflected by differences in protein accumulation (Table S3). Many CWPs were more abundant in the endosperm at 250 GDD (in particular expansins, GH9, or GH28), while accumulation of a large number of CWPs was observed at 150 GDD in the outer layers (with numerous GDSL lipases/esterases, or oxidoreductases, such as class III peroxidases) (Table S3, Figure S3).

The distribution of wheat grain CWPs into functional classes slightly differed from those of the other cell wall proteomes. It appeared that the cell wall proteomes of grass grains (wheat and *B. distachyon*) contained more proteases and less oxido-reductases than the cell wall proteomes already described (Figure 4, [24–46]). The cell wall proteome of wheat grain also stood out as having numerous proteins with predicted interaction domains compared to other cell wall proteomes, including that of the *B. distachyon* grain [24,25]. Among proteins with interaction domains, more than 50 CWPs belonged to the superfamily of the "plant invertase/PME inhibitors". Although found in all conditions, these proteins were more represented in the endosperm, and at 150 GDD. The transient expression of a fusion protein plant invertase/PME inhibitor::GFP in tobacco leaves confirmed its presence in the cell wall and in the apoplast (Figure 3). Wheat has a complex genome, with six copies of each chromosome and many nearly identical sequences scattered throughout, which could account for the large number of protein isoforms. As a matter of fact, many of the identified plant invertase/PME inhibitors present a strong homology. In wheat grain, there is a considerable number of plant invertase/PME inhibitors compared to other plant cell wall proteomes. In the cell wall proteomes of *A. thaliana* and *B. distachyon* (compilated data from *WallProtDB*), only 12 and 7 plant invertase/PME inhibitors were identified, respectively. In *B. distachyon,* they were all found in the developing grain [25]. In rice, several of the 54 predicted invertase/PME inhibitors are expressed specifically in developing endosperm [44]. PMEIs inhibit the demethylesterification of HGs by PMEs [45]. In wheat, PMEIs were suggested to play a role in plant defense mechanisms by decreasing the cell wall porosity and/or by conferring resistance against several pathogens [46,47]. Recently, two invertase inhibitor genes were identified in sugarcane, one of them encoding a protein (ShINH1) localized in the apoplast of the root, flower, stalk, and leaf. Heterologous expression of *ShINH1* confirmed its possible function as an invertase inhibitor [48]. In the same way, an apoplastic invertase inhibitor was shown to affect seed mass in soybean by regulating extracellular invertase during seed maturation [49]. Invertase inhibitor ZM-INVINH1 has been reported to bind cell wall invertase during grain development in maize [50]. Invertases belong to the GH32, GH68, and GH100 families. In our study, five GH32 that were predicted to be secreted were only found in the outer layers, which is consistent with a preferential presence of such proteins in the endosperm. On the other hand, the high amount of the plant invertase/PME inhibitors raises the possibility that they could regulate other carbohydrate metabolizing enzymes, as previously suggested [44].

Additional protein families were also well represented in the endosperm, such as GHs belonging to the GH9, GH28, and GH38 families, respectively annotated as endo-β-1,4-glucanases, polygalacturonases, and α-mannosidases. GH28 proteins were potentially active on HGs. Although present at the both developmental stages, some were more abundant at 250 GDD in the endosperm. It was already surprising to identify numerous GH28s in the cell wall proteome of the *B. distachyon* grain, at three developmental stages (9, 13, and 19 days after flowering), pectin also being poorly abundant in this plant species [25]. Although cell walls of grasses contain low level of pectins (HGs and rhamnogalacturonans), the presence of such enzymes is consistent with the remodeling of early synthesized pectins in the developing grain [14,17]. Two putative α-mannosidases belonging to the GH38 family were previously identified in the developing grain of *B. distachyon*, but no data are available concerning their fine localization [25]. Many plant α-mannosidases have been isolated and characterized from seeds and fruits. More generally located in the Golgi or endoplasmic reticulum, they are known to be involved in the maturation of *N*-glycans, and they appear to play a critical role in plant development [51,52]. Finally, the quantitative data indicated that GH9 were more abundant at 250 GDD than at 150 GDD in the endosperm. These enzymes are usually able to hydrolyze artificial soluble cellulose derivatives, such as carboxymethyl cellulose (CMC) or hydroxyethyl cellulose (HEC), and biochemical analyses have revealed their specificity for different substrates in vitro [53]. More generally, a large proportion of the endosperm CWPs were more abundant at 250 GDD than at 150 GDD, suggesting a more intensive cell wall metabolic activity in this tissue at this developmental stage of grain. This could be explained by the differentiation of the aleurone cells in addition to the expansion of the starchy endosperm cells at the beginning of storage accumulation.

Our work revealed significantly different cell wall proteomes between the endosperm and outer layers of the wheat grain, which could be largely explained by the differences of the cell wall structures between these two parts of grain. Thus, numerous protein families, highly represented in the cell wall proteome of outer layers, are known to act not only on cell wall polysaccharides but also on hydroxycinnamic acids, lignin, or cuticle formation, the last two being characteristic of the outer layers. This was the case with the GDSL lipases/esterases being highly numerous at 150 GDD. These proteins are members of a large family that display numerous functions, such as cutin synthase, cutin hydrolase, deacetylation of polysaccharides, polysaccharide hydrolysis, fucosidase, and esterification of secondary metabolites [54]. They certainly collaborate to implement the cuticle during grain development, including the association of cutin with polysaccharides and the rearrangement of cutin and polysaccharides [54]. In wheat, cuticles were observed early during grain development [14], which is consistent with the presence of GDSL lipases/esterases at 150 GDD in outer layers. The accumulation of lipid transfer proteins (LTPs) in the outer layer matches with their role in the synthesis of cuticular waxes [21,55]. However, although less numerous, their presence in the endosperm suggest another function, such as antimicrobial defense or signaling during pathogen attacks [55]. Beside the cuticle, other non-polysaccharidic components are present in cell walls, such as lignin or hydroxycinnamic acids. Feruloylated AXs and lignins were indeed detected early in developing grain [14]. Current evidence suggests that two enzyme families, class III peroxidases and laccases, are involved in lignin polymerization [56]. These proteins were identified in the outer layers of developing grain in wheat. One laccase was identified only in the outer layers and numerous class III peroxidases were found mainly in the outer layers at 150 GDD with several of them more abundant at 150 GDD than at 250 GDD. Moreover, class III peroxidases could also be involved in AX crosslinking by ferulic acid, as previously shown in maize [57].

The outer layers and endosperm exhibit distinct functions that might explain differences of the cell wall proteomes. For instance, we found eight members of the GH19 family with four of them only identified in outer layers. In addition, the GH19 found in the two parts of the grain were always more abundant in the outer layers. These proteins are involved in the biotic stress response and they are known to be defense weapons against fungi by exhibiting chitinase activity [58]. These findings correlate with the protective role of outer layers throughout grain development. Among the protein families of the PAC class, many have a higher population in the outer layers. This is the case for the DUF642 proteins, which were identified only in the outer layers at both developmental stages. The DUF642 proteins belong to a well-conserved family. They were shown to be involved in plant development [59]. They may modulate the activity of PMEs, which are more abundant in the outer layers than in the endosperm, suggesting that their function could be related to the regulation of HG modifications by interacting with the catalytic domain of PME. Their interaction with cellulose was also demonstrated in vitro but no functional studies yet confirm these interactions [60]. Pectin acetylesterases (PAEs) and AXEs were almost only found in outer layers. Acetylated AXs were detected in young wheat grain (245 GDD), but not at a mature stage (700 GDD) [61]. AXEs could participate in the deacetylation of AXs during the grain development. The deacetylation of AXs could improve the rigidity of the cell wall, as deacetylation of xylans facilitates their bonding to cellulose [62].

Other protein families, even though not specific to one developmental stage, are best represented at 150 GDD and in the outer layers, particularly the expansin, DUF538 proteins, and plastocyanins. In wheat, expansin genes were reported to be expressed at a high level during early grain expansion, with a peak of *TaExp6* transcript accumulation in the pericarp at 141 GDD, and a higher level of expression in the pericarp and endosperm at 187 GDD [7]. These results are consistent with the identification of more expansins at the cellularization step, and the presence in both the endosperm and outer layers. Moreover, the role of expansins in grain is reinforced by a recent study suggesting that expansin genes could be specifically involved in grain enlargement and may control seed growth in sunflower at the early phases of development [63]. As in wheat, several DUF538 were identified in the developing grain of *B. distachyon* [25]. Several functions have been suggested for proteins containing this domain of yet unknown function, such as their involvement in the photosynthetic system of plants by degrading chlorophyll molecules to antioxidant compounds under stress induction [64,65]. Initially characterized as chloroplastic proteins, some DUF538 proteins were shown to be in other cell compartments. Recently, a maize DUF538 protein, Glossy6, was demonstrated to be involved in cuticular wax accumulation and drought tolerance [66]. Proteins annotated as plastocyanin-like proteins, also named plantacyanins, were also preferentially found at 150 GDD and in the outer layers. They are blue copper-binding proteins belonging to the oxidoreductase class, identified in numerous cell wall proteomes, such as in stems, leaves, and the developing grain of *B. distachyon* [24,26]. Recently it was shown that the blue copper proteins from the phytocyanin family were involved in the control of panicle branching and grain weight in rice through their regulation by the microRNA miR408 [67].

Proteases are well represented in the early developmental stages, and especially at 250 GDD. Endoprotease activities were characterized in developing wheat grain [68]. Serine protease activities were characterized at early stages of grain development, with metalloproteases and Asp proteases appearing later. We identified proteases belonging to the Asp-protease, Cys–protease, or serine carboxypeptidase families, both in outer layers and the endosperm. In the endosperm, several of them were more abundant at 250 GDD than at 150 GDD, suggesting stronger metabolic activities at 250 GDD in the endosperm cell wall. Endoproteases could participate in cell wall dynamics through the maturation of cell wall remodeling enzymes, their degradation, or the release of signaling peptides. In the case of subtilases, forming a large Ser protease family, it was suggested they were involved in the regulation of PMEs in *Arabidopsis* seeds [69,70]. A subtilase specifically found in the endosperm of *Medicago truncatula* or *Pisum sativum* seeds during development was shown to influence seed weight [71]. It was also suggested that endoproteases may be involved in the degeneration of the pericarp from the differentiation stage to the grain maturation, whereas in the endosperm, they may be accumulated for protein metabolism (processing or turn-over) or further degradation of storage proteins [68].

4. Material and Methods

4.1. Plant Material

Triticum aestivum (cv Recital) was grown under natural day-length conditions in a greenhouse (INRA Le Rheu, France). Two months of vernalization were applied in a growth chamber at 8 °C before the wheat seedlings were transplanted into individual pots containing a standard potting mixture (peat RHP15 Klassman, K Klassman France, Bourgoin Jallieu, France). An Osmocote (R) Exact Tablet containing nitrogen (15%), phosphate (9%), potassium hydroxide (9%), and magnesium (3%) (Scotts International B. V., Waardenburg, The Netherlands) was added. The plants were watered daily. To harvest grains at defined developmental stages, individual ears were tagged at flowering. The temperature was recorded every day in order to follow the plant development on the basis of the cumulated temperature in growing Celsius degrees days (GDD). Plants were harvested at 150 and 250 GDD after flowering. Grains were manually dissected to separate the outer layers and endosperm and immediately stored at −80 °C.

Nicotiana tabacum plants were grown under a 16-h light/8-h dark photoperiod at a temperature of 24 °C day/18 °C night for five weeks in a growth chamber prior to *Agrobacterium tumefaciens* infiltration.

4.2. Cell Wall Enrichment and Protein Extraction

Cell wall fractionation and CWPs extraction were performed as previously described [34]. Briefly, grain samples (500 mg) were ground and centrifuged in increasing sucrose concentrations to isolate cell walls. Proteins were extracted via successive washes in a 5 mM acetate buffer at pH 4.6, supplemented with 0.2 M $CaCl_2$ or 2 M LiCl. Three biological replicates were performed for each condition (developmental stages and parts of the grain).

4.3. Mass Spectrometry Analysis

Proteins (50 µg) were briefly separated in precast gels in order to reduce the mix complexity. Each track was cut into five fragments, which were reduced using dithiothreitol (DTT), alkylated by iodoacetamide and trypsinolyzed as previously described [34], prior to mass spectrometry (MS) analysis.

The trypsin-digested samples (6 μL) were injected into the LC-MS/MS system. The MS analyses were performed using an LTQ-Orbitrap VELOS mass spectrometer (i.e., a linear ion trap quadrupole mass filter associated with an Orbitrap^{TM} analyzer, Thermo-Fisher Scientific, Bremen, Germany) coupled to a nanoscale LC system (Ultimate U3000 RSLC system, Thermo-Fisher Scientific). Chromatographic separation was performed on a reversed-phase capillary column (Acclaim® PepMapTM C18 2 μm 100 A, 75 μm i.d. × 50 cm long, Thermo-Fisher Scientific, Waltham, MA, USA) at 60 °C using a linear gradient performed between the following mobile phases: (A) 97.9% water, 2% acetonitrile (ACN), 0.1% trifluoroacetic acid (TFA); and (B) 90% ACN, 0.08% formic acid. The other parameters were as described previously [72]. Analyses were performed using a typical survey method, in which full MS scans were acquired at 60,000 resolution (full width at half-maximum, FWHM) using the Orbitrap analyzer (*m/z* 300–2000), while the CID (collision induced dissociation) spectra for the eight most intense ions were recorded in the LTQ.

4.4. Bioinformatics Treatment, Functional Annotation, and Label-Free Quantification

The LC-MS/MS raw data were processed into mzXML files and were further searched against the Uniprot database restricted to *T. aestivum* (release May 2017, 136,892 accessions) and a contaminant database including human keratin and trypsin by using the X!Tandem pipeline software (version 3.4.3 "Elastine durcie", http://pappso.inra.fr/bioinfo/xtandempipeline/), which encapsulates the X!Tandem program and supplies a graphical interface to manage results [73].

Enzymatic cleavage was declared as a tryptic digestion with one possible miscleavage event. The carbamidomethylation of Cys residues, as well as the possible oxidation of Met residues, were considered in the modification parameters of the software. Precursor mass and fragment mass tolerance were set at 5 ppm and 0.5 Da, respectively. The results obtained for the peptide identification were validated by filtering the peptides with an e-value below 10^{-2}. Proteins were identified using at least two specific peptides and an e-value score of 10^{-4}.

The *ProtAnnDB* pipeline (http://www.polebio.lrsv.ups-tlse.fr/ProtAnnDB/) was used to annotate the identified proteins [35]. To predict the subcellular localization of the proteins, three software programs were used: TargetP and SignalP, which are both included in the pipeline, and Phobius (http://phobius.sbc.su.se/). Proteins were considered to be secreted when they had no ER retention signal and displayed a predicted signal peptide from their sequence by at least two out of the three software programs. In order to identify CWPs and annotate the dataset, three functional domains prediction software programs were used: Pfam, InterPro, and Prosite, which are all included in *ProtAnnDB*. The information provided by these tools allowed for the identification of CWPs in our dataset and classified them into nine functional classes. The proteomic data of the present work was included in the *Pride* database (http://www.proteomexchange.org/; dataset identifier: PXD016258).

Spectra with confident identifications were used for label-free quantification based on extracted ion chromatograms (XICs) computed using the MassChroQ tool (MassChroQ Gui, version 2.2.2) [74]. The resulting table of peptide XICs was then processed by the Rstudio statistical environment (version 1.0.143) (Boston, MA, USA). We first applied a filter to the outlying retention times of identified peptides to prevent possible false identifications. In this scope, the standard deviation of retention times for each of them was computed through the whole dataset and a 25 s threshold was applied. In a second step, we normalized the computed XICs on runs twice to remove any engine effect with the help of the experimental TIC, and on biological replicates in order to compare lanes. More precisely, we applied a simple correction factor on each computed XIC to place 1D lanes in the same range of values. The third step filtered the peptide population. Indeed, we chose to keep only peptides that were specific to one protein, assuming that the abundances of peptides should be correlated to the abundance of the related protein. Additionally, these peptides had to be identified in two of three biological replicates within at least one condition (developmental stages or grain parts). Finally, the relative abundance of a protein in a biological replicate was computed by summing the XIC values of the selected peptides.

4.5. Statistical Analysis

Statistical analysis was performed on $\log_{10}$ transformed abundance values. To manage missing values, we replaced them with a value mimicking a minimum mass spectrometer detection threshold and by picking at random such a value in a range between the minimum value of the XIC dataset and half of this minimum.

The ade4 package (version 1.7-6, http://pbil.univ-lyon1.fr/ade4/home.php?lang=eng) was used to perform PCA analysis on scaled data. In order to determine the CWPs showing a significant effect of developmental stages or grain parts on their abundance, one-way ANOVA was used. Protein abundances were considered to be significantly different for *p*-values below 0.05 and a data subset of "significant proteins" was created from this analysis.

We used this subset, restricted to CWPs to focus on, to build heat maps with the gplots package on covariance matrix (version 3.0.1) (Figure 4B, Figure S3).

4.6. Subcellular Localization

Total RNA was extracted from wheat grains harvested from 150 to 250 GDD using an RNA kit (Qiagen, Courtaboeuf, France). RNA samples were then treated twice with a DNase buffer and purified using the RNeasy MinElute Cleanup kit (Qiagen, Dusseldorf, Germany) by following the manufacturer's instructions. Reverse transcription was carried out with 2 µg of total RNA, anchored oligo-dT, and the Transcriptor First Strand cDNA synthesis kit (Roche Applied Science, Mannheim, Germany). The binary vector pvKH18En6 was used to generate the constructs encoding the green fluorescent protein (GFP) C-terminal fusion of the three proteins of interest. The full length cDNAs encoding a pectin methylesterase (PME, A0A1D5V0T8), a PME inhibitor (PMEI, A0A1D6BK49), and a glycoside hydrolase belonging to family 152 (GH152, A0A1D6DD47) were amplified via PCR using specific oligonucleotide primers (Table S1). The corresponding amplicons were purified and cloned into the pvKH18En6 binary vector [75] using the XbaI and BamHI restriction sites leading to the upstream fusion to GFP. The construct Sec-mRFP was used as an aploplactic marker [76]. Transformation of *A. tumefaciens* and the transient expression of the constructs in tobacco leaf epidermal cells were performed as previously described [77]. Transformed *A. tumefaciens* cultures were resuspended in an infiltration buffer at $OD_{600nm} = 0.05$. Fluorescence was examined with an inverted Nikon A1 confocal laser-scanning microscope (Nikon Europe, Amsterdam, The Netherlands) at three days after infiltration. Infiltrated leaf samples were placed in a solution containing 30% glycerol 5 min before imaging with the confocal microscope to plasmolyze the cells.

4.7. Histochemistry

Wheat grains at the developmental stages of approximately 150 GDD and 250 GDD were fixed overnight in 3% (*w*/*v*) paraformaldehyde in a 0.1 M phosphate buffer at pH7.4, dehydrated in ethanol series, and embedded with LR-White resin as previously described [14]. Transverse semi-thin sections (1-µm thickness) of grains were cut with an ultracut (UC7, Leica, Nussloch, Germany) and stained with Toluidine Blue O (1% in 2.5% Na_2CO_3 for 5 min, then washed in water). Stained sections were observed using a Multizoom Macroscope (AZ100M, Nikon Europe, Amsterdam, The Netherlands) under bright-field conditions.

5. Conclusions

Our quantitative proteomic work provides an overview of CWPs of early-developing grains of wheat, both in the endosperm and the outer layers. It provides important knowledge for a better understanding of cell wall deposition and remodeling during grain development that appears essential to controlling cereal quality and yield. Certainly due to the different tissues that composed the outer layers, more CWPs were identified, reflecting more numerous protein activities than in the endosperm. Additionally, a higher level of accumulation of a significant number of CWPs was found in the endosperm at 250 GDD, revealing strong metabolic activities in endosperm cell walls at this developmental stage. Altogether, many CWPs could be involved in the cell wall polysaccharide remodeling, as well as in lignin and cutin formation in the outer layers. In addition, an unexpected number of proteins annotated as plant invertase/PME inhibitors were found in the endosperm. Further characterization is required to confirm their precise function during grain development. To date, this is the most complete cell wall proteome from a monocot species, and the first one to be performed on separated tissues of cereal grains.

Supplementary Materials: Supplementary materials can be found at http://www.mdpi.com/1422-0067/21/1/239/s1. Table S1: Sequences of oligonucleotide primers used for gene amplification. Table S2: (A) Proteins extracted from cell walls of the developing wheat grain (endosperm and outer layers) at two developmental stages and predicted to be secreted (the so-called CWPs). (B) Proteins extracted from the cell walls of the developing wheat grain (endosperm and outer layers) at two developmental stages and predicted to be intracellular. Table S3: Statistical analysis performed on the quantitative data corresponding to wheat CWPs identified at two developmental stages, both in the endosperm and outer layers. S3a. CWPs differentially accumulated in different grain parts at 150 GDD. S3b. CWPs differentially accumulated in different grain parts at 250 GDD. S3c. CWPs differentially accumulated in the endosperm at 150 and 250 GDD. S3d. CWPs differentially accumulated in the outer layers at 150 and 250 GDD. Figure S1: 1DE profiles of wheat grain proteins extracted from enriched cell wall fractions. Ten micrograms of proteins from each sample (three biological replicates (1–3) of proteins extracted from the outer layers (OL) or endosperm (E) at 150 GDD and 250 GDD have been separated by 1DE and stained with Coomassie Blue. Molecular mass markers (M) are in kDa. Figure S2: Distribution into functional classes of CWPs identified in the endosperm and the outer layers at 150 GDD and 250 GDD according to their predicted function. PACs: Proteins acting on cell wall polysaccharides. Figure S3: Heat maps of relative amounts of the CWPs identified in the endosperm or in the outer layers, and at one of two early developmental stages. En = endosperm samples; OL = outer layers samples; 150 = 150 GDD samples; 250 = 250 GDD samples.

Author Contributions: M.F.-A. coordinated the project and wrote and reviewed the manuscript; M.F.-A., F.G., C.L., and H.R. conceived and designed the experiments and edited the manuscript; M.C. and A.G. performed the experiments; A.B. performed the validation analysis; V.L. and M.C. conducted the bioinformatics and performed the related experiments; M.C. and M.F.-A. analyzed the data; E.J. reviewed and supervised the manuscript. All authors have read and agreed to the published version of the manuscript.

Funding: This research received no external funding.

Acknowledgments: The authors are thankful to INRA (National Institute of Agronomic Research) for supporting their research work. EJ wishes to thank the CNRS and the Paul Sabatier-Toulouse 3 University for supporting her research. Wheat plants were cultivated at INRA of Rheu. LC-MS/MS analyses were performed on the Platform BIBS (Biology and Structural Biochemistry) in Nantes. The authors wish to thank the following contributors: Jean-Christophe Helleisen (INRA of Rheu) for growing wheat plants; Hélène San Clemente (LRSV, UPS) for her contribution to the functional annotation of proteins; and Camille Alvarado for providing images of toluidine blue-stained cross-sections of developing wheat grains.

Conflicts of Interest: The authors declare no conflict of interest.

References

1. Stone, B.A. Cell Walls of Cereal Grains. *Cereal Foods World* **2006**, *51*, 62–65. [CrossRef]
2. Chateigner-Boutin, A.-L.; Lapierre, C.; Alvarado, C.; Yoshinaga, A.; Barron, C.; Bouchet, B.; Bakan, B.; Saulnier, L.; Devaux, M.-F.; Girousse, C.; et al. Ferulate and lignin cross-links increase in cell walls of wheat grain outer layers during late development. *Plant Sci.* **2018**, *276*, 199–207. [CrossRef] [PubMed]
3. Xiong, F.; Yu, X.R.; Zhou, L.; Wang, F.; Xiong, A.S. Structural and physiological characterization during wheat pericarp development. *Plant Cell Rep.* **2013**, *32*, 1309–1320. [CrossRef] [PubMed]
4. Xurun, Y.; Xinyu, C.; Liang, Z.; Jing, Z.; Heng, Y.; Shanshan, S.; Fei, X.; Zhong, W. Structural development of wheat nutrient transfer tissues and their relationships with filial tissues development. *Protoplasma* **2015**, *252*, 605–617. [CrossRef]
5. Evers, T.; Millar, S. Cereal grain structure and development: Some implications for quality. *J. Cereal Sci.* **2002**, *36*, 261–284. [CrossRef]
6. Cosgrove, D.J. **Relaxation** in a high-stress environment: The molecular bases of extensible cell walls and cell enlargement. *Plant Cell* **1997**, *9*, 1031–1041. [CrossRef]
7. Lizana, X.C.; Riegel, R.; Gomez, L.D.; Herrera, J.; Isla, A.; McQueen-Mason, S.J.; Calderini, D.F. Expansins expression is associated with grain size dynamics in wheat (*Triticum aestivum* L.). *J. Exp. Bot.* **2010**, *61*, 1147–1157. [CrossRef]

8. Yan, L.; Liu, Z.; Xu, H.; Zhang, X.; Zhao, A.; Liang, F.; Xin, M.; Peng, H.; Yao, Y.; Sun, Q.; et al. Transcriptome analysis reveals potential mechanisms for different grain size between natural and resynthesized allohexaploid wheats with near-identical AABB genomes. *BMC Plant Biol.* **2018**, *18*, 28. [CrossRef]
9. Saulnier, L. Types and functionality of polysaccharides in cereal Grains. Food Chemistry, Function and Analysis. In *No. 6 Cereal Grain-based Functional Foods: Carbohydrate and Phytochemical Components*; Royal Society of Chemistry: Cambridge, UK, 2019; Volume 4, pp. 54–84.
10. Anderson, J.W.; Baird, P.; Davis, R.H., Jr.; Ferreri, S.; Knudtson, M.; Koraym, A.; Waters, V.; Williams, C.L. Health benefits of dietary fiber. *Nutr. Rev.* **2009**, *67*, 188–205. [CrossRef]
11. Lovegrove, A.; Edwards, C.H.; De Noni, I.; Patel, H.; El, S.N.; Grassby, T.; Zielke, C.; Ulmius, M.; Nilsson, L.; Butterworth, P.J.; et al. Role of polysaccharides in food, digestion, and health. *Crit. Rev. Food Sci. Nutr.* **2017**, *57*, 237–253. [CrossRef]
12. Philippe, S.; Saulnier, L.; Guillon, F. Arabinoxylan and (1→3),(1→4)-β-glucan deposition in cell walls during wheat endosperm development. *Planta* **2006**, *224*, 449–461. [CrossRef] [PubMed]
13. Robert, P.; Jamme, F.; Barron, C.; Bouchet, B.; Saulnier, L.; Dumas, P.; Guillon, F. Change in wall composition of transfer and aleurone cells during wheat grain development. *Planta* **2011**, *233*, 393–406. [CrossRef] [PubMed]
14. Chateigner-Boutin, A.-L.; Bouchet, B.; Alvarado, C.; Bakan, B.; Guillon, F. The wheat grain contains pectic domains exhibiting specific spatial and development-associated distribution. *PLoS ONE* **2014**, *9*, e89620. [CrossRef] [PubMed]
15. Pellny, T.K.; Lovegrove, A.; Freeman, J.; Tosi, P.; Love, C.G.; Knox, J.P.; Shewry, P.R.; Mitchell, R.A.C. Cell walls of developing wheat starchy endosperm: Comparison of composition and RNA-Seq transcriptome. *Plant Physiol.* **2012**, *158*, 612–627. [CrossRef] [PubMed]
16. Toole, G.A.; Le Gall, G.; Colquhoun, I.J.; Nemeth, C.; Saulnier, L.; Lovegrove, A.; Pellny, T.; Wilkinson, M.D.; Freeman, J.; Mitchell, R.A.C.; et al. Temporal and spatial changes in cell wall composition in developing grains of wheat cv. Hereward. *Planta* **2010**, *232*, 677–689. [CrossRef] [PubMed]
17. Palmer, R.; Cornuault, V.; Marcus, S.E.; Knox, J.P.; Shewry, P.R.; Tosi, P. Comparative in situ analyses of cell wall matrix polysaccharide dynamics in developing rice and wheat grain. *Planta* **2015**, *241*, 669–685. [CrossRef]
18. Saulnier, L.; Guillon, F.; Chateigner-Boutin, A.L. Cell wall deposition and metabolism in wheat grain. *J. Cereal Sci.* **2012**, *56*, 91–108. [CrossRef]
19. Dervilly, G.; Saulnier, L.; Roger, P.; Thibault, J.-F. Isolation of homogeneous fractions from wheat water-soluble arabinoxylans. Influence of the structure on their macromolecular characteristics. *J. Agric. Food Chem.* **2000**, *48*, 270–278. [CrossRef]
20. Gartaula, G.; Dhital, S.; Netzel, G.; Flanagan, B.M.; Yakubov, G.E.; Beahan, C.T.; Collins, H.M.; Burton, R.A.; Bacic, A.; Gidley, M.J. Quantitative structural organisation model for wheat endosperm cell walls: Cellulose as an important constituent. *Carbohydr. Polym.* **2018**, *196*, 199–208. [CrossRef]
21. Yeats, T.H.; Rose, J.K.C. The formation and function of plant cuticles. *Plant Physiol.* **2013**, *163*, 5–20. [CrossRef]
22. Francin-Allami, M.; Alvarado, C.; Daniel, S.; Geairon, A.; Saulnier, L.; Guillon, F. Spatial and temporal distribution of cell wall polysaccharides during grain development of *Brachypodium distachyon*. *Plant Sci.* **2019**, *280*, 367–382. [CrossRef] [PubMed]
23. De la Canal, L.; Pinedo, M. Extracellular vesicles: A missing component in plant cell wall remodeling. *J. Exp. Bot.* **2018**, *69*, 4655–4658. [CrossRef] [PubMed]
24. Francin-Allami, M.; Merah, K.; Albenne, C.; Rogniaux, H.; Pavlovic, M.; Lollier, V.; Sibout, R.; Guillon, F.; Jamet, E.; Larré, C. Cell wall proteomic of *Brachypodium distachyon* grains: A focus on cell wall remodeling proteins. *Proteomics* **2015**, *15*, 2296–2306. [CrossRef] [PubMed]
25. Francin-Allami, M.; Lollier, V.; Pavlovic, M.; San Clemente, H.; Rogniaux, H.; Jamet, E.; Guillon, F.; Larré, C. Understanding the remodelling of cell walls during *Brachypodium distachyon* grain development through a sub-cellular quantitative proteomic approach. *Proteomes* **2016**, *4*, 21. [CrossRef] [PubMed]
26. Douché, T.; Clemente, H.S.; Burlat, V.; Roujol, D.; Valot, B.; Zivy, M.; Pont-Lezica, R.; Jamet, E. *Brachypodium distachyon* as a model plant toward improved biofuel crops: Search for secreted proteins involved in biogenesis and disassembly of cell wall polymers. *Proteomics* **2013**, *13*, 2438–2454. [CrossRef] [PubMed]

27. Calderan-Rodrigues, M.J.; Jamet, E.; Bonassi, M.B.C.R.; Guidetti-Gonzalez, S.; Begossi, A.C.; Setem, L.V.; Franceschini, L.M.; Fonseca, J.G.; Labate, C.A. Cell wall proteomics of sugarcane cell suspension cultures. *Proteomics* **2014**, *14*, 738–749. [CrossRef]
28. Calderan-Rodrigues, M.J.; Jamet, E.; Douché, T.; Bonassi, M.B.R.; Cataldi, T.R.; Fonseca, J.G.; San Clemente, H.; Pont-Lezica, R.; Labate, C.A. Cell wall proteome of sugarcane stems: Comparison of a destructive and a non-destructive extraction method showed differences in glycoside hydrolases and peroxidases. *BMC Plant Biol.* **2016**, *16*, 14. [CrossRef]
29. Fonseca, J.G.; Calderan-Rodrigues, M.J.; de Moraes, F.E.; Cataldi, T.R.; Jamet, E.; Labate, C.A. Cell wall proteome of sugarcane young and mature leaves and stems. *Proteomics* **2018**, *18*. [CrossRef]
30. Chen, X.Y.; Kim, S.T.; Cho, W.K.; Rim, Y.; Kim, S.; Kim, S.W.; Kang, K.Y.; Park, Z.Y.; Kim, J.Y. Proteomics of weakly bound cell wall proteins in rice calli. *J. Plant Physiol.* **2009**, *166*, 675–685. [CrossRef]
31. Jung, Y.H.; Jeong, S.H.; So, H.K.; Singh, R.; Lee, J.E.; Cho, Y.S.; Agrawal, G.K.; Rakwal, R.; Jwa, N.S. Systematic secretome analyses of rice leaf and seed callus suspension-cultured cells: Workflow development and establishment of high-density two-dimensional gel reference maps. *J. Proteome Res.* **2008**, *7*, 5187–5210. [CrossRef]
32. Cho, W.K.; Chen, X.Y.; Chu, H.; Rim, Y.; Kim, S.; Kim, S.T.; Kim, S.W.; Park, Z.Y.; Kim, J.Y. Proteomic analysis of the secretome of rice calli. *Physiol. Plant.* **2009**, *135*, 331–341. [CrossRef] [PubMed]
33. Zhou, L.; Bokhari, S.A.; Dong, C.-J.; Liu, J.-Y. Comparative proteomics analysis of the root apoplasts of rice seedlings in response to hydrogen peroxide. *PLoS ONE* **2011**, *6*, e16723. [CrossRef] [PubMed]
34. Cherkaoui, M.; Geairon, A.; Lollier, V.; Clemente, H.S.; Larré, C.; Rogniaux, H.; Jamet, E.; Guillon, F.; Francin-Allami, M. Cell wall proteome investigation of bread wheat (*Triticum aestivum*) developing grain in endosperm and outer layers. *Proteomics* **2018**, *18*. [CrossRef] [PubMed]
35. San Clemente, H.; Pont-Lezica, R.; Jamet, E. Bioinformatics as a tool for assessing the quality of sub-cellular proteomic strategies and inferring functions of proteins: Plant cell wall proteomics as a test case. *Bioinform. Biol. Insights* **2009**, *2009*, 15–28. [CrossRef] [PubMed]
36. Calderan-Rodrigues, M.J.; Guimarães Fonseca, J.; de Moraes, F.E.; Vaz Setem, L.; Carmanhanis Begossi, A.; Labate, C.A. Plant cell wall proteomics: A focus on monocot species, *Brachypodium distachyon*, *Saccharum* spp. and *Oryza sativa*. *Int. J. Mol. Sci.* **2019**, *20*, 1975. [CrossRef]
37. Duruflé, H.; Clemente, H.S.; Balliau, T.; Zivy, M.; Dunand, C.; Jamet, E. Cell wall proteome analysis of *Arabidopsis thaliana* mature stems. *Proteomics* **2017**, *17*. [CrossRef]
38. Minic, Z.; Jouanin, L. *Plant glycoside* hydrolases involved in cell wall polysaccharide degradation. *Plant Physiol. Biochem.* **2006**, *44*, 435–449. [CrossRef]
39. Albenne, C.; Canut, H.; Hoffmann, L.; Jamet, E. Plant cell wall proteins: A large body of data, but what about runaways? *Proteomes* **2014**, *17*, 224–242. [CrossRef]
40. Chateigner-Boutin, A.L.; Suliman, M.; Bouchet, B.; Alvarado, C.; Lollier, V.; Rogniaux, H.; Guillon, F.; Larré, C. *Endomembrane proteomics* reveals putative enzymes involved in cell wall metabolism in wheat grain outer layers. *J. Exp. Bot.* **2015**, *66*, 2649–2658. [CrossRef]
41. Feiz, L.; Irshad, M.; Pont-Lezica, R.F.; Canut, H.; Jamet, E. Evaluation of cell wall preparations for proteomics: A new procedure for purifying cell walls from *Arabidopsis hypocotyls*. *Plant Methods* **2006**, *2*, 10. [CrossRef]
42. Irshad, M.; Canut, H.; Borderies, G.; Pont-Lezica, R.; Jamet, E. A new picture of cell wall protein dynamics in elongating cells of *Arabidopsis thaliana*: Confirmed actors and newcomers. *BMC Plant Biol.* **2008**, *8*, 94. [CrossRef] [PubMed]
43. Hothorn, M.; Wolf, S.; Aloy, P.; Greiner, S.; Scheffzek, K. Structural insights into the target specificity of plant invertase and pectin methylesterase inhibitory proteins. *Plant Cell* **2004**, *16*, 3437–3447. [CrossRef] [PubMed]
44. French, S.R.; Abu-Zaitoon, Y.; Uddin, M.M.; Bennett, K.; Nonhebel, H.M. Auxin and cell wall invertase related signaling during rice grain development. *Plants* **2014**, *3*, 95–112. [CrossRef] [PubMed]
45. Bellincampi, D.; Camardella, L.; Delcour, J.A.; Desseaux, V.; D'Ovidio, R.; Durand, A.; Elliot, G.; Gebruers, K.; Giovane, A.; Juge, N.; et al. Potential physiological role of plant glycosidase inhibitors. *Biochim. Biophys. Acta Proteins Proteom.* **2004**, *1696*, 265–274. [CrossRef] [PubMed]
46. Hong, M.J.; Kim, D.Y.; Lee, T.G.; Jeon, W.B.; Seo, Y.W. Functional characterization of pectin methylesterase inhibitor (PMEI) in wheat. *Genes Genet. Syst.* **2010**, *85*, 97–106. [CrossRef] [PubMed]

47. Rocchi, V.; Janni, M.; Bellincampi, D.; Giardina, T.; D'Ovidio, R. Intron retention regulates the expression of pectin methyl esterase inhibitor (Pmei) genes during wheat growth and development. *Plant Biol.* **2012**, *14*, 365–373. [CrossRef] [PubMed]
48. Shivalingamurthy, S.G.; Anangi, R.; Kalaipandian, S.; Glassop, D.; King, G.F.; Rae, A.L. Identification and functional characterization of sugarcane invertase inhibitor (ShINH1): A potential candidate for reducing pre- and post-harvest loss of sucrose in sugarcane. *Front. Plant Sci.* **2018**, *9*, 598. [CrossRef]
49. Tang, X.; Su, T.; Han, M.; Wei, L.; Wang, W.; Yu, Z.; Xue, Y.; Wei, H.; Du, Y.; Greiner, S.; et al. Suppression of extracellular invertase inhibitor gene expression improves seed weight in soybean (Glycine max). *J. Exp. Bot.* **2017**, *68*, 469–482. [CrossRef]
50. Bate, N.J.; Niu, X.; Wang, Y.; Reimann, K.S.; Helentjaris, T.G. An invertase inhibitor from maize localizes to the embryo surrounding region during early kernel development. *Plant Physiol.* **2004**, *134*, 246–254. [CrossRef]
51. Hossain, M.A.; Nakamura, K.; Kimura, Y. α-mannosidase involved in turnover of plant complex type *N*-glycans in tomato (*Lycopersicum esculentum*) fruits. *Biosci. Biotechnol. Biochem.* **2009**, *73*, 140–146. [CrossRef]
52. Boisson, M. *Arabidopsis glucosidase* I mutants reveal a critical role of *N*-glycan trimming in seed development. *EMBO J.* **2001**, *20*, 1010–1019. [CrossRef] [PubMed]
53. Yoshida, K.; Komae, K. A rice family 9 glycoside hydrolase isozyme with broad substrate specificity for hemicelluloses in type II cell walls. *Plant Cell Physiol.* **2006**, *47*, 1541–1554. [CrossRef] [PubMed]
54. Bakan, B.; Marion, D. Assembly of the Cutin Polyester: From Cells to Extracellular Cell Walls. *Plants* **2017**, *6*, 57. [CrossRef] [PubMed]
55. Salminen, T.A.; Blomqvist, K.; Edqvist, J. Lipid transfer proteins: Classification, nomenclature, structure, and function. *Planta* **2016**, *244*, 971–997. [CrossRef] [PubMed]
56. Tobimatsu, Y.; Schuetz, M. Lignin polymerization: How do plants manage the chemistry so well? *Curr. Opin. Biotechnol.* **2019**, *56*, 75–81. [CrossRef] [PubMed]
57. Burr, S.J.; Fry, S.C. *Feruloylated arabinoxylans* are oxidatively cross-linked by extracellular maize peroxidase but not by horseradish peroxidase. *Mol. Plant* **2009**, *2*, 883–892. [CrossRef]
58. Pusztahelyi, T. Chitin and chitin-related compounds in plant–fungal interactions. *Mycology* **2018**, *9*, 189–201. [CrossRef]
59. Cruz-Valderrama, J.E.; Gómez-Maqueo, X.; Salazar-Iribe, A.; Zúñiga-Sánchez, E.; Hernández-Barrera, A.; Quezada-Rodríguez, E.; Gamboa-deBuen, A. Overview of the role of cell wall DUF642 proteins in plant development. *Int. J. Mol. Sci.* **2019**, *20*, 3333. [CrossRef]
60. Vázquez-Lobo, A.; Roujol, D.; Zuñiga-Sánchez, E.; Albenne, C.; Piñero, D.; de Buen, A.G.; Jamet, E. The highly conserved spermatophyte cell wall DUF642 protein family: Phylogeny and first evidence of interaction with cell wall polysaccharides in vitro. *Mol. Phylogenet. Evol.* **2012**, *63*, 510–520. [CrossRef]
61. Veličković, D.; Ropartz, D.; Guillon, F.; Saulnier, L.; Rogniaux, H. New insights into the structural and spatial variability of cell-wall polysaccharides during wheat grain development, as revealed through MALDI mass spectrometry imaging. *J. Exp. Bot.* **2014**, *65*, 2079–2091. [CrossRef]
62. Kabel, M.A.; van den Borne, H.; Vincken, J.-P.; Voragen, A.G.J.; Schols, H.A. Structural differences of xylans affect their interaction with cellulose. *Carbohydr. Polym.* **2007**, *69*, 94–105. [CrossRef]
63. Castillo, F.M.; Canales, J.; Claude, A.; Calderini, D.F. Expansin genes expression in growing ovaries and grains of sunflower are tissue-specific and associate with final grain weight. *BMC Plant Biol.* **2018**, *18*, 327. [CrossRef] [PubMed]
64. Gholizadeh, A. DUF538 protein superfamily is predicted to be chlorophyll hydrolyzing enzymes in plants. *Physiol. Mol. Biol. Plants* **2016**, *22*, 77–85. [CrossRef] [PubMed]
65. Gholizadeh, A. Chlorophyll binding ability of non-chloroplastic DUF538 protein superfamily in plants. *Proc. Natl. Acad. Sci. India Sect. B Biol. Sci.* **2018**, *88*, 967–976. [CrossRef]
66. Li, L.; Du, Y.; He, C.; Dietrich, C.R.; Li, J.; Ma, X.; Wang, R.; Liu, Q.; Liu, S.; Wang, G.; et al. The maize glossy6 gene is involved in cuticular wax deposition and drought tolerance. *J. Exp. Bot.* **2018**, *70*, 3089–3099. [CrossRef] [PubMed]
67. Zhang, J.-P.; Yu, Y.; Feng, Y.-Z.; Zhou, Y.-F.; Zhang, F.; Yang, Y.-W.; Lei, M.-Q.; Zhang, Y.-C.; Chen, Y.-Q. MiR408 regulates grain yield and photosynthesis via a phytocyanin protein. *Plant Physiol.* **2017**, *175*, 1175–1185. [CrossRef] [PubMed]
68. Dominguez, F.; Cejudo, F.J. Characterization of the endoproteases appearing during wheat grain development. *Plant Physiol.* **1996**, *112*, 1211–1217. [CrossRef]

69. Schaller, A.; Stintzi, A.; Graff, L. Subtilases—Versatile tools for protein turnover, plant development, and interactions with the environment. *Physiol. Plant.* **2012**, *145*, 52–66. [CrossRef]
70. Rautengarten, C.; Usadel, B.; Neumetzler, L.; Hartmann, J.; Büssis, D.; Altmann, T. A subtilisin-like serine protease essential for mucilage release from Arabidopsis seed coats. *Plant J.* **2008**, *54*, 466–480. [CrossRef]
71. D'Erfurth, I.; Le Signor, C.; Aubert, G.; Sanchez, M.; Vernoud, V.; Darchy, B.; Lherminier, J.; Bourion, V.; Bouteiller, N.; Bendahmane, A.; et al. A role for an endosperm-localized subtilase in the control of seed size in legumes. *New Phytol.* **2012**, *196*, 738–751. [CrossRef]
72. Suliman, M.; Chateigner-Boutin, A.-L.; Francin-Allami, M.; Partier, A.; Bouchet, B.; Salse, J.; Pont, C.; Marion, J.; Rogniaux, H.; Tessier, D.; et al. Identification of glycosyltransferases involved in cell wall synthesis of wheat endosperm. *J. Proteom.* **2013**, *78*, 508–521. [CrossRef] [PubMed]
73. Langella, O.; Valot, B.; Balliau, T.; Blein-Nicolas, M.; Bonhomme, L.; Zivy, M. X! TandemPipeline: A tool to manage sequence redundancy for protein inference and phosphosite identification. *J. Proteom. Res.* **2017**, *16*, 494–503. [CrossRef] [PubMed]
74. Valot, B.; Langella, O.; Nano, E.; Zivy, M. MassChroQ: A versatile tool for mass spectrometry quantification. *Proteomics* **2011**, *11*, 3572–3577. [CrossRef] [PubMed]
75. Batoko, H.; Zheng, H.Q.; Hawes, C.; Moore, I. A Rab1 GTPase is required for transport between the endoplasmic reticulum and golgi apparatus and for normal Golgi movement in plants. *Plant Cell* **2000**, *12*, 2201–2217. [CrossRef] [PubMed]
76. Faso, C.; Chen, Y.N.; Tamura, K.; Held, M.; Zemelis, S.; Marti, L.; Saravanan, R.S.; Hummel, E.; Kung, L.; Miller, E.; et al. A missense mutation in the arabidopsis copii coat protein sec24a induces the formation of clusters of the endoplasmic reticulum and golgi apparatus. *Plant Cell* **2009**. [CrossRef]
77. Sparkes, I.A.; Runions, J.; Kearns, A.; Hawes, C. Rapid, transient expression of fluorescent fusion proteins in tobacco plants and generation of stably transformed plants. *Nat. Protoc.* **2006**, *1*, 2019–2025. [CrossRef]

International Journal of
Molecular Sciences

MDPI

Article

Organ and Tissue-Specific Localisation of Selected Cell Wall Epitopes in the Zygotic Embryo of *Brachypodium distachyon*

Alexander Betekhtin [1,*], Anna Milewska-Hendel [2], Joanna Lusinska [1], Lukasz Chajec [3], Ewa Kurczynska [2] and Robert Hasterok [1]

1 Department of Plant Anatomy and Cytology, Faculty of Biology and Environmental Protection, University of Silesia in Katowice, 28 Jagiellonska Street, 40-032 Katowice, Poland; joanna.lusinska@wp.pl (J.L.); robert.hasterok@us.edu.pl (R.H.)

2 Department of Cell Biology, Faculty of Biology and Environmental Protection, University of Silesia in Katowice, 28 Jagiellonska Street, 40-032 Katowice, Poland; anna.milewska@us.edu.pl (A.M.-H.); ewa.kurczynska@us.edu.pl (E.K.)

3 Department of Animal Histology and Embryology, Faculty of Biology and Environmental Protection, University of Silesia in Katowice, 28 Jagiellonska Street, 40-032 Katowice, Poland; lukasz.chajec@us.edu.pl

* Correspondence: alexander.betekhtin@us.edu.pl; Tel.: +48-32-2009-484

Received: 26 January 2018; Accepted: 1 March 2018; Published: 3 March 2018

Abstract: The plant cell wall shows a great diversity regarding its chemical composition, which may vary significantly even during different developmental stages. In this study, we analysed the distribution of several cell wall epitopes in embryos of *Brachypodium distachyon* (Brachypodium). We also described the variations in the nucleus shape and the number of nucleoli that occurred in some embryo cells. The use of transmission electron microscopy, and histological and immunolocalisation techniques permitted the distribution of selected arabinogalactan proteins, extensins, pectins, and hemicelluloses on the embryo surface, internal cell compartments, and in the context of the cell wall ultrastructure to be demonstrated. We revealed that the majority of arabinogalactan proteins and extensins were distributed on the cell surface and that pectins were the main component of the seed coat and other parts, such as the mesocotyl cell walls and the radicula. Hemicelluloses were localised in the cell wall and outside of the radicula protodermis, respectively. The specific arrangement of those components may indicate their significance during embryo development and seed germination, thus suggesting the importance of their protective functions. Despite the differences in the cell wall composition, we found that some of the antibodies can be used as markers to identify specific cells and the parts of the developing Brachypodium embryo.

Keywords: Brachypodium; cell wall; nucleus; zygotic embryo

1. Introduction

The cell wall is a protective layer that is located around the cell membrane, which is found in plant, fungi, bacteria, and archaea cells. The structure and functions of the cell wall were interconnected during the process of evolution in spite of the fact that its chemical structure, and to a lesser extent, its functions are diverse in different groups of organisms. In plants, its most fundamental function is to provide structural support and protection to the cell with some additional specific functions that may occur during plant development and cell differentiation [1].

The plant cell wall is a dynamic and highly specialised network that is formed by a heterogeneous mixture of cellulose, hemicelluloses, pectins and, to some extent, proteins and phenolic compounds. The cell wall that is formed during cell division is called the primary cell wall. In many plants, as/after the cell completes its growth, additional layers of cellulose fibres are deposited on the inner surface of the primary cell wall, which produces a secondary cell wall. The cell wall composition in vascular plants is approximately 30% cellulose, 30% hemicellulose, and 35% pectins, and 1–5% structural proteins, on a dry weight basis [2,3]. However, the precise proportions of these compounds may differ significantly at different stages of plant development. For example, grass coleoptiles consist of 60–70% hemicelluloses, 20–25% celluloses, and 10% pectin substances and the grass endosperm cell wall may contain up to as much as 85% hemicelluloses. By contrast, the secondary cell walls are generally more cellulose-rich [4–7].

Brachypodium distachyon (Brachypodium) belongs to the Pooideae subfamily and is a well-established model species for the grasses. It has several features and advantages that make it useful for gaining a better understanding of the genetic, cellular and molecular biology of temperate climate zone cereals and forage crops [8]. There are many studies, which are often connected with the chemical composition of the Brachypodium cell wall [9–12]. A comparative study of the primary cell wall in the seedlings of Brachypodium, barley and wheat demonstrated similar relative levels and developmental changes of hemicelluloses [10]. Analyses of the Brachypodium proteome facilitate better understanding of the enzymes that are involved in cell wall remodelling during seed development; such research is of great importance for gaining better understanding of these processes in grasses and for finding the key components that are responsible for the size and weight of grass grains [9]. However, there is a dearth of information about the localisation of specific cell wall components at different stages of Brachypodium development.

Here, we characterise the chemical composition of the cell walls in Brachypodium embryos and describe the differences in the number of nucleoli that were observed in the cell nuclei in different parts of an embryo. We used light and transmission electron microscopy (TEM), histological and immunolocalisation techniques to analyse the distribution of selected pectins, arabinogalactan proteins (AGP), extensins, and hemicelluloses in the cell walls, internal cell compartments, and on the embryo surface.

2. Results and Discussion

2.1. The Morphological and Histological Features of Brachypodium Embryos

In their study, Wolny et al. [13] demonstrated that Brachypodium embryos are small in size, which makes their initial examination possible only by the use of a dissecting microscope. In this study, we distinguished the main parts of the embryo, such as scutellum, V scale, coleoptile, first and second leaf, shoot apex, mesocotyl, epiblast, radicula, root cap, and coleorhiza (Figure 1). The coleoptile and coleorhiza are two organs that are found exclusively in grass species [14]. A comparison of the cell nuclei in different parts of Brachypodium embryos demonstrated that the majority contained only one nucleolus (Figure 1). However, some cells of the shoot apex, mesocotyl, radicula and root cap were characterised by the presence of a round nuclei that contained two nucleoli (Figure 1; nucleoli indicated by red arrows). TEM analysis of the selected embryo parts confirmed these observations and demonstrated the presence of a centrally positioned nucleus with one or two large nucleoli as well as a high nucleus:cytoplasm ratio (Figure 2a,b). The cytoplasm of these cells was dense and contained lipid droplets and starch granules around the nucleus. Interestingly, we also found cells in the embryo with nuclei that were extended in their shape but that also contained two nucleoli (Figure 2c). The architecture of these cells is typical for the initial vascular tissue [15]. According to Verdeil et al. [16], pluripotent plant stem cells, which are located within the root and shoot meristems, are isodiametric, have a dense cytoplasm, a high nucleus:cytoplasm ratio, a fragmented vacuome, contain granules of starch, and have a spherically-shaped nucleus with one or two nucleoli. Both of the meristematic cells of the oil palm (*Elaeis guineensis*) [17], maize (*Zea mays*) [18], and onion (*Allium cepa*) [19] have similar characteristics to those described by Verdeil et al. [16], which appear to be universal for monocots.

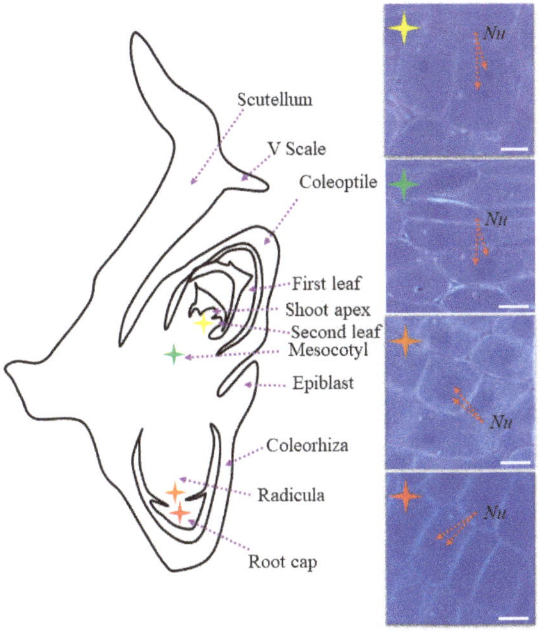

Figure 1. Schematic representation of Brachypodium embryo. Asterisks in different colours represent nuclei from histological sections of specified parts of the embryo. Red arrows mark two nucleoli. Purple arrows show respective parts of the embryo: V scale, coleoptile, first leaf, second leaf, shoot apex, mesocotyl, epiblast, coleorhiza, radicula, and root cap. Bars: 5 µm.

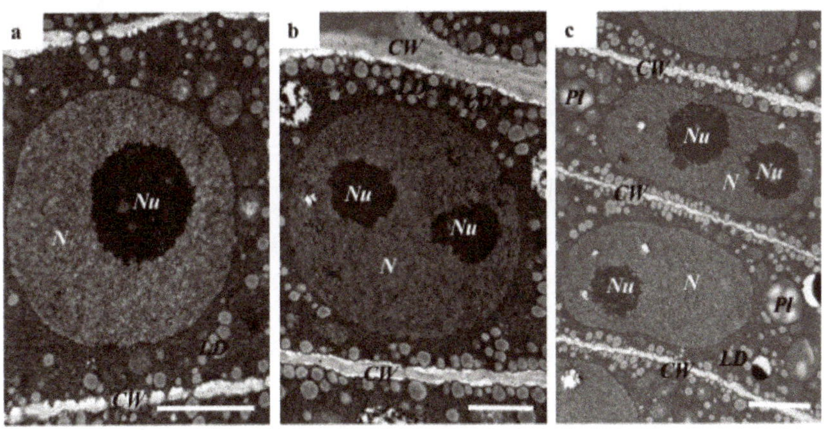

Figure 2. TEM (transmission electron microscopy) of radicula nuclei (**a**–**c**). Abbreviations: *CW*—cell wall, *LD*—lipid droplets, *N*—nucleus, *Nu*—nucleolus, *Pl*—plastid. Bars: (**a**) 3 µm; (**b**) 1.5 µm; and (**c**) 2.5 µm.

2.2. AGP, Extensin, Pectin, and Hemicellulose Epitopes in Various Tissues and Organs of Brachypodium Embryos

AGPs distribution is dynamically regulated during the ontogenesis of a plant [20]. Although we demonstrated that AGPs have a diverse localisation in Brachypodium embryos, most of them are found on the cell surface. The MAC207 antibody had an affinity to all parts of the embryo and was localised in the internal cell compartments (Figure 3a–a″). JIM8, JIM13, and JIM16 were detected mainly in outer pereclinal walls of surface embryo tissues (Figure 3b–b″,c–c″,d–d″ and Figure 4a–a″,c–c″). The JIM13 (Figure 3d–d″) and JIM16 (Figure 4c–c″) epitopes were components of the seed coat. Furthermore, the JIM16 epitope was localised in the internal cell compartments (Figure 4a–a″,b–b″). Compared to most of the analysed AGPs, the LM2 epitopes were found only in the cell wall compartments (Figure 4d–d″,e–e″). The possible functions of the AGPs on the embryo surface are to protect it from infections, contribute to plant-microbe interactions, seed germination and other important processes during plant development [20]. In their study of transcriptome and metabolome changes in *Arabidopsis thaliana* (Arabidopsis) that were connected with the seed dormancy and germination, Joosen et al. [21,22] showed that AGPs were more connected with the embryo cell walls. It is also known that AGPs are crucial in preventing infections in *Brassica napus* and *Pisum sativum* via the encystment of the *Aphanomces euteiches* zoospores, thereby inducing plant germination [23]. Van Hengel et al. [24] demonstrated that AtAGP30 is a non-classical AGP core protein in Arabidopsis, which had a root-tip specific expression in seedlings. This may imply its importance in root development or growth.

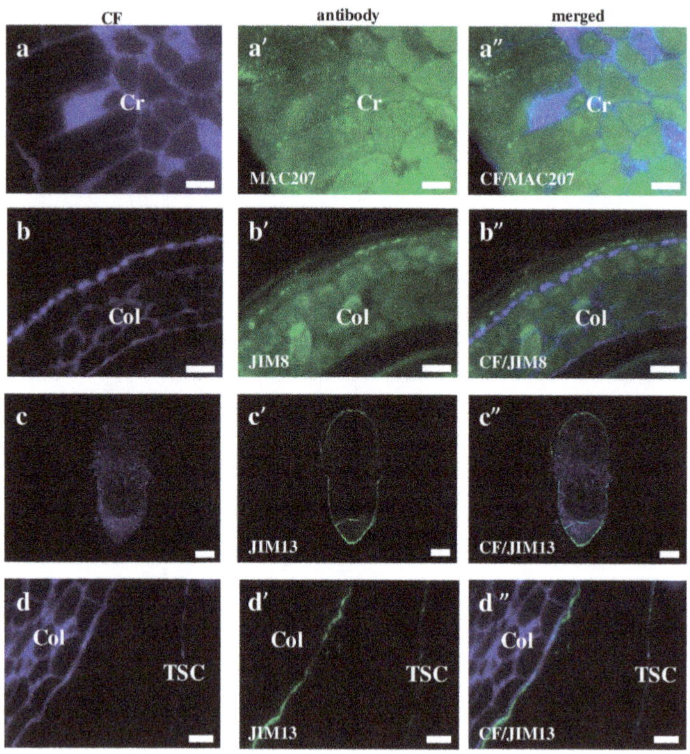

Figure 3. Immunolocalisation of arabinogalactan proteins in Brachypodium embryos. (**a–a″**) MAC207 in coleorhiza; (**b–b″**) JIM8 signal in coleoptile; (**c–c″**,**d–d″**) JIM13 in the entire embryo and the coleoptile with the seed coat, respectively. Abbreviations: CF—calcofluor, Cr—coleorhiza, Col—coleoptile, TSC—the seed coat. Bars: (**a–a″**,**b–b″**,**d–d″**) 10 µm; and (**c–c″**) 100 µm.

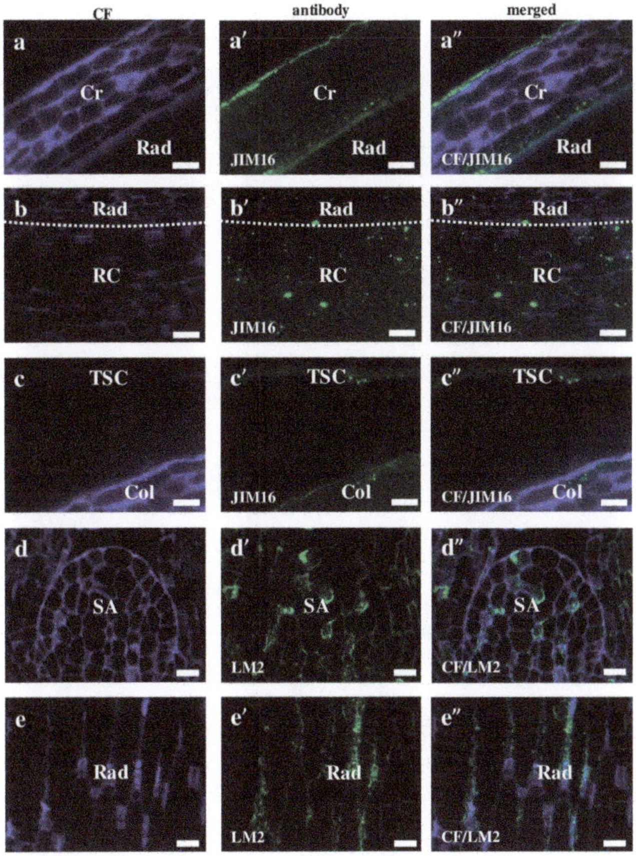

Figure 4. Immunolocalisation of arabinogalactan proteins in Brachypodium embryos. (**a–a″**,**b–b″**,**c–c″**) JIM16 with coleorhiza with radicula, radicula with root cap and coleoptile with the seed coat, respectively; Both (**d–d″**) shoot apex and (**e–e″**) radicula with LM2. Abbreviations: CF—calcofluor, Cr—coleorhiza, Col—coleoptile, Rad—radicula, RC—root cap, SA—shoot apex, and TSC—the seed coat. Bars: 10 μm.

Extensins are one of the major plant cell wall protein families found to be secreted into the cell wall. They are glycoproteins, in which about one-third of all amino acids is hydroxyproline with four monosaccharide arabinose residues attached. These compounds seem to play a strengthening and/or defensive role during the response to disease and wounding [25–27]. We observed that most of the Brachypodium extensins in this study had a surface localisation, which was similar to that of the AGPs. The signals of the LM1 antibodies were localised on the surface of the coleoptile and first leaf (Figure 5a–a″,b–b″). Furthermore, a high level of LM1 binding was observed in the internal compartments of the coleoptile epidermal cells (Figure 5b–b″; red arrows). The cell wall localisation was a specific feature of the JIM11 and JIM12 epitopes in the mesocotyl (Figure 5c–c″ and Figure 6c–c″). Signals of the first antibodies were also observed in the cell walls of the radicula (Figure 5d–d″). Additionally, strong fluorescence signals of these antibodies were also found in the inner layer of the seed coat (Figure 5e–e″). The JIM12 extensin epitope had a surface distribution on the coleoptile, but no signals of these antibodies were observed in any part of the second leaf (Figure 6a–a″,b–b″). It is still not clear whether extensins have redundant or specific functions in the different cell types of various plant tissues. The inhibition of both extensin and AGP biosynthesis by 3,4-dehydro-L-proline impedes the development of embryogenic cells and decreases the rate of embryo germination [28]. This research demonstrated that extensins appear to play a vital role in the regeneration and germination of embryos during early plant development via somatic embryogenesis. It is also known that plants increase the amount of extensins when they are under aluminium stress. Sujkowska-Rybkowska et al. [29] suggested the importance of extensins in the aluminium resistance mechanisms and demonstrated an aluminium-induced extensin accumulation in the cell walls. Additionally, some extensins protect the plant cells from pathogen attacks [30]. An analysis of the extensin transcript levels in banana (*Musa* spp.) demonstrated that these proteins mainly appear in the root cap and meristematic cells after wounding or inoculation with *Fusarium oxysporum* [30]. The level of extensin transcription in intact plants were found to be up-regulated by wounding or inoculation with a pathogen, but down-regulated by a pathogen attack on wounded plants. In general, the connection of extensins with other components of the cell wall, as well as many of their functions during the development of the cell wall architecture are not yet fully understood [26]. It is likely that more extensive studies using mutants with the knocked down genes encoding extensins would shed more light on these matters.

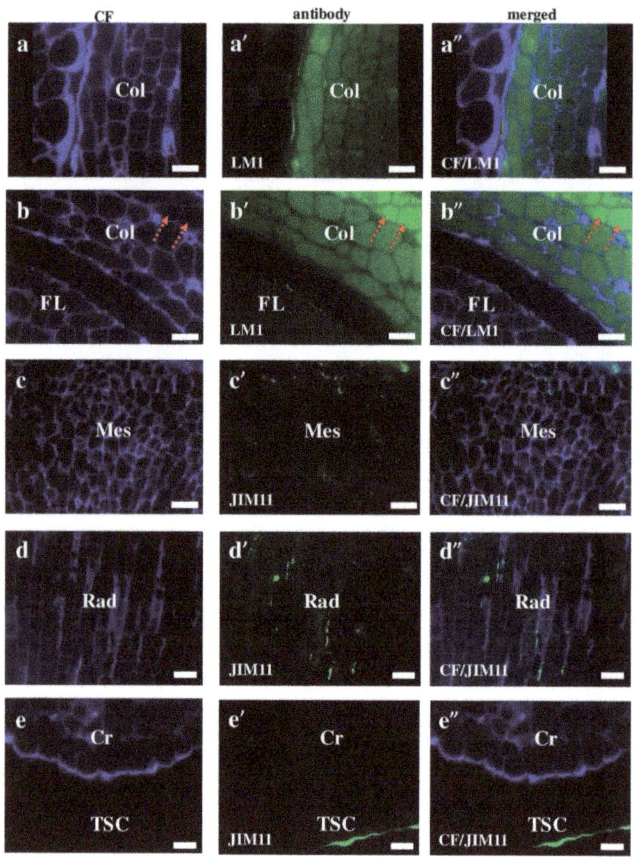

Figure 5. Immunolocalisation of extensins in Brachypodium embryos. Both (**a–a″**) coleoptile and (**b–b″**) coleoptile with first leaf with LM1; (**c–c″**,**d–d″**,**e–e″**) mesocotyl, radicula and coleorhiza (including the seed coat) with JIM11, respectively. Arrows show outer layers of coleoptile including protodermal cells (**b–b″**). Abbreviations: CF—calcofluor, Col—coleoptile, Cr—coleorhiza, FL—first leaf, Mes—mesocotyl, Rad—radicula, and TSC—the seed coat. Bars: 10 μm.

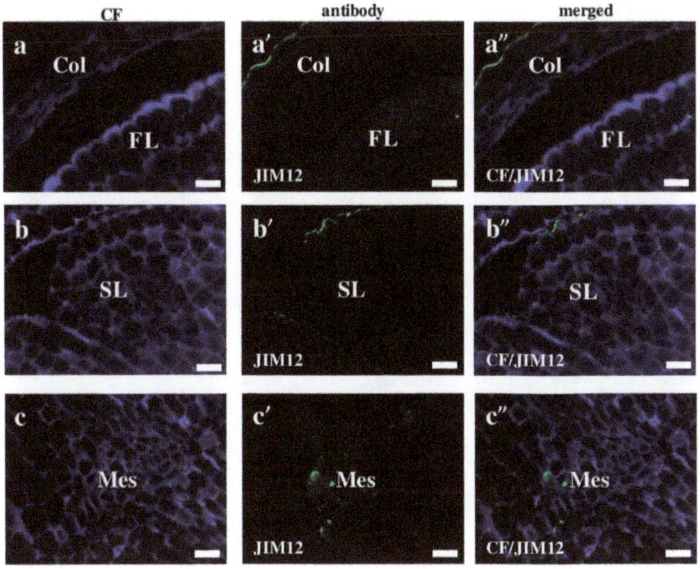

Figure 6. Immunolocalisation of extensins in Brachypodium embryos. (**a–a″**) coleoptile with first leaf; (**b–b″**) second leaf and (**c–c″**) mesocotyl. All with JIM12. Abbreviations: CF—calcofluor, Col—coleoptile, FL—first leaf, Mes—mesocotyl, and SL—second leaf. Bars: 10 µm.

Pectins are high-molecular carbohydrates with D-galacturonic acid as their main structural component of that are found in the tissues of terrestrial plants and in some algae. In some plant tissues, for example in the white part of a citrus peel, the pectin content may reach up to 30% of dry weight, while, in others, it does not exceed a fraction of a percent [31,32]. Grasses, which contain a type II cell wall, are generally pectin poor [33]. The immunolocalisation of the homogalacturonan epitopes that recognised de-methyl-esterified pectin LM19 in Brachypodium were found on the outer surface of the seed coat (Figure 7a–a″). Moreover, LM19 was localised in different embryo parts, for example, in the cell wall of the mesocotyl and radicula, respectively (Figure 7b–b″,c–c″). The LM13 epitope could be a negative marker for the epidermal cells of the first leaf of a Brachypodium embryo. These antibodies gave signals in the cell walls in almost all parts of the embryo but were absent in the epidermal cells of the first leaf (Figure 7d–d″; red arrows). A different location of LM13 was observed in the coleorhiza and radicula cells. In the coleorhiza, this epitope was found in the cell walls, while in the radicula, it had an internal cell localisation (Figure 7e–e″). In Arabidopsis, the LM13 epitope is detectable in the epidermal cell walls of the younger, more flexible regions of the inflorescence stems and is almost absent in the base of the stem [34,35]. These authors demonstrated that the contribution of the arabinan structures to the cell wall mechanical properties influences the responsiveness to the mechanical stress. Therefore, the specific lack of the LM13 signals in the epidermal cells of the first leaf of Brachypodium embryo may be related to the absence of any mechanical stress in this part of an embryo.

The fluorescence signals of LM6 against arabinans were found in both the cell wall and the internal cell compartment of the mesocotyl and radicula (Figure 8a–a″,b–b″). Furthermore, LM6 was localised on the external surface of the root cap but was absent in all other cells (Figure 8c–c″). As was demonstrated in Arabidopsis, the arabinans accumulated in developing and mature embryos but disappeared during germination and seedling establishment [36]. The early stages of Arabidopsis seed development showed a punctate distribution of this epitope compared to mature seeds, where LM6 labelling was very intense and distributed in the cell walls in the entire embryo. It was suggested that changes in the number and localisation of the arabinans might be connected with seedling germination.

In barley, at the beginning of this process, the cell walls in the coleoptile have comparably high levels of arabinose-rich pectic polysaccharides, which decrease strongly during germination [37]. The abundance of the arabinans in the cell walls of seeds is very often a characteristic feature in a wide range of species such as almond (*Prunus dulcis*) [38], rapeseed (*Brassica napus*) [39], and honey locust (*Gleditsia triacanthos*) [40].

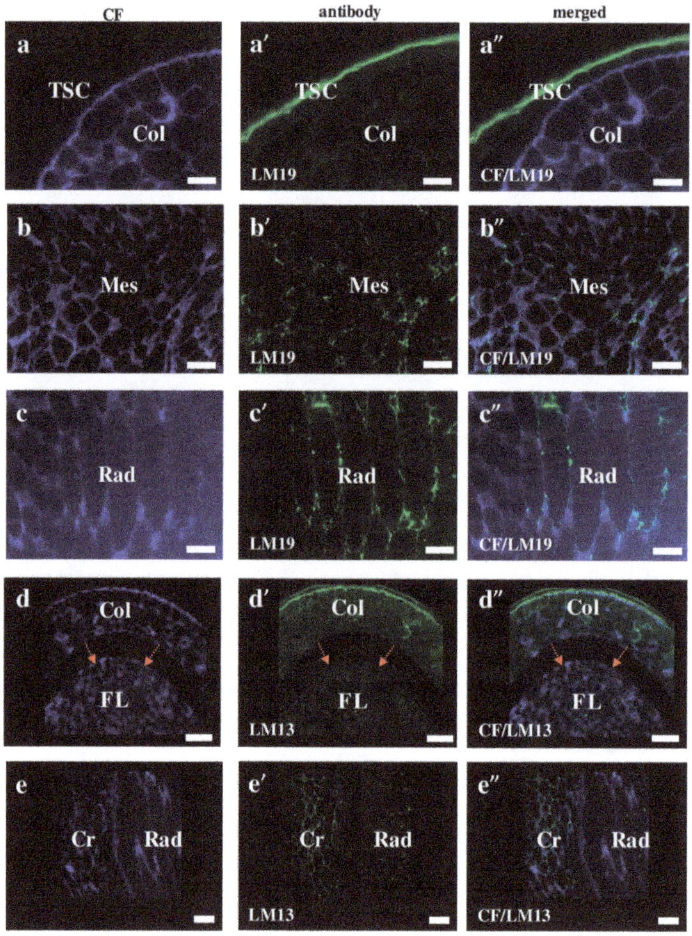

Figure 7. Immunolocalisation of pectins in Brachypodium embryos. (**a–a″**,**b–b″**,**c–c″**) Coleoptile with the seed coat, mesocotyl, and radicula are with LM19, respectively; Both (**d–d″**) coleoptile with first lead and (**e–e″**) coleorhiza with radicula with LM13. Arrows point to *FL* epidermal cells (**d–d″**). Abbreviations: CF—calcofluor, Col—coleoptile, Cr—coleorhiza, FL—first leaf, Mes—mesocotyl, Rad—radicula, and TSC—the seed coat. Bars: (**a–a″**,**b–b″**,**c–c″**) 10 μm; (**d–d″**,**e–e″**) 20 μm.

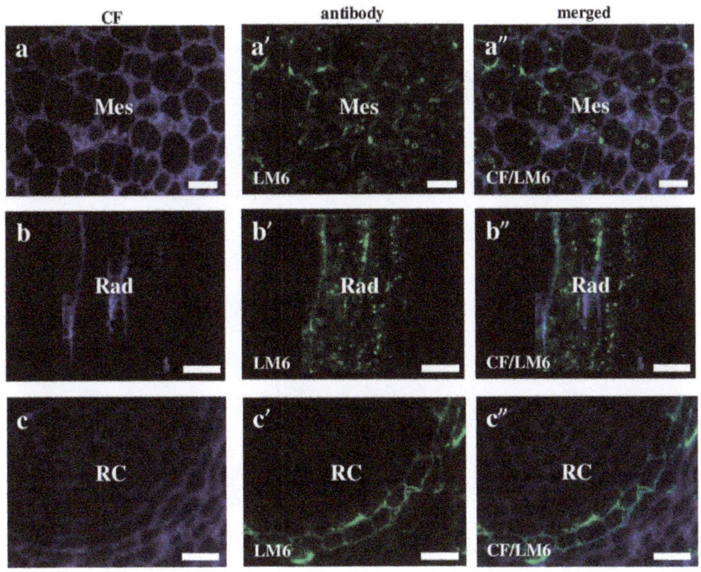

1

Figure 8. Immunolocalisation of pectins in Brachypodium embryos. (**a–a″**) Mesocotyl; (**b–b″**) radicula and (**c–c″**) root cap. All with LM6. Abbreviations: CF—calcofluor, Rad—radicula, RC—root cap, and Mes—mesocotyl. Bars: (**a–a″**,**b–b″**) 10 μm; (**c–c″**) 20 μm.

Hemicelluloses differ from cellulose in the composition of their monomers and their branched arrangement in molecules [7,41,42]. They are one of the components of the plastic matrix and impart additional strength to the cell wall, but do not hinder its growth [5]. Further, due to the ease of hydrolysis, they can also serve as reserve substances. In the Brachypodium embryo, the signals of the hemicellulose that recognise antibodies were localised in a different pattern. The signals of the LM21 antibodies were visualised in the internal cell compartments of all of the embryo parts, for example in the coleoptile and radicula (Figure 9a–a″,b–b″), while the LM25 antibodies were detected only in the radicula and root cap (Figure 9c–c″,d–d″,e–e″). Since it is difficult to determine whether the immunostaining is limited to only the cell wall, or whether it is also present outside of it using light microscopy, a TEM analysis was performed. Since this analysis was performed without immunostaining, it only showed the exact thickness of the wall (Figure S1). Thus, a calculation of the wall thickness (Figure 9c″) revealed that it was about 2 μm and that the thickness of the fluorescence "line" that was visible on Figure 9c′–c″ was about 14.2 μm. This observation indicates that the hemicellulose epitopes were also outside the radicula protodermis, which can be explained by taking into account the development of the embryo, in which, during the final stages of development, detachment of the protective tissues such as the coleorhiza and coleoptile occur [43]. The localisation of some cell wall epitopes described in this work in the cellular compartments, indicate their extensive synthesis in the organelles, such as endoplasmic reticulum and Golgi apparatus dictiosomes, which are involved in this process [28,44,45]. It is worth noting that the same antibodies (LM21 and LM25) were localised in a different way in the embryogenic callus of Brachypodium [46]. As was demonstrated there, the LM21 antibodies only gave signals in the cell junctions of the embryogenic masses and in the parenchymatous cells. The signals of LM25 were found in every cell of both the embryogenic masses and parenchymatous cells.

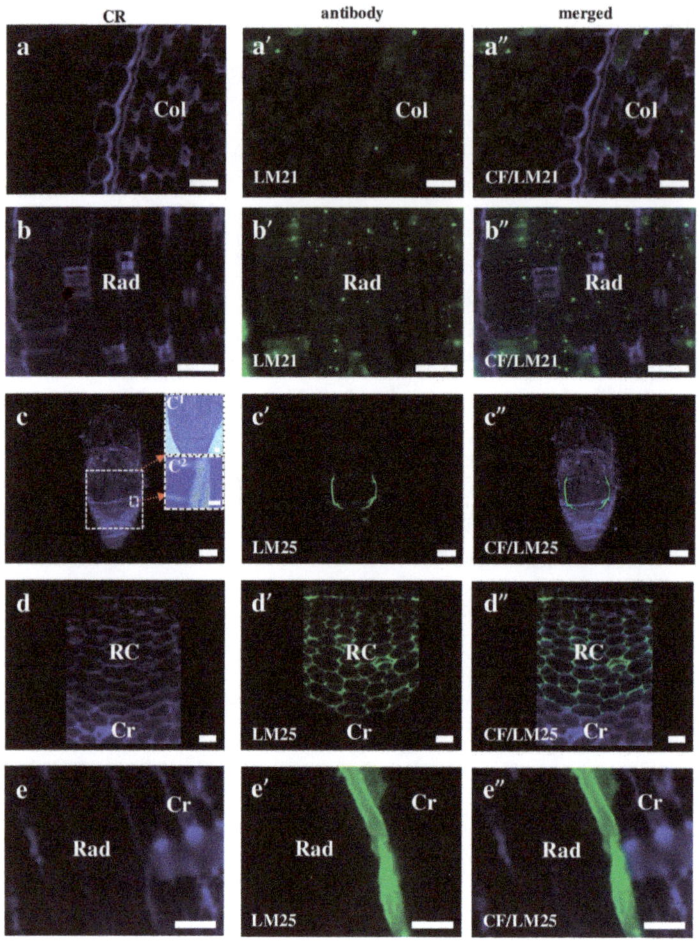

Figure 9. Immunolocalisation of hemicelluloses in Brachypodium embryos. (**a–a″**) coleoptile; (**b–b″**) radicula with LM21, respectively; (**c–c″**,**d–d″**,**e–e″**) the entire embryo, root cap with coleorhiza and radicula with coleorhiza with LM25, respectively. Arrows demonstrate insets: (c^1,c^2) histological sections of embryo, scale bars are 50 and 2 µm, respectively. Abbreviations: CF—calcofluor, Col—coleoptile, Cr—coleorhiza, Rad—radicula, and RC—root cap. Bars: (**a–a″**,**d–d″**,**e–e″**) 10 µm; (**b–b″**) 20 µm; (**c–c″**) 100 µm.

In our current study, such a high deposition of xyloglucans seems to be important as a stock of elements for embryo development. Xyloglucans are known to differentially express throughout the embryo in Arabidopsis [47]. An analysis of Arabidopsis *xyl1* mutant phenotypes, which are associated with modifications of the composition of the endosperm cell wall demonstrated altered germination characteristics, such as shorter and thicker siliques, a reduced dormancy, and an increased tolerance to germination inhibitors and thermoinhibition. This study highlighted the role of xyloglucans during seed development and germination. Xyloglucans are likely to be very important components for cell adhesion and cell separation in fruit seed coat parenchyma [48]. Research on the tomato mutant *Cnr* (colourless non-ripening), which has a pleiotropic dominant mutation with reduced cell-to-cell adhesion, demonstrated that xyloglucan polymers (galactan and mannan) that are distributed more widely can be targeted to effect cell release from a parenchyma system. It was demonstrated that a monocotyledon palm *Euterpe oleraceae* contains a highly methoxylated homogalacturonan together with small amounts of a mannoglucan. Furthermore, a type II arabinogalactan was structurally characterised [49]. The experiments in non-lignified cell walls in different monocotyledon species, using the monoclonal antibody CCRC-M1 against fucosylated xyloglucans, demonstrated the presence of this epitope in non-lignified walls. A similar pattern was also found in the palm *Phoenix canariensis*. In Zingiberales, Commelinales, and Poales this epitope was found in the phloem walls, stomatal guard, subsidiary cells, and raphide idioblasts. However, it was not found in *Tradescantia virginiana* (Commelinaceae, Commelinales), or *Zea mays* (Poaceae, Poales). On the other hand, this epitope was observed in the phloem walls of two other Poaceae species, *Lolium multiflorum* and *L. perenne*. The authors highlighted the lack of knowledge about the functions of xyloglucans in monocotyledonous plants [50]. Moreover, work on Brachypodium revealed different proteome profiles that related to the different phases of grain development, and provided information of the reorganization of cell wall constituents that occurs during these stages [9,51]. Such results indicate that changes in cell wall chemical composition are important factor involved in its assembly and remodelling on different stages of development, plant tissues, and organs, both during embryonic and post-embryonic stages of development.

3. Materials and Methods

3.1. Plant Material and Histological Procedures

Approximately 15 mature embryos of Brachypodium (29 days after fertilization), reference genotype Bd21, were isolated from seeds that were collected from plants growing in pots with soil mixed with vermiculite (3:1) in a greenhouse. Plants were grown at 20 ± 1 °C, under a 16/8 h light/dark photoperiod. To ensure synchronised flowering, approximately two-week-old plants were subjected to vernalisation for two weeks at 4 °C. The embryos were fixed in a mixture of 4% (*w*/*v*) paraformaldehyde (PFA) and 1% (*v*/*v*) glutaraldehyde (GA) in phosphate-buffered saline (PBS, pH 7.0) overnight at 4 °C. Then, they were rinsed with PBS (3 × 15 min), dehydrated in an ascending ethanol series (10%, 30%, 50%, 70%, 90%, and 100%; 2 × 30 min in each) and gradually embedded in LR White resin [52]. The material was cut into 1.5-µm thick sections using an EMUC6 ultramicrotome (Leica Microsystems, Wetzlar, Germany). Sections were collected on microscopic slides covered with poly-L-lysine. For general histology, they were stained with 0.05% (aqueous solution) Toluidine Blue O (Sigma-Aldrich, St. Louis, MO, USA) for 5 min.

3.2. Immunocytochemistry

Sections were incubated in a blocking buffer containing 2% (*v*/*v*) foetal calf serum (FCS) and 2% (*w*/*v*) bovine serum albumin (BSA) in PBS (pH 7.2) for 30 min at room temperature (RT). Next, they were incubated with specific primary monoclonal antibodies (Table 1), diluted at a 1:20 ratio in a blocking buffer (RT, minimum 1.5 h), rinsed with the blocking buffer 3 × 10 min and then incubated at RT for at least 1.5 h with the secondary antibody (Alexa Fluor 488 goat anti-rat IgG, Jackson Immuno Research Laboratories, West Grove, PA, USA) diluted 1:100 in the blocking buffer as above. After washing with the blocking buffer and PBS (3 × 10 min each), the sections were stained with 0.01% calcofluor (Sigma-Aldrich, St. Louis, MO, USA) in PBS for 5 min; then slides were thoroughly rinsed with PBS and sterile distilled water (3 × 10 min each). The drained slides were mounted in a Fluoromount (Sigma-Aldrich) antifade medium. Negative controls were performed for each antibody that was used by omitting the primary antibodies. All of the images were taken using an Axio Imager Z2 (Zeiss, Oberkochen, Germany) epifluorescent microscope equipped with an AxioCam Mrm monochromatic camera (Zeiss) with the narrow-band filters for AlexaFluor 488 and DAPI. The negative control for all of the antibodies revealed the absence of any signals that were specific for these antibodies and was representative for all of the other antibodies that were used in this study (Figure S2).

Table 1. The antibodies used for immunocytochemistry, the epitopes they recognise, and relevant references.

Antibody	Epitope	References
	Arabinogalactan proteins (AGPs)	
MAC207	β-GlcA1->3αGalA1->2Rha	[53–56]
JIM8	Arabinogalactan	[57]
JIM13	βGlcA1->3αGalA1->2Rha	[55,56,58]
JIM16	AGP glycan	[55,56,58]
LM2	β-linked GlcA	[56,59]
	Extensins	
LM1	Extensin	[60]
JIM11	Extensin	[55,61]
JIM12	Extensin	[61]
	Pectins	
LM19	α-GalA(1-4)α-GalA(1-4)α-GalA(1-4)α-GalA	[35]
LM13	α-1,5-arabinan	[62]
LM16	Processed arabinan—rhamnogalacturonan (RG)-I domain	[35]
LM6	αAra1-5αAra1-5αAra1-5αAra1-5Ara	[63]
	Hemicelluloses	
LM21	β-linked mannan polysaccharides of plant cell walls	[64]
LM25	Xyloglucan	[65]

3.3. TEM

The samples for TEM were fixed in 2.5% glutaraldehyde in a 0.1 M sodium phosphate buffer (pH 7.4) at 4 °C for 24 h. The embryos were postfixed in 1% OsO_4 (osmium tetroxide) in a 0.1 M sodium phosphate buffer for two hours at 4 °C, rinsed in the same buffer, dehydrated in an ascending series of acetone, and gradually embedded in Epon 812 (Fullam, Latham, NY, USA) [66]. Semithin sections were stained with toluidine blue and examined with a bright-field microscope. Ultrathin (70 nm) sections were cut on a Leica ultracut UCT ultramicrotome and collected on copper grids (300 mesh, Electron Microscopy Science, Hatfield, PA, USA). Sections were stained with uranyl acetate and lead citrate and examined with a Hitachi H500 TEM (Hitachi, Tokyo, Japan) at 75 kV.

4. Conclusions

In this work we traced the distribution of various cell wall epitopes in the Brachypodium zygotic embryo in mature stages of development. We have demonstrated that:

1. Among analysed AGP epitopes JIM13, JIM8, and LM2 were localised in cell walls of the embryo surface tissues, as well as some of them in the seed coat; and MAC207 and JIM16 were detected in cytoplasmic compartments.
2. Extensins are localised in the outer periclinal walls of embryo surface tissues. Our results suggest that in Brachypodium, the antibody LM13 can be used as a negative marker of epidermal cells of the embryonic leaf and that the LM25 antibody can be used as the positive marker of embryonic root epidermis cells, as well as the root cap in Brachypodium. These results demonstrate that the distribution of analysed cell wall epitopes in mature zygotic embryos of Brachypodium is tissue- and organ specific.

Without a doubt, the precise dissection of all functions of the cell wall in Brachypodium embryos requires the use of specific mutants, whose availability is limited. However, this problem should soon be overcome by the rapidly-developing techniques of site-directed mutagenesis, such as CRISPR/Cas9 that are already available for Brachypodium [67].

Supplementary Materials: The following are available online at www.mdpi.com/1422-0067/19/3/725/s1.

Acknowledgments: This work was supported by the National Science Centre, Poland (grant no. DEC-2014/14/M/NZ2/00519).

Author Contributions: Alexander Betekhtin, Anna Milewska-Hendel, Joanna Lusinska, and Lukasz Chajec performed the experiments; Alexander Betekhtin, Ewa Kurczynska, and Robert Hasterok analysed the data; the paper was written under the direction and supervision of Ewa Kurczynska and Robert Hasterok; and Alexander Betekhtin was responsible for writing this work. All authors read and approved the final manuscript.

Conflicts of Interest: The authors declare no conflict of interest.

Abbreviations

AGP	Arabinogalactan protein
BSA	Bovine serum albumin
CRISPR	Clustered regularly-interspaced short palindromic repeats
FCS	Foetal calf serum
GA	Glutaraldehyde
PBS	Phosphate-buffered saline
PFA	Paraformaldehyde
RT	Room temperature
TEM	Transmission electron microscope

References

1. Showalter, A.M. Structure and function of plant cell wall proteins. *Plant Cell* **1993**, *5*, 9–23. [CrossRef] [PubMed]
2. Ochoa-Villarreal, M.; Aispuro-Hernández, E.; Vargas-Arispuro, I.; Ángel Martínez-Téllez, M. Plant Cell Wall Polymers: Function, Structure and Biological Activity of Their Derivatives. In *Materials Science*; Gomes, A.D.S., Ed.; INTECH: Rijeka, Croatia, 2012.
3. Stafford, H.A. *Plant Cell Wall Polymers: Biogenesis and Biodegradation*; Lewis, N.G., Paice, M.G., Eds.; American Chemical Society: Washington, DC, USA, 1989; p. xii, 676p.
4. Schadel, C. Cell-Wall Hemicelluloses as Mobile Carbon Stores in Plants. Ph.D. Thesis, Universität Basel, Basel, Switzerland, 2009.
5. Cosgrove, D.J. Growth of the plant cell wall. *Nat. Rev. Mol. Cell Biol.* **2005**, *6*, 850–861. [CrossRef] [PubMed]
6. Evert, R.F. *Esau's Plant Anatomy: Meristems, Cells, and Tissues of the Plant Body: Their Structure, Function, and Development*; John Wiley & Sons, Inc.: Hoboken, NJ, USA, 2006.
7. Labavitch, J.M.; Freeman, L.E.; Albersheim, P. Structure of plant cell walls. Purification and characterization of a beta-1,4-galactanase which degrades a structural component of the primary cell walls of dicots. *J. Biol. Chem.* **1976**, *251*, 5904–5910. [PubMed]
8. Draper, J.; Mur, L.A.; Jenkins, G.; Ghosh-Biswas, G.C.; Bablak, P.; Hasterok, R.; Routledge, A.P. *Brachypodium distachyon* a new model system for functional genomics in grasses. *Plant Physiol.* **2001**, *127*, 1539–1555. [CrossRef] [PubMed]
9. Francin-Allami, M.; Merah, K.; Albenne, C.; Rogniaux, H.; Pavlovic, M.; Lollier, V.; Sibout, R.; Guillon, F.; Jamet, E.; Larre, C. Cell wall proteomic of *Brachypodium distachyon* grains: A focus on cell wall remodeling proteins. *Proteomics* **2015**, *15*, 2296–2306. [CrossRef] [PubMed]
10. Christensen, U.; Alonso-Simon, A.; Scheller, H.V.; Willats, W.G.; Harholt, J. Characterization of the primary cell walls of seedlings of *Brachypodium distachyon*-a potential model plant for temperate grasses. *Phytochemistry* **2010**, *71*, 62–69. [CrossRef] [PubMed]
11. Wang, T.; Salazar, A.; Zabotina, O.A.; Hong, M. Structure and dynamics of Brachypodium primary cell wall polysaccharides from two-dimensional ^{13}C solid-state nuclear magnetic resonance spectroscopy. *Biochemistry* **2014**, *53*, 2840–2854. [CrossRef] [PubMed]
12. Marriott, P.E.; Sibout, R.; Lapierre, C.; Fangel, J.U.; Willats, W.G.; Hofte, H.; Gomez, L.D.; McQueen-Mason, S.J. Range of cell-wall alterations enhance saccharification in *Brachypodium distachyon* mutants. *Proc. Natl. Acad. Sci. USA* **2014**, *111*, 14601–14606. [CrossRef] [PubMed]
13. Wolny, E.; Braszewska-Zalewska, A.; Hasterok, R. Spatial distribution of epigenetic modifications in *Brachypodium distachyon* embryos during seed maturation and germination. *PLoS ONE* **2014**, *9*, e101246. [CrossRef] [PubMed]

14. Tillich, H. Vergleichend morphologische Untersuchungen zur Identitat der Gramineen-Primarwurzel. *Flora* **1977**, *166*, 415–421. [CrossRef]
15. Dolan, L.; Janmaat, K.; Willemsen, V.; Linstead, P.; Poethig, S.; Roberts, K.; Scheres, B. Cellular organisation of the *Arabidopsis thaliana* root. *Development* **1993**, *119*, 71–84. [PubMed]
16. Verdeil, J.L.; Alemanno, L.; Niemenak, N.; Tranbarger, T.J. Pluripotent versus totipotent plant stem cells: Dependence versus autonomy? *Trends Plant Sci.* **2007**, *12*, 245–252. [CrossRef] [PubMed]
17. Jouannic, S.; Lartaud, M.; Herve, J.; Collin, M.; Orieux, Y.; Verdeil, J.L.; Tregear, J.W. The shoot apical meristem of oil palm (*Elaeis guineensis*; Arecaceae): Developmental progression and dynamics. *Ann. Bot.* **2011**, *108*, 1477–1487. [CrossRef] [PubMed]
18. Marín-Méndez, W.; Sanchéz-Chacón, E.; Gatica-Arias, A.M.; Ramírez Fonseca, P.; Freer-Bustamante, E.; Valdez-Melara, M. Ultrastructure and histology of organogenesis induced from shoot tips of maize (*Zea mays*, Poaceae). *Rev. Biol. Trop.* **2009**, *57*, 129–139.
19. Bennici, A.; Tani, C. Ultrastructural effects of salinity in *Nicotiana bigelovii var. bigelovii* callus and *Allium cepa* roots. *Caryologia* **2009**, *62*, 124–133.
20. Rumyantseva, N.I. Arabinogalactan proteins: Involvement in plant growth and morphogenesis. *Biochemistry* **2005**, *70*, 1073–1085. [CrossRef] [PubMed]
21. Joosen, R.V.; Arends, D.; Willems, L.A.; Ligterink, W.; Jansen, R.C.; Hilhorst, H.W. Visualizing the genetic landscape of Arabidopsis seed performance. *Plant Physiol.* **2012**, *158*, 570–589. [CrossRef] [PubMed]
22. Joosen, R.V.L.; Ligterink, W.; Dekkers, B.J.W.; Hilhorst, H.W.M. Visualization of molecular processes associated with seed dormancy and germination using MapMan. *Seed Sci. Res.* **2011**, *21*, 143–152. [CrossRef]
23. Cannesan, M.A.; Durand, C.; Burel, C.; Gangneux, C.; Lerouge, P.; Ishii, T.; Laval, K.; Follet-Gueye, M.L.; Driouich, A.; Vicre-Gibouin, M. Effect of arabinogalactan proteins from the root caps of pea and *Brassica napus* on *Aphanomyces euteiches* zoospore chemotaxis and germination. *Plant Physiol.* **2012**, *159*, 1658–1670. [CrossRef] [PubMed]
24. Van Hengel, A.J.; Roberts, K. AtAGP30, an arabinogalactan-protein in the cell walls of the primary root, plays a role in root regeneration and seed germination. *Plant J.* **2003**, *36*, 256–270. [CrossRef] [PubMed]
25. Brady, J.D.; Sadler, I.H.; Fry, S.C. Pulcherosine, an oxidatively coupled trimer of tyrosine in plant cell walls: Its role in cross-link formation. *Phytochemistry* **1998**, *47*, 349–353. [CrossRef]
26. Cannon, M.C.; Terneus, K.; Hall, Q.; Tan, L.; Wang, Y.; Wegenhart, B.L.; Chen, L.; Lamport, D.T.; Chen, Y.; Kieliszewski, M.J. Self-assembly of the plant cell wall requires an extensin scaffold. *Proc. Natl. Acad. Sci. USA* **2008**, *105*, 2226–2231. [CrossRef] [PubMed]
27. Lamport, D.T.; Kieliszewski, M.J.; Chen, Y.; Cannon, M.C. Role of the extensin superfamily in primary cell wall architecture. *Plant Physiol.* **2011**, *156*, 11–19. [CrossRef] [PubMed]
28. Xu, C.; Takac, T.; Burbach, C.; Menzel, D.; Samaj, J. Developmental localization and the role of hydroxyproline rich glycoproteins during somatic embryogenesis of banana (Musa spp. AAA). *BMC Plant Biol.* **2011**, *11*, 38. [CrossRef] [PubMed]
29. Sujkowska-Rybkowska, M.; Borucki, W. Accumulation and localization of extensin protein in apoplast of pea root nodule under aluminum stress. *Micron* **2014**, *67*, 10–19. [CrossRef] [PubMed]
30. Wu, Y.; Fan, W.; Li, X.; Chen, H.; Takac, T.; Samajova, O.; Fabrice, M.R.; Xie, L.; Ma, J.; Samaj, J.; et al. Expression and distribution of extensins and AGPs in susceptible and resistant banana cultivars in response to wounding and *Fusarium oxysporum*. *Sci. Rep.* **2017**, *7*, 42400. [CrossRef] [PubMed]
31. Szatanik-Kloc, A.; Szerement, J.; Jozefaciuk, G. The role of cell walls and pectins in cation exchange and surface area of plant roots. *J. Plant Physiol.* **2017**, *215*, 85–90. [CrossRef] [PubMed]
32. Prade, R.A.; Zhan, D.; Ayoubi, P.; Mort, A.J. Pectins, pectinases and plant-microbe interactions. *Biotechnol. Genet.* Eng. Rev. **1999**, *16*, 361–391. [CrossRef] [PubMed]
33. Carpita, N.C.; Defernez, M.; Findlay, K.; Wells, B.; Shoue, D.A.; Catchpole, G.; Wilson, R.H.; McCann, M.C. Cell wall architecture of the elongating maize coleoptile. *Plant Physiol.* **2001**, *127*, 551–565. [CrossRef] [PubMed]
34. Verhertbruggen, Y.; Marcus, S.E.; Chen, J.; Knox, J.P. Cell wall pectic arabinans influence the mechanical properties of *Arabidopsis thaliana* inflorescence stems and their response to mechanical stress. *Plant Cell Physiol.* **2013**, *54*, 1278–1288. [CrossRef] [PubMed]

35. Verhertbruggen, Y.; Marcus, S.E.; Haeger, A.; Verhoef, R.; Schols, H.A.; McCleary, B.V.; McKee, L.; Gilbert, H.J.; Knox, J.P. Developmental complexity of arabinan polysaccharides and their processing in plant cell walls. *Plant J.* **2009**, *59*, 413–425. [CrossRef] [PubMed]
36. Gomez, L.D.; Steele-King, C.G.; Jones, L.; Foster, J.M.; Vuttipongchaikij, S.; McQueen-Mason, S.J. Arabinan metabolism during seed development and germination in Arabidopsis. *Mol. Plant* **2009**, *2*, 966–976. [CrossRef] [PubMed]
37. Gibeaut, D.M.; Pauly, M.; Bacic, A.; Fincher, G.B. Changes in cell wall polysaccharides in developing barley (*Hordeum vulgare*) coleoptiles. *Planta* **2005**, *221*, 729–738. [CrossRef] [PubMed]
38. Dourado, F.; Barros, A.; Mota, M.; Coimbra, M.A.; Gama, F.M. Anatomy and cell wall polysaccharides of almond (*Prunus dulcis* D. A. Webb) seeds. *J. Agric. Food Chem.* **2004**, *52*, 1364–1370. [CrossRef] [PubMed]
39. Eriksson, I.; Andersson, R.; Westerlund, E.; Andersson, R.; Aman, P. Structural features of an arabinan fragment isolated from the water-soluble fraction of dehulled rapeseed. *Carbohydr. Res.* **1996**, *281*, 161–172. [CrossRef]
40. Navarro, D.A.; Cerezo, A.S.; Stortz, C.A. NMR spectroscopy and chemical studies of an arabinan-rich system from the endosperm of the seed of *Gleditsia triacanthos*. *Carbohydr. Res.* **2002**, *337*, 255–263. [CrossRef]
41. Wang, T.; Hong, M. Solid-state NMR investigations of cellulose structure and interactions with matrix polysaccharides in plant primary cell walls. *J. Exp. Bot.* **2016**, *67*, 503–514. [CrossRef] [PubMed]
42. Keegstra, K.; Talmadge, K.W.; Bauer, W.D.; Albersheim, P. The Structure of Plant Cell Walls: III. A Model of the Walls of Suspension-cultured Sycamore Cells Based on the Interconnections of the Macromolecular Components. *Plant Physiol.* **1973**, *51*, 188–197. [CrossRef] [PubMed]
43. Zimmermann, R.; Wolfgang, W. Pattern formation in the monocot embryo as revealed by NAM and CUC3 orthologues from *Zea mays* L. *Plant Mol. Biol.* **2005**, *58*, 669–685. [CrossRef] [PubMed]
44. Samaj, J.; Samajova, O.; Peters, M.; Baluska, F.; Lichtscheidl, I.; Knox, J.P.; Volkmann, D. Immunolocalization of LM2 arabinogalactan protein epitope associated with endomembranes of plant cells. *Protoplasma* **2000**, *212*, 186–196. [CrossRef]
45. Bourquin, V.; Nishikubo, N.; Abe, H.; Brumer, H.; Denman, S.; Eklund, M.; Christiernin, M.; Teeri, T.T.; Sundberg, B.; Mellerowicz, E.J. Xyloglucan endotransglycosylases have a function during the formation of secondary cell walls of vascular tissues. *Plant Cell* **2002**, *14*, 3073–3088. [CrossRef] [PubMed]
46. Betekhtin, A.; Rojek, M.; Milewska-Hendel, A.; Gawecki, R.; Karcz, J.; Kurczynska, E.; Hasterok, R. Spatial Distribution of Selected Chemical Cell Wall Components in the Embryogenic Callus of *Brachypodium distachyon*. *PLoS ONE* **2016**, *11*, e0167426. [CrossRef] [PubMed]
47. Sechet, J.; Frey, A.; Effroy-Cuzzi, D.; Berger, A.; Perreau, F.; Cueff, G.; Charif, D.; Rajjou, L.; Mouille, G.; North, H.M.; et al. Xyloglucan Metabolism Differentially Impacts the Cell Wall Characteristics of the Endosperm and Embryo during Arabidopsis Seed Germination. *Plant Physiol.* **2016**, *170*, 1367–1380. [CrossRef] [PubMed]
48. Ordaz-Ortiz, J.J.; Marcus, S.E.; Knox, J.P. Cell wall microstructure analysis implicates hemicellulose polysaccharides in cell adhesion in tomato fruit pericarp parenchyma. *Mol. Plant* **2009**, *2*, 910–921. [CrossRef] [PubMed]
49. Cantu-Jungles, T.M.; Iacomini, M.; Cipriani, T.R.; Cordeiro, L.M.C. Extraction and characterization of pectins from primary cell walls of edible açaí (*Euterpe oleraceae*) berries, fruits of a monocotyledon palm. *Carbohydr. Polym.* **2017**, *158*, 37–43. [CrossRef] [PubMed]
50. Brennan, M.; Harris, P.J. Distribution of fucosylated xyloglucans among the walls of different cell types in monocotyledons determined by immunofluorescence microscopy. *Mol. Plant* **2011**, *4*, 144–156. [CrossRef] [PubMed]
51. Francin-Allami, M.; Lollier, V.; Pavlovic, M.; San Clemente, H.; Rogniaux, H.; Jamet, E.; Guillon, F.; Larre, C. Understanding the remodelling of cell walls during *Brachypodium distachyon* grain development through a sub-cellular quantitative proteomic approach. *Proteomes* **2016**, *4*, 21. [CrossRef] [PubMed]
52. Skepper, J.N.; Powell, J.M. Ultrastructural immunochemistry. *CSH Protoc.* **2008**, *2008*. [CrossRef] [PubMed]
53. Bradley, D.J.; Wood, E.A.; Larkins, A.P.; Galfre, G.; Butcher, G.W.; Brewin, N.J. Isolation of monoclonal antibodies reacting with peribacteriod membranes and other components of pea root nodules containing *Rhizobium leguminosarum*. *Planta* **1988**, *173*, 149–160. [CrossRef] [PubMed]

54. Pennell, R.I.; Knox, J.P.; Scofield, G.N.; Selvendran, R.; Roberts, K. A family of abundant plasma membrane-associated glycoproteins related to the arabinogalactan proteins is unique to flowering plants. *J. Cell Biol.* **1989**, *108*, 1967–1977. [CrossRef] [PubMed]
55. Yates, E.A.; Knox, J.P. Investigations into the occurrence of plant cell surface epitopes in exudate gums. *Carbohydr. Polym.* **1994**, *24*, 281–286. [CrossRef]
56. Yates, E.A.; Valdor, J.-F.; Haslam, S.M.; Morris, H.R.; Dell, A.; Mackie, W.; Knox, J.P. Characterization of carbohydrate structural features recognized by anti-arabinogalactan-protein monoclonal antibodies. *Glycobiology* **1996**, *6*, 131–139. [CrossRef] [PubMed]
57. Pennell, R.I.; Janniche, L.; Kjellbom, P.; Scofield, G.N.; Peart, J.M.; Roberts, K. Developmental regulation of a plasma membrane arabinogalactan protein epitope in oilseed rape flowers. *Plant Cell* **1991**, *3*, 1317–1326. [CrossRef] [PubMed]
58. Knox, J.P.; Linstead, P.J.; Peart, J.; Cooper, C.; Roberts, K. Developmentally regulated epitopes of cell surface arabinogalactan proteins and their relation to root tissue pattern formation. *Plant J.* **1991**, *1*, 317–326. [CrossRef] [PubMed]
59. Smallwood, M.; Yates, E.A.; Willats, W.G.T.; Martin, H.; Knox, J.P. Immunochemical comparison of membrane-associated and secreted arabinogalactan-proteins in rice and carrot. *Planta* **1996**, *198*, 452–459. [CrossRef]
60. Smallwood, M.; Martin, H.; Knox, J.P. An epitope of rice threonine-rich and hydroxyproline-rich glycoprotein is common to cell-wall and hydrophobic plasma-membrane glycoproteins. *Planta* **1995**, *196*, 510–522. [CrossRef] [PubMed]
61. Smallwood, M.; Beven, A.; Donovan, N.; Neill, S.J.; Peart, J.; Roberts, K.; Knox, J.P. Localization of cell wall proteins in relation to the developmental anatomy of the carrot root apex. *Plant J.* **1994**, *5*, 237–246. [CrossRef]
62. Moller, I.; Marcus, S.E.; Haeger, A.; Verhertbruggen, Y.; Verhoef, R.; Schols, H.; Ulvskov, P.; Mikkelsen, J.D.; Knox, J.P.; Willats, W. High-throughput screening of monoclonal antibodies against plant cell wall glycans by hierarchical clustering of their carbohydrate microarray binding profiles. *Glycoconj. J.* **2008**, *25*, 37–48. [CrossRef] [PubMed]
63. Willats, W.G.; Marcus, S.E.; Knox, J.P. Generation of monoclonal antibody specific to (1–>5)-alpha-L-arabinan. *Carbohydr. Res.* **1998**, *308*, 149–152. [CrossRef]
64. Marcus, S.E.; Blake, A.W.; Benians, T.A.; Lee, K.J.; Poyser, C.; Donaldson, L.; Leroux, O.; Rogowski, A.; Petersen, H.L.; Boraston, A.; et al. Restricted access of proteins to mannan polysaccharides in intact plant cell walls. *Plant J.* **2010**, *64*, 191–203. [CrossRef] [PubMed]
65. Pedersen, H.L.; Fangel, J.U.; McCleary, B.; Ruzanski, C.; Rydahl, M.G.; Ralet, M.C.; Farkas, V.; von Schantz, L.; Marcus, S.E.; Andersen, M.C.; et al. Versatile high resolution oligosaccharide microarrays for plant glycobiology and cell wall research. *J. Biol. Chem.* **2012**, *287*, 39429–39438. [CrossRef] [PubMed]
66. Luft, J.H. Improvements in epoxy resin embedding methods. *J. Biophys. Biochem. Cytol.* **1961**, *9*, 409–414. [CrossRef] [PubMed]
67. O'Connor, D.L.; Elton, S.; Ticchiarelli, F.; Hsia, M.M.; Vogel, J.P.; Leyser, O. Cross-species functional diversity within the PIN auxin efflux protein family. *elife* **2017**, *6*, e31804. [CrossRef] [PubMed]

International Journal of
Molecular Sciences

MDPI

Review

Plasmodesmata-Related Structural and Functional Proteins: The Long Sought-After Secrets of a Cytoplasmic Channel in Plant Cell Walls

Xiao Han [1,†], Li-Jun Huang [2,†], Dan Feng [3,†], Wenhan Jiang [2], Wenzhuo Miu [2] and Ning Li [4,*]

1 College of Biological Science and Engineering, Fuzhou University, Fuzhou 350116, China; hanxiao@caas.cn
2 College of Forestry, Central South University of Forestry and Technology, Changsha 410004, China; nghua@126.com (L.-J.H.); jiangwenhan1999@126.com (W.J.); alic5e@163.com (W.M.)
3 Biotechnology Research Institute, Chinese Academy of Agricultural Science, Beijing 100081, China; gygzgzyx@126.com
4 State Key Laboratory of Cultivation and Protection for Non-Wood Forest Trees, Ministry of Education, Central South University of Forestry and Technology, Changsha 410004, China
* Correspondence: nli@csuft.edu.cn; Tel./Fax: +86-0731-8562-3406
† These authors contributed equally.

Received: 27 May 2019; Accepted: 13 June 2019; Published: 17 June 2019

Abstract: Plant cells are separated by cellulose cell walls that impede direct cell-to-cell contact. In order to facilitate intercellular communication, plant cells develop unique cell-wall-spanning structures termed plasmodesmata (PD). PD are membranous channels that link the cytoplasm, plasma membranes, and endoplasmic reticulum of adjacent cells to provide cytoplasmic and membrane continuity for molecular trafficking. PD play important roles for the development and physiology of all plants. The structure and function of PD in the plant cell walls are highly dynamic and tightly regulated. Despite their importance, plasmodesmata are among the few plant cell organelles that remain poorly understood. The molecular properties of PD seem largely elusive or speculative. In this review, we firstly describe the general PD structure and its protein composition. We then discuss the recent progress in identification and characterization of PD-associated plant cell-wall proteins that regulate PD function, with particular emphasis on callose metabolizing and binding proteins, and protein kinases targeted to and around PD.

Keywords: plant cell walls; plasmodesmata; callose; callose synthase

1. Introduction

Plasmodesmata (PD) span the plant cell walls and connect protoplasts of adjacent cells [1,2]. The presence of PD allows molecular communication conveniently between plant cells [3–5]. Several thousand PD form a network between fully differentiated plant cells [6], transforming individual cells into interconnected symplasts, providing channels for the intercellular exchange of electrical signaling, diffusion of small molecules, and transportation of macromolecules such as proteins, nucleic acids, and viruses [7–10]. The exchange channel provides a direct cell-to-cell cytoplasmic pathway for plant molecule transport and information transmission [11]. The PD also regulates plant metabolic activity and gene expression through the transfer of signaling chemicals, enabling plant cells to exchange information with each other to coordinate physiological processes [12–14].

Although being a functional important structure, the molecular composition of PD remains largely elusive [15,16]. PD are embedded in rigid cell walls; it is arduous to purify and study such an organelle via classical biochemical methods. High-resolution electron microscopy showed that proteinaceous particles are associated with PD structures [17]. Studies such as immunolocalization revealed that various proteins are present in PD, including kinase-like proteins [18], actin, and myosin [19], which may serve as permanent structural components. In addition, genetic and biochemical approaches applied to dissect PD-associated functional components led to the discovery of callose binding proteins and receptor-like kinase proteins in the model plants [20]. They regulate the permeability of PD and control intercellular substance transport and signal communication. PD-mediated symplasmic transport is involved in a variety of plant physiological processes that are pivotal for the development of plants of economic interest, such as shoot apex dormancy, flowering, fruit ripening, fiber elongation, and plant–microbe symbiosis [21]. Understanding PD function through characterization of PD-related proteins may have the potential for improving agronomic traits in the future.

It is generally believed that callose deposited at the PD neck region plays a regulatory role in the neck switch [22]. This is the most well-known mechanism for regulating PD permeability [23]. Callose is a polysaccharide in the form of β-1,3-glucan [24]. Callose is synthesized at the plasma membrane at the neck region of PD, squeezing the plasma membrane inwardly, thereby narrowing down the neck region, which reduces the free space that PD can use for molecular passage [25]. A large amount of evidence indicates that the change in PD permeability is due to the synthesis and hydrolysis of callose in the cell wall [26–28]. Callose synthase, β-1,3-glucanase, and PD-associated callose binding protein are involved in the regulation of callose homeostasis in PD, which in turn affects the permeability of PD channels [29,30].

2. The Formation and Structure of PD

From the perspective of origin, PD is generally formed during the mitotic cell division. In the primary formation of PD, although the protoplasm is split by the newly formed cell plate, the endoplasmic reticulum (ER) is still connected through the cell plate [31]. The presence of the ER in the channels prevents the deposition of wall-forming substances, thus maintaining the connection between the two separated cells [32]. Furthermore, the ER transforms into the PD under the pressure of the cell plate or membrane. There are also PD between non-sister cells, that is, PD can be formed between non-mitotic cells [6,33]. However, the molecular origin and process of secondary PD formation are not well understood.

Ding et al. [34] obtained a series of new PD structural details using rapid freezing and freezing replacement sample preparation techniques, high-resolution electron microscopy, and computer-aided image processing techniques. A structural model of PD was proposed based on their observations. At both ends of PD is an enlarged hole [35,36], and the center of the PD is a cylindrical body formed by the appressed ER, called a desmotubule [37]. The desmotubule is tightly constricted with little space [38]. Cytoskeletal proteins are often found in PD, such as actin and myosin [39,40]. Protein particles of about 3 nm are embedded between the appressed ER and the plasma membrane. Electron-dense radial fibrils connect the protein particles [34,41]. The protein particles on the ER membrane present a spiral or a series of circularly arranged rotations. In the cross-section, 7–9 particles can be seen [34]. The channel of the ER is considered to be the transport pathway of PD [42,43]. The two ends of the channel become smaller, called the neck region [37].

Actin filaments and callosum sphincters are two modes of controlling PD permeability. According to experimental observations, actin filaments are located near and inside the PD channel, and the polymerization state of actin changes the permeability of PD [44,45]. On the other hand, callose aggregates at the neck region narrow the space between cell wall and PD desmotubule.

3. PD-Associated Structural Proteins

As mentioned above, studies such as immunolocalization revealed that various proteins such as actin, myosin, tubulin, and calreticulin are present in PD as structural components [39,46]. Actin and myosin are components of the cell dynamic network, cytoskeleton, involved in intracellular transport [47]. Microtubule arrays consist of tubulin heterodimers which play critical roles in cell division and intracellular organization [48]. These proteins are not cell-wall proteins, but are associated with the plant cell-wall structure of PD in the context of this review (Table 1).

Table 1. Proteins associated with plasmodesmata (PD).

Gene	Protein Function	Method	Biological Function	References
Actin	Cell cytoskeleton	Immunolocalization	PD structure component	[49–51]
Myosin	Cell cytoskeleton	Immunolocalization	PD structure component	[46,51,52]
Tubulin	Cell cytoskeleton	Immunolocalization	PD structure component	[51]
GSL8 (CALS10)	Callose synthase	Genetics and cell biology	Callose deposition and auxin response	[53]
GSL12 (CALS3)	Callose synthase	Genetics and cell biology	Callose deposition and phloem development	[36]
PDBGs (AtBG_ppap)	β-1,3-glucanase	Proteomics and cell biology	Callose degradation and lateral root development	[28,54,55]
PDCBs	Callose binding protein	Proteomics and cell biology	PD localization and callose stabilization at neck region	[29]
PDLPs	Transmembrane receptor-like protein kinase	Proteomics and cell biology	SAR- and SA-induced callose accumulation	[55–59]

3.1. Actin

Actin exists in the entire PD [60]. The systematic observation of the nature and structure of PD in the epidermal cells of *Allium sativum* L. was performed using a fluorescence probe and confocal microscope combined with transmission electron microscopy and immunogold labeling [49]. White et al. [49] indirectly demonstrated the presence of actin in PD in young and mature cell walls of *Hordeum vulgare* and *Nicotiana plumbaginifolia* by colloidal gold labeling. The results effectively confirmed the presence of F-actin in PD. It is unclear how the filaments are organized within PD. They may exist in the lumen between the desmotubule and the plasma membrane, connecting the cytoskeleton between adjacent cells [45,61]. The actin filaments may also function as routes along which vesicular trafficking is possible. Treatments with actin disrupter, cytochalasin D, and actin-stabilizing fungal toxin, phalloidin, showed opposite effects on the cell-to-cell transport of a fluorescent reporter dextran; the latter chemical severely restrained the dextran movement [49,50]. These results indicate that the PD structural component may also be involved in regulating the PD permeability, with larger molecules being able to move through once actin filaments are disrupted.

3.2. Myosin

Immunochemical localization studies indicated that myosin was also an integral component of PD [49,51,52]. Radford and White's experiments indirectly demonstrated a myosin distribution on PD [52]. They used an animal myosin antibody with colloidal gold to prepare an immuno-electron microscopic sample of onions, maize, and *Arabidopsis* seedlings. They found colloidal gold-labeled particles on the cytoplasm and PD, suggesting that myosin might be present in PD. To date, myosin is confirmed to belong to 15 families, of which only some are found in plants, where the myosin found in

PD belongs to the eighth family. The effects of myosin inhibitor treatments somehow led to inconsistent results as to the role of myosin in PD [39]. How myosin regulates PD function is still under debate.

3.3. Tubulin

Blackman and Overall systemically characterized whether the cytoskeletal-related proteins were targeted to PD and found tubulin in the extracts of the PD-containing internode cell walls of corals, whereas it was not detected in the PD-independent internode cell walls, suggesting that PD may contain tubulin [51]. The role of tubulin in PD is less well characterized as compared to actin and myosin [62]. Blackman and Overall [51] speculated that tubulin may be indirectly involved in long-distance transport.

4. PD-Associated Regulatory Proteins

Callose is deposited at the cell wall near the neck of the PD and regulates the PD size exclusion limit (SEL) [22]. The PD function in plants is closely linked to callose deposition in the PD neck region. PD-associated regulatory proteins are more or less callose-related (Table 1). As mentioned above, the level of callose in PD is controlled by two antagonistic callose metabolic enzymes, callose synthase (CALS), also well known as glucan synthase-like (GSL), and β-1,3-glucanase (BG), and requires other enzymes that regulate callose stability [24,35]. In addition, the PD permeability changes involved in plant development and defense, processes involving callose, also require other proteins associated with callose [63,64].

4.1. Callose Synthases Modify Cell Wall PD to Regulate Plant Development

In the callose-dependent PD permeability, the PD channel is turned off by a high level of callose and turned on by a low level of callose, indicating that the callose level of the PD neck region is particularly crucial to PD function. The callose deposition determines PD SEL, which in turn determines their permeability and the transportation of macromolecules. Biochemical and genetic studies in barley and tobacco pollen tubes first demonstrated that GSLs produced callose [65]. This is also evidenced by studies in *Arabidopsis* [36,66]. GSLs control developmental signals by modulating the amount of callose in PD.

Plant hormones play a role in many aspects of plant physiology and development. Auxin has a gradient distribution in plants and plays an important role in plant growth and development [67]. As a small molecule, auxin can diffuse freely through the PD [68]. A recent study revealed that the auxin gradients can be maintained by controlling the PD permeability. Han et al. [53] screened all 12 *GSL* gene mutants in *Arabidopsis*, and found that the PD callose content decreased significantly only in the *gsl8/cals10* mutant. They used hypocotyls that responded to the auxin gradient distribution as experimental systems and found that the inducible *gsl8* RNA interference (RNAi) system showed lower phototropism and agravitropism in hypocotyl, which mimics auxin defects. Further studies found that a decrease in the messenger RNA (mRNA) level of *GSL8* resulted in the decrease of callose at PD, the increase of permeability of PD, and the increase of the diffusion of auxin transported by the symplast; finally, the gradient distribution of auxin was destroyed. Moreover, they proposed that the auxin gradient in plant tissues was established by the auxin–GSL8 feedback circuit associated with auxin response factor 7 (ARF7).

PD-mediated signaling plays a role in plant morphogenesis. Callose biosynthesis in the regulation of symplasmic transport by PD plays a central role in plant cell-to-cell signaling, but it is not clear how this regulation is arranged [69]. Vatén et al. [36] screened for the *gsl12/cals3* mutant in an ethyl methanesulfonate (EMS) mutagenesis library of *Arabidopsis*. Observation of the GFP::GSL12 fusion protein expressed in the leaves indicated that it was localized in PD. The gain-of-function mutation of *GSL12* leads to a transient accumulation of callose in PD in the early stage of root development, a decrease in PD SEL, a decrease in intercellular transport, and a defect in root development. This suggests that the regulatory activity of GSL12 influences the biosynthesis of callose in the cell wall surrounding the PD, which determines the pore size of the PD and regulates the transport of substances. The experiment with the *GSL12* allele (*cals3m*) showed that the PD pore size between plant tissues is controlled spatially and temporally. Using this system, they assessed the movement of the short-root (SHR) protein, microRNA165/6, and cytokinin in *Arabidopsis* roots and revealed the role of symplasmic communication in lateral root development. The above results indicate that GSL12 is a callose synthase, and that the biosynthesis of callose affects the plant developmental signals transmitted by PD. Interestingly, Yadav et al. [66] also identified GSL12 in *Arabidopsis*, and its semi-dominant mutants overproduce callose in PD.

The new cells required for plant growth and development are derived from the differentiation of stem cells. There is a group of stem cells called stem-cell niche (SCN) at the root apical meristem, and at its center are slowly dividing cells called the quiescent center (QC). It was believed that the location information provided by intercellular communication enabled QC and SCN to be maintained, but it was not verified for a long time. Liu et al. [70] developed a system that can effectively block PD-mediated communication in QC. They controllably and specifically blocked the intercellular communication between QC and neighboring cells by expressing *icals3m* (inducible form of *cals3m*) in QC. Using this system, they explored the function of intercellular communication in the SCN. They found that changes in symplastic communication between QC and adjacent cells resulted in elevated local auxin concentrations and the establishment of AP2-domain transcription factor gradients. Changes in symplasmic communication not only affect the polar transport of auxin, but also affect the biosynthesis of local auxin, which together provide local information for stem cell maintenance and play a key role in the maintenance of SCN. These studies revealed a direct link between symplasmic translocation and auxin-dependent regulation of stem-cell identity.

Slewinski et al. [71] identified a callose synthase in maize. They obtained *tie-dyed* (*TDY*) mutants of maize exhibiting green and yellow variegated leaves by EMS mutagenesis, and the yellow leaf tissue accumulated too much soluble sugar and starch. After constructing the population, they cloned the *TDY2* gene using map-based cloning, and found that *TDY2* encodes a callose synthase. Characterization of expression patterns by RT-PCR analysis, RNA in situ hybridization, and colocalization with the vascular marker, PINFORMED1a (ZmPIN1a)-YFP, revealed that *TDY2* was preferentially expressed in vascular tissues of developing young leaves. The movement experiments of ^{14}C-labeled sucrose and carboxyfluorescein diacetate (CFDA) showed that the *tdy2* leaves defected in the entry of sucrose into phloem. Solutes could not enter phloem in yellow leaf region of *tdy2* mutants, but its long-distance transport capacity was not impaired. The ultrastructure study of the *tdy2* mutant vein by transmission electron microscopy showed that the phloem and xylem cells in the yellow leaf region of *tdy2* were changed, and that the vein differentiation was incomplete, which hindered the symplasmic transport from phloem companion cells to sieve elements mediated by PD, resulting in functional defects in phloem. As to how the callose synthase TDY2 targets to PD in the cell walls, Baker et al. [72] hypothesized that TDY2 was synthesized in the ER and traveled through secretory vesicles from the Golgi apparatus to localize to PD, or that TDY2 proteins translocated through the secretory system to the plasma membrane and PD.

Song et al. [73] identified a plasmodesmal callose synthase gene *cannot reach the roof 1* (*CRR1)* that affects ovary development in rice plants. The rice *CRR1* gene is closely related to *Arabidopsis GSL8* and *GSL10*. By aniline blue staining and immunocolloidal gold analysis, the *crr1* mutation had no effect on the callose deposition at root cell plates, pericarp, and pedicel sieve plates. However, iodine staining of ovarian development showed that the starch granules were significantly less numerous than wild type (WT) in the pericarp of the *crr1* ovary, and that the distribution was uneven, indicating that the unloading from the phloem to the pericarp could be defective. RT-PCR analysis and *pCRR1*::GUS reporter transgenic plants showed that *CRR1* was specifically expressed in vascular tissues. The movement analysis of CFDA revealed that sugars were inefficiently unloaded from the lateral vascular system to the developing caryopsis. Transmission electron microscopy showed that the connection between sieve tube cells and peripheral parenchyma cells of vascular bundles was reduced, and some cells had defective cell walls, where the deposition of callose at PD was reduced. These results suggest that CRR1 may alter the transport of cell fate determinants by controlling the permeability of PD, resulting in the change of the vascular cell differentiation pattern, thereby playing a key role in determining the plant reproductive development.

4.2. Callose Hydrolases Degrade Callose to Enhance Intercellular Communication

Class I β-1,3-glucanase (GLU I) plays an important role in the defense against virus infection [74]. Iglesias et al. [75] studied the role of GLU I in PD and PD-dependent communication. They obtained a homozygous class I β-1,3-glucanase (GLU I)-deficient mutant tobacco (TAG 4.4) by antisense transformation. This mutant become insensitive to viral infections. Observation was carried out of the size of the necrotic area and the GFP expressed by a recombinant form of potato virus X (PVX). GFP showed that the spreading of the virus in this mutant was slowed down. Observation of the fluorescence of CMV's 3a movement protein in fusion with GFP (3aMP::GFP) revealed that the movement of viral movement proteins between cells was limited in TAG 4.4. Observations of fluorescein isothiocyanate (FITC)-dextrans and FITC-peptides indicated that plasmodesmal SEL was reduced in this mutant. Aniline blue staining showed that, after the loss of GLU I, the deposition of callose increased. The above results indicate that β-1,3-glucanase plays an important role in the metabolism of callose. It may promote the degradation of callose, change the permeability of PD, and regulate the movement of macromolecules through PD. Bucher et al. [76] further explored the relationship between GLU I and plant sensitivity to viruses. They studied the role of GLU I in viral migration between cells from both gain- and loss-of-function mutants. They inserted the different target fragments into the TMV-based expression vector to obtain products of different enzymatic activities. The positive insertion fragments expressed the active GLU I protein, and the necrotic area of leaves increased continuously after infection with the virus. The plants infected with viruses expressing antisense GLU I showed reciprocal effects. These experiments indicated that GLU I promoted the movement of virus between cells by hydrolyzing callose to increase PD SEL.

Levy et al. [28] identified a putative plasmodesmal associated protein β-1,3-glucanase (AtBG_ppap) in *Arabidopsis* and demonstrated that the permeability of PD is regulated by β-1,3-glucanase. The proteins in the PD-enriched fraction of *Arabidopsis* were separated by sodium dodecyl sulfate polyacrylamide gel electrophoresis (SDS-PAGE) and subjected to mass spectrometry analyses. The β-1,3-glucanase identified was named AtBG_ppap (beta-1,3-glucanase_putative PD-associated protein). Observation of transient and stable expression of GFP fusion proteins and aniline blue staining revealed that AtBG_ppap was related to the callose around PD in the cell walls. In the *Atbg_ppap* mutant, the callose level around the PD was higher, and the mobility of the GFP protein between cells was decreased, that is, the permeability of PD was lowered. The above experimental results demonstrated that AtBG_ppap is a PD-associated protein involved in plasmodesmal callose removal.

Intercellular communication regulates the function of individual cells to establish organ patterning and formation. It is long known that mobile signals are important in lateral root (LR) development. Paradoxically, the role of the intercellular channel PD in this process seems to have inadvertently not been elucidated. Recently, Benitez-Alfonso et al. [54] identified two plasmodesmal localized β-1,3-glucanase (PDBG) proteins that particularly accumulated in the LR founder tissue. Callose deposition changed correlating with changes in symplasmic transport during the course of LR development. During the formation of lateral meristems, callose amount was detected at low levels in PD. At early stages of LR development, callose accumulated to a moderate level between cells. Callose formation increased progressively when LRs developed. Such callose accumulation during LR organogenesis was regulated by two PDBGs, PDBG1 and PDBG2, which were localized in PD at early stages of lateral root primordia. In *pdbg1,2* double mutants, excessive callose deposition was detected around LR founder cells. In contrast, callose levels were significantly reduced in *PDBG1*-overexpressing plants. Consistent with callose accumulation levels, *pdbg1,2* double mutants showed a significant increase in LR formation. These results demonstrated that PD-dependent cell-wall glucanases play an important role in plant development via modulating intercellular communication.

Furthermore, Rinne et al. [77] analyzed 10 presumed cell-wall β-1,3-glucanase genes belonging to the GH17 family (glucan hydrolase family 17) in poplar, and found that different treatments, such as photoperiod, chilling, and gibberellin, regulated these genes differently. Based on this, a model of breaking bud dormancy related to PD regulated by β-1,3-glucanase was proposed. Previous studies found that flowering locus T (FT) is the primary long-distance signal that is generated at the vascular bundle by activation of Constans (CO) and transported to the apex [78]. Centroradialis-like 1 (CENL1) in poplar is an ortholog of *Arabidopsis* TFL1 (terminal flower 1)—a hypothetical signal peptide, expressed in a small region under the shoot apical meristem (SAM) [79,80]. Both FT and CENL1 move through the phloem. They enter phloem tissue via PD which connect the cytoplasm of companion cells with the sieve tube cells. When the bud is dormant, SAM closes PD with callose, and cells in SAM become disconnected from each other [80]. Chilling induces FT and affects gibberellic acid 3 (GA3) biosynthesis by upregulating the expression level of *GA3-oxidase*; then, FT and β-1,3-glucanase proteins induced by GA3 reopen the PD in dormant buds. When the climate warms up, the reopened PD enables FT and CENL1 to move to their targets to cause buds to germinate. Following the study of the mechanism of seasonal dormant axillary buds (AXBs) releasing dormancy, Rinne et al. [81] continued to study the mechanism of semi-dormant axillary buds releasing dormancy during tree branch formation. AXB dormancy breaking also involved the GA pathway, and the expression of β-1,3-glucanase gene was induced by GA. They founded that the GA synthesis gene *gibberellin 3-oxidase 2* (*GA3ox2*) expression level in AXBs was very low; thus, GA was absent in AXBs. The expression of GA receptor gene *gibberellin insensitive dwarf 1-like* (*GID1-like*) gradually increased during the maturation of semi-dormant AXBs; thus, AXBs were sensitive to GA. They found that the expression of *GA3ox2* was significantly upregulated after decapitation of the hybrid poplar; thus, the biosynthesis of GA in AXBs increased. The GH17s have CBM43 modules that position them to callose around the PD. Upregulated GH17s responded to increased GA, degraded callose in the cell wall near PD, reducing callose deposition at PD, and promoted communication between stems and buds, thereby abolishing dormancy of semi-dormant AXBs. They studied the expression patterns of GA pathway and the *GH17s* gene at different stages of bud development, summarized their expression characteristics, and demonstrated the role of PD-related β-1,3-glucanases in bud development. Conversely, photoperiod regulates seasonal growth; prior to the advent of winter, short days (SDs) during late autumn promote bud dormancy to induce growth cessation. It was found that, in hybrid aspen trees, short photoperiods act through activation of the abscisic acid (ABA) pathway [82]. Later, gene expression analysis revealed that SVL (SVP-like), an ortholog of the *Arabidopsis* floral repressor short vegetative phase (SVP), was induced by SDs in an ABA-dependent manner. SVL negatively regulated the growth promotive GA pathways and positively mediated the expression of *callose synthase 1* (*CALS1*) in the shoot apex to induce callosic plugs in the neck region of PD [83,84]. Thus, PD-associated callose synthases and glucanases in the cell walls play important roles in plant development by regulating bud dormancy and floral transition.

Previous work showed that PD is momentarily closed during the elongation of cotton fibers in the allotetraploid species *Gossypium hirsutum* [85]. Ruan et al. [86] applied fluorescence probe carboxyfluorescein (CF), aniline blue staining, and immunoelectron microscopy to cotton genotypes with different cotton fiber lengths. It was found that the closure time of cotton fiber PD varied in different cotton genotypes, which was related to the deposition and degradation of callose at the base of cotton fiber. They cloned the *GhGluc1* gene from cotton fibers and found that it specifically encoded a β-1,3-glucanase localized to the fiber cells. Moreover, they also found that callose was degraded when GhGluc1 was expressed, and GhGluc1 expression was higher in plants with shorter cotton fibers. These results indicate that GhGluc1 is a callose degrading enzyme that allows the PD to be in an open state. Callose formation at the neck region closes PD. PD closure is positively correlated with fiber length; therefore, the callose catabolic enzyme GhGluc1 plays a negative role in fiber elongation.

4.3. PD-Associated Callose Binding Proteins (PDCBs) Regulate Callose Stability

Simpson et al. [29] identified a novel PD-related protein gene, *AT5G61130*, which encodes PD callose binding protein 1 (PDCB1) with callose binding activity in *Arabidopsis*. PDCB1 fused to the YFP coding sequence driven by the CaMV *35S* or its native promoter was stably transferred into *Arabidopsis* and transiently expressed in tobacco protoplasts. The fluorescent spots were found in the epidermal cell wall of *Arabidopsis* leaf tissue. Moreover, in spongy mesophyll cells, the fluorescent points were limited to the wall-to-wall junction between adjacent cells and still retained this pattern after plasmolysis. The results indicate that PDCB1 is associated with cell walls. Immunolocalization with anti-PDCB1 antiserum indicated that PDCB1 is located at the outer neck region of the PD. Combined with aniline blue staining and confocal microscopy, they proved that PDCB1 was colocalized with callose located in PD. They also studied PDCB2 (*AT5G08000*) and PDCB3 (*AT1G18650*) belonging to this family. A homology search of the *Arabidopsis* genome revealed that PDCB2 and PDCB3 have more than 50% amino-acid sequence similarity to PDCB1. Phylogenetic analysis revealed in fact that PDCB1, -2, and -3 lack a GPI-anchor signal sequence, but contain an X8 domain adjacent to the β-1,3-glucanase catalytic domain. This is in contrast to the previously identified PD-related β-1,3-glucanase [28]. By comparing with the structure of OLE-E9, the X8 domain-containing protein with callose binding activity in olives, it was found that the PDCB1 X8 domain is very similar to the OLE-E9 domain. Gel retardation assays showed that OLE-E10, PDCB1, and PDCB2 have callose binding activity in vitro. GFP diffusion assay showed that the migration of GFP decreased with the increase of PDCB1 expression, suggesting that the deposition of callose mediated by PDCB affected the symplastic communication through PD between cells.

PDCBs contain a carbohydrate binding module family 43 (CBM43) domain (also known as an X8 domain), as well as a callose binding domain, which can stabilize callose and regulate the dynamics of callose of PD [29]. PDCB1-overexpressing plants exhibited increased callose deposition in PD and decreased intercellular movement of reporter proteins. PDCBs are also predicted to be glycosylphosphatidylinositol anchored proteins (GPI-APS). Zavaliev et al. [87] studied the effects of GPI modification on PD proteins. They selected two GPI-APs associated with the callose of PD, BG_ppap and PDCB1 in *Arabidopsis*. They also chose two GPI-APs that were not associated with PD as control, arabinogalactan protein 4 (AGP4) and lipid transfer protein 1 (LTPG1). They first verified the subcellular localization of these proteins with monomeric citrine (mCitrine) labeling and then verified the modification of GPI with mannosamine. Subsequently, they subcellular localized the full-length sequence and deleted GPI signal sequence, respectively. It was found that PD targeting of BG_ppap and PDCB1 was mediated by GPI, and that the GPI modification played a major role in protein targeting to PD; furthermore, this sorting happened before positioning.

4.4. PD-Located Receptor-Like Proteins (PDLPs) Regulate Plant Defense Responses

PDLPs are PD-localized transmembrane proteins. PDLPs consist of a short cytosolic tail and two extracellular cysteine-rich receptor-like kinase domains of unknown function (DUF26) [88,89]. In *Arabidopsis*, eight PDLPs were identified by homologous comparison [56]; PDLP1, PDLP5, and PDLP6, were later found in independent proteomic screens of PD-enriched membrane [55,57]. PDLP1 (*AT5G43980*) is the first member of this protein family identified by cell-wall proteomics in *Arabidopsis* [55]. PDLP1 belongs to the DUF26 domain family. Thomas et al. [58] developed *Arabidopsis* transgenic plants overexpressing *PDLP1* and performed GFP diffusion experiments, confirming that PDLP1 can regulate intercellular transport by regulating PD. Moreover, they found through deletion analysis that their single transmembrane domain (TMD) is required for PDLP1a targeting to PD.

When fungi and oomycetes enter host plants with a haustoria structure, they cause plant defense responses, and the haustoria structure is gradually entrapped by the callose produced by the host. Caillaud et al. [59] studied the molecular mechanism of accumulation of callose around haustoria. They infected *pPDLP1*::GUS and *pPDLP5*::GUS reporter transgenic plants with pathogens, and found that

PDLP1 was specifically expressed in cells containing haustoria and responded to pathogen infection. Observation of PDLP1::GFP showed that PDLP1 localized at PD in both uninfected and infected tissue, and PDLP1 was associated with haustoria at the early stages of pathogen invasion in infected tissue. Moreover, the signaling peptide that targets PDLP to haustoria is located at its C-terminus, allowing it to be localized to the haustorial membrane. Significantly more callose was detected in transgenic plants overexpressing *PDLP1* than in non-transgenic plants following pathogen infection. The experimental results of Caillaud et al. [59] directly demonstrate that PDLP1 contributes to the immune defense against pathogens by affecting the deposition of callose in the plant cell walls.

Pathogen invasion first leads to local infection, and then local infection produces signals that are transported to the uninfected parts of the plant, making the uninfected part resistant to subsequent enlargement of the infection. This is called systemic acquired resistance (SAR) [90–92]. Signals generated by local infection include salicylic acid (SA), azelaic acid (AzA), glycerol-3-phosphate (G3P), and reactive nitrogen species (RNS) [93,94]. Lim et al. [95] studied the transport mechanism of SAR signals and found that PDLPs were involved. They treated plants with mock solution or avirulent strains, then detected SA, AzA, and G3P in apoplastic fluids and petiole exudates representing apoplastic and symplasmic communication signals, respectively. It was found that SA was transported through the ectoplasmic pathway, while G3P and AZA were transported through the symplastic pathway mediated through cell-wall PD conduits. Evaluation of SAR in *pdlp1/pdlp5* double mutant or *PDLP5*-overexpressing plants showed that PDLP1 and PDLP5 were involved in SAR. Moreover, PDLP1 can not only affect the transmission of SAR signals by regulating the permeability of PD, but also affect the stability of SAR signal transporter azelaic acid induced 1 (AZI1) through direct protein–protein interactions [96]. Ye et al. [97] also reported that a PDLP1 homolog, PDLP8, was able to interact with the acyl-CoA-binding protein 6 (ACBP6) to stabilize ACBP6 accumulation in PD.

Various hormone signaling pathways are associated with basal immune responses. The pathways activated by SA, a defense hormone, contribute to the responses [90]. When a pathogen infects plants, SA concentration is elevated. The expression of the PD regulator *PDLP5* was upregulated resulting in the restriction of cell-to-cell movement via PD [57]. Accumulation of PDLP5 protein in PD restricts PD permeability by stimulating callose deposition at the PD neck. Most notably, Cui and Lee [98] revealed that GSL6 (CALS1) is a primary partner for PDLP5. They produced a transcriptional reporter line *pGSL6*::GUS. GUS staining experiments confirmed that the *GSL6* expression profile is most highly correlated with that of *PDLP5* in SA-treatment experiments. In *gsl6-1* mutant plants, callose deposition did not change after treatment with SA. This indicates that SA-mediated callose production requires GSL6. The results of their experiments with carboxyfluorescein, a fluorescent dye that emits fluorescence and will not retrograde once it enters the cell, also support this conclusion. Neither changes in plasmodesmal callose deposition nor carboxyfluorescein movement were detected in *gsl6-1* mutant after SA treatment. Collectively, these sets of experimental evidence establish that GSL6 is a bona fide callose synthase gene regulating SA-dependent plasmodesmal responses. Therefore, PDLP5 requires GSL6 for SA-dependent plasmodesmal regulation. Evidence of subcellular localization of GSL6 to cell-wall PD is missing. Since the *gsl10* mutant plants are gametophytic lethal [99], the *gsl10* loss-of-function plants are unavailable. The roles of *GSL10* in regulating PD permeability and plant defense responses are still not investigated. Wang et al. [100] cloned and overexpressed the extracellular domain of *Arabidopsis* PDLP5 in an insect expression system, followed by purification, crystallization, and diffraction analysis. The extracellular domain of PDLP5 includes two DUF26 domains. They collected X-ray diffraction data to 1.90 Å resolution, which indicated that PDLP5 may function as a dimer. This research provides novel understanding of the molecular mechanisms behind PD permeability regulation and plant system immunity [100]. It is still far from clear, however, how PDLPs regulate callose synthase gene expression and callose deposition in PD [101].

4.5. Other PD-Related Cell Wall Proteins and Even More to Be Identified

The proteins and lipids of eukaryotes are sequentially organized in the plane of the membrane, forming microdomains that distinguish and coordinate biological processes, ensuring specific kinetics and functions to specific regions within the cell [102–104]. Grison et al. [105] used *Arabidopsis* cultured cells and primary roots to strictly isolate plasma membrane (PM) domains of PD. Immunogold labeling and Western blotting showed that many false-positive membrane proteins, such as PMA2, CesA3, CesA6, and PIP2, did not target to PD, but previously known PDLP1 and PDCB1 were specific to PD PM [105].

A hypothetical sterol carrier protein gene, *GHSCP2D*, was isolated from elongated cotton fibers by Zhang et al. [106]. They inhibited the expression of *GHSCP2D* and observed the deposition of callose on PD, confirming that *GHSCP2D* is involved in the permeability of PD. They demonstrated that the decrease of PD permeability was due to the inhibition of the expression of the PD-targeted beta-1,3-glucanase GhPDBG3 in cotton fibers. Both downregulation of *GHSCP2D* and the application of a sterol biosynthesis inhibitor can reduce the permeability of PD, indicating that sterol also has an effect on PD.

Viruses move through PD between plant cells. Park et al. [107] studied the role of PD in viral infection and found that NbEXPA1 is a PD-specific expansin, specifically involved in cell wall relaxation. They performed a label-free quantitative proteomic analysis of PD-enriched fractions of leaves infected with TuMV and healthy controls in *Nicotiana benthamiana*. They collected in total 1070 PD protein candidates. After data processing, 100 upregulated proteins and 48 downregulated proteins were identified in the TuMV-infected samples. The observation of YFP or GFP fused with NbEXPA1 showed that the fusion protein had a dotted pattern along the cell-wall boundary, suggesting that NbEXPA1 might be located in PD. Co-localization between NbEXPA1::YFP and PDLP1::CFP and between NbEXPA1::GFP and CI::RFP indicated that NbEXPA1 was indeed a PD-localized protein. The results of proteomics and RT-PCR analysis showed that TuMV infection downregulates the gene expression and protein accumulation of NbExpA1. Colocalization of NbEXPA1::YFP and TuMV-6K2::mCherry suggests that NbEXPA1 may be involved in the movement of the virus between cells. Infection of leaves with recombinant virus expressing CFP ligated with a viral RNA polymerase NIB (CFP::NIB), combined with a yeast two-hybrid assay and BiFC analysis, suggested that NbEXPA1 acts in the viral replication complex via interaction with NIB. Silencing and overexpression of *NbEXPA1* inhibits and promotes the movement of the virus between cells, respectively. The above results indicate that NbEXPA1 is a cell-wall protein associated with PD, which affects viral gene replication and its movement through the PD between cells through interaction with NIB.

5. Further Perspectives

As the only channels for cytoplasmic communication between neighboring cells, PD are vital for the developmental process of plants. By combination of genetic, biochemical, and cell biology studies, many PD-associated proteins were identified and characterized (Figure 1). However, we still lack a fuller understanding of how the PD functions, which is largely due to the methodological difficulties in identification and characterization of PD components [108].

Increasing lines of evidence support that callose levels at the PD neck region play a central role in modulating the PD permeability and the symplasmic pathway. So far, the PD-related proteins, such as GSLs, PDGBs, PDCBs, and PDLPs, are all involved in maintaining callose homeostasis at the PD zone [28,29,36,53,56]. Further experiments are required to elucidate how transcription factors regulate the expression pattern of those genes in response to specific developmental cues and environmental stimuli. Since callose deposition and degradation in PD respond readily and rapidly to a wide range of signals, we speculate that PD-associated callose balancing proteins are mainly regulated at a post-transcriptional level. Another common feature is that those PD proteins contain transmembrane domains. Subcellular localization analysis showed that PD proteins are specifically targeted to PD embedded in the cell wall. It is also of interest to address the routes via which those proteins reach PD and more specifically the membrane of PD [109].

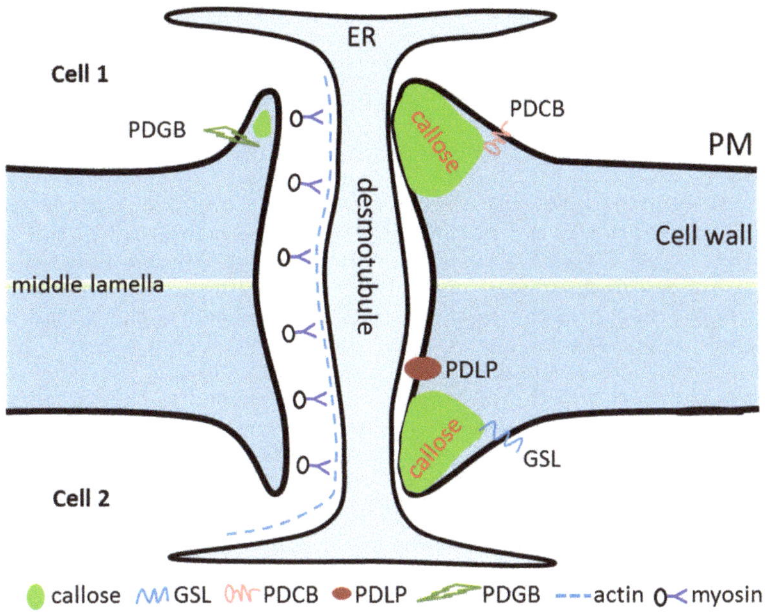

Figure 1. A simplified schematic representation of plasmodesmata (PD) structural and functional proteins. Permeability of PD in the plant cell wall is controlled by dynamic callose deposition (closed, right side) and degradation (open, left side) at the neck region of PD. ER, endoplasmic reticulum; GSL, glucan synthase-like; PM, plasma membrane; PDCB, PD-associated callose binding protein; PDGB, plasmodesmal associated β-1,3-glucanase; PDLP, PD-localized protein.

Previously, high-resolution electron microscopy and image processing methods greatly advanced our understanding of PD structure [110]. Immunochemical localization studies facilitated the detection of PD components. Indeed, our knowledge on PD is much dependent on technological progress. Despite their structural complexity, PD are important for plants and require further studies [21].

Acknowledgments: This work was supported by the Startup Foundation of Central South University of Forestry and Technology to N.L. (Numbers: 2018YJ049). We appreciate the Managing Editors and two anonymous reviewers for their corrections, suggestions, and in-depth comments, which greatly improved the manuscript.

Conflicts of Interest: The authors declare no conflicts of interest.

References

1. Lucas, W.J.; Ham, B.-K.; Kim, J.-Y. Plasmodesmata—Bridging the gap between neighboring plant cells. *Trends Cell Biol.* **2009**, *19*, 495–503. [CrossRef] [PubMed]
2. Epel, B.L. Plasmodesmata: Composition, structure and trafficking. *Plant Mol. Biol.* **1994**, *26*, 1343–1356. [CrossRef] [PubMed]
3. Lucas, W.J. Plasmodesmata: Intercellular channels for macromolecular transport in plants. *Curr. Opin. Cell Biol.* **1995**, *7*, 673–680. [CrossRef]
4. McLean, B.G.; Hempel, F.D.; Zambryski, P.C. Plant intercellular communication via plasmodesmata. *Plant Cell* **1997**, *9*, 1043–1054. [CrossRef] [PubMed]
5. Kim, J.-Y. Regulation of short-distance transport of RNA and protein. *Curr. Opin. Plant Biol.* **2005**, *8*, 45–52. [CrossRef] [PubMed]

6. Ehlers, K.; Kollmann, R. Primary and secondary plasmodesmata: Structure, origin, and functioning. *Protoplasma* **2001**, *216*, 1–30. [CrossRef]
7. Maule, A.J. Plasmodesmata: Structure, function and biogenesis. *Curr. Opin. Plant Biol.* **2008**, *11*, 680–686. [CrossRef]
8. Reagan, B.C.; Ganusova, E.E.; Fernandez, J.C.; McCray, T.N.; Burch-Smith, T.M. RNA on the move: The plasmodesmata perspective. *Plant Sci.* **2018**, *275*, 1–10. [CrossRef]
9. Han, X.; Kumar, D.; Chen, H.; Wu, S.; Kim, J.-Y. Transcription factor-mediated cell-to-cell signalling in plants. *J. Exp. Bot.* **2014**, *65*, 1737–1749. [CrossRef]
10. Lee, J.-Y. New and old roles of plasmodesmata in immunity and parallels to tunneling nanotubes. *Plant Sci.* **2014**, *221*, 13–20. [CrossRef]
11. Zambryski, P.; Crawford, K. Plasmodesmata: Gatekeepers for cell-to-cell transport of developmental signals in plants. *Annu. Rev. Cell Dev. Biol.* **2000**, *16*, 393–421. [CrossRef] [PubMed]
12. Haywood, V.; Kragler, F.; Lucas, W.J. Plasmodesmata: Pathways for protein and ribonucleoprotein signaling. *Plant Cell* **2002**, *14* (Suppl.), S303–S325. [CrossRef]
13. Otero, S.; Helariutta, Y.; Benitez-Alfonso, Y. Symplastic communication in organ formation and tissue patterning. *Curr. Opin. Plant Biol.* **2016**, *29*, 21–28. [CrossRef] [PubMed]
14. Lee, J.-Y.; Frank, M. Plasmodesmata in phloem: Different gateways for different cargoes. *Curr. Opin. Plant Biol.* **2018**, *43*, 119–124. [CrossRef] [PubMed]
15. Kim, J.-Y. Symplasmic intercellular communication through plasmodesmata. *Plants* **2018**, *7*, 23. [CrossRef]
16. Sager, R.E.; Lee, J.-Y. Plasmodesmata at a glance. *J. Cell. Sci.* **2018**, *131*, jcs209346. [CrossRef]
17. Bell, K.; Oparka, K. Imaging plasmodesmata. *Protoplasma* **2011**, *248*, 9–25. [CrossRef]
18. Lee, J.-Y.; Taoka, K.; Yoo, B.-C.; Ben-Nissan, G.; Kim, D.-J.; Lucas, W.J. Plasmodesmal-associated protein kinase in tobacco and Arabidopsis recognizes a subset of non-cell-autonomous proteins. *Plant Cell* **2005**, *17*, 2817–2831. [CrossRef]
19. Baluska, F.; Samaj, J.; Napier, R.; Volkmann, D. Maize calreticulin localizes preferentially to plasmodesmata in root apex. *Plant J.* **1999**, *19*, 481–488. [CrossRef]
20. Faulkner, C.; Maule, A. Opportunities and successes in the search for plasmodesmal proteins. *Protoplasma* **2011**, *248*, 27–38. [CrossRef]
21. Sun, Y.; Huang, D.; Chen, X. Dynamic regulation of plasmodesmatal permeability and its application to horticultural research. *Hortic. Res.* **2019**, *6*, 47. [CrossRef]
22. Wu, S.-W.; Kumar, R.; Iswanto, A.B.B.; Kim, J.-Y. Callose balancing at plasmodesmata. *J. Exp. Bot.* **2018**, *69*, 5325–5339. [CrossRef] [PubMed]
23. Amsbury, S.; Kirk, P.; Benitez-Alfonso, Y. Emerging models on the regulation of intercellular transport by plasmodesmata-associated callose. *J. Exp. Bot.* **2017**, *69*, 105–115. [CrossRef] [PubMed]
24. Chen, X.-Y.; Kim, J.-Y. Callose synthesis in higher plants. *Plant Signal. Behav.* **2009**, *4*, 489–492. [CrossRef] [PubMed]
25. De Storme, N.; Geelen, D. Callose homeostasis at plasmodesmata: Molecular regulators and developmental relevance. *Front. Plant Sci.* **2014**, *5*, 138. [CrossRef] [PubMed]
26. Heinlein, M.; Epel, B.L. Macromolecular transport and signaling through plasmodesmata. *Int. Rev. Cytol.* **2004**, *235*, 93–164. [PubMed]
27. Guenoune-Gelbart, D.; Elbaum, M.; Sagi, G.; Levy, A.; Epel, B.L. Tobacco mosaic virus (TMV) replicase and movement protein function synergistically in facilitating TMV spread by lateral diffusion in the plasmodesmal desmotubule of *Nicotiana benthamiana*. *Mol. Plant Microbe Interact.* **2008**, *21*, 335–345. [CrossRef]
28. Levy, A.; Erlanger, M.; Rosenthal, M.; Epel, B.L. A plasmodesmata-associated beta-1,3-glucanase in Arabidopsis. *Plant J.* **2007**, *49*, 669–682. [CrossRef]
29. Simpson, C.; Thomas, C.; Findlay, K.; Bayer, E.; Maule, A.J. An Arabidopsis GPI-anchor plasmodesmal neck protein with callose binding activity and potential to regulate cell-to-cell trafficking. *Plant Cell* **2009**, *21*, 581–594. [CrossRef]
30. Levy, A.; Guenoune-Gelbart, D.; Epel, B.L. Beta-1,3-glucanases: Plasmodesmal gate keepers for intercellular communication. *Plant Signal. Behav.* **2007**, *2*, 404–407. [CrossRef]
31. Knox, K.; Wang, P.; Kriechbaumer, V.; Tilsner, J.; Frigerio, L.; Sparkes, I.; Hawes, C.; Oparka, K. Putting the squeeze on plasmodesmata: A role for reticulons in primary plasmodesmata formation. *Plant Physiol.* **2015**, *168*, 1563–1572. [CrossRef] [PubMed]

32. Fitzgibbon, J.; Bell, K.; King, E.; Oparka, K. Super-resolution imaging of plasmodesmata using three-dimensional structured illumination microscopy. *Plant Physiol.* **2010**, *153*, 1453–1463. [CrossRef] [PubMed]
33. Jones, M.G.; Dropkin, V.H. Scanning electron microscopy in nematode-induced giant transfer cells. *Cytobios* **1976**, *15*, 149–161. [PubMed]
34. Ding, B.; Turgeon, R.; Parthasarathy, M.V. Substructure of freeze-substituted plasmodesmata. *Protoplasma* **1992**, *169*, 28–41. [CrossRef]
35. Zavaliev, R.; Ueki, S.; Epel, B.L.; Citovsky, V. Biology of callose (β-1,3-glucan) turnover at plasmodesmata. *Protoplasma* **2011**, *248*, 117–130. [CrossRef] [PubMed]
36. Vatén, A.; Dettmer, J.; Wu, S.; Stierhof, Y.-D.; Miyashima, S.; Yadav, S.R.; Roberts, C.J.; Campilho, A.; Bulone, V.; Lichtenberger, R.; et al. Callose biosynthesis regulates symplastic trafficking during root development. *Dev. Cell* **2011**, *21*, 1144–1155. [CrossRef] [PubMed]
37. Brecknock, S.; Dibbayawan, T.P.; Vesk, M.; Vesk, P.A.; Faulkner, C.; Barton, D.A.; Overall, R.L. High resolution scanning electron microscopy of plasmodesmata. *Planta* **2011**, *234*, 749–758. [CrossRef]
38. Tilsner, J.; Amari, K.; Torrance, L. Plasmodesmata viewed as specialised membrane adhesion sites. *Protoplasma* **2011**, *248*, 39–60. [CrossRef]
39. White, R.G.; Barton, D.A. The cytoskeleton in plasmodesmata: A role in intercellular transport? *J. Exp. Bot.* **2011**, *62*, 5249–5266. [CrossRef]
40. Kumar, D.; Kumar, R.; Hyun, T.K.; Kim, J.-Y. Cell-to-cell movement of viruses via plasmodesmata. *J. Plant Res.* **2015**, *128*, 37–47. [CrossRef]
41. Botha, C.E.J.; Hartley, B.J.; Cross, R.H.M. The ultrastructure and computer-enhanced digital image analysis of plasmodesmata at the Kranz mesophyll-bundle sheath interface of *Themeda triandra* var. *imberbis* (Retz) A. Camus in conventionally-fixed blades. *Ann. Bot.* **1993**, *72*, 255–261.
42. Wright, K.M.; Oparka, K.J. The ER Within Plasmodesmata. In *The Plant Endoplasmic Reticulum*; Robinson, D.G., Ed.; Plant Cell Monographs; Springer: Berlin/Heidelberg, Germany, 2006; pp. 279–308.
43. Gamalei, Y.V.; van Bel, A.J.E.; Pakhomova, M.V.; Sjutkina, A.V. Effects of temperature on the conformation of the endoplasmic reticulum and on starch accumulation in leaves with the symplasmic minor-vein configuration. *Planta* **1994**, *194*, 443–453. [CrossRef]
44. Cook, M.E.; Graham, L.E. Evolution of Plasmodesmata. In *Plasmodesmata: Structure, Function, Role in Cell Communication*; Van Bel, A.J.E., Van Kesteren, W.J.P., Eds.; Springer: Berlin/Heidelberg, Germany, 1999; pp. 101–117.
45. Overall, R.L.; Blackman, L.M. A model of the macromolecular structure of plasmodesmata. *Trends Plant Sci.* **1996**, *1*, 307–311. [CrossRef]
46. Chaffey, N.; Barlow, P. Myosin, microtubules, and microfilaments: Co-operation between cytoskeletal components during cambial cell division and secondary vascular differentiation in trees. *Planta* **2002**, *214*, 526–536. [CrossRef] [PubMed]
47. Vale, R.D. The molecular motor toolbox for intracellular transport. *Cell* **2003**, *112*, 467–480. [CrossRef]
48. Ehrhardt, D.W.; Shaw, S.L. Microtubule dynamics and organization in the plant cortical array. *Ann. Rev. Plant Biol.* **2006**, *57*, 859–875. [CrossRef]
49. White, R.G.; Badelt, K.; Overall, R.L.; Vesk, M. Actin associated with plasmodesmata. *Protoplasma* **1994**, *180*, 169–184. [CrossRef]
50. Ding, B.; Kwon, M.-O.; Warnberg, L. Evidence that actin filaments are involved in controlling the permeability of plasmodesmata in tobacco mesophyll. *Plant J.* **1996**, *10*, 157–164. [CrossRef]
51. Blackman, L.M.; Overall, R.L. Immunolocalisation of the cytoskeleton to plasmodesmata of *Chara corallina*. *Plant J.* **1998**, *14*, 733–741. [CrossRef]
52. Radford, J.E.; White, R.G. Localization of a myosin-like protein to plasmodesmata. *Plant J.* **1998**, *14*, 743–750. [CrossRef]
53. Han, X.; Hyun, T.K.; Zhang, M.; Kumar, R.; Koh, E.; Kang, B.-H.; Lucas, W.J.; Kim, J.-Y. Auxin-callose-mediated plasmodesmal gating is essential for tropic auxin gradient formation and signaling. *Dev. Cell* **2014**, *28*, 132–146. [CrossRef] [PubMed]
54. Benitez-Alfonso, Y.; Faulkner, C.; Pendle, A.; Miyashima, S.; Helariutta, Y.; Maule, A. Symplastic intercellular connectivity regulates lateral root patterning. *Dev. Cell* **2013**, *26*, 136–147. [CrossRef] [PubMed]

55. Fernandez-Calvino, L.; Faulkner, C.; Walshaw, J.; Saalbach, G.; Bayer, E.; Benitez-Alfonso, Y.; Maule, A. Arabidopsis plasmodesmal proteome. *PLoS ONE* **2011**, *6*, e18880. [CrossRef]
56. Amari, K.; Boutant, E.; Hofmann, C.; Schmitt-Keichinger, C.; Fernandez-Calvino, L.; Didier, P.; Lerich, A.; Mutterer, J.; Thomas, C.L.; Heinlein, M.; et al. A family of plasmodesmal proteins with receptor-like properties for plant viral movement proteins. *PLoS Pathog.* **2010**, *6*, e1001119. [CrossRef] [PubMed]
57. Lee, J.-Y.; Wang, X.; Cui, W.; Sager, R.; Modla, S.; Czymmek, K.; Zybaliov, B.; van Wijk, K.; Zhang, C.; Lu, H.; et al. A plasmodesmata-localized protein mediates crosstalk between cell-to-cell communication and innate immunity in Arabidopsis. *Plant Cell* **2011**, *23*, 3353–3373. [CrossRef] [PubMed]
58. Thomas, C.L.; Bayer, E.M.; Ritzenthaler, C.; Fernandez-Calvino, L.; Maule, A.J. Specific targeting of a plasmodesmal protein affecting cell-to-cell communication. *PLoS Biol.* **2008**, *6*, e7. [CrossRef] [PubMed]
59. Caillaud, M.-C.; Wirthmueller, L.; Sklenar, J.; Findlay, K.; Piquerez, S.J.M.; Jones, A.M.E.; Robatzek, S.; Jones, J.D.G.; Faulkner, C. The plasmodesmal protein PDLP1 localises to haustoria-associated membranes during downy mildew infection and regulates callose deposition. *PLoS Pathog.* **2014**, *10*, e1004496. [CrossRef] [PubMed]
60. Chen, C.; Zhang, Y.; Zhu, L.; Yuan, M. The actin cytoskeleton is involved in the regulation of the plasmodesmal size exclusion limit. *Plant Signal. Behav.* **2010**, *5*, 1663–1665. [CrossRef]
61. Badelt, K.; White, R.G.; Overall, R.L.; Vesk, M. Ultrastructural specializations of the cell wall sleeve around plasmodesmata. *Am. J. Plant Bot.* **1994**, *81*, 1422–1427. [CrossRef]
62. Aaziz, R.; Dinant, S.; Epel, B.L. Plasmodesmata and plant cytoskeleton. *Trends Plant Sci.* **2001**, *6*, 326–330. [CrossRef]
63. Brunkard, J.O.; Zambryski, P.C. Plasmodesmata enable multicellularity: New insights into their evolution, biogenesis, and functions in development and immunity. *Curr. Opin. Plant Biol.* **2017**, *35*, 76–83. [CrossRef] [PubMed]
64. Ganusova, E.E.; Burch-Smith, T.M. Review: Plant-pathogen interactions through the plasmodesma prism. *Plant Sci.* **2019**, *279*, 70–80. [CrossRef] [PubMed]
65. Turner, A.; Bacic, A.; Harris, P.J.; Read, S.M. Membrane fractionation and enrichment of callose synthase from pollen tubes of *Nicotiana alata* Link et Otto. *Planta* **1998**, *205*, 380–388. [CrossRef] [PubMed]
66. Yadav, S.R.; Yan, D.; Sevilem, I.; Helariutta, Y. Plasmodesmata-mediated intercellular signaling during plant growth and development. *Front. Plant Sci.* **2014**, *5*, 44. [CrossRef] [PubMed]
67. Olatunji, D.; Geelen, D.; Verstraeten, I. Control of endogenous auxin levels in plant root development. *Int. J. Mol. Sci.* **2017**, *18*, 2587. [CrossRef] [PubMed]
68. Han, X.; Kim, J.-Y. Integrating hormone- and micromolecule-mediated signaling with plasmodesmal communication. *Mol. Plant* **2016**, *9*, 46–56. [CrossRef] [PubMed]
69. Faulkner, C. Plasmodesmata and the symplast. *Curr. Biol.* **2018**, *28*, R1374–R1378. [CrossRef]
70. Liu, Y.; Xu, M.; Liang, N.; Zheng, Y.; Yu, Q.; Wu, S. Symplastic communication spatially directs local auxin biosynthesis to maintain root stem cell niche in Arabidopsis. *Proc. Natl. Acad. Sci. USA* **2017**, *114*, 4005–4010. [CrossRef]
71. Slewinski, T.L.; Baker, R.F.; Stubert, A.; Braun, D.M. *Tie-dyed2* encodes a callose synthase that functions in vein development and affects symplastic trafficking within the phloem of maize leaves. *Plant Physiol.* **2012**, *160*, 1540–1550. [CrossRef]
72. Baker, R.F.; Slewinski, T.L.; Braun, D.M. The *tie-dyed* pathway promotes symplastic trafficking in the phloem. *Plant Signal. Behav.* **2013**, *8*, e24540. [CrossRef]
73. Song, L.; Wang, R.; Zhang, L.; Wang, Y.; Yao, S. *CRR1* encoding callose synthase functions in ovary expansion by affecting vascular cell patterning in rice. *Plant J.* **2016**, *88*, 620–632. [CrossRef] [PubMed]
74. Beffa, R.; Meins, F. Pathogenesis-related functions of plant beta-1,3-glucanases investigated by antisense transformation—A review. *Gene* **1996**, *179*, 97–103. [CrossRef]
75. Iglesias, V.A.; Meins, F. Movement of plant viruses is delayed in a beta-1,3-glucanase-deficient mutant showing a reduced plasmodesmatal size exclusion limit and enhanced callose deposition. *Plant J.* **2000**, *21*, 157–166. [CrossRef] [PubMed]
76. Bucher, G.L.; Tarina, C.; Heinlein, M.; Di Serio, F.; Meins, F.; Iglesias, V.A. Local expression of enzymatically active class I beta-1, 3-glucanase enhances symptoms of TMV infection in tobacco. *Plant J.* **2001**, *28*, 361–369. [CrossRef] [PubMed]

77. Rinne, P.L.H.; Welling, A.; Vahala, J.; Ripel, L.; Ruonala, R.; Kangasjärvi, J.; van der Schoot, C. Chilling of dormant buds hyperinduces *FLOWERING LOCUS T* and recruits GA-inducible 1,3-beta-glucanases to reopen signal conduits and release dormancy in *Populus*. *Plant Cell* **2011**, *23*, 130–146. [CrossRef] [PubMed]
78. Xu, F.; Rong, X.; Huang, X.; Cheng, S. Recent advances of flowering locus T gene in higher plants. *Int. J. Mol. Sci.* **2012**, *13*, 3773–3781. [CrossRef] [PubMed]
79. Conti, L.; Bradley, D. TERMINAL FLOWER1 is a mobile signal controlling *Arabidopsis* architecture. *Plant Cell* **2007**, *19*, 767–778. [CrossRef]
80. Ruonala, R.; Rinne, P.L.H.; Kangasjarvi, J.; van der Schoot, C. *CENL1* expression in the rib meristem affects stem elongation and the transition to dormancy in *Populus*. *Plant Cell* **2008**, *20*, 59–74. [CrossRef]
81. Rinne, P.L.H.; Paul, L.K.; Vahala, J.; Kangasjärvi, J.; van der Schoot, C. Axillary buds are dwarfed shoots that tightly regulate GA pathway and GA-inducible 1,3-β-glucanase genes during branching in hybrid aspen. *J. Exp. Bot.* **2016**, *67*, 5975–5991. [CrossRef]
82. Tylewicz, S.; Petterle, A.; Marttila, S.; Miskolczi, P.; Azeez, A.; Singh, R.K.; Immanen, J.; Mähler, N.; Hvidsten, T.R.; Eklund, D.M.; et al. Photoperiodic control of seasonal growth is mediated by ABA acting on cell-cell communication. *Science* **2018**, *360*, 212–215. [CrossRef]
83. Singh, R.K.; Miskolczi, P.; Maurya, J.P.; Bhalerao, R.P. A tree ortholog of *SHORT VEGETATIVE PHASE* floral repressor mediates photoperiodic control of bud dormancy. *Curr. Biol.* **2019**, *29*, 128–133. [CrossRef] [PubMed]
84. Singh, R.K.; Maurya, J.P.; Azeez, A.; Miskolczi, P.; Tylewicz, S.; Stojkovič, K.; Delhomme, N.; Busov, V.; Bhalerao, R.P. A genetic network mediating the control of bud break in hybrid aspen. *Nat. Commun.* **2018**, *9*, 4173. [CrossRef] [PubMed]
85. Ruan, Y.L.; Llewellyn, D.J.; Furbank, R.T. The control of single-celled cotton fiber elongation by developmentally reversible gating of plasmodesmata and coordinated expression of sucrose and K+ transporters and expansin. *Plant Cell* **2001**, *13*, 47–60. [PubMed]
86. Ruan, Y.-L.; Xu, S.-M.; White, R.; Furbank, R.T. Genotypic and developmental evidence for the role of plasmodesmatal regulation in cotton fiber elongation mediated by callose turnover. *Plant Physiol.* **2004**, *136*, 4104–4113. [CrossRef] [PubMed]
87. Zavaliev, R.; Dong, X.; Epel, B.L. Glycosylphosphatidylinositol (GPI) modification serves as a primary plasmodesmal sorting signal. *Plant Physiol.* **2016**, *172*, 1061–1073.
88. Brunkard, J.O.; Runkel, A.M.; Zambryski, P.C. The cytosol must flow: Intercellular transport through plasmodesmata. *Curr. Opin. Cell Biol.* **2015**, *35*, 13–20. [CrossRef]
89. Vaattovaara, A.; Brandt, B.; Rajaraman, S.; Safronov, O.; Veidenberg, A.; Luklová, M.; Kangasjärvi, J.; Löytynoja, A.; Hothorn, M.; Salojärvi, J.; et al. Mechanistic insights into the evolution of DUF26-containing proteins in land plants. *Commun. Biol.* **2019**, *2*, 56. [CrossRef]
90. Klessig, D.F.; Choi, H.W.; Dempsey, D.A. Systemic acquired resistance and salicylic acid: Past, present, and future. *Mol. Plant Microbe Interact.* **2018**, *31*, 871–888. [CrossRef]
91. Ádám, A.L.; Nagy, Z.Á.; Kátay, G.; Mergenthaler, E.; Viczián, O. Signals of systemic immunity in plants: Progress and open questions. *Int. J. Mol. Sci.* **2018**, *19*, 1146. [CrossRef]
92. Li, N.; Han, X.; Feng, D.; Yuan, D.; Huang, L.-J. Signaling crosstalk between salicylic acid and ethylene/jasmonate in plant defense: Do we understand what they are whispering? *Int. J. Mol. Sci.* **2019**, *20*, 671. [CrossRef]
93. Yu, K.; Soares, J.M.; Mandal, M.K.; Wang, C.; Chanda, B.; Gifford, A.N.; Fowler, J.S.; Navarre, D.; Kachroo, A.; Kachroo, P. A feedback regulatory loop between G3P and lipid transfer proteins DIR1 and AZI1 mediates azelaic-acid-induced systemic immunity. *Cell Rep.* **2013**, *3*, 1266–1278. [CrossRef] [PubMed]
94. Wang, C.; El-Shetehy, M.; Shine, M.B.; Yu, K.; Navarre, D.; Wendehenne, D.; Kachroo, A.; Kachroo, P. Free radicals mediate systemic acquired resistance. *Cell Rep.* **2014**, *7*, 348–355. [CrossRef] [PubMed]
95. Lim, G.-H.; Shine, M.B.; de Lorenzo, L.; Yu, K.; Cui, W.; Navarre, D.; Hunt, A.G.; Lee, J.-Y.; Kachroo, A.; Kachroo, P. Plasmodesmata localizing proteins regulate transport and signaling during systemic acquired immunity in plants. *Cell Host Microbe* **2016**, *19*, 541–549. [CrossRef] [PubMed]
96. Lim, G.-H.; Kachroo, A.; Kachroo, P. Role of plasmodesmata and plasmodesmata localizing proteins in systemic immunity. *Plant Signal. Behav.* **2016**, *11*, e1219829. [CrossRef] [PubMed]
97. Ye, Z.-W.; Chen, Q.-F.; Chye, M.-L. *Arabidopsis thaliana* acyl-CoA-binding protein ACBP6 interacts with plasmodesmata-located protein PDLP8. *Plant Signal. Behav.* **2017**, *12*, e1359365. [CrossRef] [PubMed]

98. Cui, W.; Lee, J.-Y. Arabidopsis callose synthases CalS1/8 regulate plasmodesmal permeability during stress. *Nat. Plants* **2016**, *2*, 16034. [CrossRef] [PubMed]
99. Huang, L.; Chen, X.-Y.; Rim, Y.; Han, X.; Cho, W.K.; Kim, S.-W.; Kim, J.-Y. Arabidopsis glucan synthase-like 10 functions in male gametogenesis. *J. Plant Physiol.* **2009**, *166*, 344–352. [CrossRef] [PubMed]
100. Wang, X.; Zhu, P.; Qu, S.; Zhao, J.; Singh, P.K.; Wang, W. Ectodomain of plasmodesmata-localized protein 5 in Arabidopsis: Expression, purification, crystallization and crystallographic analysis. *Acta Crystallogr. F-Struct. Biol. Commun.* **2017**, *73*, 532–535. [CrossRef] [PubMed]
101. Cheval, C.; Faulkner, C. Plasmodesmal regulation during plant-pathogen interactions. *New Phytol.* **2018**, *217*, 62–67. [CrossRef] [PubMed]
102. Sonnino, S.; Prinetti, A. Lipids and membrane lateral organization. *Front. Physiol.* **2010**, *1*, 153. [CrossRef] [PubMed]
103. Jarsch, I.K.; Konrad, S.S.A.; Stratil, T.F.; Urbanus, S.L.; Szymanski, W.; Braun, P.; Braun, K.-H.; Ott, T. Plasma membranes are subcompartmentalized into a plethora of coexisting and diverse microdomains in *Arabidopsis* and *Nicotiana benthamiana*. *Plant Cell* **2014**, *26*, 1698–1711. [CrossRef] [PubMed]
104. Iswanto, A.B.B.; Kim, J.-Y. Lipid raft, regulator of plasmodesmal callose homeostasis. *Plants* **2017**, *6*, 15. [CrossRef] [PubMed]
105. Grison, M.S.; Brocard, L.; Fouillen, L.; Nicolas, W.; Wewer, V.; Dörmann, P.; Nacir, H.; Benitez-Alfonso, Y.; Claverol, S.; Germain, V.; et al. Specific membrane lipid composition is important for plasmodesmata function in Arabidopsis. *Plant Cell* **2015**, *27*, 1228–1250. [CrossRef] [PubMed]
106. Zhang, Z.; Ruan, Y.-L.; Zhou, N.; Wang, F.; Guan, X.; Fang, L.; Shang, X.; Guo, W.; Zhu, S.; Zhang, T. Suppressing a putative sterol carrier gene reduces plasmodesmal permeability and activates sucrose transporter genes during cotton fiber elongation. *Plant Cell* **2017**, *29*, 2027–2046. [CrossRef] [PubMed]
107. Park, S.-H.; Li, F.; Renaud, J.; Shen, W.; Li, Y.; Guo, L.; Cui, H.; Sumarah, M.; Wang, A. NbEXPA1, an α-expansin, is plasmodesmata-specific and a novel host factor for potyviral infection. *Plant J.* **2017**, *92*, 846–861. [CrossRef] [PubMed]
108. Oparka, K.J. Signalling via plasmodesmata—The neglected pathway. *Semin. Cell Biol.* **1993**, *4*, 131–138. [CrossRef]
109. Tilsner, J.; Nicolas, W.; Rosado, A.; Bayer, E.M. Staying tight: Plasmodesmal membrane contact sites and the control of cell-to-cell connectivity in plants. *Ann. Rev. Plant Biol.* **2016**, *67*, 337–364. [CrossRef]
110. Schubert, V. Super-resolution microscopy—Applications in plant cell research. *Front. Plant Sci.* **2017**, *8*, 531. [CrossRef]

International Journal of
Molecular Sciences

MDPI

Article

Arabidopsis Trichome Contains Two Plasma Membrane Domains with Different Lipid Compositions Which Attract Distinct EXO70 Subunits

Zdeňka Kubátová [1], Přemysl Pejchar [1,2], Martin Potocký [1,2], Juraj Sekereš [2], Viktor Žárský [1,2] and Ivan Kulich [1,*]

[1] Department of Experimental Plant Biology, Faculty of Science, Charles University, 12800 Prague, Czech Republic
[2] Institute of Experimental Botany, Czech Academy of Sciences, 165 02 Prague, Czech Republic
* Correspondence: kulich@natur.cuni.cz

Received: 21 June 2019; Accepted: 1 August 2019; Published: 3 August 2019

Abstract: Plasma membrane (PM) lipid composition and domain organization are modulated by polarized exocytosis. Conversely, targeting of secretory vesicles at specific domains in the PM is carried out by exocyst complexes, which contain EXO70 subunits that play a significant role in the final recognition of the target membrane. As we have shown previously, a mature *Arabidopsis* trichome contains a basal domain with a thin cell wall and an apical domain with a thick secondary cell wall, which is developed in an EXO70H4-dependent manner. These domains are separated by a cell wall structure named the Ortmannian ring. Using phospholipid markers, we demonstrate that there are two distinct PM domains corresponding to these cell wall domains. The apical domain is enriched in phosphatidic acid (PA) and phosphatidylserine, with an undetectable amount of phosphatidylinositol 4,5-bisphosphate (PIP_2), whereas the basal domain is PIP_2-rich. While the apical domain recruits EXO70H4, the basal domain recruits EXO70A1, which corresponds to the lipid-binding capacities of these two paralogs. Loss of EXO70H4 results in a loss of the Ortmannian ring border and decreased apical PA accumulation, which causes the PA and PIP_2 domains to merge together. Using transmission electron microscopy, we describe these accumulations as a unique anatomical feature of the apical cell wall—radially distributed rod-shaped membranous pockets, where both EXO70H4 and lipid markers are immobilized.

Keywords: cell wall; EXO70; exocyst complex; phosphatidic acid; phosphatidylinositol 4,5-bisphosphate; phospholipids; plasma membrane domains; polar exocytosis; trichome

---

1. Introduction

Despite their unicellularity, *Arabidopsis* trichomes grow into extraordinarily shaped and precisely polarized structures, which makes them a potent model of cell polarization and morphogenesis [1]. According to [2], trichome development can be divided into six stages. During the first stage, several rounds of endoreduplication lead to an increased DNA content and, later, to a remarkably large trichome cell. The second and third stages are crucial for trichome shaping because they involve oriented growth above the surface and two or three branching events. During the fourth and fifth stages, the branches elongate significantly and the trichome reaches its final size and shape. When the growth is complete, the trichome enters the last stage of its development, which is maturation of the cell wall. In this stage, the trichome is divided into two markedly different domains—the basal domain, with a thin cell wall, and the apical domain, with an extremely thick cell wall. This latter secondary cell wall (SCW) consists of at least two distinct layers—the outer, cellulose-rich layer and the inner, callose-rich layer [3]. The inner layer is autofluorescent and silicified in a callose-dependent manner [4].

The apical cell wall domain also contains surface papillae, which accumulate cuticular waxes that may be different from those in other epidermal cells [5]. Basal and apical trichome domains are separated by a ring-shaped, callose-rich structure named the Ortmannian ring [3].

The development of this cell wall is dependent on the EXO70H4 exocyst subunit, which is necessary for callose synthase delivery to the plasma membrane (PM) [4]. EXO70H4 is a subunit of the exocyst complex, which is a eukaryotic protein complex involved in the tethering of post-Golgi secretory vesicles, which carry membrane and cell wall components to the PM [6]. The exocyst is composed of eight different subunits forming a functional complex, including Sec3, Sec5, Sec6, Sec8, Sec10, Sec15, Exo70, and Exo84 (for review, see [7,8]). Exo70 and Sec3 have a special position in the complex, as both of these subunits are able to directly bind to the target membrane lipids [9–11], thus regulating where the secretion will occur [12]. In yeast and mammals, the Exo70 subunit recruits the rest of the exocyst complex to the PM via a specific interaction of the EXO70 C-terminus with phosphatidylinositol 4,5-bisphosphate (PIP_2) [9,13]. Sec3 is then responsible for initiating the binary SNARE (Soluble NSF Attachment Protein Receptor) complex [14]. The crosstalk of phospholipids with their protein partners and its importance in plant cell polarity determination and membrane traffic regulation was summarized in [15].

The *Arabidopsis* genome contains 23 EXO70 paralogs [16–18]. This multiplicity of EXO70 subunits in plants led to the hypothesis that EXO70s may have divergent lipid-binding properties and thus regulate exocytosis in distinct PM domains within a single cell [19]. So far, there is one documented example of such domain separation observed for NtEXO70A1 and NtEXO70B1 paralogs in *Nicotiana tabacum* pollen tubes, where EXO70A1 was more apically localized than EXO70B1 [20]. There is also apparent functional specialization among the EXO70 paralogs, as neither EXO70A1, B1, nor any other of 18 tested paralogs can complement the *exo70H4-1* loss-of-function mutant phenotype [4]. In this study, we document the differential localization of EXO70H4 and EXO70A1 in the mature trichome and show that EXO70H4 is required for the development and separation of two PM domains with different compositions of signaling phospholipids and distinct abilities to attract other EXO70 members.

2. Results

2.1. Mature *Arabidopsis* Trichomes Contain Two Distinct Lipid Domains and EXO70H4 is Involved in Their Formation

To investigate the distribution of the PM lipids in the *Arabidopsis* trichome, we observed localization of several fluorescent phospholipid markers. We focused on phosphatidylinositol-4-phosphate (PI4P) [21], phosphatidylinositol 4,5-bisphosphate (PIP_2) [21], phosphatidic acid (PA) [22,23], and phosphatidylserine (PS) membrane lipid markers [23]. While the PIP_2 marker was localized almost exclusively to the trichome base beneath the Ortmannian ring (Figure 1), the PI4P and PA markers were distributed evenly around the whole PM, with their signals enhanced at the Ortmannian ring and in intramural pockets throughout the apical domain (Figure 1). However, the PI4P marker signal inside these intramural pockets was weaker than the PA marker signal, which is highly accumulated there (Figure 1). The identities of these pockets will be described in more detail in Section 2.3. The PS marker was mostly visible at the apical membrane domain (Figure 1 and Supplemental Video S1). No signal was observed at the base, but a weak signal may have remained undetected due to a strong cytoplasmic background of this marker line. Thus, we concluded that a mature *Arabidopsis* trichome contains two distinct membrane domains with different lipid compositions.

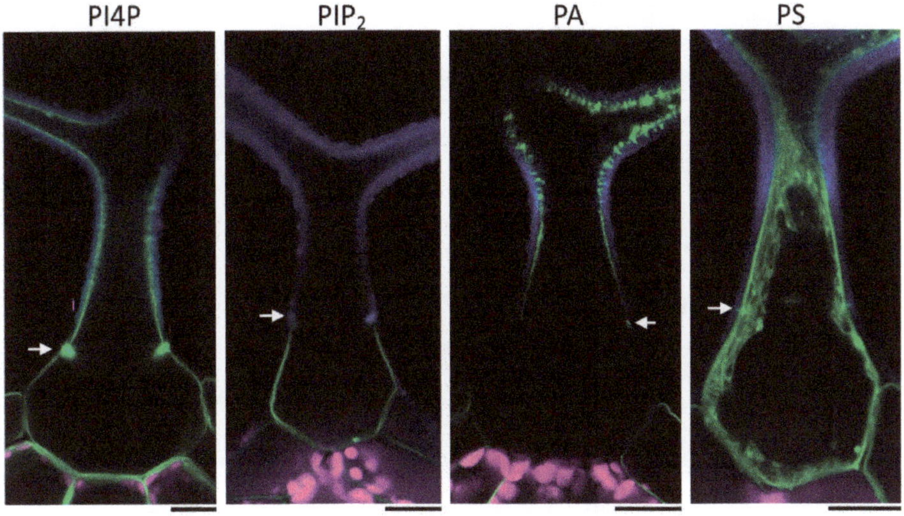

Figure 1. Representative images of different phospholipid markers in wild type mature trichome. PI4P—phosphatidylinositol-4-phosphate; PIP_2—phosphatidylinositol 4,5-bisphosphate; PA—phosphatidic acid; PS—phosphatidylserine. Blue—cell wall autofluorescence; magenta—chlorophyll autofluorescence; green—mCitrine or YFP (yellow fluorescent protein). White arrows point at the Ortmannian ring. Scale bars = 20 µm.

Next, we wanted to investigate when these two domains differentiated during trichome development. For this, we used a double marker line expressing a PA marker tagged with mCitrine and a PIP_2 marker tagged with mCherry (mCH), provided by the authors of [23]. In the wild type (WT) background, the distribution of the lipid markers was consistent with previous observations (Figure 2a). Distinct domains only appeared at stage 6 of trichome development, along with the establishment of the Ortmannian ring. In the WT trichomes of stage 4 (elongation), where the Ortmannian ring was not formed yet, PM domains were not visible (Figure 2b). The PIP_2 marker was evenly distributed all around the PM and the PA marker was not bound to the membrane at all (Figure 2b), suggesting that PA is not present in young trichome PM. Later, we introduced both lipid markers into the *exo70H4-1* mutant background, which was unable to finish the cell wall maturation [3]. Here, the membrane domains lacked a sharp border and were poorly visible (Figure 2c). We concluded that the establishment of the apical and basal membrane domains occurred while the EXO70H4-dependent SCW layer in the apical domain was formed, which depended on the SCW formation.

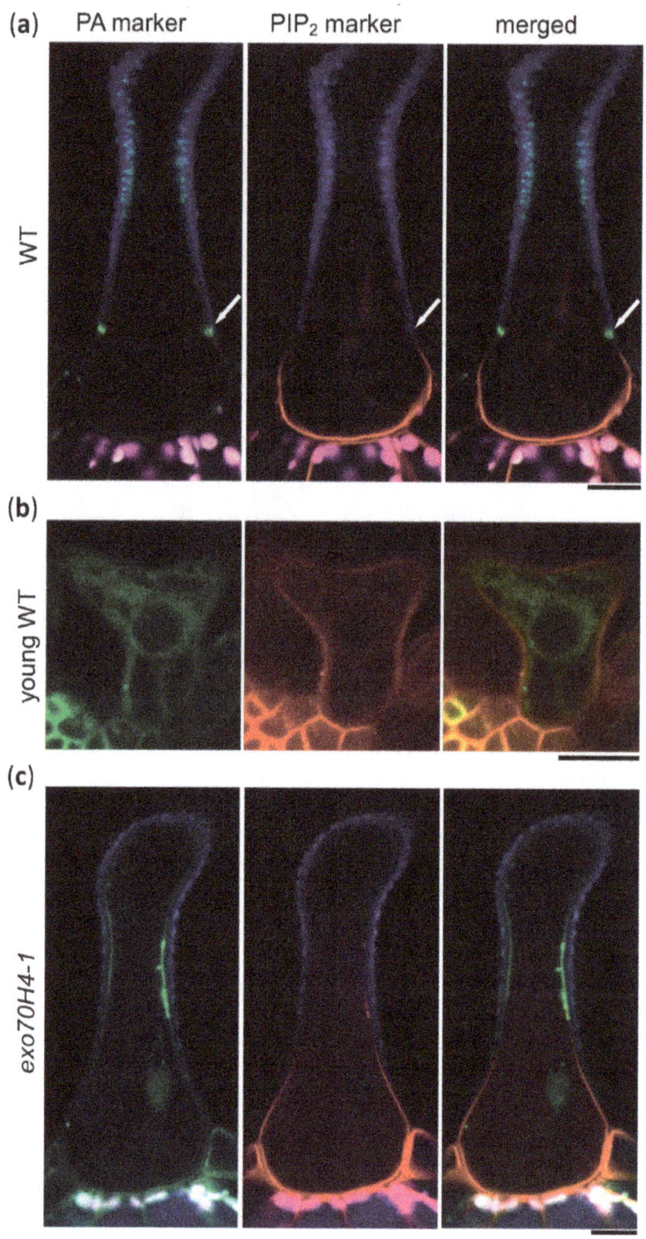

Figure 2. Lipid markers in WT and *exo70H4-1* trichomes. (**a**) Colocalization of PA and PIP_2 markers in WT trichome; (**b**) colocalization of PA and PIP_2 markers in WT trichome in the elongation stage; (**c**) colocalization of PA and PIP_2 markers in an *exo70H4-1* mutant background. Blue—cell wall autofluorescence; magenta—chlorophyll autofluorescence; green—YFP; red—mCherry. White arrows point at the Ortmannian ring. Scale bars = 20 μm.

2.2. Trichome Apical and Basal Plasma Membrane Domains Recruit Different EXO70 Proteins

To address the biological relevance of the trichome lipid domain distribution, we observed multiple EXO70 proteins under the control of the EXO70H4 promoter. EXO70A1 was previously found to bind to PIP_2 [24] and to localize to the PIP_2-rich region of the pollen tube [20]. Corresponding with its lipid affinity, EXO70A1 localized preferentially to the basal trichome domain and, in many cases, the apical trichome domain was completely devoid of EXO70A1, although there was a certain degree of variability and it was often seen also within the Ortmannian ring area (Figure 3a, Supplemental Figure S1). As reported previously [4], a lack of EXO70H4 resulted in EXO70A1 localizing all around the trichome PM (Figure 3a). To verify that EXO70A1 is natively present in *Arabidopsis* trichomes, we also generated a reporter construct of the EXO70A1 promoter fused to 2xGFP (green fluorescent protein). This revealed that EXO70A1 was indeed expressed in the mature trichome (Figure 3b).

EXO70H4 is known to localize to the apical trichome domain and to the Ortmannian ring [4]. To correlate its localization together with EXO70A1, we co-transformed EXO70H4p::mCHERRY-EXO70H4 (mCH-EXO70H4) with EXO70H4p::GFP-EXO70A1 (GFP-EXO70A1) In this case, the EXO70 isoforms localized to distinct and mostly non-overlapping PM domains that corresponded to the apical and basal trichome domains (Figure 3c).

We also investigated the localization of the remaining EXO70s, which we observed previously in the *exo70H4-1* mutant background [4]. This time, we expressed the constructs in a WT background to see their localization when it was unaffected by the *exo70H4-1* mutation. Eight paralogs were tested—EXO70A1, A2, B1, C1, D2, F1, H7, and H8. The only paralogs found to localize to the PM in the WT background in addition to EXO70A1 were EXO70A2 (localized similarly to EXO70A1) and EXO70H8. EXO70H8 strongly accumulated at the PA-rich domain in the apical part of the trichome, including Ortmannian ring and cell wall ingrowths. Although EXO70H8 mimicked EXO70H4 localization in the WT trichome, it was not capable of functionally complementing the *exo70H4-1* mutant phenotype in our previous cross-complementation study [4]. The other EXO70 paralogs that were tested remained in the cytoplasm or nucleus, as in the previous experiments in the *exo70H4-1* background.

Since the lipid-binding capacities of EXO70H4 have not yet been described, we performed a protein–lipid overlay assay with in vitro translated HA-tagged EXO70H4. This revealed a clear affinity to PS and conceivably PA, but no apparent binding to PIP_2 (Figure 3d). This corresponded well with the colocalization of EXO70H4 with PA and PS markers in the apical domain of the mature trichome. Based on these and previously published data, we concluded that different EXO70 proteins exhibit a specific capacity to bind membrane lipids and thus are recruited to distinct PM domains, contributing to the biogenesis of different cell wall domains within a single plant cell.

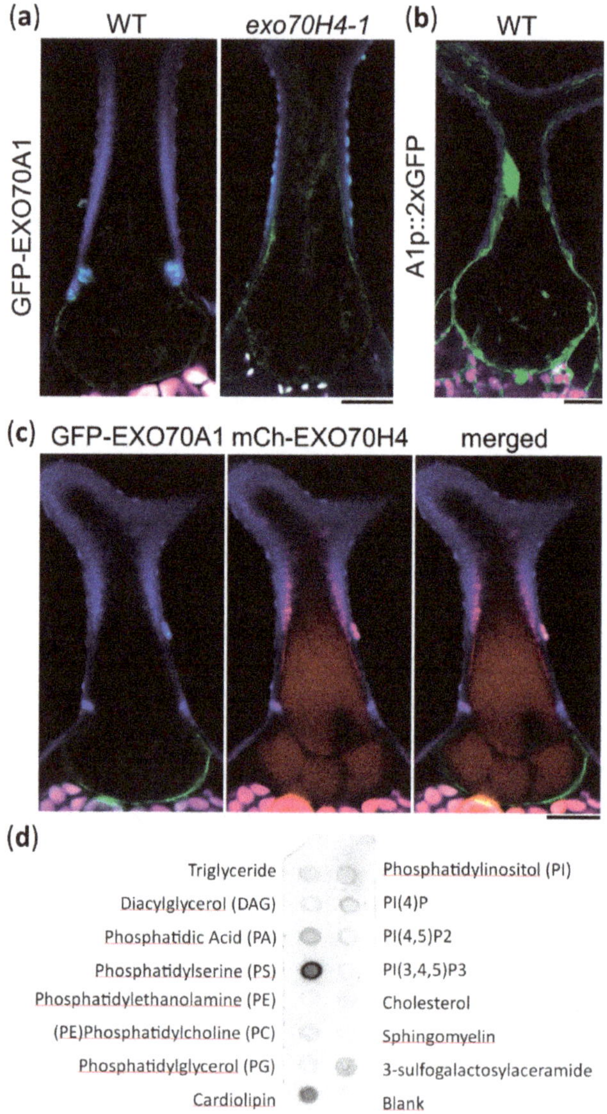

Figure 3. Trichome plasma membrane domains recruit different EXO70 proteins. (**a**) EXO70H4p::GFP-EXO70A1 (GFP-EXO70A1) preferentially localizes to the basal trichome domain in WT. This preference is lost in the *exo70H4-1* mutant. (**b**) EXO70A1p::GFP:GFP (A1p::2xGFP) expression marker in a WT trichome. (**c**) Colocalization of GFP-EXO70A1 with EXO70H4p::mCherry-EXO70H4 (mCh-EXO70H4). (**d**) Protein–lipid overlay assay of EXO70H4. Blue—cell wall autofluorescence; magenta—chlorophyll autofluorescence; green—GFP; red—mCherry. Scale bars = 20 μm.

2.3. The Apical Cell Wall Contains Entrapped Membranous Pockets

While observing the apical trichome domain, we noticed that a large portion of the signal came from within the cell wall. This was true for all of the constructs which localized to the apical membrane domain, including markers of PA, PS, PI4P, mCh-EXO70H4, GFP-EXO70H8, and occasionally also EXO70A1 and EXO70A2 with a weak signal. This strange intramural localization directed our further investigation. The signal within the SCW was distributed in a pattern remarkably similar to the callose deposits shown before [3,4]. The radially distributed rays of signal were visibly embedded within the cell wall and were well-apparent on fluorescence microscopy optical cuts of matured branches together with a wrinkled PM–SCW interface (Figure 4a,b). To be sure that this was indeed the fluorophore signal and not cell wall autofluorescence, we performed lambda scans in plants expressing GFP-EXO70H4, mCh-EXO70H4, YFP-PA marker, and untransformed control. In all of these cases, the expected emission peak was observed (Supplemental Figure S2), with no such peaks in the negative control. To check if these intramural signals were still connected to the rest of the cytoplasm, we performed a fluorescence recovery after photobleaching (FRAP) experiment on the PI4P marker, which revealed that there was no detectable recovery of the signal, unlike in the case of the signal bleached within the apparent continuum of the PM (Supplemental Figure S3). We therefore hypothesized that the intramural signal may come from the PM and cell interior, being physically entrapped within the cell wall pockets.

To investigate the SCW structure in more detail, we used transmission electron microscopy (TEM). This revealed internal cell wall structures, supporting our observations of oblong traversing pockets with light microscopy. TEM cross-sections of mature WT trichome branches displayed SCW-transpassing structures, probably aggregates of entrapped secretory lipid membranes, proteins, and the cell interior, often arranged in a radial pattern of concentric transpassing channels (Figure 4c, left). These intramural pockets were obviously separated from the continuum of cytoplasm (Figure 4c, middle), but their PM origin was apparent from images where pieces of electron-dense, non-cell wall materials were embedded into the cell wall (Figure 4c, right).

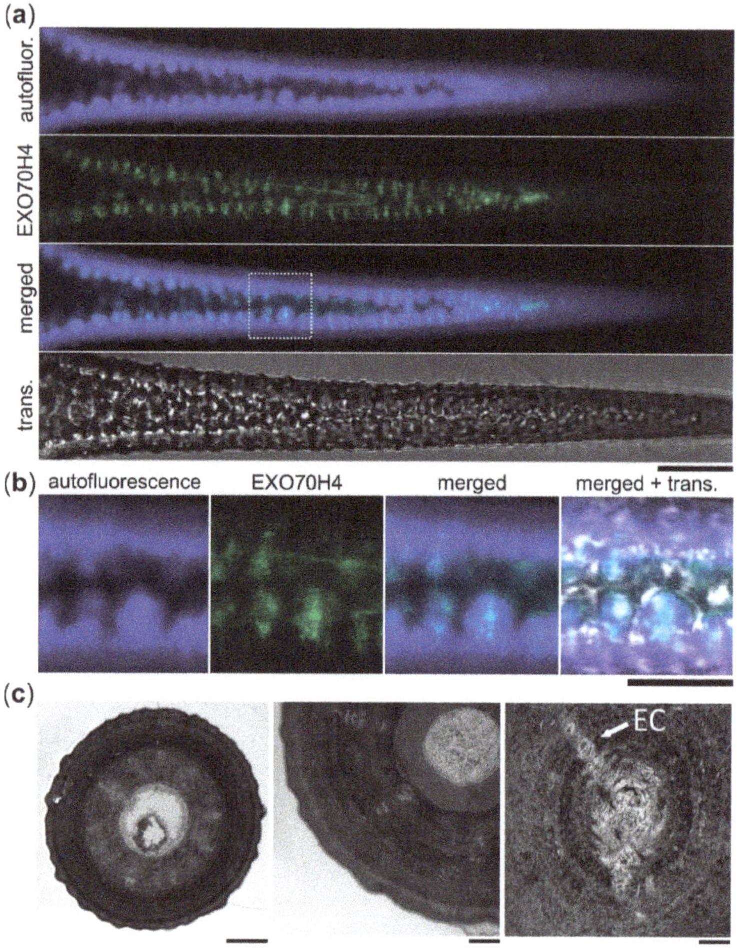

Figure 4. Plasma membrane proteins reside within the trichome apical cell wall. (**a**) GFP-EXO70H4 (EXO70H4) signal, cell wall autofluorescence (autofluor.), and transmission channel (trans.) at the branch of a mature WT trichome. The dotted square represents area enlarged in (**b**). Scale bar = 20 µm. (**b**) Detail of (**a**). Scale bar = 10 µm. (**c**) Left—TEM image of a cross-section of a trichome branch. Entrapped cell interior (EC) is visible as concentric rays. Scale bar = 3 µm. Middle—detail of entrapped cell interior obviously separated from the cytoplasm. Scale bar = 1 µm. Right—a detail of one concentric ray. Scale bar = 200 nm.

3. Discussion

We showed that the mature *Arabidopsis* trichome contains, along with two cell wall domains, two distinct PM domains that differ in their phospholipid composition and also in their ability to recruit different EXO70 proteins. EXO70A1 is recruited to the basal trichome domain, which is PIP_2-rich and contains a thin, pectin-rich cell wall [25]. EXO70H4 is recruited to the apical domain, which is PS- and PA-rich and has a thick, autofluorescent cell wall [3].

There are a number of different examples of cell types having different phospholipid domains and many of them can be linked with differential cell wall deposition. Pollen grains also accumulate PIP_2 at the site of the future aperture, which is an area marked with a thin cell wall [26]. Similarly, the pollen tube accumulates PIP_2 at its tip, while PS and PA are in the shank [22,23]. This corresponds with the apical cell wall of the pollen tube being almost solely made of secreted pectins, while callose and cellulose are deposited in the shank [27]. This domain separation in the pollen tube is further marked by differential membrane localization of EXO70A1 and EXO70B1 in tobacco [20]. In contrast to the pollen tube, trichomes are extremely large and static structures. The border of the two membrane domains appears to be very sharp and defined by the Ortmannian ring. A lack of a sharp border between these two domains and a partial loss of polarity in the *exo70H4-1* mutant suggests that there is a positive feedback loop and that EXO70s have a role in domain development.

The observed lack of PIP_2 and abundance of PA in the apical domain could also imply that, in the apical domain, PIP_2 is metabolized to PA. In general, PA is formed through two pathways: First, by the direct hydrolysis of structural phospholipids by phospholipase D and second, through the consecutive actions of phospholipase C and diacylglycerol kinase, where the hydrolysis of PIP_2 produces inositol-1,4,5-trisphosphate (released into the cytosol) and diacylglycerol within the PM, which is quickly phosphorylated at the membrane into PA by diacylglycerol kinase. PA produced by phospholipase Dα1 is a crucial signaling lipid, mediating the abscisic acid response in guard cells, where its role in NADPH oxidase-mediated reactive oxygen species (ROS) production was clearly demonstrated [28]. Both PA and phospholipase Dα1 have previously been linked with ROS production [29]. NADPH oxidase-mediated ROS production acts in the plant defense response [30,31]. In plants, PA is also produced in response to several stress factors, including pathogen attack [32]. EXO70H4 is also induced by *flagellin 22* in epidermal pavement cells [4]. The apical cell wall domain of the trichome is also remarkably similar to the defense papillae, as they both accumulate callose deposits, phenolic compounds [33], and an extracellular signal of membrane proteins (such as SYP121—SYNTAXIN OF PLANTS 121 [34]). We therefore suggest that the mechanism and domain organization in the trichome may be a manifestation of a general response to pathogen attack, which runs constitutively in the trichome.

It was shown that binding of PA to its protein interactors is enhanced by a negative curvature stress and that a complex membrane lipid composition strongly influences lipid–protein interactions [35]. Surprisingly, EXO70 by itself is able to induce negative membrane curvature, specifically through the homodimerization mechanism, as was demonstrated in mammals [36]. Together, these biophysical properties could induce PM deformation sufficient to form extreme membrane curvatures, leading to the formation of membranous pockets within the SCW. Another factor contributing to the pocket formation may be the cuticular wax migration across the cell wall. The apical trichome domain displays several unique features, including surface papillae formation. Papillae start to form at stage 5 of trichome development (expansion) and continue developing during stage 6 (maturation). At these stages, the apical cell wall is already quite thick (>2 μm) [37]. Papillae are little bumps filled with lipophilic cuticular substances. These waxes differ from those of other epidermal cells by a high content of C35+ alkanes [5]. How these compounds get through the thick cell wall to the surface is not understood. Membranous pockets may drift along the migration routes of the cuticular waxes through the inner layer of the SCW. This feature may be very useful for the study of cuticular wax migration across the cell wall. Another intriguing possibility is that formation of these membranous pockets within the SCW is directly linked to the mechanism of callose deposition during SCW biogenesis—a possibility we aim to study.

4. Materials and Methods

4.1. Plant Material

Plants were grown in Jiffy soil pellets in standard growth chamber conditions (long day 16 h:8 h, 100–120 μM photosynthetically active radiation m^{-2} s^{-1}). The *exo70H4-1* mutant line was described previously [3]. As WT control, outcrossed WT plants were used. Seeds of PI4P and PIP_2 markers were obtained from Yvon Jaillais and are described in detail in [21]. From this set, we used line P21Y (mCitrine-2xPH FAPP PI4P binding domain), line P24Y (mCitrine-2xPH^{PLC} PIP_2 binding domain), and line P24R (mCherry-2xPH^{PLC} PIP_2 binding domain). Two different PA markers were used in this study. In Figure 1, it was YFP:NES-2xSpo20p (cloning described below) and in Figure 2 and Supplemental Figure S2, it was a double marker line expressing a PA marker tagged with mCitrine (PAY) and a PIP_2 marker tagged with mCherry (P24R, mentioned above), provided by [23]. PS marker cloning is described below.

4.2. Confocal Microscopy

Confocal microscopy images were taken on Zeiss LSM880 with C-Apochromat 403/1.2 W Korr FCS M27 objective [GFP (488): 508–540 nm, chlorophyll (488) 650–721 nm, cell wall autofluorescence (405) 426–502, mCherry (561) 597–641]. Images were processed using the Fiji platform [38].

4.3. Trichome Isolation and TEM

Leaves of 4-week-old plants were collected and incubated for 3 h in falcon tubes in a solution of acetic acid:ethanol (1:3) and then washed three times with deionized water. Washed leaves were transferred to a solution of 150 mM KH_2PO_4 pH 9.5 and incubated overnight with shaking at 150–180 rpm. Released trichomes were collected by centrifugation (1 min, 1000 G, no break). For TEM, isolated trichomes were fixed for 24 h in 2.5% (*v*/*v*) glutaraldehyde in 0.1 M cacodylate buffer (pH 7.2) at 4 °C and postfixed in 2% (*w*/*v*) OsO4 in the same buffer. Fixed samples were dehydrated through an ascending ethanol and acetone series and embedded in Epon–Araldite.

4.4. Cloning and Stable Transgenic Line Preparation

For cloning of the expression reporter construct EXO70A1p::GFP:GFP, the multisite gateway approach was used. The EXO70A1 promoter (1 kb upstream, EXO70A1 prom for and EXO70A1 prom rev listed in Supplemental Table S1 were used) was subcloned into pDONORP4-P1r and sequenced using M13 primers. GFP constructs—pEN-L1-F-L2 and pEN-R2-GFP-L3,0—and destination vector pB7m34GW were obtained from [39]. An expression clone was assembled from these by a multisite reaction using Gateway LR clonase (Thermo Fischer, Waltham, MA, USA). All EXO70H4p::GFP-EXO70XY constructs used in this study were previously described in [4].

To prepare the construct for the PA marker (pUBQ::YFP:NES-2xSpo20p PABD), first NES-2xSpo20p was amplified from YFP:Spo20p-PABD [22] and cloned together with the ubiquitin (UBQ) promoter and YFP into the binary vector pHD71 (kindly provided by Dr. Benedikt Kost).

To prepare the construct for the PS marker (pUBQ::YFP:C2^{LACT}), YFP-C2^{LACT} from Lat52:YFP-C2^{LACT} [22] was cloned together with the UBQ promoter into the binary vector pHD71; the primers used are listed in Supplemental Table S1.

Final constructs were electroporated into competent cells of *Agrobacterium tumefaciens*, strain GV3101. Col-0 wild type plants were transformed by the floral dip method [40] and transformants were selected on kanamycin plates or soil by BASTA spraying (150 mg/L of glufosinate-ammonium). No fewer than five individual transformants were observed in each experiment and at least two biological replicates were made.

4.5. Protein–Lipid Overlay Assay

The protein–lipid overlay assay was performed using N-terminally HA-tagged EXO70H4 and echelon lipid strips. First, the EXO70H4 coding sequence was cloned into a pTNT vector (Promega, Fitchburg, WI, USA), optimized for in vitro coupled transcription and translation, using primers listed in Supplemental Table S1. The cloning was performed in two steps. First, the EXO70H4 coding sequence was amplified with the use of the primers EXO70H4-forward and EXO70H4-reverse. Subsequently, a megaprimer containing the HA-tag sequence and Kozak consensus sequence facilitating efficient translation initiation was used together with EXO70H4-reverse primer. The product was cloned into the pTNT vector via SalI and NotI restriction sites.

The construct was used as a template for in vitro coupled transcription and translation reactions using TNT® SP6 High-Yield Wheat Germ Protein Expression System (Promega), according to the manufacturer's instructions, in a total volume of 50 µL. To verify protein expression and stability, 10 µL of the yield was used in Western blotting. Since the protein was tagged on the N-terminus, Western blot analysis would have detected products of prematurely terminated translation, but none were observed (see Supplemental Figure S4).

The remaining 40 µL of the reaction product was used for the protein–lipid overlay assay. The echelon lipid strip (Elcheon Biosciences, Salt Lake City, UT, USA) was blocked for one hour in blocking solution (50 mM Tris, 150 mM NaCl, 0.05% Tween 20, pH 7.6 + 3% bovine serum albumine). After three washes in wash solution (50 mM Tris, 150 mM NaCl, 0.05% Tween 20, pH 7.6), the strip was incubated for two hours with primary mouse anti-HA antibody (Sigma-Aldrich, St. Louis, MO, USA) diluted 1:1000 in blocking solution. After three washes with wash solution, the strip was subsequently incubated with secondary anti-mouse antibody conjugated to horseradish peroxidase (Promega), diluted 1:20,000 in blocking solution. Finally, the strip was washed three times with wash solution and incubated for three minutes with Amersham ECL Prime Western Blotting Detection Reagent (GE Healthcare, Chicago, IL, USA) and imaged with Bio-Rad Laboratories (Hercules, CA, USA) Chemidoc gel imaging system. The experiment was repeated twice with consistent results.

Supplementary Materials: Supplementary Materials can be found at http://www.mdpi.com/1422-0067/20/15/3803/s1.

Author Contributions: Conceptualization, I.K. and V.Z.; validation, Z.K. and I.K.; investigation, Z.K., J.S. and I.K.; resources, P.P. and M.P.; writing—original draft preparation, Z.K. and I.K.; writing—review and editing, P.P., M.P., J.S. and V.Ž.; visualization, Z.K. and I.K.; supervision, I.K. and V.Z; funding acquisition, I.K.

Funding: This project was financed by Czech Science Foundation grant 18-12579S. Microscopy was performed in the Laboratory of Confocal and Fluorescence Microscopy co-financed by the European Regional Development Fund and the state budget of the Czech Republic (project no. CZ.1.05/4.1.00/16.0347). Part of income of V.Ž. is supported by the Ministry of Education, Youth and Sports of CR from European Regional Development Fund-Project "Centre for Experimental Plant Biology": No. CZ.02.1.01/0.0/0.0/16_019/0000738.

Acknowledgments: We would like to thank Yvon Jaillais for providing us with the lipid markers, Patrick Moxon for cloning EXO70A1 promoter, Jachym Metlicka for help with image preparation, and our technician, Marta Cadyova, for running the laboratory.

Conflicts of Interest: The authors declare no conflict of interest.

Abbreviations

FRAP	fluorescence recovery after photobleaching
mCh	mCherry
PA	phosphatidic acid
PI4P	phosphatidylinositol-4-phosphate
PIP_2	phosphatidylinositol 4,5-bisphosphate
PM	plasma membrane
PS	phosphatidylserine
SCW	secondary cell wall
TEM	transmission electron microscopy
WT	wild type

References

1. Szymanski, D.B.; Lloyd, A.M.; Marks, M.D. Progress in the molecular genetic analysis of trichome initiation and morphogenesis in Arabidopsis. *Trends Plant Sci.* **2000**, *5*, 214–219. [CrossRef]
2. Hülskamp, M. Plant trichomes: A model for cell differentiation. *Nat. Rev. Mol. Cell Biol.* **2004**, *5*, 471–480. [CrossRef] [PubMed]

3. Kulich, I.; Vojtíková, Z.; Glanc, M.; Ortmannová, J.; Rasmann, S.; Žárský, V. Cell wall maturation of Arabidopsis trichomes is dependent on exocyst subunit EXO70H4 and involves callose deposition. *Plant Physiol.* **2015**, *168*, 120–131. [CrossRef] [PubMed]
4. Kulich, I.; Vojtíková, Z.; Sabol, P.; Ortmannová, J.; Neděla, V.; Tihlaříková, E.; Žárský, V. Exocyst Subunit EXO70H4 Has a Specific Role in Callose Synthase Secretion and Silica Accumulation. *Plant Physiol.* **2018**, *176*, 2040–2051. [CrossRef] [PubMed]
5. Hegebarth, D.; Buschhaus, C.; Joubès, J.; Thoraval, D.; Bird, D.; Jetter, R. Arabidopsis ketoacyl-CoA synthase 16 (KCS16) forms C/C acyl precursors for leaf trichome and pavement surface wax. *Plant Cell Environ.* **2017**, *40*, 1761–1776. [CrossRef] [PubMed]
6. TerBush, D.R.; Maurice, T.; Roth, D.; Novick, P. The Exocyst is a multiprotein complex required for exocytosis in Saccharomyces cerevisiae. *EMBO J.* **1996**, *15*, 6483–6494. [CrossRef]
7. Hsu, S.-C.; TerBush, D.; Abraham, M.; Guo, W. The exocyst complex in polarized exocytosis. *Int. Rev. Cytol.* **2004**, *233*, 243–265.
8. Lepore, D.M.; Martínez-Núñez, L.; Munson, M. Exposing the Elusive Exocyst Structure. *Trends Biochem. Sci.* **2018**, *43*, 714–725. [CrossRef]
9. Liu, J.; Zuo, X.; Yue, P.; Guo, W. Phosphatidylinositol 4,5-Bisphosphate Mediates the Targeting of the Exocyst to the Plasma Membrane for Exocytosis in Mammalian Cells. *Mol. Biol. Cell* **2007**, *18*, 4483–4492. [CrossRef]
10. Zhang, X.; Orlando, K.; He, B.; Xi, F.; Zhang, J.; Zajac, A.; Guo, W. Membrane association and functional regulation of Sec3 by phospholipids and Cdc42. *J. Cell Biol.* **2008**, *180*, 145–158. [CrossRef]
11. Bendezú, F.O.; Vincenzetti, V.; Martin, S.G. Fission yeast Sec3 and Exo70 are transported on actin cables and localize the exocyst complex to cell poles. *PLoS ONE* **2012**, *7*, e40248. [CrossRef]
12. Luo, G.; Zhang, J.; Guo, W. The role of Sec3p in secretory vesicle targeting and exocyst complex assembly. *Mol. Biol. Cell* **2014**, *25*, 3813–3822. [CrossRef]
13. He, B.; Xi, F.; Zhang, X.; Zhang, J.; Guo, W. Exo70 interacts with phospholipids and mediates the targeting of the exocyst to the plasma membrane. *EMBO J.* **2007**, *26*, 4053–4065. [CrossRef]
14. Yue, P.; Zhang, Y.; Mei, K.; Wang, S.; Lesigang, J.; Zhu, Y.; Dong, G.; Guo, W. Sec3 promotes the initial binary t-SNARE complex assembly and membrane fusion. *Nat. Commun.* **2017**, *8*, 14236. [CrossRef]
15. Sekereš, J.; Pleskot, R.; Pejchar, P.; Žárský, V.; Potocký, M. The song of lipids and proteins: Dynamic lipid-protein interfaces in the regulation of plant cell polarity at different scales. *J. Exp. Bot.* **2015**, *66*, 1587–1598. [CrossRef]
16. Elias, M.; Drdova, E.; Ziak, D.; Bavlnka, B.; Hala, M.; Cvrckova, F.; Soukupova, H.; Zarsky, V. The exocyst complex in plants. *Cell Biol. Int.* **2003**, *27*, 199–201. [CrossRef]
17. Synek, L.; Schlager, N.; Eliás, M.; Quentin, M.; Hauser, M.-T.; Zárský, V. AtEXO70A1, a member of a family of putative exocyst subunits specifically expanded in land plants, is important for polar growth and plant development. *Plant J.* **2006**, *48*, 54–72. [CrossRef]
18. Cvrčková, F.; Grunt, M.; Bezvoda, R.; Hála, M.; Kulich, I.; Rawat, A.; Zárský, V. Evolution of the land plant exocyst complexes. *Front. Plant Sci.* **2012**, *3*, 159. [CrossRef]
19. Zárský, V.; Cvrcková, F.; Potocký, M.; Hála, M. Exocytosis and cell polarity in plants–exocyst and recycling domains. *New Phytol.* **2009**, *183*, 255–272. [CrossRef]
20. Sekereš, J.; Pejchar, P.; Šantrůček, J.; Vukašinović, N.; Žárský, V.; Potocký, M. Analysis of Exocyst Subunit EXO70 Family Reveals Distinct Membrane Polar Domains in Tobacco Pollen Tubes. *Plant Physiol.* **2017**, *173*, 1659–1675. [CrossRef]
21. Simon, M.L.A.; Platre, M.P.; Assil, S.; van Wijk, R.; Chen, W.Y.; Chory, J.; Dreux, M.; Munnik, T.; Jaillais, Y. A multi-colour/multi-affinity marker set to visualize phosphoinositide dynamics in Arabidopsis. *Plant J.* **2014**, *77*, 322–337. [CrossRef]
22. Potocký, M.; Pleskot, R.; Pejchar, P.; Vitale, N.; Kost, B.; Zárský, V. Live-cell imaging of phosphatidic acid dynamics in pollen tubes visualized by Spo20p-derived biosensor. *New Phytol.* **2014**, *203*, 483–494. [CrossRef]
23. Platre, M.P.; Noack, L.C.; Doumane, M.; Bayle, V.; Simon, M.L.A.; Maneta-Peyret, L.; Fouillen, L.; Stanislas, T.; Armengot, L.; Pejchar, P.; et al. A Combinatorial Lipid Code Shapes the Electrostatic Landscape of Plant Endomembranes. *Dev. Cell* **2018**, *45*, 465–480. [CrossRef]
24. Wu, C.; Tan, L.; van Hooren, M.; Tan, X.; Liu, F.; Li, Y.; Zhao, Y.; Li, B.; Rui, Q.; Munnik, T.; et al. Arabidopsis EXO70A1 recruits Patellin3 to the cell membrane independent of its role as an exocyst subunit. *J. Integr. Plant Biol.* **2017**, *59*, 851–865. [CrossRef]

25. Zhang, X.; Oppenheimer, D.G. A simple and efficient method for isolating trichomes for downstream analyses. *Plant Cell Physiol.* **2004**, *45*, 221–224. [CrossRef]
26. Lee, B.H.; Weber, Z.T.; Zourelidou, M.; Hofmeister, B.T.; Schmitz, R.J.; Schwechheimer, C.; Dobritsa, A.A. Arabidopsis Protein Kinase D6PKL3 Is Involved in the Formation of Distinct Plasma Membrane Aperture Domains on the Pollen Surface. *Plant Cell* **2018**, *30*, 2038–2056. [CrossRef]
27. Ferguson, C.; Teeri, T.T.; Siika-aho, M.; Read, S.M.; Bacic, A. Location of cellulose and callose in pollen tubes and grains of Nicotiana tabacum. *Planta* **1998**, *206*, 452–460. [CrossRef]
28. Zhang, Y.; Zhu, H.; Zhang, Q.; Li, M.; Yan, M.; Wang, R.; Wang, L.; Welti, R.; Zhang, W.; Wang, X. Phospholipase dalpha1 and phosphatidic acid regulate NADPH oxidase activity and production of reactive oxygen species in ABA-mediated stomatal closure in Arabidopsis. *Plant Cell* **2009**, *21*, 2357–2377. [CrossRef]
29. Sang, Y.; Cui, D.; Wang, X. Phospholipase D and phosphatidic acid-mediated generation of superoxide in Arabidopsis. *Plant Physiol.* **2001**, *126*, 1449–1458. [CrossRef]
30. Torres, M.A.; Dangl, J.L. Functions of the respiratory burst oxidase in biotic interactions, abiotic stress and development. *Curr. Opin. Plant Biol.* **2005**, *8*, 397–403. [CrossRef]
31. Torres, M.A.; Dangl, J.L.; Jones, J.D.G. Arabidopsis gp91phox homologues AtrbohD and AtrbohF are required for accumulation of reactive oxygen intermediates in the plant defense response. *Proc. Natl. Acad. Sci. USA* **2002**, *99*, 517–522. [CrossRef]
32. Testerink, C.; Munnik, T. Phosphatidic acid: A multifunctional stress signaling lipid in plants. *Trends Plant Sci.* **2005**, *10*, 368–375. [CrossRef]
33. Bélanger, R.R. *The Powdery Mildews: A Comprehensive Treatise*; Amer Phytopathological Society: Amer, India, 2002.
34. Assaad, F.F.; Qiu, J.-L.; Youngs, H.; Ehrhardt, D.; Zimmerli, L.; Kalde, M.; Wanner, G.; Peck, S.C.; Edwards, H.; Ramonell, K.; et al. The PEN1 syntaxin defines a novel cellular compartment upon fungal attack and is required for the timely assembly of papillae. *Mol. Biol. Cell* **2004**, *15*, 5118–5129. [CrossRef]
35. Putta, P.; Rankenberg, J.; Korver, R.A.; van Wijk, R.; Munnik, T.; Testerink, C.; Kooijman, E.E. Phosphatidic acid binding proteins display differential binding as a function of membrane curvature stress and chemical properties. *Biochim. Biophys. Acta* **2016**, *1858*, 2709–2716. [CrossRef]
36. Zhao, Y.; Liu, J.; Yang, C.; Capraro, B.R.; Baumgart, T.; Bradley, R.P.; Ramakrishnan, N.; Xu, X.; Radhakrishnan, R.; Svitkina, T.; et al. Exo70 generates membrane curvature for morphogenesis and cell migration. *Dev. Cell* **2013**, *26*, 266–278. [CrossRef]
37. Marks, M.D.; Gilding, E.; Wenger, J.P. Genetic interaction between glabra3-shapeshifter and siamese in Arabidopsis thaliana converts trichome precursors into cells with meristematic activity. *Plant J.* **2007**, *52*, 352–361. [CrossRef]
38. Schindelin, J.; Arganda-Carreras, I.; Frise, E.; Kaynig, V.; Longair, M.; Pietzsch, T.; Preibisch, S.; Rueden, C.; Saalfeld, S.; Schmid, B.; et al. Fiji: An open-source platform for biological-image analysis. *Nat. Methods* **2012**, *9*, 676–682. [CrossRef]
39. Karimi, M.; Bleys, A.; Vanderhaeghen, R.; Hilson, P. Building blocks for plant gene assembly. *Plant Physiol.* **2007**, *145*, 1183–1191. [CrossRef]
40. Clough, S.J.; Bent, A.F. Floral dip: A simplified method for Agrobacterium-mediated transformation of Arabidopsis thaliana. *Plant J.* **1998**, *16*, 735–743. [CrossRef]

International Journal of
Molecular Sciences

MDPI

Article

Hydroxyproline-Rich Glycoproteins as Markers of Temperature Stress in the Leaves of *Brachypodium distachyon*

Artur Pinski [1], Alexander Betekhtin [1,*], Katarzyna Sala [2], Kamila Godel-Jedrychowska [2], Ewa Kurczynska [2,*] and Robert Hasterok [1]

1 Department of Plant Anatomy and Cytology, Faculty of Biology and Environmental Protection, University of Silesia in Katowice, 40-032 Katowice, Poland; apinski@us.edu.pl (A.P.); robert.hasterok@us.edu.pl (R.H.)

2 Department of Cell Biology, Faculty of Biology and Environmental Protection, University of Silesia in Katowice, 40-032 Katowice, Poland; katarzyna.sala@us.edu.pl (K.S.); kgodel@us.edu.pl (K.G.-J.)

* Correspondence: alexander.betekhtin@us.edu.pl (A.B.); ewa.kurczynska@us.edu.pl (E.K.); Tel.: +48-32-2009-484 (A.B.); +48-32-2009-447 (E.K.)

Received: 7 May 2019; Accepted: 23 May 2019; Published: 25 May 2019

Abstract: Plants frequently encounter diverse abiotic stresses, one of which is environmental thermal stress. To cope with these stresses, plants have developed a range of mechanisms, including altering the cell wall architecture, which is facilitated by the arabinogalactan proteins (AGP) and extensins (EXT). In order to characterise the localisation of the epitopes of the AGP and EXT, which are induced by the stress connected with a low (4 °C) or a high (40 °C) temperature, in the leaves of *Brachypodium distachyon*, we performed immunohistochemical analyses using the antibodies that bind to selected AGP (JIM8, JIM13, JIM16, LM2 and MAC207), pectin/AGP (LM6) as well as EXT (JIM11, JIM12 and JIM20). The analyses of the epitopes of the AGP indicated their presence in the phloem and in the inner bundle sheath (JIM8, JIM13, JIM16 and LM2). The JIM16 epitope was less abundant in the leaves from the low or high temperature compared to the control leaves. The LM2 epitope was more abundant in the leaves that had been subjected to the high temperatures. In the case of JIM13 and MAC207, no changes were observed at the different temperatures. The epitopes of the EXT were primarily observed in the mesophyll and xylem cells of the major vascular bundle (JIM11, JIM12 and JIM20) and no correlation was observed between the presence of the epitopes and the temperature stress. We also analysed changes in the level of transcript accumulation of some of the genes encoding EXT, EXT-like receptor kinases and AGP in the response to the temperature stress. In both cases, although we observed the upregulation of the genes encoding AGP in stressed plants, the changes were more pronounced at the high temperature. Similar changes were observed in the expression profiles of the EXT and EXT-like receptor kinase genes. Our findings may be relevant for genetic engineering of plants with increased resistance to the temperature stress.

Keywords: arabinogalactan proteins; *Brachypodium distachyon*; cell wall; extensins; immunohistochemistry; leaf; RT-qPCR; temperature stress

1. Introduction

Plant growth and productivity are compromised by various abiotic stresses, among which are high and low temperature stress. Even short periods of temperature stress may significantly decrease the yield, especially when it occurs during the crucial stages of plant development [1]. As was predicted, heat waves and other extreme temperature events are to become more intense, frequent and long-lasting because of global climate change [2]. Thus, thermal stresses must be better understood in the context of the response and adaptation of plants, which may enable crops with improved thermotolerance to be obtained and bred [3].

Brachypodium distachyon L. Beauv. (Brachypodium), which is a member of the Pooideae subfamily, is a wild annual grass species that is widespread in the regions of the Mediterranean basin, Western Europe, the Middle East, south-west Asia, north-east Africa, North and South America and Australia. It is closely related to many temperate zone key cereals, including wheat, barley, rye, oats and various forage grasses [4]. Due to its numerous favourable biological features such as a relatively small nuclear genome that ranges from 270 Mb to 350 Mb (depending on the methodology that is used), small stature, self-fertility, a life cycle of less than four months and undemanding growth requirements, *B. distachyon* constitutes an excellent model species [5].

Low-temperature stress results in the downregulation of many photosynthesis-related proteins and, at the same time, the upregulation of the proteins that are involved in reactive oxygen species (ROS) scavenging, redox adjustment, cytoskeletal rearrangements, cryoprotection and cell wall remodelling [6]. Similar results are observed in plants that are stressed by a high temperature [7]. Though the cell wall structure is not primarily altered under heat stress, numerous studies have indicated various changes in its architecture. In low temperature stress, changes in the cell wall rigidity may be an important factor in thermotolerance. Changes in the cell wall are more pronounced in roots because they are more sensitive to temperature stresses than the aerial parts of a plant, though the adverse effect of such stress on leaves directly affects plant productivity. Alterations in the cell wall in response to temperature stress concern cellulose and hemicelluloses biosynthesis, pectin modifications by pectin methylesterases, lignin biosynthesis and changes in the abundance of hydroxyproline-rich glycoproteins (HRGP) [8].

HRGP are usually divided into three complex multigene families: (i) arabinogalactan proteins (AGP); (ii) extensins (EXT); and (iii) proline-rich proteins [9]. AGP are further divided into four sub-families according to their polypeptide core: classical AGP, lysine-rich AGP (Lys-rich AGP), arabinogalactan peptides (AG peptides) and fasciclin-like AGP (FLA) [10]. Typically, AGP are strongly O-glycosylated and most of them have glycosylphosphatidylinositol (GPI) anchors that attach the proteins to the plasma membrane, though some of them can be released into the wall matrix *via* GPI cleavage [11]. In connection with their abundance, ubiquitous presence and localisation, AGP play a crucial role in various biological processes such as cell division, cellular communication, programmed cell death, organ abscission, plant-microbe interactions, plant defence and growth as well as in the reproductive processes [12–14]. A decrease in the amount of AGP has also been linked with the loss of embryogenic potential in callus cultures of *B. distachyon* [15]. Despite many studies on the role of AGP in plant development, our understanding of their role in the reaction of the plant to temperature stress is still quite limited. Recent studies have shown that temperature stress strongly affects the distribution and content of AGP in the stigma and ovule of *Solanum lycopersicum* as well as in banana leaves and roots, which may indicate that AGP are differentially regulated in the response to temperature stress and that their expression and distribution is tissue specific [16–18].

Based on a bioinformatic analysis, EXT were divided into seven classes: classical, short, leucine-rich repeat extensins (LRX), proline-rich extensin-like receptor kinases (PERK), formin-homolog EXT (FH EXT), chimeric and long chimeric EXT. EXT are characterised by the presence of serine, which is followed by three to five proline residues. These prolines are hydroxylated and glycosylated [19]. EXT are known to play important roles in the response to wounding and pathogen infections [20]. This family was also indicated as playing an important role in root-microbe interactions [14,21]. A study on a *B. distachyon* callus showed that one of the chimeric EXT could be considered to be a good marker for embryogenic cells [15]. A chimeric leucine-rich repeat/extensin, LRX1, was shown to be required for root hair morphogenesis in *Arabidopsis thaliana* [22]. However, information on the synthesis and location of extensins in response to temperature stress is scarce.

Thus, the aim of this work was to investigate any changes in the distribution of the epitopes of AGP and EXT in *B. distachyon* leaves through an immunostaining analysis. This approach enabled the distribution of these epitopes and the changes in their leaves that had been stressed by a high or low temperature to be determined. We also determined the level of transcript accumulation of selected genes encoding EXT, EXT-like receptor kinases, and FLA in the leaves of *B. distachyon* that had been stressed by a high or low temperature using RT-qPCR.

2. Results

2.1. Distribution of the Epitopes of AGP and EXT in Leaves in Response to Temperature Stress

The distribution of the epitopes of AGP (JIM8, JIM13, JIM16, LM2 and MAC207), pectin/AGP (LM6) and EXT (JIM11, JIM12 and JIM20) in the leaves of *B. distachyon* under normal (21 °C) and thermal stress conditions (4 and 40 °C) was determined. Considering the phenotype, we observed no visible changes induced by the thermal stress. The general anatomy of a *B. distachyon* leaf is shown in Figure 1. In order to present the results clearly, only the antibodies for which changes in their localisation or the intensity of the fluorescence signal in different temperature conditions were observed, are presented in the main text. Figures A1–A5 show the localisation of the remaining epitopes, in which no changes were identified in their response to temperature treatment.

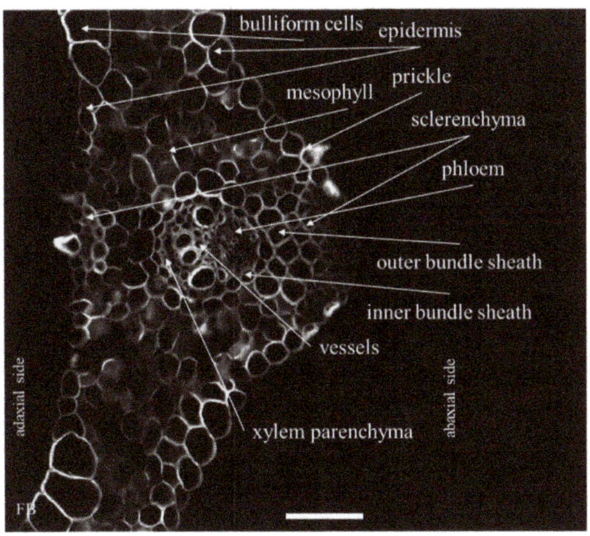

Figure 1. A cross-section of a *B. distachyon* leaf through the major vascular bundle (nomenclature according to Botha [23]) that had been stained with a fluorescent brightener (FB). Scale bar: 50 µm.

The occurrence of the epitopes of AGP was mostly associated with the vascular bundle. The JIM8 epitope was present in the walls and cellular compartments of the inner bundle sheath cells and phloem (Figure 2D–F). This epitope also occurred in the sclerenchyma fibres that were located next to the vascular bundle (Figure A1A–C) or were developing at the edge of the leaf blade (Figure A1D–F). The presence and spatial distribution of the JIM8 epitope were diverse at different temperatures. The JIM8 epitope was less represented in the leaves that were growing at a low temperature (Figure 2A–C). However, in the leaves that had been subjected to a temperature of 40 °C, an increase in the intensity of fluorescence signal was observed in the walls of phloem cells compared to the leaves that were growing at a low temperature (Figure 2G–I). JIM13 was found at the same locations as the JIM8 epitope (Figure A2A–F) and was additionally detected in the intercellular compartments (under intercellular compartments we define the localisation of epitope within the cytoplasm endomembrane system or organelles that are associated with the biosynthesis and secretion pathway to the wall, however these are not visible on the light microscope level [24]) of the prickles (Figure A2G–I). There were no changes in the distribution of the JIM13 epitope or in the intensity of the fluorescence signal between the analysed temperatures. The JIM16 epitope in the control leaves was present in a low amount in the intercellular compartments of the inner and outer bundle sheath cells and phloem as well as in the xylem parenchyma (Figure 3D–F). This epitope was not detected in the xylem parenchyma in the leaves that were growing at 40 °C (Figure 3G–I) and in the leaves from 4 °C, the presence of this epitope was not detected (apart from single dots in the vascular bundle cells; Figure 3A–C). Another AGP epitope, LM2, occurred in the cellular compartments of bundle sheath, phloem and xylem parenchyma in the control leaves (Figure 4D–F). At a low temperature (4 °C), the occurrence of this epitope was very low (Figure 4A–C), while in the leaves that were growing at a high temperature, it was more abundant (Figure 4G–I) compared to the control plants (Figure 4D–F). Additionally, LM2, was detected in the intercellular compartments and/or walls of the epidermis and bulliform cells in the leaves from plants that had been subjected to a high temperature (Figure 5A–I). A signal in the mesophyll cells was detected only in the leaves that were growing at 40 °C (Figure 5G–I). The MAC207 epitope was present in large amounts in the intercellular compartments and/or walls of the phloem cells, mesophyll cells and bulliform cells (Figure A3A–F). At all of the temperatures, its fluorescence had a similar cellular distribution and intensity. The LM6 epitope was detected abundantly in the phloem, xylem parenchyma and, in lower amounts, in the cellular compartments and/or walls of the outer bundle sheath (Figure 6D–F). The fluorescence signal of this epitope in the leaves that were growing at 40 °C was more intense compared to the other temperature treatments (Figure 6G–I vs. Figure 6A–F). Outside the vascular bundle, LM6 was present in the cell walls and/or in the intercellular compartments of the mesophyll cells (Figure A4A–C).

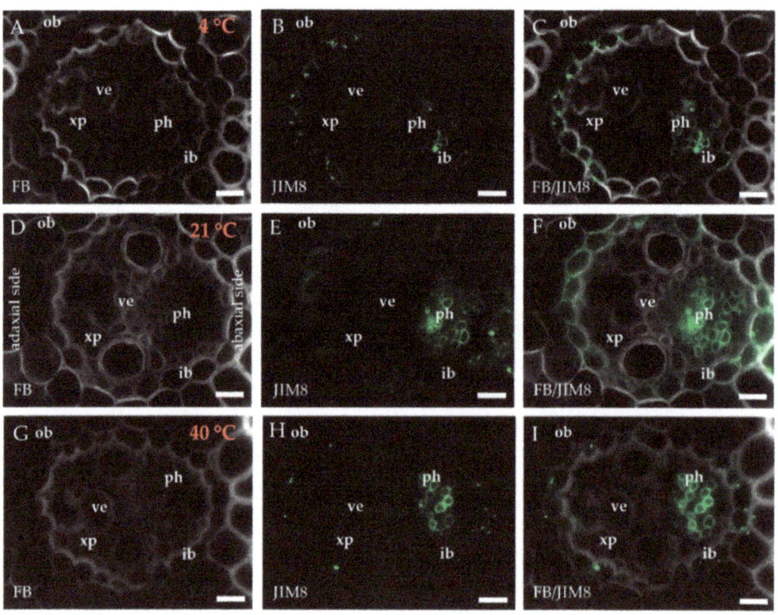

Figure 2. Immunolocalisation of the JIM8 epitope (**A–I**) in cross-sections of the *B. distachyon* leaves, (**A–I**): through the major vascular bundle. (**A–C**): 4 °C; (**D–F**): 21 °C; (**G–I**): 40 °C. Abbreviations: FB—fluorescent brightener, ib—inner bundle sheath, ob—outer bundle sheath, ph—phloem, ve—vessels, xp—xylem parenchyma. The green colour shows epitope occurrence. Scale bars: 10 μm.

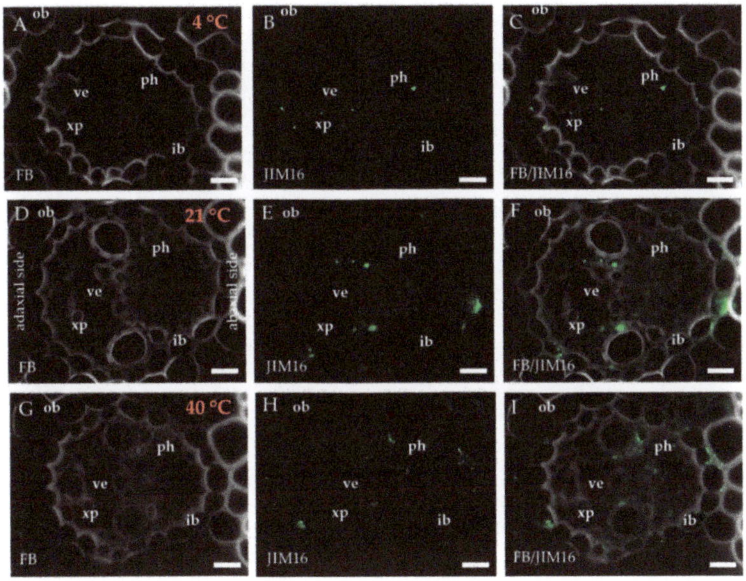

Figure 3. Immunolocalisation of the JIM16 epitope (**A–I**) in cross-sections of the *B. distachyon* leaves, (A-I): through the major vascular bundle. (**A–C**): 4 °C; (**D–F**): 21 °C; (**G–I**): 40 °C. Abbreviations: FB—fluorescent brightener, ib—inner bundle sheath, ob—outer bundle sheath, ph—phloem, ve—vessels, xp—xylem parenchyma. The green colour shows epitope occurrence. Scale bars: 10 μm.

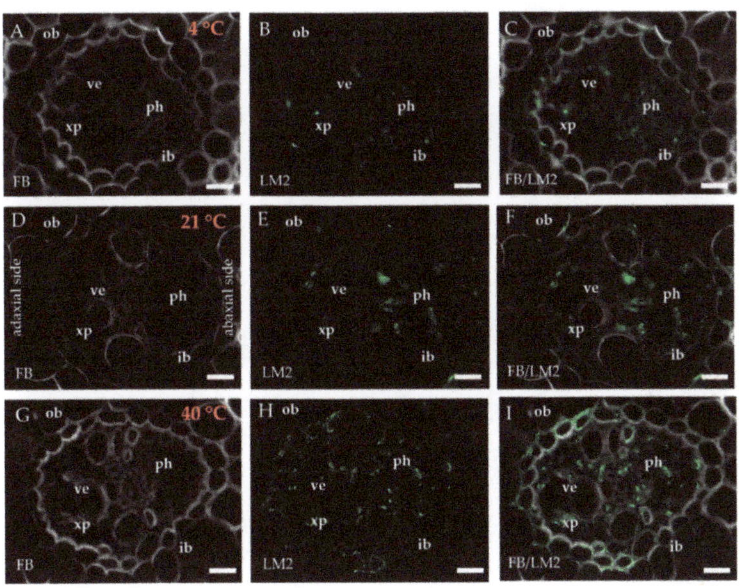

Figure 4. Immunolocalisation of the LM2 epitope (**A–I**) in cross-sections of the *B. distachyon* leaves, (**A–I**): through the major vascular bundle. (**A–C**): 4 °C; (**D–F**): 21 °C; (**G–I**): 40 °C. Abbreviations: FB—fluorescent brightener, ib—inner bundle sheath, ob—outer bundle sheath, ph—phloem, ve—vessels, xp—xylem parenchyma. The green colour shows epitope occurrence. Scale bars: 10 μm.

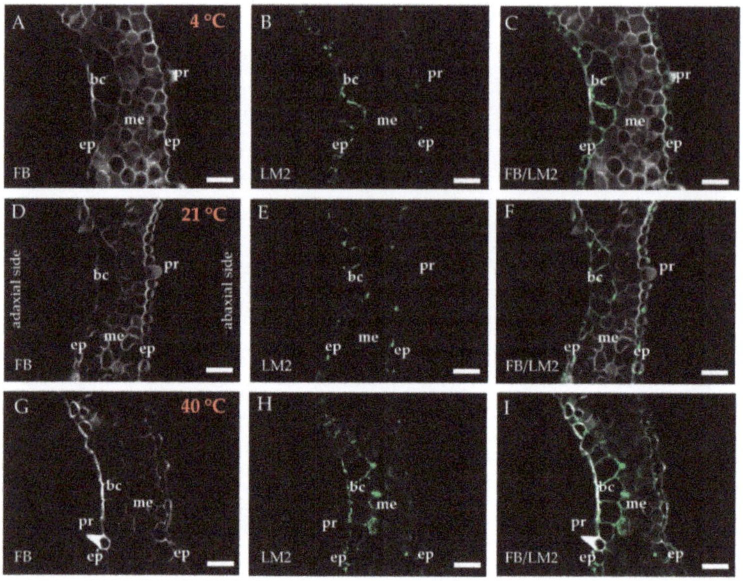

Figure 5. Immunolocalisation of the LM2 epitope (**A–I**) in cross-sections of the *B. distachyon* leaves, (**A–I**): through the mesophyll and bulliform cells. (**A–C**): 4 °C; (**D–F**): 21 °C; (**G–I**): 40 °C. Abbreviations: bc—bulliform cells, ep—epidermis, FB—fluorescent brightener, me—mesophyll, pr—prickle. The green colour shows epitope occurrence. Scale bars: 20 μm.

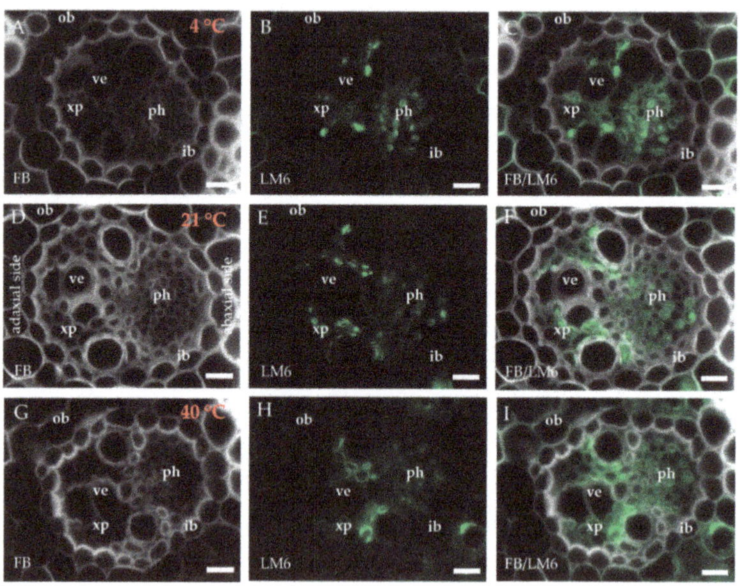

Figure 6. Immunolocalisation of the LM6 epitope (**A–I**) in cross-sections of the *B. distachyon* leaves, (**A–I**): through the major vascular bundle. (**A–C**): 4 °C; (**D–F**): 21 °C; (**G–I**): 40 °C. Abbreviations: FB—fluorescent brightener, ib—inner bundle sheath, ob—outer bundle sheath, ph—phloem, ve—vessels, xp—xylem parenchyma. The green colour shows epitope occurrence. Scale bars: 10 μm.

All three extensin epitopes that are recognised by the JIM11, JIM12 and JIM20 antibodies were mostly associated with the mesophyll cells. The JIM11 epitope was present only in the mesophyll cell walls (Figure A5A–C). In addition to occurring in the mesophyll cell walls (Figure A5D–F), JIM12 was also found in the walls of some of the outer bundle sheath cells and vessels (Figure A5G–I). The occurrence of the JIM20 epitope was similar to JIM12 (Figure A5J–O), but had an additional location in the walls and/or cellular compartments of the phloem (Figure A5M–O). All three extensin epitopes occurred abundantly and there were no differences in their distribution or in the intensity of the fluorescence signal among the temperatures that were analysed.

2.2. Analysis of the Level of Transcript Accumulation of the Genes Encoding the FLA, EXT and EXT-Like Receptor Kinases

In this study, we determined the level of transcript accumulation of five different genes encoding *FLA* (*Bradi4g34420, Bradi2g00220, Bradi5g18950, Bradi3g39740* and *Bradi2g60270*). The transcript accumulation levels of *Bradi4g34420* and *Bradi2g00220* increased in both temperatures, 4 and 40 °C, compared to the control conditions (Figure 7A). The increase in expression of the *Bradi4g34420* gene at 40 °C (4.7-fold) was higher than at 4 °C (1.9-fold) (Figure 7A). In the case of the *Bradi5g18950* gene, the expression at 4 °C was approximately the same as in the control, while its expression at 40 °C was 6-fold higher (Figure 7A). A similar pattern of expression was found for the *Bradi3g39740* gene, though there was only a slight (1.7-fold) increase in its expression at 4 °C (Figure 7B). Interestingly, there was a dramatic increase (28-fold) in the expression of this gene at 40 °C. Notably, the expression of the *Bradi2g60270* gene was only detectable in the leaves at 40 °C. Generally, temperature stress resulted in a higher level of transcript accumulation of *FLA*, though the increase was more pronounced at 40 °C.

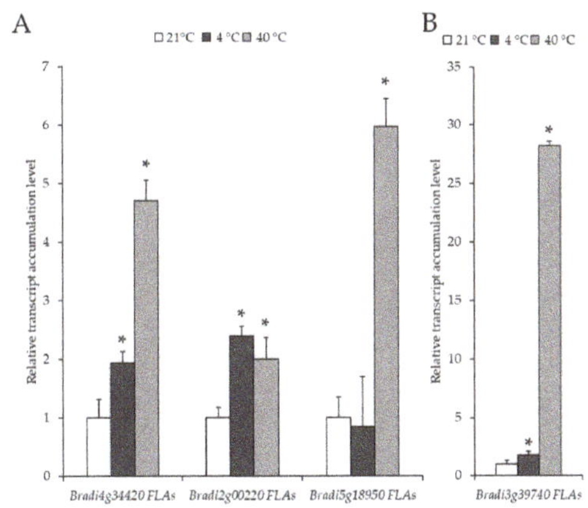

Figure 7. Relative level of transcript accumulation of the fasciclin-like AGP (*FLA*) genes: (**A**) *Bradi4g34420*, *Bradi2g00220* and *Bradi5g18950* and (**B**) *Bradi3g39740*. The relative expression levels were normalised to an internal control (*Bradi1g32860*, gene encoding ubiquitin) and calibrated to the control. Asterisks indicate significant differences from the control using the Student's t-test ($p < 0.05$; mean ± SD, $n = 3$).

The level of transcript accumulation of nine different genes encoding EXT and EXT-like receptor kinases were also determined. Each gene was assigned to a group based on its structure: *FH EXT* (*Bradi1g22980, Bradi3g59780* and *Bradi4g03720*), *chimeric EXT* (*Bradi4g11250* and *Bradi3g12902*) and *PERK EXT* (*Bradi2g00900, Bradi2g49240, Bradi1g07010* and *Bradi3g31967*). The level of transcript accumulation of the *Bradi1g22980* gene in the treated plants was not statistically different from the control (Figure 8A). The level of transcript accumulation of two other *FH EXT*, *Bradi3g59780* and *Bradi4g03720* was only statistically higher at 40 °C (Figure 8A). Conversely, the level of transcript accumulation of the *chimeric*

EXT, *Bradi4g11250*, increased significantly at 4 °C, though there was no clear difference in its expression for the *Bradi3g12902* gene (Figure 8B). When considering *PERK*, a higher transcript accumulation of the *Bradi2g00900* gene at 4 °C and a higher level of transcript accumulation of the *Bradi2g49240* gene at 4 °C and 40 °C was determined (Figure 8C). Intriguingly, the expression of the other *PERK* gene, *Bradi3g31967*, was only observed in the temperature-stressed samples (Figure 8D). The distribution of all of the epitopes together with changes in the level of transcript accumulation of the analysed genes are summarised in Figure 9.

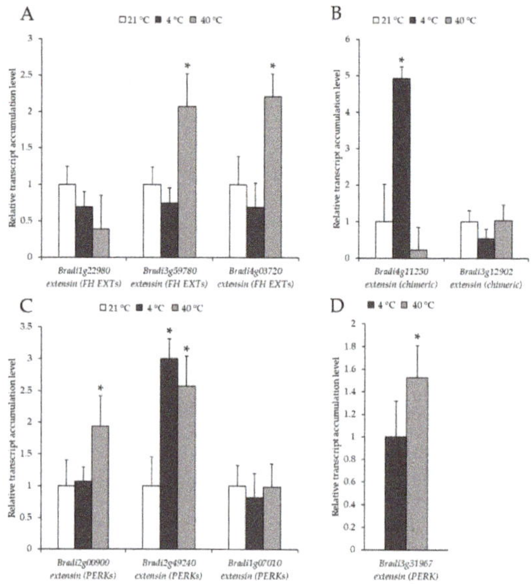

Figure 8. Relative level of transcript accumulation of the extensins (EXT) genes: (**A**) formin-homolog (FH) EXT: *Bradi1g22980, Bradi3g59780, Bradi4g03720*, (**B**) chimeric EXT: *Bradi4g11250, Bradi3g12902*, (**C**) proline-rich extensin-like receptor kinase (PERK): *Bradi2g00900, Bradi2g49240, Bradi1g07010* and (**D**) *Bradi3g31967*. The relative expression levels were normalised to an internal control (*Bradi1g32860*, gene encoding ubiquitin) and calibrated to the control. Asterisks indicate significant differences from the control using the Student's t-test ($p < 0.05$; mean ± SD, $n = 3$).

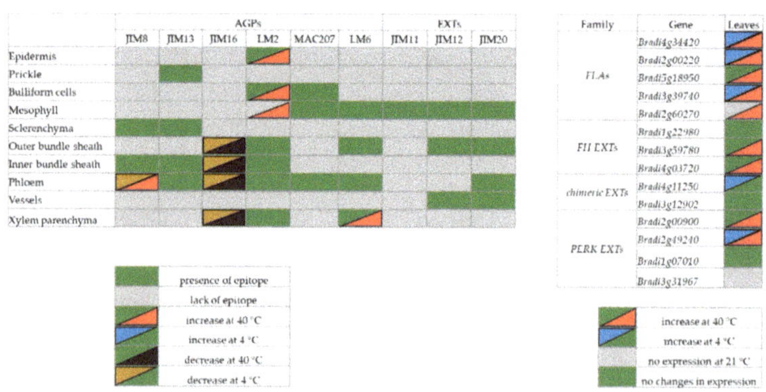

Figure 9. Consolidated results of the distribution of the epitopes EXT and AGP in the leaves of *B. distachyon* and changes in the level of transcript accumulation of the analysed genes.

3. Discussion

Although immunohistochemical analyses are widely used to study changes in the chemical components of the cell wall during different developmental processes, in vivo and in vitro information concerning the presence and distribution of the cell wall proteins in leaves that have been subjected to biotic and/or abiotic stresses are scarce and remain largely unexplored [16,18,25]. Previous studies have primarily focused on the differential expression of AGP in response to temperatures stresses in roots and seedlings; however, the involvement of AGP in the response to temperature stress has rarely been studied in leaves [16]. For example, the transient appearance of two AGP proteins in *Triticum aestivum* in response to a low temperature were observed, thus indicating their involvement in the activation of the plant cell defence [16,26]. In transgenic *A. thaliana* plants, non-classical AGP improved the freezing tolerance of seedlings [27]. Temperature stress is one of the factors that limit plant growth and productivity [2,28]. Therefore, data showing changes in the distribution of individual cell wall components, particularly AGP, in connection with temperature stress, are particularly important as the results can be used in genetic engineering for stress tolerance [29].

In the present study, while the distribution of the epitopes of the analysed AGP and EXT was primarily observed in the major vascular bundles (nomenclature according to Botha [23]) and sclerenchyma cells, in the case of the LM2 and MAC207 epitopes, they were present in the mesophyll and epidermal cells, especially in the bulliform cells. Generally, the results for *B. distachyon* presented here are similar to those that have been described for banana leaves in terms of the distribution of epitopes in leaf tissues [18]. In banana, the JIM8 epitope increased in abundance at lower temperatures, thus indicating its role in the tolerance to strong chilling stress [18]. In the pistils of *Solanum lycopersicum* cv Micro-Tom, a high temperature strongly affected the distribution of the JIM8 epitope, which decreased in the stigma and ovule [17]. These varied results with respect to this and other epitopes mean that further intensive studies are necessary. Moreover, such results may indicate that the changes in the chemical composition of the cell walls in response to temperature stress are species-specific. In banana leaves, the presence of the JIM16 epitope was higher in a tolerant cultivar at low temperature stress, thus suggesting that these epitopes may be involved in determining the tolerance of banana to temperature stress [18]. Among the analysed epitopes, the LM2 epitope was detected in most of the leaf tissues and an increase of the LM2 epitope as a response to high temperature stress was observed. A similar distribution was detected in banana leaves, in which this epitope was found in the phloem, bundle sheath, mesophyll and epidermal cells. Low temperature treatment increased the abundance of this epitope in banana [18]. In the leaves of *Tilia x euchlora*, the LM2 epitope was present in the epidermis, hypodermis and parenchyma cells, although as a response to salt stress [25]. Such results may indicate that this wall epitope may be a marker of the plant response to diverse stresses. The abundant presence of the LM6 antibody in *B. distachyon* leaves was found. As LM6 antibody exhibits a high affinity to the (1-5)-α-L-arabinosyl residues, it detects the (1-5)-α-L-arabinan (a pectin rhamnogalacturonan I side chain), however, it can also bind to some AGP (http://www.plantprobes.net/index.php). Thus, the increase in the fluorescence signal in the xylem parenchyma of the temperature-stressed plants may not necessarily indicate changes in the presence of AGP. The function of the arabinan side chains is not well understood and their roles are postulated to be an involvement in the rehydration of the cell wall and flexibility [30–33]. The more abundant presence of the wall components that are recognised by LM6 antibody, especially in the cytoplasmic compartments, that were observed in our study indicate that leaves react to temperature stress by synthesising and depositing (1-5)-α-L-arabinans into the cell walls.

As has been shown in previous studies, analyses of the distribution and changes in the signal intensity of epitopes can be compared with the expression profiles of the genes encoding the proteins that are targeted by these antibodies [15,34]. In this work, we focused on the FLA that have been implicated in modulation of signalling upstream of cell wall polymer biosynthesis, remodelling, as well as in the stress response as one of the AGP sub-families [35–37]. We found an increase in the level of transcript accumulation of the *FLA* in response to temperature stress, which concurs with the immunohistochemistry observations that have been made for the LM2 antibody. It is worth noting that the increase in the level of transcript accumulation of these genes was more pronounced at the high temperature. However, in wheat, four genes encoding *FLA* had a decreased level of transcript accumulation in response to low temperature stress [38]. Similarly, two other *FLA* genes (*OsFLA1* and *OsFLA4*) were downregulated by cold stress in rice [39]. This may reflect intrinsic differences between the analysed species or might be the result of fragmentariness of the conducted experiments, which only focus on a few of the numerous *FLA* genes that are present in a genome. The *Bradi5g18950* gene, which was analysed in our work, exhibited a higher level of transcript accumulation at the high temperature and it was previously shown to be upregulated in a 30-day-old callus that was characterised by an increased embryogenic potential. Conversely, the *Bradi3g39740* gene was linked with a gradual loss of embryogenic potential in *B. distachyon* [15]. Temperature stresses that are induced by low or high temperatures inhibit water uptake, which immediately leads to a slowing down of leaf growth. This observation was correlated with a loss of turgor in leaf cells and an adjustment of the osmoticum, which enables cells to regain turgor [40]. AGP are well known for their water-holding properties [41]. For example, in the resurrection plants, side chains of pectin are highly enriched in arabinose-rich polymers, including AGP. Their presence can prevent water loss during desiccation [8]. In *Coffea arabica* plants that had been subjected to heat stress, there were extensive changes in the cell wall of the leaves. The plants accumulated a higher content of arabinose and galactose, which may suggest that the response of coffee leaves to heat stress is related to type II arabinogalactans and pectins. Moreover, during heat stress, the palisade parenchyma cells were more separated and thinner relative to the control, which resulted in a decreased thickness of the leaves [42]. It has been hypothesised that the organs that are susceptible to water loss such as leaves increase the thickness of the cell walls, thereby limiting desiccation through the production of specific molecules such as AGP [16]. Our results seem to support this hypothesis.

Similar to the temperature stress, salt stress results in a decrease of available water due to a reduction in osmotic potential of the soil solution, which leads to a water deficit [8]. AGP have also been shown to play an important part in the salt stress response and an upregulation of *AGP* in salt-adapted tobacco BY-2 cell cultures was observed. It has been proposed that AGP act as a possible sodium carrier via vesicle trafficking from the apoplast to the vacuoles in salt-adapted tobacco BY-2 cells [43]. A significant upregulation of the *FLA* genes in the salt stress response was observed in the roots of *Populus trichocarpa* [44]. Moreover, AGP were found to act as pectin plasticisers [8,45]. As was shown for an *FLA sos5* (*salt-overly sensitive*) mutant of *A. thaliana*, it exhibits a root-swelling phenotype under salt stress [46]. Further studies showed that the SOS5 protein mediates adherence via its interaction with the cell wall pectin [47]. *At-FLA4* is one of the *FLA* genes in *A. thaliana* that encodes the predicted lipid-anchored glycoprotein and it was shown to positively regulate cell wall biosynthesis and root growth by modulating abscisic acid signalling. Moreover, an *At-fla4* mutant was found to be sensitive to the salt stress [37]. It has been suggested that *At-fla4* might interact with the pectin network via the covalent or non-covalent interactions of its glycans. *At-fla4* may mechanically link pectin with the *AtFei1* and *AtFei2* receptor kinases and the plasma membrane, thus contributing to some biophysical properties such as swelling and interpolymer connectivity [48]. As was indicated by another inactivated mutant of *A. thaliana*, the *FLA1* gene is involved in the early events of lateral root and shoot development in tissue cultures [49]. Considering the salt stress response, it is possible that the upregulation of *FLA* and the increased signal intensity of some of the epitopes of AGP during temperature stress may link with other cell wall polymers such as pectins and thus may modulate the signalling pathways. As has been shown by a number of studies on various species, the pectin content increases during cold stress. Conversely, it decreases during heat stress [8].

An immunohistochemical analysis of the distribution of the *EXT* using the JIM11, JIM12 and JIM20 antibodies showed the presence of all of these epitopes in the mesophyll and the JIM12 and JIM20 epitopes in the vessels. Although no immunohistochemical studies targeting the EXT using the JIM11, JIM12 and JIM20 antibodies were done in the leaves, the distribution of these epitopes has been widely studied in the callus embryos and roots [20,50–53]. For example, changes in the signal intensity of the JIM11 and JIM12 antibodies were connected with a gradual loss of embryogenic potential in *B. distachyon* callus cultures [15]. Zhang, et al. [54] showed that the EXT that are recognised by the JIM20 antibody were present in the pollen tubes and transmitting tissue of *Nicotiana tabacum* and that the application of hydroxyproline synthesis inhibitor, 3,4-dehydro-L-proline, decreased pollen tube growth. However, studies dedicated to the role of EXT in abiotic stress and especially temperature stress are still scarce. In our study, while we did not observe any changes in the selected epitopes of the EXT in the mesophyll, outer bundle sheath, phloem or vessels at the level of the immunohistochemical analyses, we found an increase of *EXT* and *EXT-like receptor kinase* level of transcript accumulation in the response to temperature stresses, especially during the high temperature stress. This may be partially explained by the fact that we used only three antibodies that bind to the EXT, but it is possible that the application of other anti-EXT antibodies could reveal some changes. Additionally, the effectiveness of immunohistochemical analyses is limited, since it does not provide sufficient resolution and does not focus on individual genes, as is the case of RT-qPCR-based analyses. Changes in the *EXT* gene expression greatly depended on the class of extensins that were being analysed as was shown for *A. thaliana* plants that had been subjected to low temperature stress [55]. Another transcriptomic analysis of the *A. thaliana* response to cold stress showed the downregulation of one of the *PERK* genes [56]. In our experiment, we found that the *PERK extensin*, *Bradi3g31967*, was only expressed in the stressed leaves. The genes that belong to this class have been found in the apical dominance, floral organ defects and root cell elongation [57]. An increased accumulation level of the *PERK4* gene transcript in *A. thaliana* was observed in response to abscisic acid, which is a key regulator of abiotic stress tolerance in plants [58,59]. *PERK1* mRNA from *Brassica napus* was shown to be dramatically and rapidly accumulated in response to wounding and moderately accumulated in response to infection by the fungal pathogen *Sclerotinia sclerotiorum* [60]. Interestingly, the LRX proteins were found to regulate salt tolerance in *A. thaliana*. A triple mutant in the *LRX* genes exhibited a severe salt hypersensitivity and these genes were determined to be an important sensor of the cell wall integrity signals [61]. Moreover, recent studies have hinted at the role of EXT in the plant defence against phytopathogens as well as in interactions with beneficial microorganisms [14,21,62]. EXT have also been implicated in aluminium resistance and its accumulation in the cell walls of pea roots was observed [63]. The changes in the plant cell wall in response to temperature stress are diverse and not only include AGP and EXT, but also alterations in cellulose, hemicellulose, pectin and lignin biosynthesis [8]. Further investigations into the changes in cell wall proteomes could unravel the involvement of other proteins in the stress response because the proteome of the *B. distachyon* cell walls is complex and consists of at least 594 proteins [64–68].

4. Materials and Methods

4.1. Plant Material

The plants of the *B. distachyon* reference genotype Bd21 that were used is this experiment were cultivated in pots that had been filled with soil mixed with vermiculite at a ratio of 3:1 in a greenhouse. The seeds of *B. distachyon* genotype Bd21 (accession number: PI 254867) were sourced from the collection held by the United States Department of Agriculture—National Plant Germplasm System. The plants were grown in the greenhouse under a 16 h/8 h light/dark photoperiod at 21 ± 1 °C and were illuminated by lamps emitting white light at an intensity of 10 000 lx. For the low temperature stress, the plants were incubated at 4 °C for 24 h and for the high temperature stress, the plants were incubated at 40 °C for 24 h in growing chambers [69]. Plants at the fourth stage of principal growth according to the Hong, et al. [70] were used in this experiment. This stage is referred to as booting and is characterised by the emergence of the head at the top of the growing shoot. The flag leaf was harvested and used to isolate the RNA and to perform the RT-qPCR analysis. For the immunohistochemistry analysis, the middle part of the leaf was collected because this permitted clear observations of the major vascular bundle, epidermis, bulliform cells, mesophyll and sclerenchyma.

4.2. Sample Preparation

To determine the chemistry of the cell wall, a set of monoclonal antibodies against the specific cell wall epitopes of the AGP (antibodies JIM8, JIM13, JIM16, LM2, MAC207), pectin/AGP (LM6) and EXT (JIM11, JIM12 and JIM20) (Plant Probes, Leeds, UK) were used. The references and information on the antibodies are shown in Table 1. The leaves were excised, fixed and embedded in Steedman's wax [20,71]. Transverse sections of the leaf blade (7 μm thick) were cut using a HYRAX M40 rotary microtome (Zeiss, Oberkochen, Germany) and collected on microscopic slides coated with poly-L-lysine (Menzel Gläser, Braunscheig, Germany).

4.3. Immunohistochemistry

The sections were de-waxed and rehydrated in an ethanol series (three times in 100, 90 and 50% ethanol in phosphate buffered saline PBS, *v*/*v*, each for 10 min) and PBS (10 min) [71]. The detailed procedure for immunochemical analysis and histological section observation was as previously described [20]. The slides were stained with 0.01% (*w*/*v*) fluorescent brightener 28 (FB) (Sigma-Aldrich, St. Louis, MO, USA) in PBS, which was used to visualise cell walls due to its affinity to cellulose. Two biological replicates were performed with at least eight sections for each replicate.

4.4. RT-qPCR

In order to characterise the level of transcript accumulation of the selected genes, RT-qPCR was performed using a LightCycler® 480 SYBR Green I Master in a LightCycler® 480 Real-Time PCR System (Roche, Basel, Switzerland). The total RNA was isolated from the leaves of *B. distachyon*. The primers used in this research are shown in Table A1. The genes encoding extensins with their division into classes were as previously described [19]. The *FLA* genes were selected based on the annotation found in the Phytozome database (https://phytozome.jgi.doe.gov/pz/portal.html). The detailed procedure for RT-qPCR was as in Betekhtin, et al. [72]. Briefly, the isolated RNA were treated with the DNase (QIAGEN, Hilden, Germany), and subsequently used for first-strand cDNA generation. Samples were run in the LightCycler® 480 Real-Time PCR System (Roche, Basel, Switzerland). The PCR conditions were as follow: 5 min at 95 °C, 45 cycles of 10 s at 95 °C, 20 s at 60 °C and 10 s at 72 °C with signal acquisition. Ubiquitin was used as the reference gene and analysis was performed using the $2^{-\Delta\Delta CT}$ method. The significant differences between the samples and control were calculated using the Student's *t*-test.

Table 1. The antibodies that were used for the immunocytochemistry, the epitopes they recognise and the relevant references.

Antibody	Epitope	References
	AGP	
JIM8	Arabinogalactan	[73]
JIM13	(β)GlcA1->3(α)GalA1->2Rha	[74–76]
JIM16	AGP glycan	[74–76]
LM2	β-linked GlcA	[75,77]
MAC207	(β)GlcA1->3(α)GalA1->2Rha	[74,75,78,79]
	Pectin/AGP	
LM6	(1-5)-α-L-arabinosyl residues, can also bind to some AGP	[80,81]
	EXT	
JIM11	Extensin	[74,82]
JIM12	Extensin	[82]
JIM20	Extensin	[82]

5. Conclusions

In our work, we demonstrated changes in the abundance and diversified expression of the epitopes of the AGP genes encoding FLA, EXT and EXT-like receptor kinases in the leaves of *B. distachyon* in response to temperature stress.

The main findings are as follows:

1. An increase in the JIM8 signal in the walls of phloem cells at 40 °C and a decrease at 4 °C.
2. An increase in the abundance of the LM2 epitope in the leaves of plants that had been subjected to the high temperature.
3. A decrease in the JIM16 signal intensity at 4 and 40 °C.
4. The upregulation of some *FLA*, *EXT* and *EXT-like* receptor *kinases* genes in response to temperature stress (4 and 40 °C).
5. The expression of the *PERK EXT* gene *Bradi3g31967* only in the leaves under low and high temperature stress.

To summarise, our results extend the knowledge about the presence of these epitopes in connection with temperature stress in *B. distachyon*. A precise dissection of the functions of AGP and EXT in response to abiotic stresses and to temperature stress, among others, requires the use of specific mutants whose availability is still limited. However, recent developments in the site-directed mutagenesis techniques such as CRISPR/Cas9 should allow the selective targeting of these genes, which may be helpful to better understand their roles.

Author Contributions: Conceptualisation, A.P., A.B.; methodology, A.P., A.B., K.S., K.G.-J.; validation, A.P., K.S., A.B.; formal analysis, A.P., K.S., A.B.; investigation, A.P., K.S., A.B.; resources, R.H., E.K.; data curation, A.P., A.B., K.S.; writing—original draft preparation, A.P., A.B., K.S., E.K.; writing—review and editing, A.P., A.B., K.S., E.K., R.H.; visualisation, A.P.; supervision, A.B., R.H.; funding acquisition, R.H., E.K.

Funding: This research was funded by the National Science Centre Poland (grant DEC-2014/14/M/NZ2/00519).

Conflicts of Interest: The authors declare no conflict of interest.

Abbreviations

AG Peptides	Arabinogalactan peptides
AGP	Arabinogalactan proteins
EXT	Extensins
FH EXT	Formin-homolog EXT
FLA	Fasciclin-like AGP
GPI	Glycosylphosphatidylinositol
HRGP	Hydroxyproline-rich glycoproteins
LRX	Leucine-rich repeat extensins
Lys-rich AGP	Lysine-rich arabinogalactan proteins
PBS	Phosphate-buffered saline
PERK	Proline-rich extensin-like receptor kinases
ROS	Reactive oxygen species
RT-qPCR	Reverse transcription-quantitative polymerase chain reaction

Appendix A

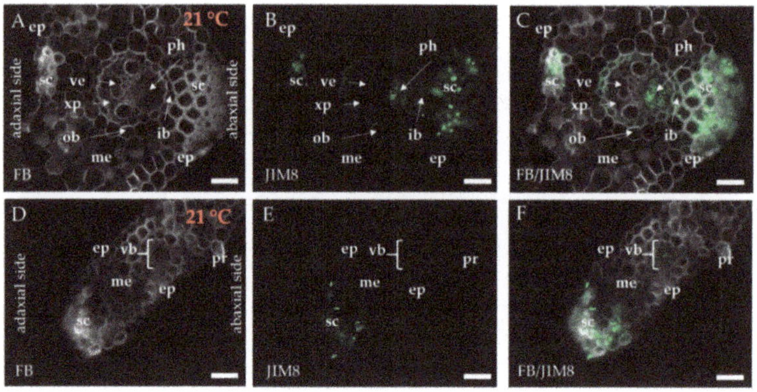

Figure A1. Immunolocalisation of the JIM8 epitope (**A–F**) in cross-sections of *B. distachyon* leaves (21 °C), (**A–C**): through the major vascular bundle; (**D–F**): through the edge of the leaf blade. The white arrows point to the respective parts of the leaf. Abbreviations: ep—epidermis, FB—fluorescent brightener, ib—inner bundle sheath, me—mesophyll, ob—outer bundle sheath, ph—phloem, pr—prickle, sc—sclerenchyma, vb—vascular bundle, ve—vessels, xp—xylem parenchyma. The green colour shows epitope occurrence. Scale bars: 20 μm.

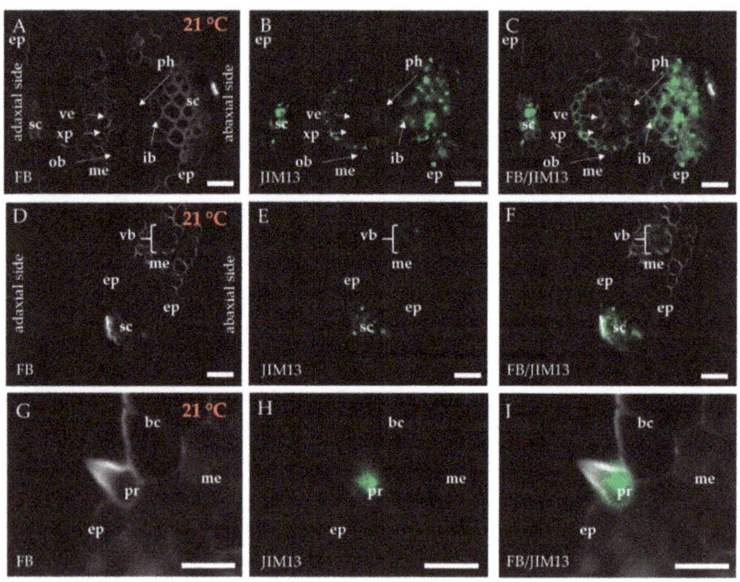

Figure A2. Immunolocalisation of the JIM13 epitope (**A–I**) in cross-sections of *B. distachyon* leaves (21 °C), (**A–C**): through the major vascular bundle; (**D–F**): through the edge of the leaf blade; (**G–I**): prickle. The white arrows point to the respective parts of the leaf. Abbreviations: bc—bulliform cells, ep—epidermis, FB—fluorescent brightener, ib—inner bundle sheath, me—mesophyll, ob—outer bundle sheath, ph—phloem, pr—prickle, sc—sclerenchyma, vb—vascular bundle, ve—vessels, xp—xylem parenchyma. The green colour shows epitope occurrence. Scale bars: 20 μm.

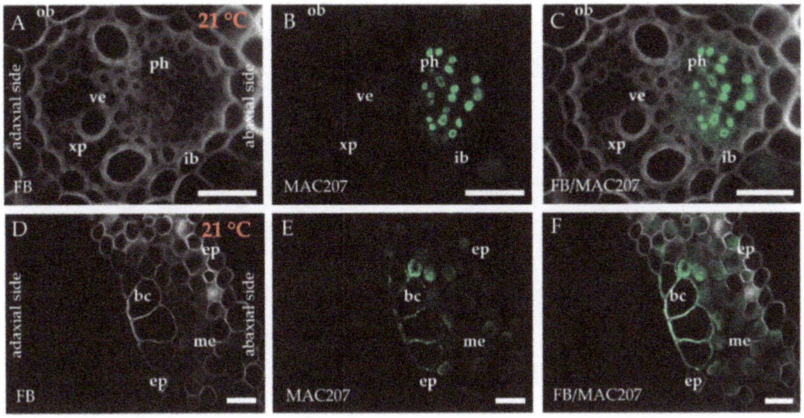

Figure A3. Immunolocalisation of the MAC207 epitope in cross-sections of *B. distachyon* leaves (21 °C), (**A–C**): through the major vascular bundle; (**D–F**): through the mesophyll and bulliform cells. Abbreviations: ep—epidermis, b—bulliform cells, FB—fluorescent brightener, ib—inner bundle sheath, me—mesophyll, ob—outer bundle sheath, ph—phloem, ve—vessels, xp—xylem parenchyma. The green colour shows epitope occurrence. Scale bars: 20 μm.

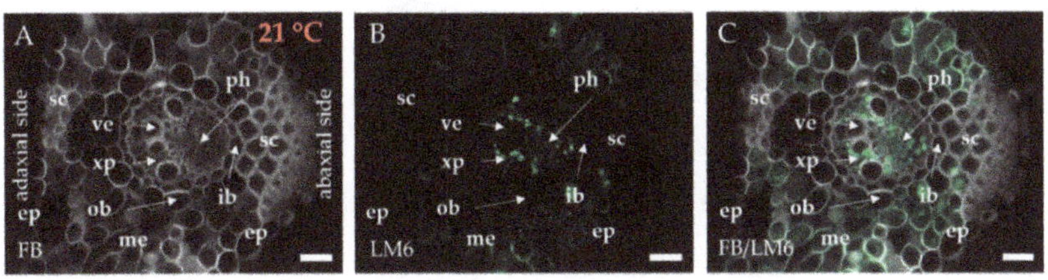

Figure A4. Immunolocalisation of the LM6 epitope (**A–C**) in cross-sections of *B. distachyon* leaves (21 °C), through the major vascular bundle. The white arrows point to the respective parts of the leaf. Abbreviations: ep—epidermis, FB—fluorescent brightener, ib—inner bundle sheath, me—mesophyll, o—outer bundle sheath, ph—phloem, sc—sclerenchyma ve—vessels, xp—xylem parenchyma. The green colour shows epitope occurrence. Scale bars: 20 μm.

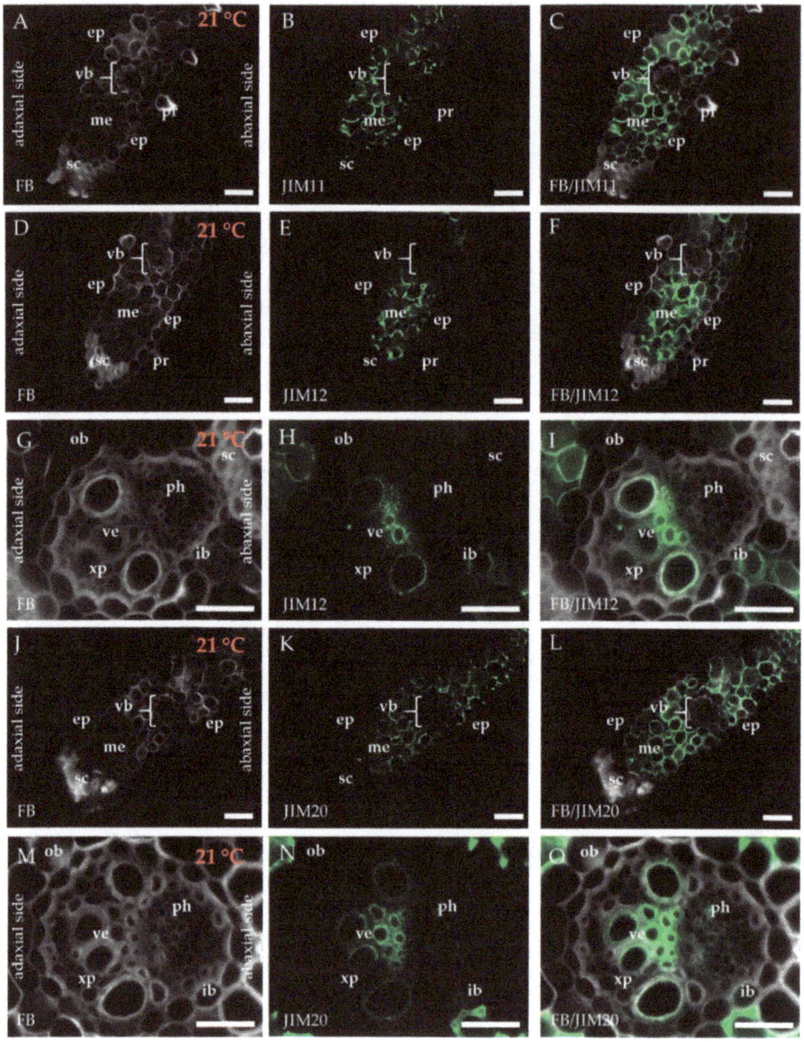

Figure A5. Immunolocalisation of the JIM11 (**A–I**), JIM12 (**D–I**) and JIM20 (**J–O**) epitopes in cross-sections of *B. distachyon* leaves (21 °C), (**A–F**,**J–L**): through the edge of the leaf blade; (**G–I**,**M–O**): through the major vascular bundle. Abbreviations: bc—bulliform cells, ep—epidermis, FB—fluorescent brightener, ib—inner bundle sheath, me—mesophyll, ob—outer bundle sheath, p— phloem, pr—prickle, sc—sclerenchyma, vb—vascular bundle, ve—vessels, xp—xylem parenchyma. The green colour shows epitope occurrence. Scale bars: 20 μm.

Table A1. The oligonucleotide primers that were used for the RT-qPCR reaction with relevant descriptions of the genes.

Genes	Description of the Genes	Primer Sequence (5′-3′)
Bradi1g32860	*ubiquitin*	pF-GAGGGTGGACTCCTTTTGGA pR-TCCACACTCCACTTGGTGCT
	***EXT* and *EXT*-like receptor kinase**	
Bradi4g11250	*extensin (chimeric EXT)*	pF-GCGACTGCGACAATGATGTG pR-ACCCCTTGCTAAGCCCTCTA
Bradi3g12902	*extensin (chimeric EXT)*	pF-CATCTGGACCTGCCAATGGT pR-TCCCAGTTTTGGAGTCTCGC
Bradi3g59780	*formin-homolog extensin (FH EXT)*	pF-GATGAATGCCGGAACAGCAC pR-GTGGAGAAGAGTGGTGCCTC
Bradi4g03720	*formin-homolog extensin (FH EXT)*	pF-GAAGCAGATTGAGGCCGAGA pR-CGCGCCTCCATCTTTTGATT
Bradi1g22980	*formin-homolog extensin (FH EXT)*	pF-CAGCAGAGCCTGTTGCTTGAC pR-TTCTAGGTTTCCGTGCATGAGT
Bradi2g49240	*proline-rich extensin-like receptor kinase (PERK)*	pF-TTCTCAGCCGTTGGGAGATG pR-GGAAGGTCCCCAAAGTCTCG
Bradi1g07010	*proline-rich extensin-like receptor kinase (PERK)*	pF-CCTCCACGGTAAAGGGCTG pR-GATCCGTGGATGGCAGTCTT
Bradi3g31967	*proline-rich extensin-like receptor kinase (PERK)*	pF-CCGTCGCCATTAAGAATCTGC pR-GATTCTTGTGCCGAACTCGC
Bradi2g00900	*proline-rich extensin-like receptor kinase (PERK)*	pF-TAACTTTGAGGCACAGGTTGCT pR-AGCCATGTATCCAAAAGTCCCC
	FLA	
Bradi2g00220	*fasciclin-like arabinogalactan protein*	pF-AGCTCAACAGCTCCCAGAC pR-CGAAAGCGAGTTGAGCGTG
Bradi5g18950	*fasciclin-like arabinogalactan protein*	pF-AATAAAGGGAAGTCACCGTCGC pR-CCGTTCTTCTTGTCATGGACCT
Bradi4g34420	*fasciclin-like arabinogalactan protein*	pF-CACATCCTCCAGATGCACGTC pR-CCGGACTCCTGGAACATGG
Bradi3g39740	*fasciclin-like arabinogalactan protein*	pF-GTACTATTCCCTGGCGGAGTTC pR-CCATGTTGTCGGTGAGGTTGAG
Bradi2g60270	*fasciclin-like arabinogalactan protein*	pF-AGCAGAGCAATCCTCTAGTAGC pR-TGGGTTCTTCTCGCCATTGTTA

References

1. Barlow, K.M.; Christy, B.P.; O'Leary, G.J.; Riffkin, P.A.; Nuttall, J.G. Simulating the impact of extreme heat and frost events on wheat crop production: A review. *Field Crops Res.* **2015**, *171*, 109–119. [CrossRef]
2. Hatfield, J.L.; Prueger, J.H. Temperature extremes: Effect on plant growth and development. *Weather Clim. Extrem.* **2015**, *10*, 4–10. [CrossRef]
3. Ohama, N.; Sato, H.; Shinozaki, K.; Yamaguchi-Shinozaki, K. Transcriptional regulatory network of plant heat stress response. *Trends Plant Sci.* **2017**, *22*, 53–65. [CrossRef]
4. Garvin, D.F.; Gu, Y.Q.; Hasterok, R.; Hazen, S.P.; Jenkins, G.; Mockler, T.C.; Mur, L.A.J.; Vogel, J.P. Development of genetic and genomic research resources for *Brachypodium distachyon*, a new model system for grass crop research. *Crop Sci.* **2008**, *48*, S-69. [CrossRef]
5. International Brachypodium Initiative. Genome sequencing and analysis of the model grass *Brachypodium distachyon*. *Nature* **2010**, *463*, 763–768. [CrossRef] [PubMed]
6. Janmohammadi, M.; Zolla, L.; Rinalducci, S. Low temperature tolerance in plants: Changes at the protein level. *Phytochemistry* **2015**, *117*, 76–89. [CrossRef]

7. Bita, C.E.; Gerats, T. Plant tolerance to high temperature in a changing environment: Scientific fundamentals and production of heat stress-tolerant crops. *Front. Plant Sci.* **2013**, *4*, 273. [CrossRef]
8. Le Gall, H.; Philippe, F.; Domon, J.M.; Gillet, F.; Pelloux, J.; Rayon, C. Cell wall metabolism in response to abiotic stress. *Plants* **2015**, *4*, 112–166. [CrossRef] [PubMed]
9. Johnson, K.L.; Cassin, A.M.; Lonsdale, A.; Bacic, A.; Doblin, M.S.; Schultz, C.J. Pipeline to identify hydroxyproline-rich glycoproteins. *Plant Physiol.* **2017**, *174*, 886–903. [CrossRef] [PubMed]
10. Pereira, A.M.; Pereira, L.G.; Coimbra, S. Arabinogalactan proteins: Rising attention from plant biologists. *Plant Reprod.* **2015**, *28*, 1–15. [CrossRef] [PubMed]
11. Showalter, A.M.; Basu, D. Extensin and arabinogalactan-protein biosynthesis: Glycosyltransferases, research challenges, and biosensors. *Front. Plant Sci.* **2016**, *7*, 814. [CrossRef] [PubMed]
12. Su, S.; Higashiyama, T. Arabinogalactan proteins and their sugar chains: Functions in plant reproduction, research methods, and biosynthesis. *Plant Reprod.* **2018**, *31*, 67–75. [CrossRef]
13. Baetz, U.; Martinoia, E. Root exudates: The hidden part of plant defense. *Trends Plant Sci.* **2014**, *19*, 90–98. [CrossRef] [PubMed]
14. Pinski, A.; Betekhtin, A.; Hupert-Kocurek, K.; Mur, L.A.J.; Hasterok, R. Defining the genetic basis of plant–endophytic bacteria interactions. *Int. J. Mol. Sci.* **2019**, *20*, 1947. [CrossRef] [PubMed]
15. Betekhtin, A.; Rojek, M.; Nowak, K.; Pinski, A.; Milewska-Hendel, A.; Kurczynska, E.; Doonan, J.H.; Hasterok, R. Cell wall epitopes and endoploidy as reporters of embryogenic potential in *Brachypodium distachyon* callus culture. *Int. J. Mol. Sci.* **2018**, *19*, 3811. [CrossRef] [PubMed]
16. Mareri, L.; Romi, M.; Cai, G. Arabinogalactan proteins: Actors or spectators during abiotic and biotic stress in plants? *Plant Biosyst.* **2018**, *153*, 173–185. [CrossRef]
17. Mareri, L.; Faleri, C.; Romi, M.; Mariani, C.; Cresti, M.; Cai, G. Heat stress affects the distribution of JIM8-labelled arabinogalactan proteins in pistils of *Solanum lycopersicum* cv Micro-Tom. *Acta Physiol. Plant* **2016**, *38*, 184. [CrossRef]
18. Yan, Y.; Takac, T.; Li, X.; Chen, H.; Wang, Y.; Xu, E.; Xie, L.; Su, Z.; Samaj, J.; Xu, C. Variable content and distribution of arabinogalactan proteins in banana (*Musa* spp.) under low temperature stress. *Front. Plant Sci.* **2015**, *6*, 353. [CrossRef]
19. Liu, X.; Wolfe, R.; Welch, L.R.; Domozych, D.S.; Popper, Z.A.; Showalter, A.M. Bioinformatic identification and analysis of extensins in the plant kingdom. *PLoS ONE* **2016**, *11*, e0150177. [CrossRef]
20. Betekhtin, A.; Rojek, M.; Milewska-Hendel, A.; Gawecki, R.; Karcz, J.; Kurczynska, E.; Hasterok, R. Spatial distribution of selected chemical cell wall components in the embryogenic callus of *Brachypodium distachyon*. *PLoS ONE* **2016**, *11*, e0167426. [CrossRef]
21. Castilleux, R.; Plancot, B.; Ropitaux, M.; Carreras, A.; Leprince, J.; Boulogne, I.; Follet-Gueye, M.L.; Popper, Z.A.; Driouich, A.; Vicre, M. Cell wall extensins in root-microbe interactions and root secretions. *J. Exp. Bot.* **2018**, *69*, 4235–4247. [CrossRef] [PubMed]
22. Baumberger, N.; Ringli, C.; Keller, B. Systematic identification of novel protein domain families associated with nuclear functions. *Genome Res.* **2002**, *12*, 47–56.
23. Botha, C.E. A tale of two neglected systems-structure and function of the thin- and thick-walled sieve tubes in monocotyledonous leaves. *Front. Plant Sci.* **2013**, *4*, 297. [CrossRef]
24. Gawecki, R.; Sala, K.; Kurczyńska, E.U.; Świątek, P.; Płachno, B.J. Immunodetection of some pectic, arabinogalactan proteins and hemicellulose epitopes in the micropylar transmitting tissue of apomictic dandelions (*Taraxacum*, Asteraceae, Lactuceae). *Protoplasma* **2016**, *254*, 657–668. [CrossRef] [PubMed]
25. Milewska-Hendel, A.; Baczewska, A.H.; Sala, K.; Dmuchowski, W.; Brągoszewska, P.; Gozdowski, D.; Jozwiak, A.; Chojnacki, T.; Swiezewska, E.; Kurczynska, E. Quantitative and qualitative characteristics of cell wall components and prenyl lipids in the leaves of *Tilia* x *euchlora* trees growing under salt stress. *PLoS ONE* **2017**, *12*, e0172682. [CrossRef]
26. Garaeva, L.D.; Pozdeeva, S.A.; Timofeeva, O.A.; Khokhlova, L.P. Cell-wall lectins during winter wheat cold hardening. *Russ. J. Plant. Physiol.* **2006**, *53*, 746–750. [CrossRef]
27. Gong, S.Y.; Huang, G.Q.; Sun, X.; Li, P.; Zhao, L.L.; Zhang, D.J.; Li, X.B. GhAGP31, a cotton non-classical arabinogalactan protein, is involved in response to cold stress during early seedling development. *Plant. Biol.* **2012**, *14*, 447–457. [CrossRef] [PubMed]
28. Kang, Y.; Khan, S.; Ma, X. Climate change impacts on crop yield, crop water productivity and food security—A review. *Prog. Nat. Sci.* **2009**, *19*, 1665–1674. [CrossRef]

29. Wang, W.; Vinocur, B.; Altman, A. Plant responses to drought, salinity and extreme temperatures: Towards genetic engineering for stress tolerance. *Planta* **2003**, *218*, 1–14. [CrossRef]
30. Larsen, F.H.; Byg, I.; Damager, I.; Diaz, J.; Engelsen, S.B.; Ulvskov, P. Residue specific hydration of primary cell wall potato pectin identified by solid-state 13C single-pulse MAS and CP/MAS NMR spectroscopy. *Biomacromolecules* **2011**, *12*, 1844–1850. [CrossRef] [PubMed]
31. Ha, M.A.; Vietor, R.J.; Jardine, G.D.; Apperley, D.C.; Jarvis, M.C. Conformation and mobility of the arabinan and galactan side-chains of pectin. *Phytochemistry* **2005**, *66*, 1817–1824. [CrossRef]
32. Tenhaken, R. Cell wall remodeling under abiotic stress. *Front. Plant. Sci.* **2014**, *5*, 771. [CrossRef]
33. Moore, J.P.; Farrant, J.M.; Driouich, A. A role for pectin-associated arabinans in maintaining the flexibility of the plant cell wall during water deficit stress. *Plant. Signal. Behav.* **2014**, *3*, 102–104. [CrossRef]
34. Betekhtin, A.; Pinski, A.; Milewska-Hendel, A.; Kurczynska, E.; Hasterok, R. Stability and instability processes in the calli of *Fagopyrum tataricum* that have different morphogenic potentials. *Plant. Cell Tissue Organ. Cult.* **2019**, 1–15. [CrossRef]
35. Xue, H.; Seifert, G.J. Fasciclin like arabinogalactan protein 4 and respiratory burst oxidase homolog d and F independently modulate abscisic acid signaling. *Plant. Signal. Behav.* **2015**, *10*, e989064. [CrossRef]
36. Johnson, K.L.; Jones, B.J.; Bacic, A.; Schultz, C.J. The fasciclin-like arabinogalactan proteins of Arabidopsis. A multigene family of putative cell adhesion molecules. *Plant. Physiol.* **2003**, *133*, 1911–1925. [CrossRef]
37. Seifert, G.J.; Xue, H.; Acet, T. The *Arabidopsis thaliana FASCICLIN LIKE ARABINOGALACTAN PROTEIN 4* gene acts synergistically with abscisic acid signalling to control root growth. *Ann. Bot.* **2014**, *114*, 1125–1133. [CrossRef]
38. Faik, A.; Abouzouhair, J.; Sarhan, F. Putative fasciclin-like arabinogalactan-proteins (FLA) in wheat (*Triticum aestivum*) and rice (*Oryza sativa*): Identification and bioinformatic analyses. *Mol. Genet. Genom.* **2006**, *276*, 478–494. [CrossRef]
39. Ma, H.; Zhao, J. Genome-wide identification, classification, and expression analysis of the arabinogalactan protein gene family in rice (*Oryza sativa* L.). *J. Exp. Bot.* **2010**, *61*, 2647–2668. [CrossRef]
40. Nahar, K.; Hasanuzzaman, M.; Ahamed, K.U.; Hakeem, K.R.; Ozturk, M.; Fujita, M. Plant responses and tolerance to high temperature stress: Role of exogenous phytoprotectants. In *Crop Production and Global Environmental Issues*; Hakeem, K.R., Ed.; Springer International Publishing: Cham, The Netherlands, 2015; pp. 385–435.
41. Ellis, M.; Egelund, J.; Schultz, C.J.; Bacic, A. Arabinogalactan-proteins: Key regulators at the cell surface? *Plant. Physiol.* **2010**, *153*, 403–419. [CrossRef]
42. Lima, R.B.; dos Santos, T.B.; Vieira, L.G.; Ferrarese Mde, L.; Ferrarese-Filho, O.; Donatti, L.; Boeger, M.R.; Petkowicz, C.L. Heat stress causes alterations in the cell-wall polymers and anatomy of coffee leaves (*Coffea arabica* L.). *Carbohydr. Polym.* **2013**, *93*, 135–143. [CrossRef]
43. Olmos, E.; Garcia De La Garma, J.; Gomez-Jimenez, M.C.; Fernandez-Garcia, N. Arabinogalactan proteins are involved in salt-adaptation and vesicle trafficking in tobacco by-2 cell cultures. *Front. Plant Sci.* **2017**, *8*, 1092. [CrossRef]
44. Zang, L.; Zheng, T.; Chu, Y.; Ding, C.; Zhang, W.; Huang, Q.; Su, X. Genome-wide analysis of the fasciclin-like arabinogalactan protein gene family reveals differential expression patterns, localization, and salt stress response in *Populus*. *Front. Plant Sci.* **2015**, *6*, 1140. [CrossRef] [PubMed]
45. Lamport, D.T.; Kieliszewski, M.J.; Showalter, A.M. Salt stress upregulates periplasmic arabinogalactan proteins: Using salt stress to analyse AGP function. *New Phytol.* **2006**, *169*, 479–492. [CrossRef] [PubMed]
46. Shi, H.; Kim, Y.; Guo, Y.; Stevenson, B.; Zhu, J.K. The Arabidopsis SOS5 locus encodes a putative cell surface adhesion protein and is required for normal cell expansion. *Plant. Cell* **2002**, *15*, 19–32. [CrossRef] [PubMed]
47. Griffiths, J.S.; Tsai, A.Y.; Xue, H.; Voiniciuc, C.; Sola, K.; Seifert, G.J.; Mansfield, S.D.; Haughn, G.W. SALT-OVERLY SENSITIVE5 mediates Arabidopsis seed coat mucilage adherence and organization through pectins. *Plant Physiol.* **2014**, *165*, 991–1004. [CrossRef] [PubMed]
48. Seifert, G.J. Fascinating fasciclins: A surprisingly widespread family of proteins that mediate interactions between the cell exterior and the cell surface. *Int. J. Mol. Sci.* **2018**, *19*, 1628. [CrossRef] [PubMed]

49. Johnson, K.L.; Kibble, N.A.; Bacic, A.; Schultz, C.J. A fasciclin-like arabinogalactan-protein (FLA) mutant of *Arabidopsis thaliana*, *fla1*, shows defects in shoot regeneration. *PLoS ONE* **2011**, *6*, e25154. [CrossRef]
50. Wu, Y.; Fan, W.; Li, X.; Chen, H.; Takac, T.; Samajova, O.; Fabrice, M.R.; Xie, L.; Ma, J.; Samaj, J.; et al. Expression and distribution of extensins and AGPs in susceptible and resistant banana cultivars in response to wounding and *Fusarium oxysporum*. *Sci. Rep.* **2017**, *7*, 42400. [CrossRef]
51. Casero, P.J.; Casimiro, I.; Knox, J.P. Occurrence of cell surface arabinogalactan-protein and extensin epitopes in relation to pericycle and vascular tissue development in the root apex of four species. *Planta* **1998**, *204*, 252–259. [CrossRef]
52. Zhang, X.; Ren, Y.; Zhao, J. Roles of extensins in cotyledon primordium formation and shoot apical meristem activity in *Nicotiana tabacum*. *J. Exp. Bot.* **2008**, *59*, 4045–4058. [CrossRef] [PubMed]
53. Xu, C.; Takáč, T.; Burbach, C.; Menzel, D.; Šamaj, J. Developmental localization and the role of hydroxyproline rich glycoproteins during somatic embryogenesis of banana (*Musa* spp. AAA). *BMC Plant. Biol.* **2011**, *11*, 38. [CrossRef]
54. Zhang, X.; Ma, H.; Qi, H.; Zhao, J. Roles of hydroxyproline-rich glycoproteins in the pollen tube and style cell growth of tobacco (*Nicotiana tabacum* L.). *J. Plant. Physiol.* **2014**, *171*, 1036–1045. [CrossRef] [PubMed]
55. Seki, M.; Narusaka, M.; Ishida, J.; Nanjo, T.; Fujita, M.; Oono, Y.; Kamiya, A.; Nakajima, M.; Enju, A.; Sakurai, T.; et al. Monitoring the expression profiles of 7000 Arabidopsis genes under drought, cold and high-salinity stresses using a full-length cDNA microarray. *Plant. J.* **2002**, *31*, 279–292. [CrossRef] [PubMed]
56. Kreps, J.A.; Wu, Y.; Chang, H.S.; Zhu, T.; Wang, X.; Harper, J.F. Transcriptome changes for Arabidopsis in response to salt, osmotic, and cold stress. *Plant. Physiol.* **2002**, *130*, 2129–2141. [CrossRef]
57. Hwang, Y.; Lee, H.; Lee, Y.S.; Cho, H.T. Cell wall-associated ROOT HAIR SPECIFIC 10, a proline-rich receptor-like kinase, is a negative modulator of Arabidopsis root hair growth. *J. Exp. Bot.* **2016**, *67*, 2007–2022. [CrossRef]
58. Dar, N.A.; Amin, I.; Wani, W.; Shafiq, A.; Shikari, A.B.; Wani, S.H.; Masoodi, K.Z. Abscisic acid: A key regulator of abiotic stress tolerance in plants. *Plant. Gene* **2017**, *11*, 106–111. [CrossRef]
59. Bai, L.; Zhang, G.; Zhou, Y.; Zhang, Z.; Wang, W.; Du, Y.; Wu, Z.; Song, C.P. Plasma membrane-associated proline-rich extensin-like receptor kinase 4, a novel regulator of Ca signalling, is required for abscisic acid responses in *Arabidopsis thaliana*. *Plant. J.* **2009**, *60*, 314–327. [CrossRef] [PubMed]
60. Silva, N.F.; Goring, D.R. The proline-rich, extensin-like receptor kinase-1 (PERK1) gene is rapidly induced by wounding. *Plant. Mol. Biol.* **2002**, *50*, 667–685. [CrossRef]
61. Zhao, C.; Zayed, O.; Yu, Z.; Jiang, W.; Zhu, P.; Hsu, C.C.; Zhang, L.; Tao, W.A.; Lozano-Duran, R.; Zhu, J.K. Leucine-rich repeat extensin proteins regulate plant salt tolerance in Arabidopsis. *Proc. Natl. Acad. Sci. USA* **2018**, *115*, 13123–13128. [CrossRef]
62. Irizarry, I.; White, J.F. *Bacillus amyloliquefaciens* alters gene expression, ROS production and lignin synthesis in cotton seedling roots. *J. Appl. Microbiol.* **2018**, *124*, 1589–1603. [CrossRef]
63. Sujkowska-Rybkowska, M.; Borucki, W. Accumulation and localization of extensin protein in apoplast of pea root nodule under aluminum stress. *Micron* **2014**, *67*, 10–19. [CrossRef]
64. Francin-Allami, M.; Merah, K.; Albenne, C.; Rogniaux, H.; Pavlovic, M.; Lollier, V.; Sibout, R.; Guillon, F.; Jamet, E.; Larre, C. Cell wall proteomic of *Brachypodium distachyon* grains: A focus on cell wall remodeling proteins. *Proteomics* **2015**, *15*, 2296–2306. [CrossRef]
65. Calderan-Rodrigues, M.J.; Guimarães Fonseca, J.; de Moraes, F.E.; Vaz Setem, L.; Carmanhanis Begossi, A.; Labate, C.A. Plant cell wall proteomics: A focus on monocot species, *Brachypodium distachyon*, *Saccharum* spp. and *Oryza sativa*. *Int. J. Mol. Sci.* **2019**, *20*, 1975. [CrossRef] [PubMed]
66. Douche, T.; San Clemente, H.; Burlat, V.; Roujol, D.; Valot, B.; Zivy, M.; Pont-Lezica, R.; Jamet, E. *Brachypodium distachyon* as a model plant toward improved biofuel crops: Search for secreted proteins involved in biogenesis and disassembly of cell wall polymers. *Proteomics* **2013**, *13*, 2438–2454. [CrossRef] [PubMed]
67. Francin-Allami, M.; Lollier, V.; Pavlovic, M.; San Clemente, H.; Rogniaux, H.; Jamet, E.; Guillon, F.; Larre, C. Understanding the remodelling of cell walls during *Brachypodium distachyon* grain development through a sub-cellular quantitative proteomic approach. *Proteomes* **2016**, *4*, 21. [CrossRef]
68. Zhang, M.; Chen, G.X.; Lv, D.W.; Li, X.H.; Yan, Y.M. N-linked glycoproteome profiling of seedling leaf in *Brachypodium distachyon* L. *J. Proteome Res.* **2015**, *14*, 1727–1738. [CrossRef] [PubMed]

69. Priest, H.D.; Fox, S.E.; Rowley, E.R.; Murray, J.R.; Michael, T.P.; Mockler, T.C. Analysis of global gene expression in *Brachypodium distachyon* reveals extensive network plasticity in response to abiotic stress. *PLoS ONE* **2014**, *9*, e87499. [CrossRef]
70. Hong, S.Y.; Park, J.H.; Cho, S.H.; Yang, M.S.; Park, C.M. Phenological growth stages of *Brachypodium distachyon*: Codification and description. *Weed Res.* **2011**, *51*, 612–620. [CrossRef]
71. Sala, K.; Malarz, K.; Barlow, P.W.; Kurczyńska, E.U. Distribution of some pectic and arabinogalactan protein epitopes during *Solanum lycopersicum* (L.) adventitious root development. *BMC Plant. Biol.* **2017**, *17*, 25. [CrossRef] [PubMed]
72. Betekhtin, A.; Milewska-Hendel, A.; Chajec, L.; Rojek, M.; Nowak, K.; Kwasniewska, J.; Wolny, E.; Kurczynska, E.; Hasterok, R. 5-azacitidine induces cell death in a tissue culture of *Brachypodium distachyon*. *Int. J. Mol. Sci.* **2018**, *19*, 1806. [CrossRef] [PubMed]
73. Pennell, R.I.; Janniche, L.; Kjellbom, P.; Scofield, G.N.; Peart, J.M.; Roberts, K. Developmental regulation of a plasma membrane arabinogalactan protein epitope in oilseed rape flowers. *Plant. Cell* **1991**, *3*, 1317–1326. [CrossRef] [PubMed]
74. Yates, E.A.; Knox, J.P. Investigations into the occurrence of plant cell surface epitopes in exudate gums. *Carbohydr. Polym.* **1994**, *24*, 281–286. [CrossRef]
75. Yates, E.A.; Valdor, J.F.; Haslam, S.M.; Morris, H.R.; Dell, A.; Mackie, W.; Knox, J.P. Characterization of carbohydrate structural features recognized by anti-arabinogalactan-protein monoclonal antibodies. *Glycobiology* **1996**, *6*, 131–139. [CrossRef]
76. Knox, J.P.; Linstead, P.J.; Peart, J.; Cooper, C.; Roberts, K. Developmentally regulated epitopes of cell surface arabinogalactan proteins and their relation to root tissue pattern formation. *Plant. J.* **1991**, *1*, 317–326. [CrossRef] [PubMed]
77. Smallwood, M.; Yates, E.A.; Willats, W.G.T.; Martin, H.; Knox, J.P. Immunochemical comparison of membrane-associated and secreted arabinogalactan-proteins in rice and carrot. *Planta* **1996**, *198*, 452–459. [CrossRef]
78. Bradley, D.J.; Wood, E.A.; Larkins, A.P.; Galfre, G.; Butcher, G.W.; Brewin, N.J. Isolation of monoclonal antibodies reacting with peribacteroid membranes and other components of pea root nodules containing *Rhizobium leguminosarum*. *Planta* **1988**, *173*, 149–160. [CrossRef] [PubMed]
79. Pennell, R.I.; Knox, J.P.; Scofield, G.N.; Selvendran, R.R.; Roberts, K. A family of abundant plasma membrane-associated glycoproteins related to the arabinogalactan proteins is unique to flowering plants. *J. Cell. Biol.* **1989**, *108*, 1967–1977. [CrossRef]
80. Willats, W.G.T.; Marcus, S.E.; Knox, J.P. Generation of a monoclonal antibody specific to (1→5)-α-L-arabinan. *Carbohydr. Res.* **1998**, *308*, 149–152. [CrossRef]
81. El-Tantawy, A.A.; Solis, M.T.; Da Costa, M.L.; Coimbra, S.; Risueno, M.C.; Testillano, P.S. Arabinogalactan protein profiles and distribution patterns during microspore embryogenesis and pollen development in *Brassica napus*. *Plant. Reprod.* **2013**, *26*, 231–243. [CrossRef] [PubMed]
82. Smallwood, M.; Beven, A.; Donovan, N.; Neill, S.J.; Peart, J.; Roberts, K.; Knox, J.P. Localization of cell wall proteins in relation to the developmental anatomy of the carrot root apex. *Plant. J.* **1994**, *5*, 237–246. [CrossRef]

International Journal of
Molecular Sciences

MDPI

Article

Characterization of the *XTH* Gene Family: New Insight to the Roles in Soybean Flooding Tolerance

Li Song [1,2,*], Babu Valliyodan [2], Silvas Prince [2,3], Jinrong Wan [2] and Henry T. Nguyen [2,*]

1 Institutes of Agricultural Science and Technology Development, Joint International Research Laboratory of Agriculture and Agri-Product Safety, Co-Innovation Center for Modern Production Technology of Grain Crops, Yangzhou University, Yangzhou 225009, China
2 National Center for Soybean Biotechnology and Division of Plant Sciences, University of Missouri, Columbia, MO 65211, USA; valliyodanb@missouri.edu (B.V.); sjprince@noble.org (S.P.); wanj@missouri.edu (J.W.)
3 Noble Research Institute, 2510 Sam noble Pkwy, Ardmore, OK 73401, USA
* Correspondence: songli@yzu.edu.cn (L.S.); nguyenhenry@missouri.edu (H.T.N.)

Received: 26 June 2018; Accepted: 6 September 2018; Published: 11 September 2018

Abstract: Xyloglucan endotransglycosylases/hydrolases (XTHs) are a class of enzymes involved in the construction and remodeling of cellulose/xyloglucan crosslinks and play an important role in regulating cell wall extensibility. However, little is known about this class of enzymes in soybean. Here, 61 soybean *XTH* genes (*GmXTHs*) were identified and classified into three subgroups through comparative phylogenetic analysis. Genome duplication greatly contributed to the expansion of *GmXTH* genes in soybean. A conserved amino acid motif responsible for the catalytic activity was identified in all GmXTHs. Further expression analysis revealed that most *GmXTHs* exhibited a distinct organ-specific expression pattern, and the expression level of many *GmXTH* genes was significantly associated with ethylene and flooding stress. To illustrate a possible role of *XTH* genes in regulating stress responses, the *Arabidopsis AtXTH31* gene was overexpressed in soybean. The generated transgenic plants exhibited improved tolerance to flooding stress, with a higher germination rate and longer roots/hypocotyls during the seedling stage and vegetative growth stages. In summary, our combined bioinformatics and gene expression pattern analyses suggest that *GmXTH* genes play a role in regulating soybean stress responses. The enhanced soybean flooding tolerance resulting from the expression of an *Arabidopsis* XTH also supports the role of *XTH* genes in regulating plant flooding stress responses.

Keywords: *Glycine max*; *XTH* gene family; transgenic soybean; plant genome; plant hormone; flooding; root plasticity

1. Introduction

Xyloglucan endotransglycosylases/hydrolases (XET/XTHs also named XTHs) are classified as glycoside hydrolase family 16 (available online: http://www.cazy.org/fam/GH16.html) and play an important role in organ elongation by modifying xyloglucan chains or catalyzing the hydrolysis of xyloglucan [1–4]. Several studies have emphasized the significant role of this gene family in the regulation of cell wall extensibility. Overexpression of the *BcXTH1* gene from *Brassica campestris* enhanced stem elongation in *Arabidopsis* by promoting cell expansion and elongation [5]. Increased *SlXET1* activity affects the xyloglucan structure during the fruit ripening and softening process in transgenic tomato fruit [6]. Overexpression of *GhXTH1* in cotton loosens and elongates cell wall fibers due to cleavage down the xyloglucan-cellulose chains [7,8]. *PtxtXET16-34* is strongly expressed in primary-walled xylem. Transgenic hybrid Aspen analysis indicated that wood cell expansion and xyloglucan content were affected when the *PtxtXET16-34* gene was overexpressed [9]. *AtXTH31* regulates cell wall xyloglucan content, and *AtXTH21* influences the development of primary roots by regulating the deposition of cellulose and the thickness of the cell wall in *Arabidopsis* [10,11].

It was reported that the XTH enzyme activity may vary with changes in environmental conditions (i.e., abiotic stresses), and that hormones play important roles in tuning XTH activity during plant development. XET activity in the maize primary root elongation zone contributes to cell wall loosening at low water potential, which is partly regulated by abscisic acid (ABA) [12,13]. Low water potential decreased the XET activity in the hypocotyl elongation zone of dark-grown soybean [14]. A decrease in XET activity was also reported in the basal 5–10 mm of maize primary roots treated with polyethylene glycol solution, which reduced the cell wall extensibility and cell elongation in that region [15]. *CaXTH3* overexpressing transgenic *Arabidopsis* showed improved tolerance to water deficit and less tolerance to high salinity compared to wild type [16]. *OsXTH8*, which is highly expressed in the vascular bundles of leaf sheath and young nodal roots in rice, was upregulated by gibberellic acid. Repressed growth in transgenic rice was associated with knocking down the expression of *OsXTH8* [17]. Loss of function of *AtXTH31* reduces sensitivity to ABA and faster germination in *Arabidopsis* [18]. Downregulation of *XTH8* and *XTH31* is also responsible for the reduced leaf cell expansion of the *Arabidopsis siz1* mutant in an SA-dependent manner [19]. The homologue of *GmXTH16* in maize is induced by flooding and ethylene and is associated with aerenchyma development [20].

As the third most cultivated crop worldwide, soybean provides protein, oil, and plant natural products for human and animal consumption, but its production is limited by environmental constraints [21–23]. Among the major abiotic stresses, soybean is particularly sensitive to flooding stress, as plant growth and grain yields are markedly reduced in flooded soil [24]. The downregulation of gene expression related to cell wall metabolism, cellulose synthesis, and cell wall degradation caused by flooding indicates that cell wall biosynthesis is inhibited by flooding [25]. Therefore, functional characterisation of the *XTH* gene family in soybean will be very useful for revealing the mechanism of soybean flooding resistance. However, little is known about soybean *XTHs*, except for the report showing that *pBRU1* (*GmXTH16*) is involved in brassinosteroid-regulated soybean epicotyls elongation [26]. The availability of soybean genome sequences and comparative analysis of the XTH gene family across plant species provide an excellent opportunity to explore the XTH diversity in soybean. In this study, genome-wide analyses of soybean *XTHs* were performed, including phylogenetic analysis, chromosomal distribution, structural, and conserved motif (DExDxEFLG) analysis. Then, a comparative analysis of the XTH gene family transcriptome in soybean tissues was performed. Further functional validation of the roles of *AtXTH31* was executed in transgenic soybean during the early seedling stage under flooding stress.

2. Results

2.1. Genome-Wide Identification of Soybean XTH Family and Phylogenetic Relationship

A total of 61 *GmXTHs* were identified with a blasting core value over 100 by using the AtXTH31 amino acid sequence as query. They were designated *GmXTH1* through *GmXTH61* according to their positions on chromosomes 1 to 20 (Supplemental Table S1). The putative proteins encoded by these *GmXTHs* document the conserved structural features of the XTHs.

An unrooted phylogenetic analysis was constructed using a total of 123 full-length XTH protein sequences from *Arabidopsis* (AtXTH), rice (OsXTH), and soybean (GmXTH). According to the NJ phylogenetic tree, 46 *GmXTH* genes were clustered in group I/II, and 15 members were classified in group III (Figure 1). Two major groups have been previously classified based on the sequence similarity as group I/II and group III in *Arabidopsis* and rice [27,28]. Phylogenetic analysis revealed that the soybean *GmXTH* gene family was expanded widely in contrast to *Arabidopsis* and rice, which may correspond to the larger genome of soybean. Group III is further divided into two subgroups: group IIIA (red) and group IIIB (pink), which contains eight and seven *GmXTH* genes, respectively.

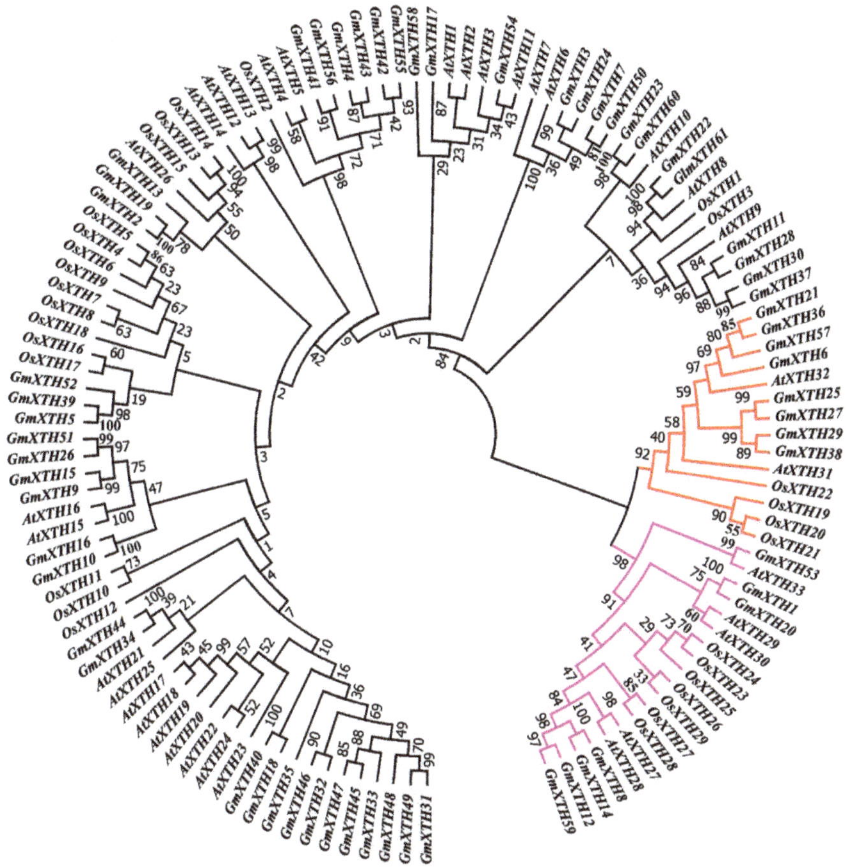

Figure 1. An unrooted phylogenetic tree for *AtXTH*, *OsXTH*, and *GmXTH* genes. A phylogenetic tree was constructed using the neighbor-joining method implemented in MEGA7. The number beside the branches represents bootstrap values based on 1000 replications. The XTH members are classified into three subfamilies. Genes from groups I/II and III are shown in the black and red/pink lines, respectively. Group III was designated group IIIA (red) and group IIIB (pink).

2.2. Chromosomal Location and Duplication Process of GmXTHs

Genome localization analysis revealed that *GmXTHs* were widely dispersed across 19 of the 20 chromosomes (Figure 2). Most *GmXTH* genes were distributed on the chromosome arms, except for two genes (*GmXTH40* and *GmXTH53*) which are located in the heterochromatin regions around the centromeric repeats. Chromosome 13 contained the largest number (eight *XTH* genes), followed by seven on chromosome 17 and five on chromosome 8. No *XTH* genes were found on chromosome 4, and the remaining chromosomes each contained one to four *XTH* genes.

Here, we found that tandem duplication or segmental duplication was involved in the expansion of the *GmXTH* gene family (Supplementary Table S3). Mapping the *XTH* genes to their chromosome physical positions (Figure 3) revealed that many *XTH* genes were clustered together, suggesting that they might be the result of tandem duplication events. For example, *GmXTH31, 32, 33, 34,* and *35* on chromosome 13 and *GmXTH44, 45, 46, 47, 48,* and *49* on chromosome 17 come from tandem duplication. *GmXTH41/42* pair and *GmXTH55/56* pair come from tandem duplication. In addition, most of the segmental duplications of *GmXTH* genes occurred approximately 13 Million years ago when Glycine-specific duplication occurred in the soybean genome (Schmutz et al., 2010, [29] Supplementary Table S3). The Ka/Ks ratio for each segmentally duplicated gene pair varied from 0.06 to 0.46. This analysis suggests that all mutations in paralogous *GmXTH* genes are neutral or disadvantageous, as their Ka/Ks ratios were less than 1. In general, tandem and segmental duplication contributed to *XTH* gene expansion in soybean. Only seven genes (*GmXTH11*, *GmXTH13*, *GmXTH17*, *GmXTH28*, *GmXTH53*, *GmXTH54*, and *GmXTH58*) were not involved in duplication events.

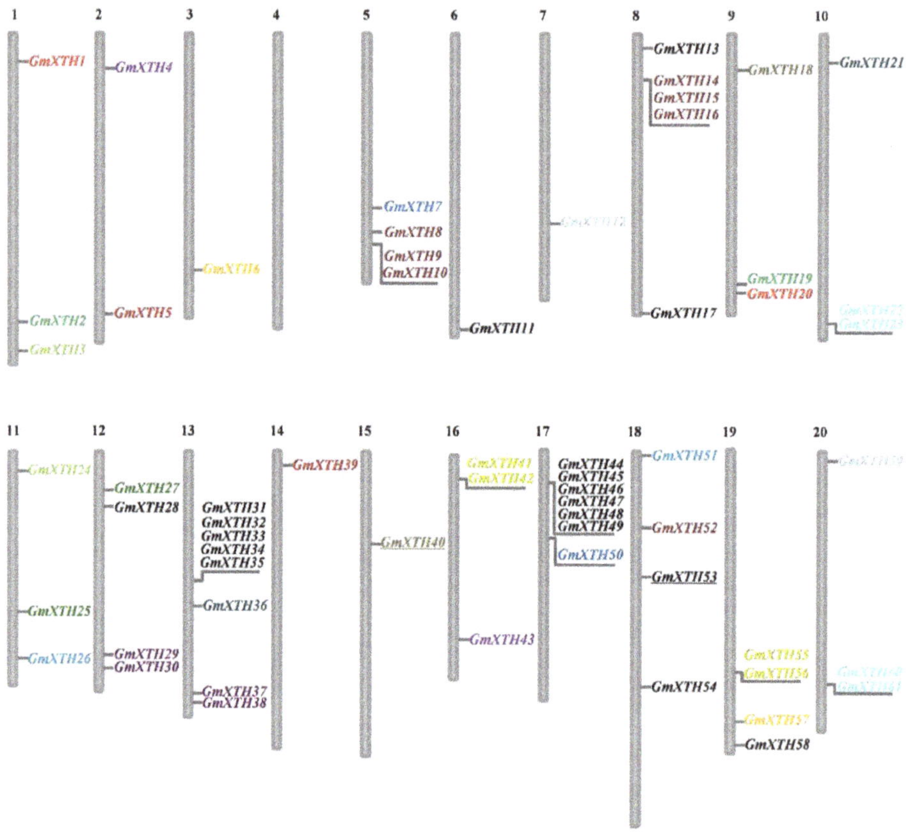

Figure 2. Chromosomal location of 61 *GmXTH* genes along soybean's 20 chromosomes. Physical map showed the distribution of the *Glycine max XTH* genes along the 20 chromosomes with colors indicating duplicated gene pairs. The chromosome number is indicated at the bottom.

2.3. Gene Structure and Conserved Protein Motif Analysis

To gain further insights into the evolutionary relationships among *GmXTH* genes, the exon/intron structures were predicted based on the alignment of coding sequence (CDS) sequences with corresponding genomic DNA sequences. As illustrated in Supplementary Figure S1, all members of the *GmXTH* family contain three or four exons. Several genes showed a different structure; for example, *GmXTH44* has an extremely long 5′ untranslated region (UTR), and *GmXTH39*, *GmXTH5*, and *GmXTH58* have no 5′ and 3′ UTRs.

All GmXTH proteins contained the conserved amino acid motif DExDxEFLG (Figure 3A), which is predicted to be responsible for the catalytic activity [30,31]. Therefore, GmXTHs reported here may cut/rejoin xyloglucan chains or catalyze the hydrolysis of xyloglucan. Compared with the genes of group III, this conserved motif could be extended upstream and downstream with a more conserved motif among *Arabidopsis*, rice, and soybean (Figure 3B).

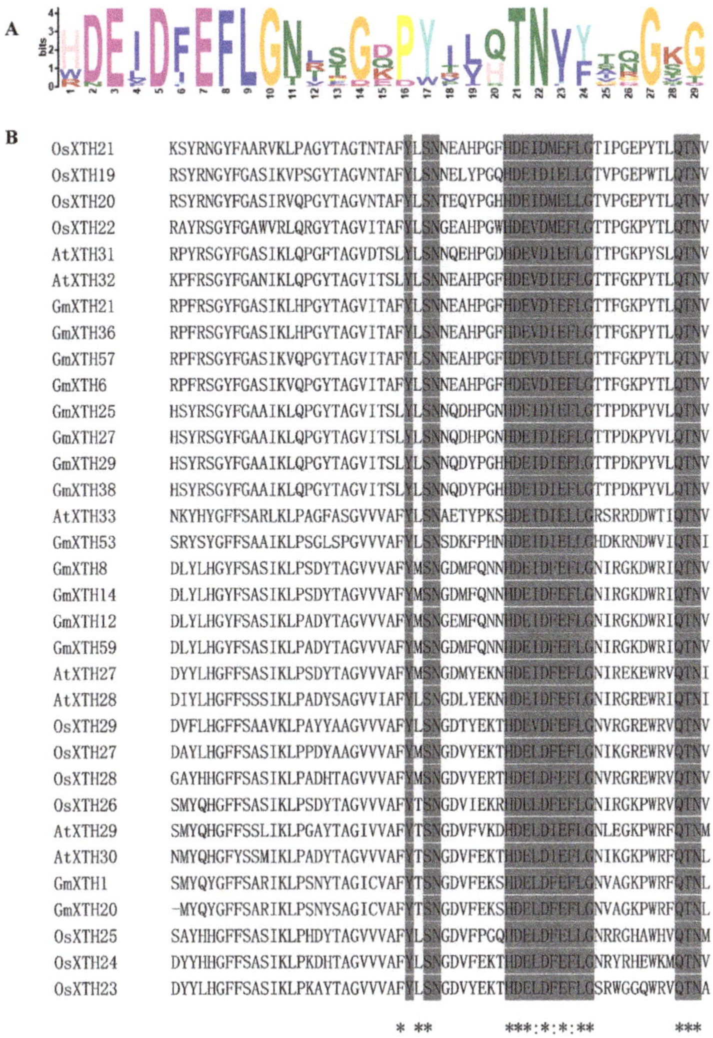

Figure 3. Conserved protein motifs in soybean XTHs. (**A**) Motifs in XTH protein sequences of 61 GmXTH identified with the MEME tool. (**B**) Alignment of the putative-site amino acid residues in group III XTH proteins from *Arabidopsis*, rice, and soybean constructed with the CLUSTALW2 program. Amino acid residues that are identical within this motif are indicated by gray shading. "*" means that the residues or nucleotides in that column are identical in all sequences in the alignment. ":" means that conserved substitutions have been observed.

2.4. GmXTHs Show an Organ-Specific Expression Pattern

To determine the expression patterns of *GmXTH* genes, we used publicly available genome-wide transcriptome data of soybean organs as a resource [32]. As shown in Figure 4, *GmXTH* genes are broadly expressed in various soybean organs. However, most of the *GmXTHs* exhibit distinct tissue-specific expression patterns. For example, fourteen *GmXTHs* (*XTH4, 41, 53, 10, 27, 17, 48, 45, 47,*

33, 2, 19, 16, and *61*) were highly expressed in roots, eighteen *GmXTHs* (*XTH9, 38, 34, 44, 8, 35, 20, 42, 40, 52, 5, 39, 1, 18, 29, 31, 43,* and *56*) were highly expressed in flowers, thirteen *GmXTHs* (*XTH60, 12, 14, 23, 6, 46, 49, 57, 3, 51, 26, 59,* and *25*) were highly expressed in stems, and three *GmXTHs* (*XTH 36, 13,* and *22*) highly expressed in seeds, whereas expression levels were relatively low in other organs. In contrast, several organs only contained one or a few genes that showed specific expression patterns. For example, *GmXTH54* was the only gene highly expressed in leaves compared with other organs, and *GmXTH55* was the only gene highly expressed in root hair compared with other organs.

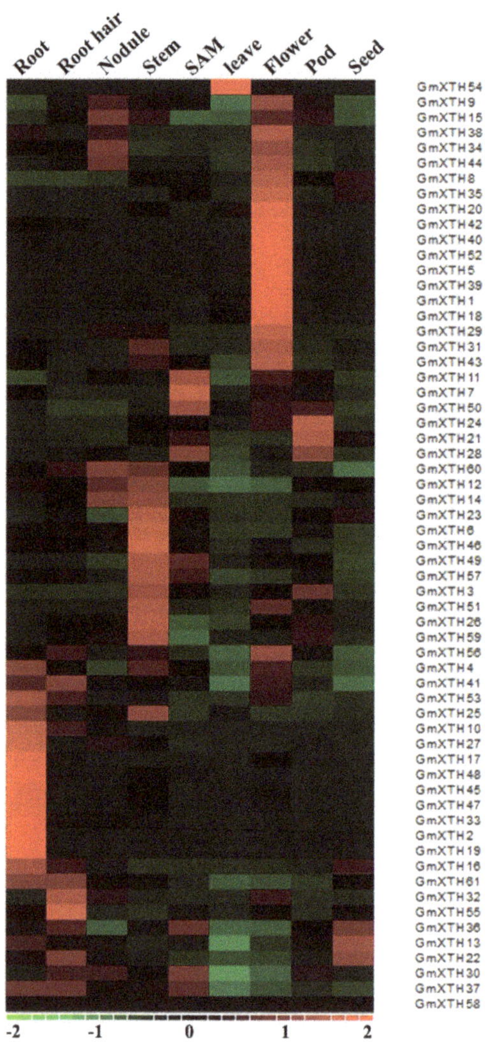

Figure 4. Heatmap of the expression profiles of the *GmXTH* gene family in nine organs. Relative organ expression levels of *GmXTHs* based on RNA-seq data were used to construct the heatmap. The expression values (Reads Per Kilobase Million) were median-cantered and normalized for each gene in different tissues before transforming to color scale. The color bar at the bottom shows the range of expression values from highest expression level (red) to lowest expression level (green), and 0 is the median expression level (Black). SAM (Shoot Apical Meristem).

No sequence reads were found for *GmXTH58* in any of the soybean organs included in the study, indicating that *GmXTH58* is probably a pseudogene or expressed under special conditions or at specific developmental stages. No *GmXTH* gene showed specifically higher expression level in nodules. Similar expression patterns were found for some phylogenetically paired genes. For example, *GmXTH25* and *GmXTH27* from group IIIA were highly expressed in roots, and *GmXTH29* and *GmXTH38* from group IIIA were highly expressed in flowers. *GmXTH14*, *GmXTH12*, and *GmXTH59* from group IIIB were highly expressed in stems.

2.5. Expression Patterns of GmXTHs Correlated with Flooding Stress

Identification of the regulatory elements indicated that one 4-bp oxygen-deficiency response element (S000493 GTAC) was significantly enriched in most *GmXTH* promoter regions, except *GmXTH23* and *GmXTH55* (Supplementary Table S4). Most *GmXTH* genes contain more than 3 response elements. The main effect of flooding is hypoxia, which reduces submerged plant normal growth and nutrient/water uptake [33]. These results suggest that the *GmXTH* gene family may play important roles during soybean sustained flooding stress.

To further assess whether the expression profiles of *GmXTH* genes are changed under flooding stress, soybean seedlings (Williams 82 genotype) were exposed to flooding conditions for 7 days [34], and their leaves were collected for gene expression analysis by RNA-seq method. Ten out of sixty-one genes were significantly regulated under flooding stress (Figure 5A), and all genes were located in group I/II. Then, the gene expression levels of twenty genes from group I/II were assessed in soybean hypocotyl and root organs under 2 days of flooding stress by quantitative Real-Time PCR (qRT-PCR) method. The results indicated that *GmXTH32* and *GmXTH34*, together with the marker genes GmADH2 and GmPDC1 [35,36], were predominantly upregulated by flooding stress, and all others showed higher expression levels than the untreated control, except four genes that were slightly downregulated (Figure 5B).

Ethylene will quickly accumulate under flooding conditions, and many regulators of submergence respond to ethylene [33,37]. To further investigate the relationship between the expression level of *GmXTH* genes and ethylene, we monitored the abundance of 61 *GmXTH* transcripts in soybean root tissue exposed to 50 μM 1-Aminocyclopropane-1-carboxylic acid (ACC) for 5 h by RNA-seq method (Wang and Nguyen et al., University of Missouri, Columbia, MO. Williams 82, 2015). We found that twenty-three *GmXTH* genes were significantly regulated under ACC treatment (Figure 5). Among them, *GmXTH18* was highly induced, and *GmXTH29* was highly down-regulated. Interestingly, *GmXTH25*, *27*, *28*, and *29* (group IIIA) were downregulated. We also found the *GmXTH40* was oppositely adjusted by flooding in leaf and root tissues. *GmXTH32* and *GmXTH34* were significantly induced by flooding in roots but not in leaves. Interestingly, a similar expression pattern was found for the tandem or segmental duplicated gene pairs located on chr13 and chr17. For example, *GmXTH31* and *GmXTH49* are both upregulated, and *GmXTH18* and *GmXTH40* are both downregulated by flooding stress in the root (Figure 5).

2.6. Stable Transgenic Soybean with Overexpression of AtXTH31

To further characterize how *XTH* genes affect root morphology and maintain physiological response under flooding conditions during soybean seedling development, *AtXTH31* (under the control of the AtMyb2 promoter) was transformed into soybean (cultivar Maverick). Over twenty independent transgenic events were generated, and six events were first identified through the general PCR method. Subsequently, the copy number variation of the transgene in those events (T0 generation) was inferred using the Q3D PCR method. Three events (Code: ND-30-2, ND-30-9 and ND-30-11) contained one copy number of the transgene, and one event (Code: ND-30-12) showed two copies of inserts (Figure 6A). However, the other two events showed a very low copy number (less than 0.5), which indicates that those events are chimeras (transgenic plant or plant part that is a mixture of two or more genetically different types of cells). Homozygous T3 transgenic soybean lines from the above events were obtained and confirmed using herbicide-resistance segregation analysis. An ND-30-12 single copy insert line was chosen from segregated T1 generation by the Q3D PCR method.

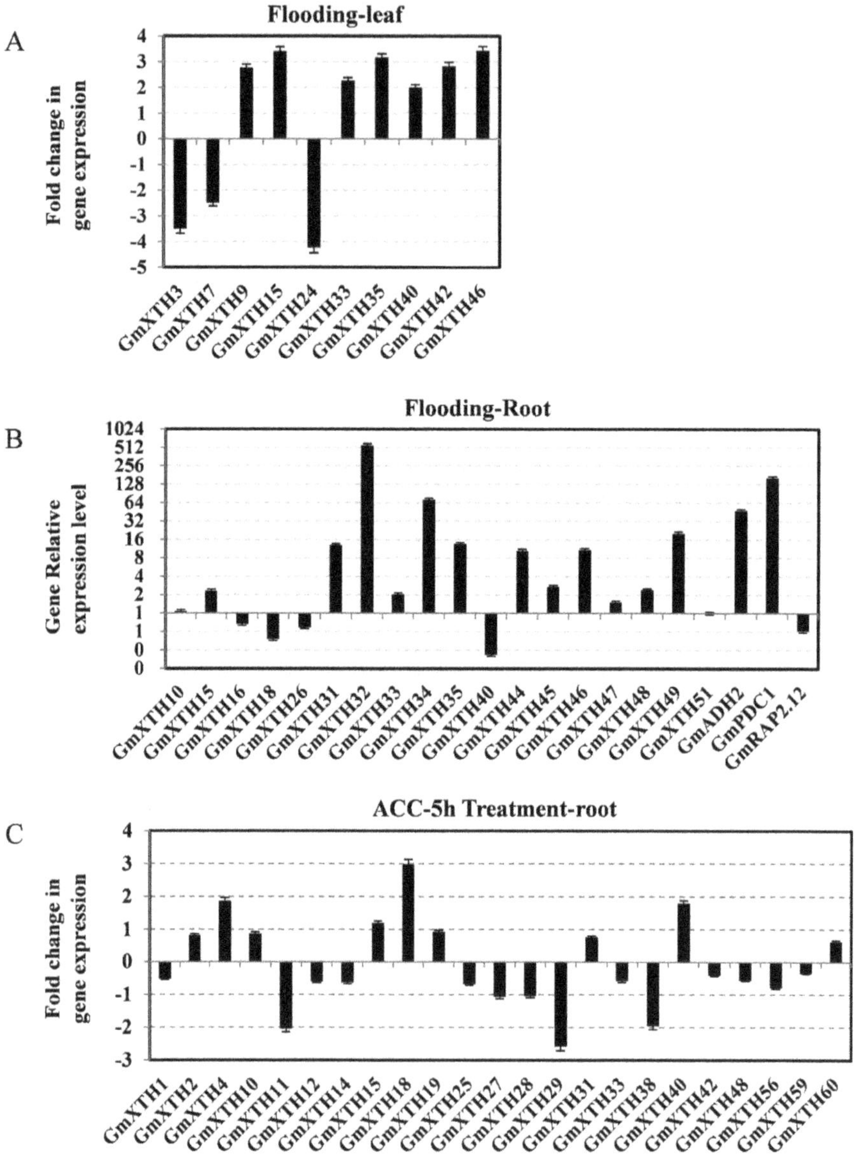

Figure 5. Expression patterns of individual *XTH* genes in response to flooding and ACC treatment studied by qRT-PCR or RNA-seq. (**A**) Expression pattern of individual *XTH* genes in response to 24 h flooding treatment in root and hypocotyl organs of two-day-old soybean seedlings. (**B**,**C**) Organ-specific expression analysis showed that most *XTH* genes were unregulated by the ACC in the root tissue, but there was no significant difference in the aerial parts of the three-week-old soybean. Three flooding-related homologous marker genes in soybean were studied as positive controls. Fold change (Log2) of relative gene expression (Actin (Glyma.18G290800)) of soybean was used as the normalization control.

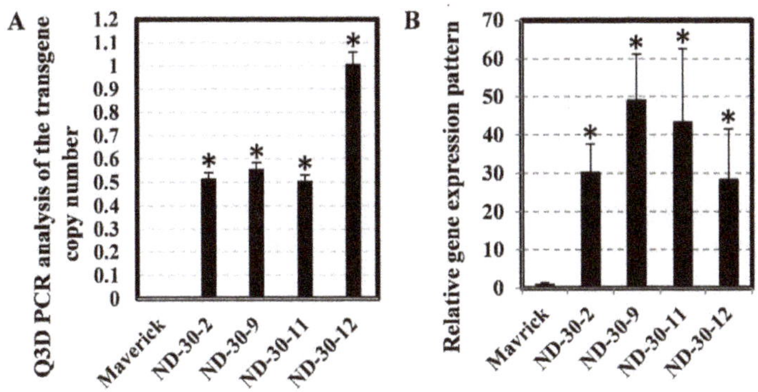

Figure 6. Calculation of transgene AtXTH31 copy numbers and relative expression levels in four transgenic events. (**A**) Ratios of copy number between *AtXTH31* and lectin gene (Glyma.02G009600) were determined by digital PCR in soybean T0 transgenic generation. Soybean transgenic plants contained a single insert copy when the ratio value was equal to 0.5 and two insert copies when the ratio value was equal to 1. (**B**) The relative expression of *AtXTH31* in T3 homozygous transgenic soybean roots determined by qRT-PCR. The relative levels of transcripts were normalized to the soybean actin gene (Glyma.18G290800). Bars represent mean values of three biological replicates (standard error). * indicates significantly different at $p < 0.05$ as tested by Fisher's least significant difference. Non-transgenic Maverick soybean as a control and MYB2:AtXTH31 transgenic soybean lines ND-30-2, ND-30-9, ND-30-11, and ND-30-12 with overexpression of *AtXTH31* were studied.

The transcript abundance of *AtXTH31* in different T3 homozygous transgenic soybean lines was investigated using qRT-PCR. Lines ND-30-2 and ND-30-12 had approximately 28-fold increases compared to the non-transgenic control, whereas lines ND-30-9 and ND-30-11 had moderately high (between 43-fold to 48-fold) increases (Figure 6B).

2.7. *Transgenic Soybean Exhibits Tolerance to Flooding during the Germination Stage*

In a comparison of the flooding tolerance between the control and transgenic lines, the germination rate of each line was counted first. As shown in Figure 7, all transgenic soybean lines had longer roots and hypocotyls (range from 45 mm to 51 mm) than the control (38 mm) after 5 days of flooding (Figure 7A). Four transgenic homozygous T3 transgenic lines with varied transgene expression conferred a range of tolerance with an increase in germination rate after 7 days of flooding, which ranged from 40% to 58% and two lines (ND-30-2 and ND-30-9) showed significant increase (Figure 7B). The length of roots and hypocotyls in the two lines (ND-30-9 and ND-30-11) was significantly greater than that in the control (Figure 7C). These results indicated that overexpressed AtXTH31 in soybean induced higher germination rate, and enhanced root/hypocotyl elongation compared with susceptible parent Maverick.

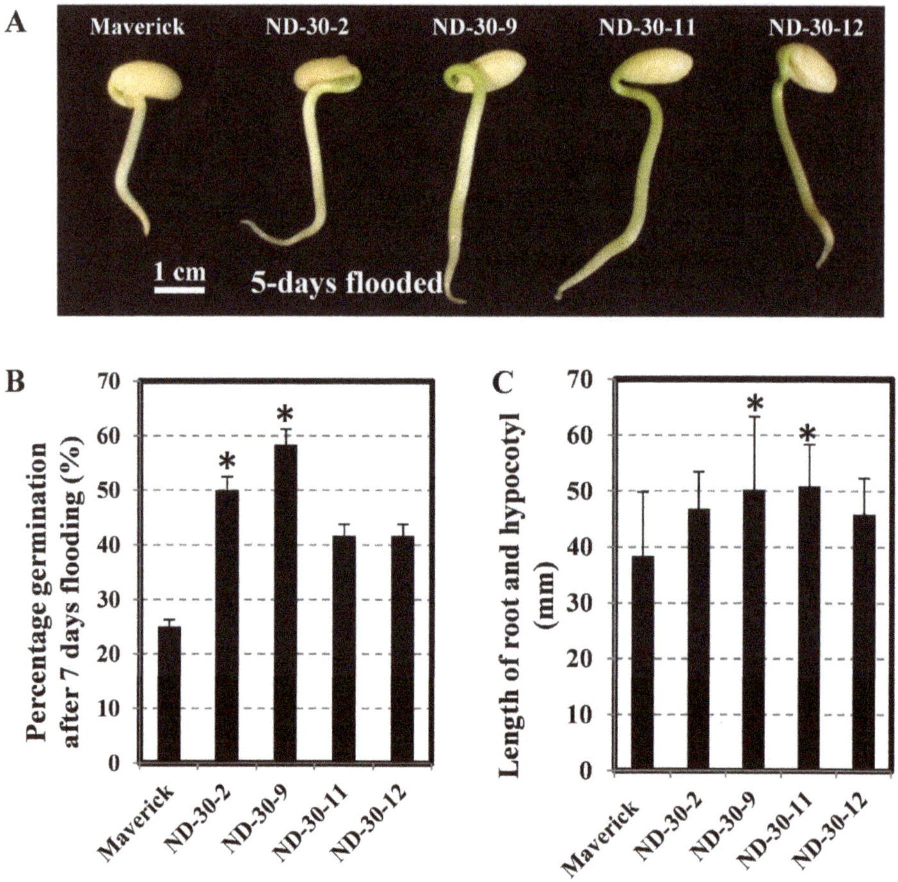

Figure 7. Soybean *AtXTH31* transgenic plants show an enhanced germination ratio and elongated root and hypocotyl under flooding conditions. (**A**) Two-day-old seedlings were flooded with water for 5 days. Bar indicates 1 cm. (**B**) The germination rate of transgenic and wild-type plants under 7 days of flooding. (**C**) Length of roots and hypocotyls of flooded Maverick soybean and transgenic seedlings. ($n \geq 30$). * indicates differences between the maverick and transgenic soybean ($p < 0.05$).

2.8. Transgenic Soybean Exhibits Tolerance to Flooding during the Vegetative Stage

Similarly, the ability of transgenic seedlings (V1 stage) to withstand flooding was then investigated. Seedlings were grown up to the V1 stage and flooded for 14 days (Figure 8A). We found that the primary root of transgenic plants was longer than that of non-transgenic controls under flooding conditions. Except for ND-30-2, the roots of all other three transgenic plants showed a significant increase in length (Figure 8B). In addition, two transgenic events (ND-30-2 and ND-30-9) showed great number of lateral roots and tertiary root tips as a response to flooding. Meanwhile, we observed that aerial root formation significantly increased in other two lines than the wild type (Figure 8E). Thus, ectopic expression of AtXTH31 in soybean could promote root development under flooding conditions and provide enhanced tolerance to flooding stress.

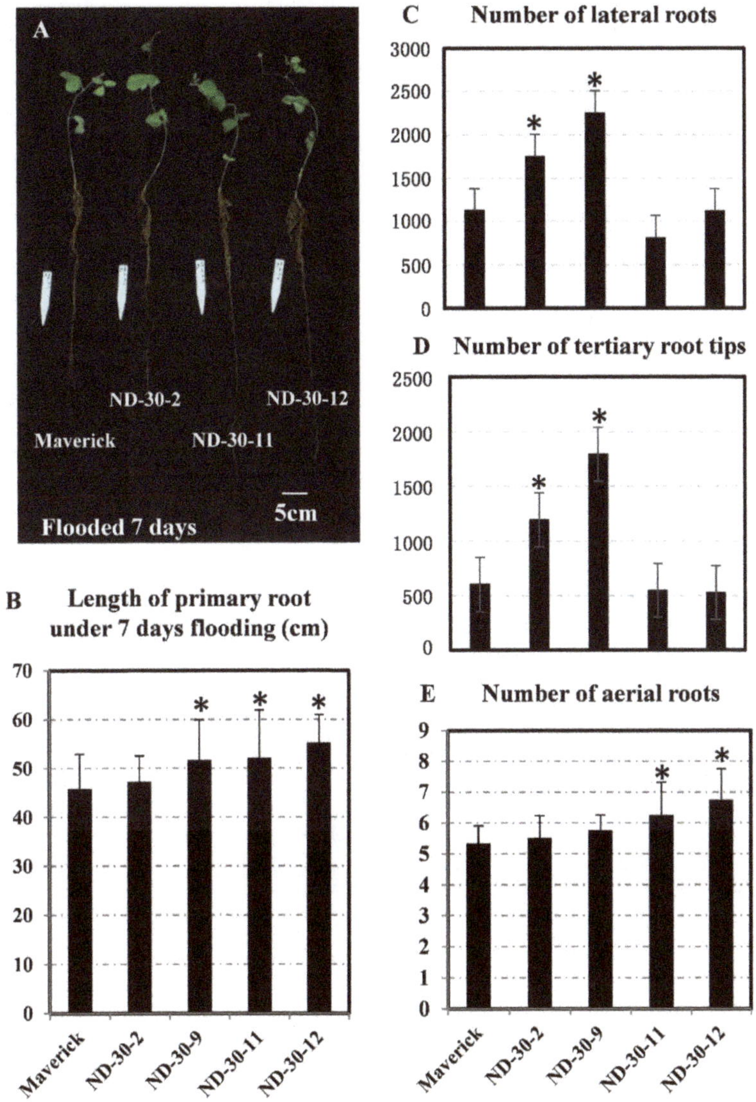

Figure 8. Soybean AtXTH31 transgenic plants show an enhanced flooding tolerance phenotype by promoting adventitious roots, lateral roots, tertiary root tips, and elongated primary roots. (**A**) Flooding effects on soybean seedlings. The V1 stage seedlings were flooded with water for 7 days. Bar indicates 5 cm. (**B**) Length of primary root compared between transgenic and control soybean plants under 7 days of flooding conditions. (**C–E**) The effects of flooding on number of lateral roots, tertiary root tips and adventitious roots per transgenic plant. ($n \geq 20$). * indicates differences between the maverick and transgenic soybean ($p < 0.05$).

3. Discussion

The *XTH* gene family has been identified in several plant species, including *Arabidopsis* [27], rice [28], barley [38], poplar [39], tomato [40], and bryophyte [41]. In this study, we report the identification and characterization of the soybean *XTH* gene family and make a comparison to

Arabidopsis and rice. Expression pattern analysis suggested that *GmXTHs* may play an important role under flooding stress. Transgenic soybean plants overexpressing *AtXTH31* showed an increase in tolerance to flooding, along with the increased aerial root number and elongated primary root length.

3.1. Charaterization of GmXTHs Gene Family

Although the role of plant XTHs in regulating cell wall extensibility is well known, there is limited information on the XTH gene family size and the evolutionary relationships between XTH genes in soybean. Previous phylogenetic studies showed that XTHs form three groups in Arabidopsis and rice [27,28]. The number of *GmXTH* genes identified was substantially higher than in *Arabidopsis* and rice, and clustered into three groups. Further, the evolutionary mechanism analysis suggested that GmXTHs family expanded partly due to segmental and tandem duplication events. These duplication events further contributed to the conserved protein motif and gene structure. These *GmXTHs* genes displayed differential expression patterns either between different organs or under flooding stress. For example, *GmXTH22* showed the highest expression level in root hair and seeds, but *GmXTH23* exhibited the highest expression level in stems. *GmXTH29* and *GmXTH30* showed the highest and lower expression levels in flower tissue, respectively. The expression patterns of these paralogous pairs suggest that *GmXTH* gene family might have undergone sub-functionalization or neofunctionalization during the subsequent evolution process. Considering these facts, the characterization of *GmXTH* gene family provides valuable information on the evolution of the *XTH* soybean gene family and underlines basis for future research.

3.2. The Expression Patterns of GmXTHs Were Regulated by Flooding and Ethylene

Flooding causes severe production losses in soybean [42,43] through inhibition of stem and root growth, decreased photosynthesis, and induced leaf abscission and premature fruit drop [20,44,45]. Therefore, investigating gene expression patterns of the *GmXTH* gene family can help us advance the fundamental understanding of how soybean adjusts to flooding stress during growth and development.

In this study, we found that the expression pattern of *GmXTHs* may confer precise regulation with temporal, spatial, and environmental conditions. One of the main effects of flooding is the deprivation of oxygen from plant roots, and low oxygen will increase the synthesis of ethylene [46]. Ethylene production was higher in soybean waterlogging-tolerant lines than in susceptible lines [47]. It has been reported that the expression of *XTH* genes is associated with shoot elongation, which is promoted by ethylene in arrowhead tubers [48]. Several *AtXTH* genes were differentially regulated during ACC-induced inhibition of *Arabidopsis* root cell elongation [49]. Ethylene increases *XTH* and *EXPANSIN7* (*EXP7*) expression in *Arabidopsis* roots [50]. Accordingly, investigation of *GmXTH* expression levels under ACC treatment will provide more hints to further gene functional analysis. In this study, we found that 23 *GmXTH* genes were significantly regulated by ethylene in soybean roots, indicating that the hormone ethylene plays important roles in GmXTH-mediated cell wall remodeling during flooding stress. However, further analysis is needed to reveal the relationship between hormone ACC and cell wall remodeling by regulating *XTH* transcription level in soybean.

3.3. The Biological Function of AtXTH31 in Soybean Root Development Under Flooding Stress

It was reported that the elongation of soybean roots was suppressed in the first 24 h and then significantly retarded after 48 h under flooding stress, which indicated that the flooding responses in the initial stages are critical for soybean growth and survival [51]. The XTH activity in the hypocotyl elongation zone of dark-grown soybean decreases when the root is exposed to low water potential [14], which directly indicates that XTH may be involved in the abiotic stress response. In this paper, the phenotypes of transgenic soybean plants carrying *AtXTH31* gene on seedling growth under flooding conditions were studied. *AtXTH31* belongs to subgroup IIIA, which was predicted to exhibit hydrolase activity with higher activity in vitro than XET activity [10,52]. *AtXTH31* exhibits a root-specific expression pattern and is involved in *Arabidopsis* cell wall modification and cell elongation [27]. The *xth31* mutant shows slower root elongation [10]. Therefore, the *AtXTH31* gene was selected for heterologous overexpression in soybean. Here we found that transgenic soybean's ability to produce more adventitious roots and longer primary roots corresponded to an increase in tolerance under flooding stress during early seedling development. It was reported that soybean root length was positively correlated with waterlogging tolerance in soybean germplasm lines [53,54]. Waterlogging-tolerant soybean lines normally developed more adventitious roots than waterlogging-susceptible lines [47,55]. Clearly, our results indicate that XTH-mediated cell wall adjustment may play a critical role in the adaptation of plants to flooding stress, and *AtXTH31* could be a useful candidate gene to develop flooding tolerance lines through molecular transgenic breeding methods. However, the corresponding tolerance mechanisms demand further investigation.

3.4. Digital PCR Provides a Simple and Accurate Method for Soybean Transgene Copy Number Analysis

Detection and quantification of transgene copy numbers are very important in characterizing transgenic plants. Recently, the application of digital PCR for the precise analysis of transgene copy numbers in crops has been reported in an array of crops [56–58]. This technology accelerates molecular breeding workflow in transgenic plants, enhances data quality in characterizing transgenes, and finally benefits the environment. However, no reports have been available on soybean for detecting copy number variations by the application of digital PCR until now.

In our study, digital PCR technology was applied to validate the transgenic *AtXTH31* copy numbers using T0 plants. Here, the digital PCR method provided more accurate results than those provided by Southern blotting and classical PCR. For example, chimeric plants (a plant or plant part that is a mixture of two or more genetically different types of cells) can be easily identified through the ratio of copy numbers. Furthermore, the dPCR method was continuously used to identify homozygous plants in the T1 generation, which saved the experiment time and no need to evaluate the homozygous lines through calculating the segregation rate of T2. In particular, the dPCR method was successfully applied to choose single-copy insert transgenic lines through analysis of copy number variation in the T1 generation. In summary, the dPCR method provides a very useful technical support for the transgenic soybean research community.

4. Materials & Methods

4.1. Identification, Chromosomal Location, and Structural Organization of GmXTH Family Members in Glycine Max

All sequence information for genes and proteins was retrieved by searching Phytozome v10.3 database (available online: http://www.phytozome.net) with a BLASTP algorithm using the *AtXTH31* amino acid sequence. The chromosome location of each *GmXTH* was obtained from Phytozome. The exon/intron organizations of *GmXTHs* were visualized with the Gene Structure Display Server program ([59] GSDS: available online: http://gsds.cbi.pku.edu.cn/).

4.2. Protein Sequence Alignment, Phylogenetic Analysis, and Gene Duplications of GmXTH *Genes*

Multiple sequence alignments were constructed using ClustalW2 (available online: http://www.ebi.ac.uk/Tools/clustalw2/index.html). Subsequently, a phylogenetic tree was constructed using the neighbor-joining method and implemented using the MEGA7 software tool [60]. The reliability of an inferred tree was confirmed with bootstrap analysis performed with 1000 replications. The evolutionary distances were computed using the Poisson correction method [61] and are in the units of the number of amino acid substitutions per site. A total of 123 coding sequences from *Arabidopsis*, rice, and soybean were collected for phylogenetic analysis. All positions containing gaps and missing data were eliminated. The Ks (synonymous substitutions per synonymous site) and Ka (non-synonymous substitutions per non-synonymous site) values were extracted from the Plant Genome Duplication Database (PGDD: available online: http://chibba.agtec.uga.edu/duplication/), and these were used for calculating the approximate dates of duplication events. The date of duplication events was subsequently estimated according to the equation $T = Ks/2\lambda$, in which the mean synonymous substitution rate (λ) for soybean is 6.1×10^{-9} [62].

4.3. Plant Growth, Hormonal/Flooding Treatments, and Tissue Collection

Soybean cultivar Williams 82 was used for gene expression pattern analysis. For ethylene precursor 1-aminocyclopropane-1-carboxylic acid (ACC) treatment, plants were grown in 4-gallon pots (Greenhouse megastore, USA) containing a 3:1 mixture of turface and sand in a growth chamber under the conditions of 28/20 °C day/night temperature, 14/10 h light/dark photoperiod, 800 μmol m^{-2} s^{-1} light intensity and 60% humidity. Two-week-old plants were sprayed with 50 μM ACC (or a mock solution without ACC) on the leaf, irrigated with 2 l of hormone solution in a plastic case and then incubated for 5 h before root tissue collection.

For flooding stress treatment, two-day-old seedlings cultivated on quartz sand were completely submerged in 700 ml of water for 5 days at 25 °C with a light/dark cycle (600 μmol m^{-2} s^{-1}, 16 h light/8 h dark). The water level was kept at 2 cm above the quartz sand surface, and control seedlings were grown with a water level below the quartz sand surface. Roots and hypocotyls were collected from soybean seedlings. The collected tissues were frozen immediately in liquid nitrogen and stored at −80 °C. All samples were collected in biological triplicate.

For waterlogging treatment, transgenic and control seeds were sown in cones filled with turface and sand in the ratio of 2:1. The cones with plants at the V1 stage were kept inside a tub and flooded above the soil level (>2 cm high) for 14 days. The greenhouse temperatures (24~26 °C) were exactly the same during day and night. The heating system turned on when temperatures were below 23.8 °C. A passive cooling ridge vent opened when temperatures were above 26.6 °C. Active cooling fans turned on when temperatures were above 29.4 °C. The shade was set to stay open all the time, and the HID lights were set to be on at all levels between 5 am and 7 pm.

4.4. Promoter Analysis

The DNA sequences 2000 bp upstream of the translation start site (ATG) were extracted from the soybean genome, and the presence and abundance of the known cis-elements were analyzed with the help of the program SOGO (available online: https://sogo.dna.affrc.go.jp/cgi-bin/sogo.cgi?lang=en&pj=640&action=page&page=newplace).

4.5. Expression Profiling Using RNA-seq Datasets

The RNA-seq data generated by Libault et al. [32] from nine different soybean tissues (William 82 genotype) including flowers, leaves, nodules, pods, roots, root hairs, seeds, shoot apical meristems, and stems were used to analyze expression patterns of *GmXTHs* members. Chen et al. [34] generated RNA-seq data using soybean (Williams 82 genotype) leaf tissue under flooding stress. Briefly, flooding stress was imposed at the soybean V4 stage (four unfolded trifoliate leaves) by placing the pots into a larger pot filled up to a water level of 4 cm above the soil surface for 7 days.

4.6. RNA Extraction for Expression Pattern Analysis

The frozen samples were ground to powder in liquid nitrogen with a mortar and pestle. Approximately 100 mg tissue samples were used for RNA extraction using a RNeasy Plant mini kit (Cat# 74904, Qiagen, Valencia, CA, USA) according to the manufacturer's protocol. On-Column DNA digestion performed by following the RNase-Free DNase kit (Cat#79254, Qiagen) manufacturer's protocol. The quality and quantity of RNA were assessed using a Nanodrop®1000 spectrophotometer (Thermo Scientific, Wilmington, DE, USA).

4.7. Quantitative RT-PCR Analysis

A total of 2 μg RNA from each sample was reverse-transcribed to cDNA in 20 μL reaction volume using RNA to cDNA Ecopry™ Premix (Double primed) cDNA Synthesis Kit (Cat# 639549, Clontech, Foster City, CA, USA) according to the manufacturer's protocol. PCR was performed in a 10 μL reaction volume using the Maxima SYBR Green/ROX qPCR Master Mix (Cat# K0223, Thermo, USA) on the ABI7900HT detection system machine (ABI PRISM® 7900HT, Foster City, CA, USA). The results from three biological replicates and two technical replicates were used for data analysis. The PCR conditions were as follows: 50 °C for 2 min, 95 °C for 10 min, and 40 cycles of 95 °C for 15 s, and 60 °C for 1 min. To normalize the gene expression levels, the actin (Glyma.18G290800) gene was used as an internal control. All novel primers were designed using the Primer3 web interface (available online: http://frodo.wi.mit.edu/primer3/ [63,64]. The primer sequences are listed in Supplementary Table S2.

4.8. Construction of the pZY101-AtXTH31 Vector, Agrobacterium-mediated Soybean (Glycine max) Transformation and Progeny Segregation Analysis

The gene-specific primer pair 5′-CATGCCATGGATGGCTTTGTCTCTTATCTTTC-3′ and 5′-CATGCCATGGCTAACATTCTGGTGTTTGGG-3′ was designed to isolate the full-length CDS of *AtXTH31* from *Arabidopsis*. The PCR product (902 bp) was cloned into the pCR4-TOPO vector, and the positive plasmid was fully sequenced with M13 sequencing primers. The *AtXTH31* gene sequence was inserted into the pCNSH.131. AtMyb2p-Gus vector, which contained the Myb2 promoter. Finally, the whole cassette contained a promoter, and the gene sequence was moved into the pZY101-Asc binary vector. An improved *Agrobacterium*-mediated transformation of the soybean cotyledonary node system [65] was performed using the elite genotype "Maverick". To determine the segregation of gene of interest and selectable marker gene, at least 30 plants from each T0 event were screened using leaf paint (100 mg/L glufosinate, Sigma, St. Louis, MO, USA) analysis carried out for the T0 generation. T2 progeny from the T1 generation was similarly analyzed to identify homozygous T1 lines for subsequent study.

4.9. DNA Extraction and Quantification and PCR Confirmation of Transgenes

DNA was extracted from the transgenic plant leaves (mixed leaf tissue) using CTAB methods [66]. DNA concentrations and quality were initially estimated using a Nanodrop spectrometer (Thermo Fisher Scientific, Waltham, MA, USA) and then estimated using a QuantiT dsDNA HS Kit (Invitrogen, Thermo Fisher Scientific, Waltham, MA, USA). The concentrations from the Qubit assays were used to quantify the DNA input in each PCR reaction. Primers were designed to detect the *AtXTH31* gene of interest and the bar gene as the selectable marker gene (Supplementary Table S2). PCR conditions were as follows: 95 °C hot start 30 s, followed by 36 cycles of 95 °C denaturing 10 s, 55 °C annealing 10 s, and 72 °C extension 1 min, followed by 72 °C final extension 10 min. PCR products were analyzed on an agarose gel, and events were considered transgenic if they displayed an approximately 800 bp band for the gene of interest and a 500 bp band for the bar gene. Four positive transgenic events were obtained and used for further phenotype analysis.

4.10. TaqMan Assays and QuantStudio 3D Digital PCR Analysis for Soybean AtXTH31 Transgenic Copy Number Variation

The following equipment and chemicals were used from Applied Biosystems (Waltham, MA, USA), Thermo Fisher Scientific, USA: QuantStudio™ 3D Imager (Cat#: PN4489084), QuantStudio™ 3D Loader (Cat#: PN4482592), Dual Flat Block GeneAmpR PCR System 9700 (Cat#: PN4428235), Tilt Base & chip adapters (Cat#: PN4486414 and 4485513), QuantStudio™ 3D Digital PCR 20K Chips (Cat#: PN4485507), and QuantStudio™ 3D Digital PCR Master Mix (Cat#: PN4485718). The probe was designed and synthesized by Life Technology Company. The dPCR reaction volume was 20 μL and contained 10 μL 2× TaqMan 3D mix, 1 μL 20× FAM labeled primers and probe, 1 μL 20× VIC labeled primers and probe, 1 μL DNA samples (40 ng/ μL), and 7 μL nuclease-free water. In total, 14.5 μL each reaction product was loaded on the chips. Data analysis was conducted using QuantStudio™ 3D Analysis Suite™ Cloud Software as described previously [55]. The designed probes could only amplify transgene *AtXTH31* to ensure that no soybean homologous genes were detected. The ratio of the copy number of *AtXTH31* with lectin in the same soybean material was calculated as follows: (copies/µL of the *AtXTH31* transgene)/(copies/µL of the lectin gene Glyma.02G009600) in the same PCR reaction product. Soybean transgenic plants contained a single insert copy when the ratio value was equal to 0.5 and two insert copies when the ratio value was equal to 1. The ratio value was less than 0.5 when chimeric transgenic plants were found. The primers and probe sequences are shown in Supplementary Table S2.

5. Conclusions

In conclusion, our results showed that the soybean genome contains 61 *XTH* genes, the largest family of XTH proteins characterized in any organism to date. The results of phylogenetic analysis and chromosome location/structure provide an overview of the soybean *GmXTH* gene family. The results of the segmental and tandem duplication during expansion of the *GmXTH* gene family provide a genome-wide evolutionary overview. The results of conserved amino acid motif analysis and expression pattern analysis further provide insight into their putative function. Additionally, functional analysis of *AtXTH31* in a heterologous system suggests that the higher germination rate and longer roots/hypocotyls induced by the increased XTH activity may be responsible for the flooding tolerance of transgenic plants. Further comprehensive experiments may be required to elucidate the cellular locations and functions to understand the biological role of XTHs in soybean.

Supplementary Materials: Supplementary materials can be found at http://www.mdpi.com/1422-0067/19/9/2705/s1. Supplementary Table S1: Information related to the 61 genes homologous to XTH genes in the soybean genome; Supplementary Table S2: Oligonucleotide primer sequences used in this work; Supplementary Table S3: Estimate of the dates of the segmental duplication events of the soybean XTH gene pairs; Supplementary Figure S1: Exon-intron structure of soybean *XTHs*; Supplementary Table S4: Search for cis-elements in the GmXTH gene promoters.

Author Contributions: L.S. designed the experiments, analyzed the data, and prepared the manuscript. B.V. worked on the gene clone and vector construction. S.P. worked on the transgenic soybean phenotype analysis. J.W. analyzed the digital PCR data. H.T.N. conceived and supervised the project. All authors have read, revised, and approved the manuscript.

Acknowledgments: We acknowledge Missouri Soybean Merchandising Council, USA and Yang Zhou Science and Technology Bureau (YZ2018156), China for financial support and Plant Transformation Core Facility at the University of Missouri for soybean transformation. We thank Thermo Scientific (Life Technology) for supporting the Q3D PCR equipment. We thank Caifu Chen, Yalei Wu, Pius Brzoska, and David Keys for digital PCR technical support. We thank Yongqin Wang and Chenglin Chai for helping with plant hormone treatment. We thank Mackensie Murphy and Raymond Mutava for their help on the transgenic plant flooding assay.

Conflicts of Interest: The authors declare no conflict of interest.

References

1. Cantarel, B.L.; Coutinho, P.M.; Rancurel, C.; Bernard, T.; Lombard, V.; Henrissat, B. The Carbohydrate-Active EnZymes database (CAZy): An expert resource for Glycogenomics. *Nucleic Acids Res.* **2009**, D233–D238. [CrossRef] [PubMed]
2. Fry, S.C.; Smith, R.C.; Renwick, K.F.; Martin, D.J.; Hodge, S.K.; Matthews, K.J. Xyloglucan endotransglycosylase, a new wall-loosening enzyme activity from plants. *Biochem. J.* **1992**, *282*, 821–828. [CrossRef] [PubMed]
3. Thompson, J.E.; Fry, S.C. Restructuring of wall-bound xyloglucan by transglycosylation in living plant cells. *Plant J.* **2001**, *26*, 23–34. [CrossRef] [PubMed]
4. Van Sandt, V.S.; Suslov, D.; Verbelen, J.P.; Vissenberg, K. Xyloglucan endotransglucosylase activity loosens a plant cell wall. *Ann. Bot.* **2007**, *100*, 1467–1473. [CrossRef] [PubMed]
5. Shin, Y.K.; Yum, H.; Kim, E.S.; Cho, H.; Gothandam, K.M.; Hyun, J.; Chung, Y.Y. BcXTH1, a Brassica campestris homologue of Arabidopsis XTH9, is associated with cell expansion. *Planta* **2006**, *224*, 32–41. [CrossRef] [PubMed]

6. Miedes, E.; Herbers, K.; Sonnewald, U.; Lorences, E.P. Overexpression of a cell wall enzyme reduces xyloglucan depolymerization and softening of transgenic tomato fruits. *J. Agric. Food Chem.* **2010**, *58*, 5708–5713. [CrossRef] [PubMed]
7. Lee, J.; Burns, T.H.; Light, G.; Sun, Y.; Fokar, M.; Kasukabe, Y.; Allen, R.D. Xyloglucan endotransglycosylase/hydrolase genes in cotton and their role in fiber elongation. *Planta* **2010**, *232*, 1191–1205. [CrossRef] [PubMed]
8. Shao, M.Y.; Wang, X.D.; Ni, M.; Bibi, N.; Yuan, S.N.; Malik, W.; Zhang, H.P.; Liu, Y.X.; Hua, S.J. Regulation of cotton fiber elongation by xyloglucan endotransglycosylase/hydrolase genes. *Genet. Mol. Res.* **2011**, *10*, 3771–3782. [CrossRef] [PubMed]
9. Nishikubo, N.; Takahashi, J.; Roos, A.A.; Derba-Maceluch, M.; Piens, K.; Brumer, H.; Teeri, T.T.; Stålbrand, H.; Mellerowicz, E.J. Xyloglucan endo-transglycosylase-mediated xyloglucan rearrangements in developing wood of hybrid aspen. *Plant Physiol.* **2011**, *155*, 399–413. [CrossRef] [PubMed]
10. Zhu, X.F.; Shi, Y.Z.; Lei, G.J.; Fry, S.C.; Zhang, B.C.; Zhou, Y.H.; Braam, J.; Jiang, T.; Xu, X.Y.; Mao, C.Z.; et al. XTH31, encoding an in vitro XEH/XET-active enzyme, regulates aluminum sensitivity by modulating in vivo XET action, cell wall xyloglucan content, and aluminum binding capacity in Arabidopsis. *Plant Cell* **2012**, *24*, 4731–4747. [CrossRef] [PubMed]
11. Liu, Y.B.; Lu, S.M.; Zhang, J.F.; Liu, S.; Lu, Y.T. A xyloglucan endotransglucosylase/hydrolase involves in growth of primary root and alters the deposition of cellulose in Arabidopsis. *Planta* **2007**, *226*, 1547–1560. [CrossRef] [PubMed]
12. Wu, Y.; Spollen, W.G.; Sharp, R.E.; Hetherington, P.R.; Fry, S.C. Root Growth Maintenance at Low Water Potentials (Increased Activity of Xyloglucan Endotransglycosylase and Its Possible Regulation by Abscisic Acid). *Plant Physiol.* **1994**, *106*, 607–615. [CrossRef] [PubMed]
13. Moore, J.P.; Vicré-Gibouin, M.; Farrant, J.M.; Driouich, A. Adaptations of higher plant cell walls to water loss: Drought vs desiccation. *Physiol. Plant.* **2008**, *134*, 237–245. [CrossRef] [PubMed]
14. Wu, Y.; Jeong, B.R.; Fry, S.C.; Boyer, J.S. Change in XET activities, cell wall extensibility and hypocotyl elongation of soybean seedlings at low water potential. *Planta* **2005**, *220*, 593–601. [CrossRef] [PubMed]
15. Pritchard, J.; Hetherington, P.R.; Fry, S.C.; Tomos, A.D. Xyloglucan endotransglycosylase activity, microfibril orientation and the profiles of cell wall properties along growing regions of maize roots. *J. Exp. Bot.* **1993**, *44*, 1281–1289. [CrossRef]
16. Cho, S.K.; Kim, J.E.; Park, J.A.; Eom, T.J.; Kim, W.T. Constitutive expression of abiotic stress-inducible hot pepper CaXTH3, which encodes a xyloglucan endotransglucosylase/hydrolase homolog, improves drought and salt tolerance in transgenic Arabidopsis plants. *FEBS Lett.* **2006**, *580*, 3136–3144. [CrossRef] [PubMed]
17. Jan, A.; Yang, G.; Nakamura, H.; Ichikawa, H.; Kitano, H.; Matsuoka, M.; Komatsu, S. Characterization of a xyloglucan endotransglucosylase gene that is up-regulated by gibberellin in rice. *Plant Physiol.* **2004**, *136*, 3670–3681. [CrossRef] [PubMed]
18. Endo, A.; Tatematsu, K.; Hanada, K.; Duermeyer, L.; Okamoto, M.; Yonekura-Sakakibara, K.; Saito, K.; Toyoda, T.; Kawakami, N.; Kamiya, Y.; et al. Tissue-specific transcriptome analysis reveals cell wall metabolism, flavonol biosynthesis and defense responses are activated in the endosperm of germinating Arabidopsis thaliana seeds. *Plant Cell Physiol.* **2012**, *53*, 16–27. [CrossRef] [PubMed]
19. Miura, K.; Lee, J.; Miura, T.; Hasegawa, P.M. SIZ1 controls cell growth and plant development in Arabidopsis through salicylic acid. *Plant Cell Physiol.* **2010**, *51*, 103–113. [CrossRef] [PubMed]
20. Saab, I.N.; Sachs, M.M. A flooding-induced xyloglucan endo-transglycosylase homolog in maize is responsive to ethylene and associated with aerenchyma. *Plant Physiol.* **1996**, *112*, 385–391. [CrossRef] [PubMed]
21. Stacey, G.; Vodkin, L.; Parrott, W.A.; Shoemaker, R.C. National science foundation-sponsored workshop report. Draft plan for soybean genomics. *Plant Physiol.* **2004**, *135*, 59–70. [CrossRef] [PubMed]
22. Manavalan, L.P.; Guttikonda, S.K.; Tran, L.S.; Nguyen, H.T. Physiological and molecular approaches to improve drought resistance in soybean. *Plant Cell Physiol.* **2009**, *50*, 1260–1276. [CrossRef] [PubMed]
23. Valliyodan, B.; Ye, H.; Song, L.; Murphy, M.; Shannon, J.G.; Nguyen, H.T. Genetic diversity and genomic strategies for improving drought and waterlogging tolerance in soybeans. *J. Exp. Bot.* **2017**, *68*, 1835–1849. [CrossRef] [PubMed]
24. Ye, H.; Song, L.; Chen, H.; Valliyodan, B.; Ali, L.; Vuong, T.; Wu, C.; Orlowski, J.; Buckley, B.; Chen, P.; et al. A Major Natural Genetic Variation Associated with Root System Architecture and Plasticity Improves Waterlogging Tolerance and Yield in Soybean. *Plant Cell Environ.* **2018**, *41*, 2169–2182. [CrossRef] [PubMed]

25. Nanjo, Y.; Maruyama, K.; Yasue, H.; Yamaguchi-Shinozaki, K.; Shinozaki, K.; Komatsu, S. Transcriptional responses to flooding stress in roots including hypocotyl of soybean seedlings. *Plant Mol. Biol.* **2011**, *77*, 129–144. [CrossRef] [PubMed]
26. Zurek, D.M.; Clouse, S.D. Molecular cloning and characterization of a brassinosteroid-regulated gene from elongating soybean (*Glycine max* L.) epicotyls. *Plant Physiol.* **1994**, *104*, 161–170. [CrossRef] [PubMed]
27. Yokoyama, R.; Nishitani, K. A comprehensive expression analysis of all members of a gene family encoding cell-wall enzymes allowed us to predict cis-regulatory regions involved in cell-wall construction in specific organs of Arabidopsis. *Plant Cell Physiol.* **2001**, *42*, 1025–1033. [CrossRef] [PubMed]
28. Yokoyama, R.; Rose, J.K.; Nishitani, K. A surprising diversity and abundance of xyloglucan endotransglucosylase/hydrolases in rice. Classification and expression analysis. *Plant Physiol.* **2004**, *134*, 1088–1099. [CrossRef] [PubMed]
29. Schmutz, J.; Cannon, S.B.; Schlueter, J.; Ma, J.; Mitros, T.; Nelson, W.; Hyten, D.L.; Song, Q.; Thelen, J.J.; Cheng, J.; et al. Genome sequence of the palaeopolyploid soybean. *Nature* **2010**, *463*, 178–183. [CrossRef] [PubMed]
30. Planas, A.; Juncosa, M.; Lloberas, J.; Querol, E. Essential catalytic role of Glu134 in endo-β-1,3-1,4-D-glucan 4-glucanohydrolase from *B. licheniformis* as determined by site-directed mutagenesis. *FEBS Lett.* **1992**, *308*, 141–145. [CrossRef]
31. Juncosa, M.; Pons, J.; Dot, T.; Querol, E.; Planas, A. Identification of active site carboxylic residues in Bacillus licheniformis 1,3-1,4-beta-D-glucan 4-glucanohydrolase by site-directed mutagenesis. *J. Biol. Chem.* **1994**, *269*, 14530–14535. [PubMed]
32. Libault, M.; Farmer, A.; Joshi, T.; Takahashi, K.; Langley, R.J.; Franklin, L.D.; He, J.; Xu, D.; May, G.; Stacey, G. An integrated transcriptome atlas of the crop model Glycine max, and its use in comparative analyses in plants. *Plant J.* **2010**, *63*, 86–99. [CrossRef] [PubMed]
33. Tamang, B.G.; Magliozzi, J.O.; Maroof, M.A.S.; Fukao, T. Physiological and transcriptomic characterization of submergence and reoxygenation responses in soybean seedlings. *Plant, Cell Environ.* **2014**, *37*, 2350–2365. [CrossRef] [PubMed]
34. Chen, W.; Yao, Q.; Patil, G.B.; Agarwal, G.; Deshmukh, R.K.; Lin, L.; Wang, B.; Wang, Y.; Prince, S.J.; Song, L.; et al. Identification and Comparative Analysis of Differential Gene Expression in Soybean Leaf Tissue under Drought and Flooding Stress Revealed by RNA-Seq. Front. *Plant Sci.* **2016**, *7*. [CrossRef] [PubMed]
35. Komatsu, S.; Deschamps, T.; Hiraga, S.; Kato, M.; Chiba, M.; Hashiguchi, A.; Tougou, M.; Shimamura, S.; Yasue, H. Characterization of a novel flooding stress-responsive alcohol dehydrogenase expressed in soybean roots. *Plant Mol. Biol.* **2011**, *77*, 309–322. [CrossRef] [PubMed]
36. Kürsteiner, O.; Dupuis, I.; Kuhlemeier, C. The pyruvate decarboxylase1 gene of Arabidopsis is required during anoxia but not other environmental stresses. *Plant Physiol.* **2003**, *132*, 968–978. [CrossRef] [PubMed]
37. Hattori, Y.; Nagai, K.; Furukawa, S.; Song, X.J.; Kawano, R.; Sakakibara, H.; Wu, J.; Matsumoto, T.; Yoshimura, A.; Kitano, H.; et al. The ethylene response factors SNORKEL1 and SNORKEL2 allow rice to adapt to deep water. *Nature* **2009**, *460*, 1026–1030. [CrossRef] [PubMed]
38. Strohmeier, M.; Hrmova, M.; Fischer, M.; Harvey, A.J.; Fincher, G.B.; Pleiss, J. Molecular modeling of family GH16 glycoside hydrolases: Potential roles for xyloglucan transglucosylases/hydrolases in cell wall modification in the poaceae. *Protein Sci.* **2004**, *13*, 3200–3213. [CrossRef] [PubMed]
39. Geisler-Lee, J.; Geisler, M.; Coutinho, P.M.; Segerman, B.; Nishikubo, N.; Takahashi, J.; Aspeborg, H.; Djerbi, S.; Master, E.; Andersson-Gunnerås, S.; et al. Poplar carbohydrate-active enzymes. Gene identification and expression analyses. *Plant Physiol.* **2006**, *140*, 946–962. [CrossRef] [PubMed]
40. Saladié, M.; Rose, J.K.; Cosgrove, D.J.; Catalá, C. Characterization of a new xyloglucan endotransglucosylase/hydrolase (XTH) from ripening tomato fruit and implications for the diverse modes of enzymic action. *Plant J.* **2006**, *47*, 282–295. [CrossRef] [PubMed]
41. Yokoyama, R.; Uwagaki, Y.; Sasaki, H.; Harada, T.; Hiwatashi, Y.; Hasebe, M.; Nishitani, K. Biological implications of the occurrence of 32 members of the XTH (xyloglucan endotransglucosylase/hydrolase) family of proteins in the bryophyte Physcomitrella patens. *Plant J.* **2010**, *64*, 645–656. [CrossRef] [PubMed]
42. Linkemer, G.; Board, J.E.; Musgrave, M.E. Waterlogging effects on growth and yield components in late-planted soybean. *Crop Sci.* **1998**, *38*, 1576–1584. [CrossRef] [PubMed]

43. Wuebker, E.F.; Mullen, R.E.; Koehler, K. Flooding and temperature effects on soybean germination. *Crop Sci.* **2001**, *41*, 1857–1861. [CrossRef]
44. Vartapetian, B.B.; Jackson, M.B. Plant adaptations to anaerobic stress. *Ann. Bot.* **1997**, *79*, 3–20. [CrossRef]
45. Komatsu, S.; Yamamoto, R.; Nanjo, Y.; Mikami, Y.; Yunokawa, H.; Sakata, K. A comprehensive analysis of the soybean genes and proteins expressed under flooding stress using transcriptome and proteome techniques. *J. Proteome Res.* **2009**, *8*, 4766–4778. [CrossRef] [PubMed]
46. Jackson, M.B.; Colmer, T.D. Response and adaptation by plants to flooding stress. *Ann. Bot.* **2005**, *96*, 501–505. [CrossRef] [PubMed]
47. Kim, Y.H.; Hwang, S.J.; Waqas, M.; Khan, A.L.; Lee, J.H.; Lee, J.D.; Nguyen, H.T.; Lee, I.J. Comparative analysis of endogenous hormones level in two soybean (*Glycine max* L.) lines differing in waterlogging tolerance. *Front. Plant Sci.* **2015**, *6*, 714. [CrossRef] [PubMed]
48. Ookawara, R.; Satoh, S.; Yoshioka, T.; Ishizawa, K. Expression ofα-expansin and xyloglucan endotransglucosylase/hydrolase genes associated with shoot elongation enhanced by anoxia, ethylene and carbon dioxide in arrowhead (*Sagittaria pygmaea* Miq.) tubers. *Ann. Bot.* **2005**, *96*, 693–702. [CrossRef] [PubMed]
49. Markakis, M.N.; De Cnodder, T.; Lewandowski, M.; Simon, D.; Boron, A.; Balcerowicz, D.; Doubbo, T.; Taconnat, L.; Renou, J.P.; Höfte, H.; et al. Identification of genes involved in the ACC-mediated control of root cell elongation in Arabidopsis thaliana. *BMC Plant Biol.* **2012**, *12*, 208. [CrossRef] [PubMed]
50. Cho, H.T.; Cosgrove, D.J. Regulation of root hair initiation and expansin gene expression in Arabidopsis. *Plant Cell* **2002**, *14*, 3237–3253. [CrossRef] [PubMed]
51. Hashiguchi, A.; Sakata, K.; Komatsu, S. Proteome analysis of early-stage soybean seedlings under flooding stress. *J. Proteome Res.* **2009**, *8*, 2058–2069. [CrossRef] [PubMed]
52. Baumann, M.J.; Eklöf, J.M.; Michel, G.; Kallas, A.M.; Teeri, T.T.; Czjzek, M.; Brumer, H. 3rd. Structural evidence for the evolution of xyloglucanase activity from xyloglucan endo-transglycosylases: Biological implications for cell wall metabolism Fry. *Plant Cell* **2007**, *9*, 1947–1963. [CrossRef] [PubMed]
53. Sakazono, S.; Nagata, T.; Matsuo, R.; Kajihara, S.; Watanabe, M.; Ishimoto, M.; Shimamura, S.; Harada, K.; Takahashi, R.; Mochizuki, T. Variation in root development response to flooding among 92 soybean lines during early growth stages. *Plant Prod. Sci.* **2014**, *17*, 228–236. [CrossRef]
54. Jitsuyama, Y. Morphological root responses of soybean to rhizosphere hypoxia reflect waterlogging tolerance. *Can. J. Plant Sci.* **2015**, *95*, 999–1005. [CrossRef]
55. Valliyodan, B.; Van Toai, T.T.; Alves, J.D.; de Fátima, P.; Goulart, P.; Lee, J.D.; Fritschi, F.B.; Rahman, M.A.; Islam, R.; Shannon, J.G.; et al. Expression of root-related transcription factors associated with flooding tolerance of soybean (*Glycine max*). *Int. J. Mol. Sci.* **2014**, *15*, 17622–17643. [CrossRef] [PubMed]
56. Corbisier, P.; Bhat, S.; Partis, L.; Xie, V.R.; Emslie, K.R. Absolute quantification of genetically modified MON810 maize (*Zea mays* L.) by digital polymerase chain reaction. *Anal. Bioanal. Chem.* **2010**, *396*, 2143–2150. [CrossRef] [PubMed]
57. Wan, J.R.; Song, L.; Wu, Y.L.; Brzoska, P.; Keys, D.; Chen, C.F.; Valliyodan, B.; Shannon, J.G.; Nguyen, T.H. Application of Digital PCR in the Analysis of Transgenic Soybean Plants. *Adv. Biosci. Biotechnol.* **2016**, *7*, 403–417. [CrossRef]
58. Collier, R.; Dasgupta, K.; Xing, Y.P.; Hernandez, B.T.; Shao, M.; Rohozinski, D.; Kovak, E.; Lin, J.; de Oliveira, M.L.P.; Stover, E.; et al. Accurate measurement of transgene copy number in crop plants using droplet digital PCR. *Plant J.* **2017**, *90*, 1014–1025. [CrossRef] [PubMed]
59. Hu, B.; Jin, J.; Guo, A.Y.; Zhang, H.; Luo, J.; Gao, G. GSDS 2.0: An upgraded gene feature visualization server. *Bioinformatics* **2015**, *3*, 1296–1297. [CrossRef] [PubMed]
60. Kumar, S.; Stecher, G.; Tamura, K. MEGA7: Molecular Evolutionary Genetics Analysis version 7.0 for bigger datasets. *Mol. Biol. Evol.* **2016**, *33*, 1870–1874. [CrossRef] [PubMed]
61. Zuckerkandl, E.; Pauling, L. Evolutionary divergence and convergence in proteins. In *Evolving Genes and Proteins*; Bryson, V., Vogel, H.J., Eds.; Academic Press: New York, NY, USA, 1965; pp. 97–166.
62. Lynch, M.; Conery, J.S. The evolutionary fate and consequences of duplicate genes. *Science* **2000**, *290*, 1151–1155. [CrossRef] [PubMed]
63. Koressaar, T.; Remm, M. Enhancements and modifications of primer design program Primer3. *Bioinformatics* **2007**, *23*, 1289–1291. [CrossRef] [PubMed]
64. Untergasser, A.; Cutcutache, I.; Koressaar, T.; Ye, J.; Faircloth, B.C.; Remm, M.; Rozen, S.G. Primer3-new capabilities and interfaces. *Nucleic Acids Res.* **2012**, *40*, E115. [CrossRef] [PubMed]

65. Zeng, P.; Vadnais, D.; Zhang, Z.; Polacco, J. Refined glufosinate selection in Agrobacterium-mediated transformation of soybean [*Glycine max* (L.) *Merr.*]. *Plant Cell Rep.* **2004**, *22*, 478–482. [CrossRef] [PubMed]
66. Vuong, T.D.; Sleper, D.A.; Shannon, J.G.; Nguyen, H.T. Novel quantitative trait loci for broad-based resistance to soybean cyst nematode (*Heterodera glycines Ichinohe*) in soybean PI 567516C. *Theor. Appl. Genet.* **2010**, *121*, 1253–1266. [CrossRef] [PubMed]

International Journal of
Molecular Sciences

MDPI

Review

Cell Wall Proteins Play Critical Roles in Plant Adaptation to Phosphorus Deficiency

Weiwei Wu, Shengnan Zhu, Qianqian Chen, Yan Lin, Jiang Tian and Cuiyue Liang *

Root Biology Center, State Key Laboratory for Conservation and Utilization of Subtropical Agro-Bioresources, South China Agricultural University, Guangzhou 510642, China; weiweiwu2016@163.com (W.W.); shnzhu@163.com (S.Z.); qian2016200202@163.com (Q.C.); anlynlin@126.com (Y.L.); jtian@scau.edu.cn (J.T.)

* Correspondence: liangcy@scau.edu.cn; Tel.: +86-20-8528-0156; Fax: +86-20-8528-1829

Received: 2 August 2019; Accepted: 14 October 2019; Published: 23 October 2019

Abstract: Phosphorus is one of the mineral nutrient elements essential for plant growth and development. Low phosphate (Pi) availability in soils adversely affects crop production. To cope with low P stress, remodeling of root morphology and architecture is generally observed in plants, which must be accompanied by root cell wall modifications. It has been documented that cell wall proteins (CWPs) play critical roles in shaping cell walls, transmitting signals, and protecting cells against environmental stresses. However, understanding of the functions of CWPs involved in plant adaptation to P deficiency remains fragmentary. The aim of this review was to summarize advances in identification and functional characterization of CWPs in responses to P deficiency, and to highlight the critical roles of CWPs in mediating root growth, P reutilization, and mobilization in plants.

Keywords: cell wall protein; phosphate starvation; Pi mobilization; roots

1. Introduction

Phosphorus (P) is one of the mineral nutrients essential for plant development and growth, and serves as a major structural component and functions directly/indirectly in a set of biological and metabolic processes such as energy transmission, membrane and nucleotide synthesis, photosynthesis, and signal transduction [1]. However, due to its strong fixation and binding ability with organic matter and other minerals in soils, phosphate (Pi) bioavailability is often limiting, particularly in the highly-weathered acid soils in tropical and subtropical areas [2]. Therefore, low Pi availability becomes a major factor inhibiting crop production on arable lands, especially on acid soils [2,3]. To cope with low P stress on soils, plants have evolved a series of adaptive strategies to increase their Pi acquisition and utilization efficiency [4,5]. In plant roots, these multiple adaptive strategies to P deficiency mainly include: (i) using alternative metabolism pathways to reduce ATP consumption, (ii) stimulating root proliferation and enhancing symbiotic association with arbuscular mycorrhiza fungi to explore a more extensive volume of soil for Pi, (iii) enhancing root-mediated organic acid and purple acid phosphatase (PAP) exudation to remobilize sparingly-soluble P sources, and (iv) up-regulation of Pi transporters to enhance Pi uptake and translocation [1,4–7].

As part of the apoplast, which constitutes the extracellular space outside the plasma membrane, the cell wall is generally believed to play essential roles in cell division, enlargement, and differentiation. Meanwhile, the plant cell wall could also function as an important P pool for Pi reutilization when plants encounter P deficiency [8]. Therefore, many physiological and developmental changes of plant responses to Pi starvation are functionally related to cell wall modifications, and these modifications mainly determined by cell wall proteins (CWPs) [8–12]. For example, pectin content and the activities of pectin methylesterase (PMEs) in rice cell wall were found to be increased under low P conditions [11,12]. It was suggested that the carboxylate group (–COO–) in homogalacturonans can trap PO_4^{3-} by forming a –COO–Fe–PO_4^{3-} likage. The increasing amount of pectins in low P conditions might be able to provide more –COO–, which will bind Fe more tightly and subsequently release the trapped PO_4^{3-} [8]. Therefore, the coordinated increase of PME activities under low P conditions has been suggested to provided more negative charges for the cell wall, which facilitates the release of cell-wall-bound Pi for reutilization [11–13].

With the aid of genetic, molecular, and proteomic analysis, numerous CWPs have been identified and functionally characterized [14–20]. In particular, great progress has been achieved in clarifying the function of CWPs in plant responses to various abiotic stresses [19,21–24]. Moreover, complex regulatory networks underlying CWP responses to stresses have been documented [19,25–27]. This review focuses on the characterization of Pi-starvation-responsive CWPs mediating plant adaptation to P deficiency. We briefly summarize the current methods for purification and identification of CWPs in plants. Subsequently, the role of CWPs in controlling plant adaptation to P deficiency has been discussed, especially for expansins, proline-rich proteins (PRPs), oxidoreductases, and purple acid phosphatases.

2. Extraction and Identification of CWPs in Responses to Pi Starvation

Proteins localized in cell walls are ubiquitous in plants. Early studies characterized CWPs as proteins unusually rich in one or two amino acids, containing highly repetitive sequence domains, and glycosylated to varying degrees [14]. The most common types of CWPs include hydroxyproline-rich glycoproteins (HRGPs) or extensins, the arabinogalactan proteins (AGPs), the glycine-rich proteins (GRPs), the proline-rich proteins (PRPs), and chimeric proteins that contain extensin-like domains [14,28]. With assistances of the development of molecular, genetic, and proteomic technologies, a large number of other CWPs have been identified and the classical CWPs are understood to harbor several properties, including containing an N-terminal cleavage signal peptide, but lacking the ER-retention C-terminal motif KDEL/HDEL and transmembrane domains [18,20,29–32].

According to their binding ability to cell walls, plant CWPs can be classified into three categories: labile proteins, weakly bound proteins, and strongly bound proteins [18,30]. Due to these properties, different extraction approaches have been separately developed and widely applied to isolate diverse CWPs from plants [18,24,30,33–36]. Feiz et al. (2006) developed an efficient method for extracting weakly bound CWPs, in which several sequential steps are conducted, including plant sample homogenization in low ionic strength acid buffer, purification of cell walls through increasing density cushions, extensive washing of cell walls with a low ionic strength acid buffer to remove cytosolic protein contamination, and using $CaCl_2$ and high concentration of LiCl to extract CWPs instead of detergents. Compared to other available methods of CWP extraction, this method increases the proportion of the potential CWPs while decreasing the intracellular protein contamination [33]. This method has been successfully and widely used to extract CPWs from a broad range of plants, such as potato (*Solanum tuberosum*), flax (*Linum usitatissimum*), and soybean [37–39]. However, it could not be avoided losing about 15% CWPs during cell wall purification and protein extraction because of the complexity of cell wall components and the diversity of CWP properties in plants [37]. Therefore, more efficient methods for CWP purification and isolation are yet to be designed.

Compared to CWP extraction, the approaches for separation, identification, and quantification of CWPs are better developed [40–42]. With the aid of various proteomics techniques, CWPs have been widely identified in different organs of many plant species, including Medicago (*Medicago truncatula*) [17,18,30,36,43–46], maize (*Zea mays*) [47], chickpea (*Cicer arietinum*) [48], rice (*Oryza sativa*) [34,49–52], potato [53], flax [54], and sugarcane (*Saccharum officinarum*) cell suspension cultures [55]. These identified CWPs can be functionally classified as proteases, structural proteins, proteins functioning in carbohydrate metabolism, proteins acting as oxidoreductases, proteins associated with lipid metabolism, proteins related to signaling transduction, proteins with predicted interaction domains, miscellaneous proteins, and unknown function proteins [18]. The diverse biological functions of CWPs indicate sophisticated modulations of the cell wall during plant development.

In addition, proteomics techniques have been widely used to identify CWPs differentially accumulated in responses to abiotic stresses in plants. For instance, Zhu et al (2007) [56] used 2D electrophoresis to separate the CWPs from water-deficit maize primary roots, and subsequently applied time-of-flight mass spectrometry to identify the differentially accumulated proteins. The results showed that a total of 152 water-deficit-responsive CWPs were identified, which were functionally categorized into proteins related to reactive oxygen species (ROS) metabolism, proteins involved in defense repsonse, hydrolases, proteins associated with carbohydrate metabolism, and other/unknown proteins [56]. Similar proteomic studies were also conducted to quantify and characterize CWPs in plant responses to a set of stresses, such as dehydration stress in maize [56], chickpea [57], and rice [58], flooding stress in both wheat (*Triticum aestivum*) [22] and soybean (*Glycine max*) [59], and hydrogen peroxide stress in rice [60]. The results indicate that stress-responsive CWPs mediate multiple processes, and thus regulate the responses of cell elongation and expansion under abiotic stress conditions.

Despite the cumulative studies on the identification of differentially accumulated CWPs in response to abiotic stresses, few studies have been conducted to globally identify CWPs in responses to nutrient deficiencies using proteomic approaches, particularly for P deficiency. Recently, a pioneer study was conducted to identify Pi-starvation-responsive and weakly bound CWPs in soybean roots through proteomic analysis [39]. The CWPs were extracted from soybean roots following the methods developed by Feiz et al. [33] and were subsequently analyzed by isobaric tag in relative and absolute quantitation (iTRAQ) proteomics techniques. A total of 71 differentially accumulated proteins were identified, among which, 53 proteins (74.6%) were predicated to be secreted and were considered cell-wall-associated proteins by bioinformatic analysis. These proteins were predicted to be involved in carbohydrate metabolism, oxido-reduction, protein modification and turnover, signal transduction, miscellaneous functions, and unknown functions [39]. The results strongly suggest that CWPs may regulate the complex responses in soybean roots under low P conditions. However, except for the purple acid phosphatase1 like (PAP1-like) protein, which was subsequently characterized, the functions of all other CWPs in soybean adaptation to low P stress remain unclear [39].

3. CWPs Related to Root Growth in Responses to Pi Starvation

Changes of root morphology and architecture have been widely observed in plants in response to Pi starvation, which extensively increases root proliferation for soil exploration [4,61]. The modulation of root morphology and architecture is highly related to root cell wall synthesis and remodeling, which mainly relies on coordinated regulation of CWPs' enzymatic activities. However, differences of root growth in response to Pi starvation among various plant species could lead to more complex regulatory networks. For example, in P-deficient Arabidopsis, growth of primary root is arrested, while growth of lateral roots and root hairs is promoted [62]. For other plant species, such as rice and soybean, both the primary root and lateral (adventitious) root length are increased by short-term Pi starvation [39,63]. Moreover, white lupine (*Lupinus albus*) develops large numbers of cluster roots under Pi starvation conditions [64]. Thus, different responses of root morphology in various plant species should be also taken into consideration when functions of CWPs are characterized in plant acclimatization to low P stress.

So far, the functions of most CWPs in mediating plant root growth in responses to Pi starvation remain largely unknown, except for a small set of CWPs including expansins, PRPs, and some oxidoreductases [65–69].

3.1. Expansin

Expansins have been widely identified in plant cell walls, and are believed to loosen the cell wall by breaking the hydrogen bonds between hemicellulose and cellulose microfibrils in a pH-dependent manner, thus enabling turgor-driven cell expansion [70,71]. Since increases in cell size are considered to be essential for plant root growth, it is generally assumed that expansins play a critical role in mediating root growth and development. In plants, expansins are classified into four sub-families, including α-expansin or expansin A (EXPA), β-expansin or expansin B (EXPB), expansin-like A (EXPLA), and expansin-like B (EXPLB) [71,72].

Since the first expansin protein was identified in cucumber (*Cucumis sativus*) hypocotyls [73], diverse biological functions of expansin members have been widely found to be associated with environmental stresses in plants [71]. However, the roles of expansins in plant adaptation to P deficiency has attracted little attention. A β-expansin gene from soybean, *GmEXPB2*, was the first one found to regulate root growth in responses to Pi starvation [65]. It was reported that *GmEXPB2* was highly induced by Pi starvation. Transgenic Arabidopsis with *GmEXPB2* overexpression exhibited increased root cell division and elongation, which subsequently enhanced plant growth and Pi uptake [65]. A follow-up report provided supporting evidence that transgenic soybean with *GmEXPB2* overexpression also had higher plant dry weight and P content compared to wild-type plants when gown in calcareous soils [66]. Similar results were reported in wheat coleoptiles: that multiple expansins (five α- and nine β-expansin genes) were up- or down-regulated in responses to Pi application [74]. Overexpression of one expansin member, *TaEXPB23*, significantly increased the number of lateral roots in transgenic tobacco (*Nicotiana tabacum*) under both high and low P conditions [74]. These results strongly suggest the involvement of expansins in mediating root growth in responses to P deficiency.

Besides enhancing tap or lateral root elongation, root hair growth is also regulated by expansins. In soybean, GmEXPB2 was suggested to play roles in the process of soybean root hair development basing on the observation that overexpression of *GmEXPB2* dramatically increased root hair density and size of the root hair zone [75]. Similar functions were also reported for a group of expansins in other plants [71]. For example, AtEXPA7 in Arabidopsis and OsEXPA17 in rice have been documented to be required for root hair elongation [76,77], while HvEXPB1 in barley has been demonstrated to be root-hair-specific and to play roles in root hair formation [78,79]. Since increased root hair density and length is considered to improve Pi uptake efficiency in plants [80,81], root hair growth and development regulated by expansins presents a good example in which CWPs play a critical role in mediating root growth in responses to low P stress.

3.2. Proline-Rich Proteins

PRPs is a protein family with Pro- and Hyp-rich amino acid sequences. Following the identification of the first PRP in carrot (*Daucus carrota*) storage roots [82,83], a number of PRPs have been characterized in many plant species such as soybean [84–89], bean (*Phaseolus vulgaris*) [90], Medicago [91], pea (*Pisum sativum*) [86,92], maize [93], tomato (*Lycopersicon esculentum*) [94–96], and Arabidopsis [97–99]. PRPs have multiple functions, and are assumed to participate in plant development and growth, nodule formation, pathogen infection, and abiotic stress and wounding responses [83,86,90–92,100–103].

Although the roles of PRPs in response to biotic and abiotic stresses have been well documented, little information is available on their functions in plant adaptation to P deficiency. Recently, a SRPP protein, which is partially homologous to PRPs but lacks the proline-rich domain in its N-terminal region, was reported to function in Arabidopsis responses to Pi starvation [69]. It was found that SRPP is localized in cell walls, and its transcripts are specifically accumulated in both root hairs and fruits. Furthermore, loss function of *SRPP* results in shorter root hair length and reduced root hair viability under Pi starvation conditions, strongly suggesting that SRPP is involved in the regulation of root hair growth and development under low P conditions [69]. Although functions of other PRPs in plant adaptation to Pi starvation remain largely unknown, it could be hypothesized that they might play a role in plants' adaptation to Pi starvation, because transcripts of other *PRP* or *PRP*-like genes were found to be up-regulated by Pi starvation through transcriptomic analysis in plants such as Arabidopsis [104].

3.3. Oxidoreductases

A number of CWPs associated with oxido-reduction have been identified and functionally characterized in plant responses to Pi starvation [39,67]. A typical example is the cell-wall-localized ferroxidase LPR1 (Low Phosphate Root1), which has been well documented to play a critical role in cell-wall-mediated primary root responses to Pi starvation in Arabidopsis [67]. During Pi limitation, LPR1 modulates Fe deposition in cell walls of the root apical meristem and elongation zone, and thus leads to an oxidative environment for enhancing reactive oxygen species (ROS) generation. Meanwhile, increased accumulation of ROS in the apoplast subsequently induces callose deposition in cell walls and finally impairs SHORT ROOT protein movement and the maintenance of stem cell niche in root apical meristem [67]. On the other hand, ROS may also enhance the activity of class III peroxidases in the cell wall during Pi starvation, which could use H_2O_2 as a co-substrate and stiffen cell walls by catalyzing the cross-link between cell wall components [68,105]. However, since lateral root responses to Pi starvation are quite different from those of the primary roots, it still remains unclear whether cell wall modifications mediated by the LPR1-modulated ROS pathway are applicable to lateral root development under low P conditions.

Meanwhile, with the aid of transcriptomic and proteomics analysis, Pi-starvation-responsive peroxidases have been observed in plants [39,106,107]. For example, proteomics analysis of CWPs in soybean roots revealed that abundance of five peroxidase isoforms was either increased or decreased by Pi starvation, indicating a possible role of these peroxidases in soybean cell wall modification in response to low P stress [39]. Similar results were also obtained in cotton (*Gossypium hirsutum*): that 180 peroxidases were predicted to be secreted into the apoplast [107]. Among them, the expression patterns of 18 peroxidase genes were significantly influenced by P deficiency [107]. The results indicate complex regulation and diverse functions of peroxidases in responses to Pi starvation, suggesting that it is worth to investigating their functions in plant cell wall modification under low P conditions.

In addition, a cell-wall-targeted berberine bridge enzyme (BBE)-like protein was demonstrated to be regulated by Pi starvation in soybean roots [39]. It was revealed that protein abundance of the BBE-like protein was increased by about 6-fold under low P conditions, strongly suggesting that BBE-like protein might participate in root responses to Pi starvation [39]. The BBE protein in plants was first functionally characterized in poppy (*Eschscholzia californica*) in which it oxidates the N-methyl group of (S)-reticuline, yielding (S)-scoulerine [108]. In recent years, numbers of genes encoding BBE-like enzymes have been identified in plants, but functions of BBE-like proteins in plants are largely unexplored. In Arabidopsis, a total of 28 *BBE*-like genes have been identified [109]. The transcripts of *AtBBE*-like genes were found by microarray assay at different root developmental stages, such as lateral root initiation, root elongation and maturation [109,110]. Moreover, four members of this group, including AT1G01980, AT4G20840, AT1G11770, and AT4G20830, exhibit the ability to oxidize oligogalacturonides, thus enhancing resistance to fungi [111]. Additionally, oligogalacturonides have been demonstrated to play a role in root elongation by inducing the oxidative burst in alfalfa (*Medicago sativa*), suggesting that BBE-like proteins might have the potential role of controlling root growth by mediated the dynamic equilibrium of oligogalacturonides in cell walls [112]. However, the functions of BBE-like proteins in plant root development in responses to Pi starvation remain enigmatic.

4. Purple Acid Phosphatase Functions in Cell Wall Synthesis and Pi Mobilization

Another plant adaptation strategy to low P stress is to increase the accumulation of extracellular acid phosphatase (APase), which facilitates the scavenging of Pi from organophosphate compounds in apoplasts [4,113].

PAPs belong to the APase family, which exhibits distinct biochemical properties, such as purple color in aqueous solution, activity insensitive to inhibition by L-tartrate, and a bimetallic active site [113]. The plant PAP family is composed of a large group of members and displays non-specific APase activities, which catalyze the hydrolysis of Pi from a wide range of organophosphates, with optimum reaction pH below 7.0. Although diverse functions of plant PAPs have been observed in plants, most plant PAPs are suggested to play a critical role in P scavenging and recycling in plants, especially for extracellular PAPs [113,114].

A group of cell-wall-targeted PAPs has been well characterized in plants, including Arabidopsis, soybean, and tobacco [113]. In Arabidopsis, 14 PAPs have been identified in cell wall proteomes (WallProtDB, http://www.polebio.lrsv.ups-tlse.fr/WallProtDB/). Among them, four PAPs, including AtPAP10, AtPAP12, AtPAP25, and AtPAP26, have been functionally characterized in Arabidopsis responses to Pi starvation [115–119]; AtPAP12 and AtPAP26 contribute to over 70% of the total APase activity secreted by roots or targeted to shoot cell walls of Arabidopsis under low P conditions [115,117]. Furthermore, an *atpap12/atpap26* double mutant exhibited impaired growth when organic P (e.g., glycerol-3-phospahte and herring sperm DNA) was applied as the sole P source, strongly suggesting that cell-wall-associated AtPAP12 and/or AtPAP26 participates in recycling Pi from endogenous phosphomonoesters that leak into cell walls [118,120]. Similarly, AtPAP10 is predominantly localized to the surface of root epidermal cells and functions in the acclimation of Arabidopsis to Pi deprivation through activating organic P in solid MS medium [116,117].

Recently, an iTRAQ proteomics analysis was conducted to identify CWPs with differential accumulations in soybean roots in response to Pi starvation [39]. Among differential CWPs, a cell wall-localized PAP, GmPAP1-like, was identified to be up-regulated by Pi starvation. Furthermore, *GmPAP1-like* overexpression resulted in significant increases of both relative growth and P content in soybean hairy roots when deoxy-ribonucleotide triphosphate (dNTP) was supplied as the sole external P source, strongly suggesting that cell-wall-targeted GmPAP1-like participates in extracellular organic-P utilization in plants [39].

However, unlike GmPAP1-like, AtPAP12, and AtPAP26, a cell-wall-targeted PAP, AtPAP25, exhibited high phosphatase activity against several phosphoproteins and phosphoamino acids in vitro. Moreover, mutation of *AtPAP25* led to a reduced level of transcripts encoding *At4*, *AtPPCK1*, *AtRNS1*, *AtPAP12*, and *AtPAP17*, which were involved in Pi starvation responses. Therefore, it was suggested that AtPAP25 might function as a phosphoprotein phosphatase and play important signaling roles during Pi deprivation [119]. Similarly, a cell wall localized PAP, NtPAP12, was purified in tobacco [121]. Furthermore, it was found that NtPAP12 could dephosphorylate several cell-wall-bound enzymes, such as α-xylosidase and β-glucosidase, in vitro, suggesting that NtPAP12 might be involved in the regulation of cell wall biosynthesis [122].

Therefore, it seems that functional dissection of cell-wall-targeted PAPs will be a complex task because functions vary among PAPs owing to their respective protein structures, accumulation patterns, and a wide variety of potential substrates. Furthermore, considering that several substrates of plant PAPs have been suggested to be signaling molecules, such as ATP, it is reasonable to hypothesize that PAP could influence signaling molecule metabolism, and thus participate in the complex signaling network in plants. Therefore, identification and characterization of cell-wall-targeted PAPs will be helpful in elucidating the functions of PAP in plant acclimation to P deficiency [113].

5. Conclusions and Perspectives

Great progress has been made in revealing the important roles of CWPs in the modification of cell wall properties, which is crucial for the meticulous modulation of plant roots in response to P deficiency. For example, LPR1-mediated callose deposition results in the inhibition of primary root growth under Pi-starvation conditions in Arabidopsis [67]. However, the primary root growth responses of other plant species are quite different from those of Arabidopsis [39,63]; whether the same regulatory pathway also exists in other plant species remains for further studies to determine. Moreover, the plant cell wall is a complex matrix, mainly consisting of cellulose microfibrills and other wall components, such as pectins, hemicelluloses, callose, and lignin, depending on the cell wall type [14,27,123]. When plants were subjected to Pi starvation, the cell wall components were changed, thus changing the cell wall's physical properties, and subsequently influencing cell expansion [8,68]. Therefore, to understand the roles of CWPs in root modulation under low P conditions, future studies are expected to investigate how CWPs concomitantly modulate the biochemistry and physiological properties of the cell wall components.

In order to elucidate the functions of CWPs in cell wall adaptation to Pi starvation, identification of CWPs with differential accumulation is critical. However, few proteomic studies have been conducted to identify CWPs in response to Pi starvation, thus limiting the understanding of the functions of CWPs in plant adaptation to Pi starvation. Meanwhile, since covalently bound proteins are poorly extracted, the full coverage of plant cell wall proteomes remains challenging [37]. A better extraction protocol remains to be developed to increase this coverage.

Taken together, the studies referenced here have demonstrated that CWPs play important roles in plant responses to Pi starvation. It can be anticipated that increased knowledge in this field will provide important theoretical bases for engineering crops with increased P efficiency.

Funding: This research was funded by the National Natural Science Foundation of China (31672220 and 31872164), Nature Science Foundation of Guangdong Province (2015A030306034), Integrated Demonstration of Key Techniques for the Industrial Development of Featured Crops in Rocky Desertification Areas of Yunnan-Guangxi-Guizhou Provinces, National Key Research and Development Program of China (2016YFD0100700), the High Level Talents Special Support Plan of Guangdong Province (2015TX01N042 and 2015TQ01N078), Key R&D Program of Guangdong Province (2019B020219003) and the Research Team Project of the Natural Science Foundation of Guangdong Province (2016A030312009).

Conflicts of Interest: The authors declare no conflict of interest.

References

1. Plaxton, W.C.; Lambers, H. Phosphorus metabolism in plants. *Annu. Plant Rev.* **2015**, *48*, 3–15.
2. Kochian, L.V.; Hoekenga, O.A.; Pineros, M.A. How do crop plants tolerate acid soils? Mechanisms of aluminum tolerance and phosphorous efficiency. *Annu. Rev. Plant Biol.* **2004**, *55*, 459–493. [CrossRef] [PubMed]
3. Vance, C.P.; Stone, C.J.; Allan, D.L. Phosphorus acquisition and use: Critical adaptations by plants for securing a nonrenewable resource. *New Phytol.* **2003**, *157*, 423–447. [CrossRef]
4. Liang, C.; Wang, J.; Zhao, J.; Tian, J.; Liao, H. Control of phosphate homeostasis through gene regulation in crops. *Curr. Opin. Plant Biol.* **2014**, *21*, 59–66. [CrossRef]
5. Ham, B.K.; Chen, J.; Yan, Y.; Lucas, W.J. Insights into plant phosphate sensing and signaling. *Curr. Opin. Biotechnol.* **2018**, *49*, 1–9. [CrossRef]
6. Chiou, T.J.; Lin, S.I. Signaling network in sensing phosphate availability in plants. *Annu. Rev. Plant Biol.* **2011**, *62*, 185–206. [CrossRef]
7. Wu, P.; Shou, H.; Xu, G.; Lian, X. Improvement of phosphorus efficiency in rice on the basis of understanding phosphate signaling and homeostasis. *Curr. Opin. Plant Biol.* **2013**, *16*, 205–212. [CrossRef]
8. Zhu, X.F.; Wang, Z.W.; Wan, J.X.; Sun, Y.; Wu, Y.R.; Li, G.X.; Shen, R.F.; Zheng, S.J. Pectin enhances rice (*Oryza sativa*) root phosphorus remobilization. *J. Exp. Bot.* **2014**, *66*, 1017–1024. [CrossRef]
9. Cramer, G.R.; Urano, K.; Delrot, S.; Pezzotti, M.; Shinozaki, K. Effects of abiotic stress on plants: A systems biology perspective. *BMC Plant Biol.* **2011**, *11*, 163. [CrossRef]
10. Hoehenwarter, W.; Monchgesang, S.; Neumann, S.; Majovsky, P.; Abel, S.; Muller, J. Comparative expression profiling reveals a role of the root apoplast in local phosphate response. *BMC Plant Biol.* **2016**, *16*, 106. [CrossRef]
11. Zhu, C.X.; Zhu, X.F.; Hu, A.Y.; Wang, C.; Wang, B.; Dong, X.Y.; Shen, R.F. Differential effects of nitrogen forms on cell wall phosphorus remobilization are mediated by nitric oxide, pectin content, and phosphate transporter expression. *Plant Physiol.* **2016**, *171*, 1407–1417. [CrossRef] [PubMed]
12. Zhu, X.F.; Zhao, X.S.; Wu, Q.; Shen, R.F. Abscisic acid is involved in root cell wall phosphorus remobilization independent of nitric oxide and ethylene in rice (*Oryza sativa*). *Ann. Bot.* **2018**, *121*, 1361–1368.
13. Zhu, X.; Li, S.; Pan, S.; Xin, X.; Gu, Y. CSI1, PATROL1, and exocyst complex cooperate in delivery of cellulose synthase complexes to the plasma membrane. *Proc. Natl. Acad. Sci. USA* **2018**, *115*, E3578–E3587. [CrossRef] [PubMed]
14. Cassab, G.I. Plant cell wall proteins. *Annu. Rev. Plant Physiol. Plant Mol. Biol.* **1998**, *49*, 281–309. [CrossRef] [PubMed]

15. Borderies, G.; Jamet, E.; Lafitte, C.; Rossignol, M.; Jauneau, A.; Boudart, G.; Monsarrat, B.; Esquerre-Tugaye, M.T.; Boudet, A.; Pont-Lezica, R. Proteomics of loosely bound cell wall proteins of *Arabidopsis thaliana* cell suspension cultures: A critical analysis. *Electrophoresis* **2003**, *24*, 3421–3432. [CrossRef] [PubMed]
16. Lee, S.J.; Saravanan, R.S.; Damasceno, C.M.; Yamane, H.; Kim, B.D.; Rose, J.K. Digging deeper into the plant cell wall proteome. *Plant Physiol. Biochem.* **2004**, *42*, 979–988. [CrossRef] [PubMed]
17. Bayer, E.M.; Bottrill, A.R.; Walshaw, J.; Vigouroux, M.; Naldrett, M.J.; Thomas, C.L.; Maule, A.J. *Arabidopsis* cell wall proteome defined using multidimensional protein identification technology. *Proteomics* **2006**, *6*, 301–311. [CrossRef]
18. Jamet, E.; Albenne, C.; Boudart, G.; Irshad, M.; Canut, H.; Pont-Lezica, R. Recent advances in plant cell wall proteomics. *Proteomics* **2008**, *8*, 893–908. [CrossRef]
19. Novakovic, L.; Guo, T.; Bacic, A.; Sampathkumar, A.; Johnson, K.L. Hitting the wall-sensing and signaling pathways involved in plant cell wall remodeling in response to abiotic stress. *Plants* **2018**, *7*, 89. [CrossRef]
20. Calderan-Rodrigues, M.J.; Guimarães Fonseca, J.; Edgar de Moraes, F.; Setem, L.V.; Begossi, A.C.; Labate, C.A. Plant cell wall proteomics: A focus on monocot species, *Brachypodium distachyon*, *Saccharum* spp. and *Oryza sativa*. *Int. J. Mol. Sci.* **2019**, *20*, 1975. [CrossRef]
21. Skirycz, A.; Inze, D. More from less: Plant growth under limited water. *Curr. Opin. Biotechnol.* **2010**, *21*, 197–203. [CrossRef] [PubMed]
22. Kong, F.J.; Oyanagi, A.; Komatsu, S. Cell wall proteome of wheat roots under flooding stress using gel-based and LC MS/MS-based proteomics approaches. *Biochim. Biophys. Acta* **2010**, *1804*, 124–136. [CrossRef] [PubMed]
23. Ranjan, A.; Pandey, N.; Lakhwani, D.; Dubey, N.K.; Pathre, U.V.; Sawant, S.V. Comparative transcriptomic analysis of roots of contrasting *Gossypium herbaceum* genotypes revealing adaptation to drought. *BMC Genomics* **2012**, *13*, 680. [CrossRef] [PubMed]
24. Komatsu, S.; Yanagawa, Y. Cell wall proteomics of crops. *Front. Plant Sci.* **2013**, *4*, 17. [CrossRef]
25. Yang, J.L.; Zhu, X.F.; Peng, Y.X.; Zheng, C.; Li, G.X.; Liu, Y.; Shi, Y.Z.; Zheng, S.J. Cell wall hemicellulose contributes significantly to aluminum adsorption and root growth in *Arabidopsis*. *Plant Physiol.* **2011**, *155*, 885–892. [CrossRef]
26. Liu, H.; Ma, Y.; Chen, N.; Guo, S.; Liu, H.; Guo, X.; Chong, K.; Xu, Y. Overexpression of stress-inducible OsBURP16, the beta subunit of polygalacturonase 1, decreases pectin content and cell adhesion and increases abiotic stress sensitivity in rice. *Plant Cell Environ.* **2014**, *37*, 1144–1158. [CrossRef]
27. Ogden, M.; Hoefgen, R.; Roessner, U.; Persson, S.; Khan, G.A. Feeding the walls: How does nutrient availability regulate cell wall composition? *Int. J. Mol. Sci.* **2018**, *19*, 2691. [CrossRef]
28. Carpita, N.; McCann, M.; Griffing, L.R. The plant extracellular matrix: News from the cell's frontier. *Plant Cell* **1996**, *8*, 1451–1463.
29. Chivasa, S.; Ndimba, B.K.; Simon, W.J.; Robertson, D.; Yu, X.L.; Knox, J.P.; Bolwell, P.; Slabas, A.R. Proteomic analysis of the *Arabidopsis thaliana* cell wall. *Electrophoresis* **2002**, *23*, 1754–1765. [CrossRef]
30. Jamet, E.; Canut, H.; Boudart, G.; Pont-Lezica, R.F. Cell wall proteins: A new insight through proteomics. *Trends Plant Sci.* **2006**, *11*, 33–39. [CrossRef]
31. Clemente, H.S.; Pont-Lezica, R.; Jamet, E. Bioinformatics as a tool for assessing the quality of sub-cellular proteomic strategies and inferring functions of proteins: Plant cell wall proteomics as a test case. *Bioinforma. Biol. Insights* **2009**, *3*, 15–28. [CrossRef]
32. Albenne, C.; Canut, H.; Jamet, E. Plant cell wall proteomics: The leadership of *Arabidopsis thaliana*. *Front. Plant Sci.* **2013**, *4*, 111. [CrossRef] [PubMed]
33. Feiz, L.; Irshad, M.; Pont-Lezica, R.F.; Canut, H.; Jamet, E. Evaluation of cell wall preparations for proteomics: A new procedure for purifying cell walls from *Arabidopsis* hypocotyls. *Plant Methods* **2006**, *2*, 10. [CrossRef] [PubMed]
34. Cho, W.K.; Hyun, T.K.; Kumar, D.; Rim, Y.; Chen, X.Y.; Jo, Y.; Kim, S.; Lee, K.W.; Park, Z.Y.; Lucas, W.J.; et al. Proteomic analysis to identify tightly-bound cell wall protein in rice calli. *Mol. Cells* **2015**, *38*, 685–696. [CrossRef] [PubMed]
35. Ghahremani, M.; Stigter, K.A.; Plaxton, W. Extraction and characterization of extracellular proteins and their post-translational modifications from *Arabidopsis thaliana* suspension cell cultures and seedlings: A critical review. *Proteomes* **2016**, *4*, 25. [CrossRef] [PubMed]

36. Duruflé, H.; Clemente, H.S.; Balliau, T.; Zivy, M.; Dunand, C.; Jamet, E. Cell wall proteome analysis of Arabidopsis thaliana mature stems. *Proteomics* **2017**, *17*, 1600449. [CrossRef] [PubMed]
37. Albenne, C.; Canut, H.; Hoffmann, L.; Jamet, E. Plant cell wall proteins: A large body of data, but what about runaways? *Proteomes* **2014**, *17*, 224–242. [CrossRef]
38. Elagamey, E.; Narula, K.; Sinha, A.; Aggarwal, P.R.; Ghosh, S.; Chakraborty, N.; Chakraborty, S. Extracellular matrix proteome and phosphoproteome of potato reveals functionally distinct and diverse canonical and non-canonical proteoforms. *Proteomes* **2016**, *4*, 20. [CrossRef]
39. Wu, W.W.; Lin, Y.; Liu, P.D.; Chen, Q.Q.; Tian, J.; Liang, C.Y. Association of extracellular dNTP utilization with a *GmPAP1*-like protein identified in cell wall proteomic analysis of soybean roots. *J. Exp. Bot.* **2018**, *69*, 603–617. [CrossRef]
40. Cheng, F.Y.; Blackburn, K.; Lin, Y.M.; Goshe, M.B.; Williamson, J.D. Absolute protein quantification by LC/MS(E) for global analysis of salicylic acid-induced plant protein secretion responses. *J. Proteome Res.* **2009**, *8*, 82–93. [CrossRef]
41. Thelen, J.J.; Peck, S.C. Quantitative proteomics in plants: Choices in abundance. *Plant Cell* **2007**, *19*, 3339–3346. [CrossRef] [PubMed]
42. Zhu, M.; Dai, S.; McClung, S.; Yan, X.; Chen, S. Functional differentiation of *Brassica napus* guard cells and mesophyll cells revealed by comparative proteomics. *Mol. Cell. Proteomics* **2009**, *8*, 752–766. [CrossRef] [PubMed]
43. Watson, B.S.; Lei, Z.; Dixon, R.A.; Sumner, L.W. Proteomics of *Medicago sativa* cell walls. *Phytochemistry* **2004**, *65*, 1709–1720. [CrossRef]
44. Soares, N.C.; Francisco, R.; Ricardo, C.P.; Jackson, P.A. Proteomics of ionically bound and soluble extracellular proteins in *Medicago truncatula* leaves. *Proteomics* **2007**, *7*, 2070–2082. [CrossRef] [PubMed]
45. Minic, Z.; Jamet, E.; Negroni, L.; Arsene der Garabedian, P.; Zivy, M.; Jouanin, L. A sub-proteome of *Arabidopsis thaliana* mature stems trapped on concanavalin A is enriched in cell wall glycoside hydrolases. *J. Exp. Bot.* **2007**, *58*, 2503–2512. [CrossRef]
46. Zhang, Y.; Giboulot, A.; Zivy, M.; Valot, B.; Jamet, E.; Albenne, C. Combining various strategies to increase the coverage of the plant cell wall glycoproteome. *Phytochemistry* **2011**, *72*, 1109–1123. [CrossRef]
47. Zhu, J.; Chen, S.; Alvarez, S.; Asirvatham, V.S.; Schachtman, D.P.; Wu, Y.; Sharp, R.E. Cell wall proteome in the maize primary root elongation zone. I. Extraction and identification of water-soluble and lightly ionically bound proteins. *Plant Physiol.* **2006**, *140*, 311–325. [CrossRef]
48. Bhushan, D.; Pandey, A.; Chattopadhyay, A.; Choudhary, M.K.; Chakraborty, S.; Datta, A.; Chakraborty, N. Extracellular matrix proteome of chickpea (*Cicer arietinum* L.) illustrates pathway abundance, novel protein functions and evolutionary perspect. *J. Proteome Res.* **2006**, *5*, 1711–1720. [CrossRef]
49. Jung, Y.H.; Jeong, S.H.; Kim, S.H.; Singh, R.; Lee, J.E.; Cho, Y.S.; Agrawal, G.K.; Rakwal, R.; Jwa, N.S. Systematic secretome analyses of rice leaf and seed callus suspension-cultured cells: Workflow development and establishment of high-density two-dimensional gel reference maps. *J. Proteome Res.* **2008**, *7*, 5187–5210. [CrossRef]
50. Chen, X.Y.; Kim, S.T.; Cho, W.K.; Rim, Y.; Kim, S.; Kim, S.W.; Kang, K.Y.; Park, Z.Y.; Kim, J.Y. Proteomics of weakly bound cell wall proteins in rice calli. *J. Plant Physiol.* **2009**, *166*, 675–685. [CrossRef]
51. Cho, W.K.; Chen, X.Y.; Chu, H.; Rim, Y.; Kim, S.; Kim, S.T.; Kim, S.W.; Park, Z.Y.; Kim, J.Y. Proteomic analysis of the secretome of rice calli. *Physiol. Plant* **2009**, *135*, 331–341. [CrossRef] [PubMed]
52. Lin, F.; Williams, B.J.; Thangella, P.A.V.; Ladak, A.; Schepmoes, A.A.; Olivos, H.J.; Zhao, K.; Callister, S.J.; Bartley, B.E. Proteomics coupled with metabolite and cell wall profiling reveal metabolic processes of a developing rice stem internode. *Front. Plant Sci.* **2017**, *8*, 1134. [CrossRef] [PubMed]
53. Lim, S.; Chisholm, K.; Coffin, R.H.; Peters, R.D.; Al-Mughrabi, K.I.; Wang-Pruski, G.; Pinto, D.M. Protein profiling in potato (*Solanum tuberosum* L.) leaf tissues by differential centrifugation. *J. Proteome Res.* **2012**, *11*, 2594–2601. [CrossRef] [PubMed]
54. Chabi, M.; Goulas, E.; Leclercq, C.C.; de Waele, I.; Rihouey, C.; Cenci, U.; Day, A.; Blervacq, A.S.; Neutelings, G.; Duponchel, L.; et al. A cell wall proteome and targeted cell wall analyses provide novel information on hemicellulose metabolism in Flax. *Mol. Cell. Proteomics* **2017**, *16*, 1634–1651. [CrossRef]
55. Calderan-Rodrigues, M.J.; Jamet, E.; Bonassi, M.B.; Guidetti-Gonzalez, S.; Begossi, A.C.; Setem, L.V.; Franceschini, L.M.; Fonseca, J.G.; Labate, C.A. Cell wall proteomics of sugarcane cell suspension cultures. *Proteomics* **2014**, *14*, 738–749. [CrossRef]

56. Zhu, J.; Alvarez, S.; Marsh, E.L.; Lenoble, M.E.; Cho, I.J.; Sivaguru, M.; Chen, S.; Nguyen, H.T.; Wu, Y.; Schachtman, D.P.; et al. Cell wall proteome in the maize primary root elongation zone. II. Region-specific changes in water soluble and lightly ionically bound proteins under water deficit. *Plant Physiol.* **2007**, *145*, 1533–1548. [CrossRef]
57. Bhushan, D.; Pandey, A.; Choudhary, M.K.; Datta, A.; Chakraborty, S.; Chakraborty, N. Comparative proteomics analysis of differentially expressed proteins in chickpea extracellular matrix during dehydration stress. *Mol. Cell. Proteomics* **2007**, *6*, 1868–1884. [CrossRef]
58. Pandey, A.; Rajamani, U.; Verma, J.; Subba, P.; Chakraborty, N.; Datta, A.; Chakraborty, S.; Chakraborty, N. Identification of extracellular matrix proteins of rice (*Oryza sativa* L.) involved in dehydration-responsive network: A proteomic approach. *J. Proteome Res.* **2010**, *9*, 3443–3464. [CrossRef]
59. Komatsu, S.; Kobayashi, Y.; Nishizawa, K.; Nanjo, Y.; Furukawa, K. Comparative proteomics analysis of differentially expressed proteins in soybean cell wall during flooding stress. *Amino Acids* **2010**, *39*, 1435–1449. [CrossRef]
60. Zhou, L.; Bokhari, S.A.; Dong, C.J.; Liu, J.Y. Comparative proteomics analysis of the root apoplasts of rice seedlings in response to hydrogenperoxide. *PLoS ONE* **2011**, *6*, e16723.
61. Kochian, L.V. Root architecture. *J. Integr. Plant Biol.* **2016**, *58*, 190–192. [CrossRef] [PubMed]
62. Gruber, B.D.; Giehl, R.F.; Friedel, S.; von Wiren, N. Plasticity of the *Arabidopsis* root system under nutrient deficiencies. *Plant Physiol.* **2013**, *163*, 161–179. [CrossRef] [PubMed]
63. Gamuyao, R.; Chin, J.H.; Pariasca-Tanaka, J.; Pesaresi, P.; Catausan, S.; Dalid, C.; Slamet-Loedin, I.; Tecson-Mendoza, E.M.; Wissuwa, M.; Heuer, S. The protein kinase Pstol1 from traditional rice confers tolerance of phosphorus deficiency. *Nature* **2012**, *488*, 535–539. [CrossRef] [PubMed]
64. Valdés-López, O.; Hernandez, G. Transcriptional regulation and signaling in phosphorus starvation: What about legumes? *J. Integr. Plant Biol.* **2008**, *50*, 1213–1222. [CrossRef]
65. Guo, W.; Zhao, J.; Li, X.; Qin, L.; Yan, X.; Liao, H. A soybean beta-expansin gene *GmEXPB2* intrinsically involved in root system architecture responses to abiotic stresses. *Plant J.* **2011**, *66*, 541–552. [CrossRef]
66. Zhou, J.; Xie, J.; Liao, H.; Wang, X. Overexpression of beta-expansin gene *GmEXPB2* improves phosphorus efficiency in soybean. *Physiol. Plant* **2014**, *150*, 194–204. [CrossRef]
67. Müller, J.; Toev, T.; Heisters, M.; Teller, J.; Moore, K.L.; Hause, G.; Dinesh, D.C.; Bürstenbinder, K.; Abel, S. Iron-dependent callose deposition adjusts root meristem maintenance to phosphate availability. *Dev. Cell* **2015**, *33*, 216–230. [CrossRef]
68. Balzergue, C.; Dartevelle, T.; Godon, C.; Laugier, E.; Meisrimler, C.; Teulon, J.M.; Creff, A.; Bissler, M.; Brouchoud, C.; Hagege, A.; et al. Low phosphate activates STOP1-ALMT1 to rapidly inhibit root cell elongation. *Nat. Commun.* **2017**, *8*, 15300. [CrossRef]
69. Tanaka, N.; Uno, H.; Okuda, S.; Gunji, S.; Ferjani, A.; Aoyama, T.; Maeshima, M. SRPP, a cell wall protein is involved in development and protection of seeds and root hairs in *Arabidopsis thaliana*. *Plant Cell Physiol.* **2017**, *58*, 760–769. [CrossRef]
70. McQueen-Mason, S.; Cosgrove, D.J. Disruption of hydrogen bonding between plant cell wall polymers by proteins that induce wall extension. *Proc. Natl. Acad. Sci. USA* **1994**, *91*, 6574–6578. [CrossRef]
71. Marowa, P.; Ding, A.; Kong, Y. Expansins: Roles in plant growth and potential applications in crop improvement. *Plant Cell Rep.* **2016**, *35*, 949–965. [CrossRef] [PubMed]
72. Lee, Y.; Choi, D.; Kende, H. Expansins: Ever-expanding numbers and functions. *Curr. Opin. Plant Biol.* **2001**, *4*, 527–532. [CrossRef]
73. Cosgrove, D.J.; Durachko, D.M. Autolysis and extension of isolated walls from growing cucumber hypocotyls. *J. Exp. Bot.* **1994**, *45*, 1711–1719. [CrossRef] [PubMed]
74. Han, Y.Y.; Zhou, S.; Chen, Y.H.; Kong, X.; Xu, Y.; Wang, W. The involvement of expansins in responses to phosphorus availability in wheat, and its potentials in improving phosphorus efficiency of plants. *Plant Physiol. Biochem.* **2014**, *78*, 53–62. [CrossRef] [PubMed]
75. Li, X.; Zhao, J.; Tan, Z.; Zeng, R.; Liao, H. *GmEXPB2*, a cell wall beta-Expansin, affects soybean nodulation through modifying root architecture and promoting nodule formation and development. *Plant Physiol.* **2015**, *169*, 2640–2653.
76. Lin, C.; Choi, H.S.; Cho, H.T. Root hair-specific *EXPANSIN A7* is required for root hair elongation in *Arabidopsis*. *Mol. Cells* **2011**, *31*, 393–397. [CrossRef]

77. Yu, Z.M.; Bo, K.; He, X.W.; Lv, S.L.; Bai, Y.H.; Ding, W.N.; Chen, M.; Cho, H.-T.; Wu, P. Root hair-specific expansins modulate root hair elongation in rice. *Plant J.* **2011**, *66*, 725–734.
78. Kwasniewski, M.; Szarejko, I. Molecular cloning and characterization of beta-expansin gene related to root hair formation in barley. *Plant Physiol.* **2006**, *141*, 1149–1158. [CrossRef]
79. Won, S.K.; Choi, S.B.; Kumari, S.; Cho, M.; Lee, S.H.; Cho, H.T. Root hair-specific *EXPANSIN B* genes have been selected for *Graminaceae* root hairs. *Mol. Cells.* **2010**, *30*, 369–376. [CrossRef]
80. Niu, Y.F.; Chai, R.S.; Jin, G.L.; Wang, H.; Tang, C.X.; Zhang, Y.S. Responses of root architecture development to low phosphorus availability: A review. *Ann. Bot.* **2013**, *112*, 391–408. [CrossRef]
81. Bhosale, R.; Giri, J.; Pandey, B.K.; Giehl, R.F.H.; Hartmann, A.; Traini, R.; Truskina, J.; Leftley, N.; Hanlon, M.; Swarup, K.; et al. A mechanistic framework for auxin dependent Arabidopsis root hair elongation to low external phosphate. *Nat. Commun.* **2018**, *9*, 1818. [CrossRef] [PubMed]
82. Chen, J.; Varner, J.E. Isolation and characterization of cDNA clones for carrot extensin and a proline-rich 33-kDa protein. *Proc. Natl. Acad. Sci. USA* **1985**, *82*, 4399–4403. [CrossRef] [PubMed]
83. Tierney, M.L.; Wiechert, J.; Pluymers, D. Analysis of the expression of extensin and p33-related cell wall proteins. *Mol. Genet. Genomics* **1988**, *211*, 393–399. [CrossRef]
84. Hong, J.C.; Nagao, R.T.; Key, J.L. Developmentally regulated expression of soybean proline-rich cell wall protein genes. *Plant Cell* **1989**, *1*, 937–943. [PubMed]
85. Francisco, K.S.; Tierney, M.L. Isolation and characterization of a proline-rich cell wall protein from soybean seedlings. *Plant Physiol.* **1990**, *94*, 1897–1902. [CrossRef] [PubMed]
86. Van de Wiel, C.; Scheres, B.; Franssen, H.; van Lierop, M.J.; van Lammeren, A.; van Kammen, A.; Bisseling, T. The early nodulin transcript *ENOD2* is located in the nodule parenchyma (inner cortex) of pea and soybean root nodules. *EMBO J.* **1990**, *9*, 1–7. [CrossRef] [PubMed]
87. Lindstrom, J.T.; Vodkin, L.O. A soybean cell wall protein is affected by seed color genotype. *Plant Cell* **1991**, *3*, 561–571.
88. Wyatt, R.E.; Nagao, R.T.; Key, J.L. Patterns of soybean proline-rich protein gene expression. *Plant Cell* **1992**, *4*, 99–110.
89. Chen, L.; Jiang, B.; Wu, C.; Sun, S.; Hou, W.; Han, T. *GmPRP2* promoter drives root-preferential expression in transgenic *Arabidopsis* and soybean hairy roots. *BMC Plant Biol.* **2014**, *14*, 245. [CrossRef]
90. Sheng, J.; D'Ovidio, R.; Mehdy, M.C. Negative and positive regulation of a novel proline-rich protein mRNA by fungal elicitor and wounding. *Plant J.* **1991**, *1*, 345–354. [CrossRef]
91. Wilson, R.C.; Long, F.; Maruoka, E.M.; Cooper, J.B. A new proline-rich early nodulin from *Medicago truncatula* is highly expressed in nodule meristematic cells. *Plant Cell* **1994**, *6*, 1265–1275. [PubMed]
92. Scheres, B.; Van De Wiel, C.; Zalensky, A.; Horvath, B.; Spaink, H.; Van Eck, H.; Zwartkruis, F.; Wolters, A.M.; Gloudemans, T.; Van Kammen, A.; et al. The *ENOD12* gene product is involved in the infection process during the pea-Rhizobium interaction. *Cell* **1990**, *60*, 281–294. [CrossRef]
93. Jose-Estanyol, M.; Ruiz-Avila, L.; Puigdomenech, P. A maize embryo-specific gene encodes a proline-rich and hydrophobic protein. *Plant Cell* **1992**, *4*, 413–423. [CrossRef] [PubMed]
94. Salts, Y.; Wachs, R.; Gruissem, W.; Barg, R. Sequence coding for a novel proline-rich protein preferentially expressed in young tomato fruit. *Plant Mol. Biol.* **1991**, *17*, 149–150. [CrossRef] [PubMed]
95. Santino, C.G.; Stanford, G.L.; Conner, T.W. Developmental and transgenic analysis of two tomato fruit enhanced genes. *Plant Mol. Biol.* **1997**, *33*, 405–416. [CrossRef] [PubMed]
96. Bucher, M.; Brunner, S.; Zimmermann, P.; Zardi, G.I.; Amrhein, N.; Willmitzer, L.; Riesmeier, J.W. The expression of an extensin-like protein correlates with cellular tip growth in tomato. *Plant Physiol.* **2002**, *128*, 911–923. [CrossRef]
97. Fowler, T.J.; Bernhardt, C.; Tierney, M.L. Characterization and expression of four proline-rich cell wall protein genes in Arabidopsis encoding two distinct subsets of multiple domain proteins. *Plant Physiol.* **1999**, *121*, 1081–1092. [CrossRef]
98. Bernhardt, C.; Tierney, M.L. Expression of AtPRP3, a proline-rich structural cell wall protein from Arabidopsis, is regulated by cell-type-specific developmental pathways involved in root hair formation. *Plant Physiol.* **2000**, *122*, 705–714. [CrossRef]
99. Boron, A.K.; Van Orden, J.; Nektarios Markakis, M.; Mouille, G.; Adriaensen, D.; Verbelen, J.P.; Hofte, H.; Vissenberg, K. Proline-rich protein-like PRPL1 controls elongation of root hairs in *Arabidopsis thaliana*. *J. Exp. Bot.* **2014**, *65*, 5485–5495. [CrossRef]

100. Ebener, W.; Fowler, T.J.; Suzuki, H.; Shaver, J.; Tierney, M.L. Expression of *DcPRP1* is linked to carrot storage root formation and is induced by wounding and auxin treatment. *Plant Physiol.* **1993**, *101*, 259–265. [CrossRef]
101. Suzuki, H.; Wagner, T.; Tierney, M.L. Differential expression of two soybean (*Glycine max* L.) proline-rich protein genes after wounding. *Plant Physiol.* **1993**, *101*, 1283–1287. [CrossRef] [PubMed]
102. Wu, P.; Ma, L.; Hou, X.; Wang, M.; Wu, Y.; Liu, F.; Deng, X.W. Phosphate starvation triggers distinct alterations of genome expression in Arabidopsis roots and leaves. *Plant Physiol.* **2003**, *132*, 1260–1271. [CrossRef] [PubMed]
103. Nawaz, G.; Han, Y.; Usman, B.; Liu, F.; Qin, B.; Li, R. Knockout of *OsPRP1*, a gene encoding proline-rich protein, confers enhanced cold sensitivity in rice (*Oryza sativa* L.) at the seedling stage. *Biotech* **2019**, *9*, 254. [CrossRef] [PubMed]
104. Jorge, E.; Salazar, H.; Wolfgang, S. An inventory of nutrient-responsive genes in Arabidopsis root hairs. *Front. Plant Sci.* **2016**, *7*, 237.
105. Fry, S.C. Oxidative scission of plant cell wall polysaccharides by ascorbate-induced hydroxyl radicals. *Biochem. J.* **1998**, *332*, 507–515. [CrossRef] [PubMed]
106. Hammond, J.P.; Bennett, M.J.; Bowen, H.C.; Broadley, M.R.; Eastwood, D.C.; May, S.T.; Rahn, C.; Swarup, R.; Woolaway, K.E.; White, P.J. Changes in gene expression in Arabidopsis shoots during phosphate starvation and the potential for developing smart plants. *Plant Physiol.* **2003**, *132*, 578–596. [CrossRef]
107. Duan, P.; Wang, G.; Chao, M.; Zhang, Z.; Zhang, B. Genome-wide identification and analysis of Class III peroxidases in allotetraploid cotton (*Gossypium hirsutum* L.) and their responses to PK deficiency. *Genes* **2019**, *10*, 473. [CrossRef]
108. Winkler, A.; Hartner, F.; Kutchan, T.M.; Glieder, A.; Macheroux, P. Biochemical evidence that berberine bridge enzyme belongs to a novel family of flavoproteins containing a bi-covalently attached FAD cofactor. *J. Biol. Chem.* **2006**, *281*, 21276–21285. [CrossRef]
109. Winter, D.; Vinegar, B.; Nahal, H.; Ammar, R.; Wilson, G.V.; Provart, N.J. An "Electronic Fluorescent Pictograph" browser for exploring and analyzing large-scale biological data sets. *PLoS ONE* **2007**, *2*, 718. [CrossRef]
110. Daniel, B.; Pavkov-Keller, T.; Steiner, B.; Dordic, A.; Gutmann, A.; Nidetzky, B.; Sensen, C.W.; van der Graa, E.; Wallner, S.; Gruber, K.; et al. Oxidation of monolignols by members of the berberine bridge enzyme family suggests a role in plant cell wall metabolism. *J. Biol. Chem.* **2015**, *290*, 18770–18781. [CrossRef]
111. Benedetti, M.; Verrascina, I.; Pontiggia, D.; Locci, F.; Mattei, B.; De Lorenzo, G.; Cervone, F. Four Arabidopsis berberine bridge enzyme-like proteins are specific oxidases that inactivate the elicitor-active oligogalacturonides. *Plant J.* **2018**, *94*, 260–273. [CrossRef] [PubMed]
112. Camejo, D.; Martí, M.C.; Jiménez, A.; Cabrera, J.C.; Olmos, E.; Sevilla, F. Effect of oligogalacturonides on root length, extracellular alkalinization and O_2−-accumulation in alfalfa. *Plant Physiol.* **2011**, *168*, 566–575. [CrossRef] [PubMed]
113. Tian, J.T.; Liao, H. *Annual Plant Reviews Book Series; Volume 48, Phosphorus Metabolism in Plants III. P-Deprivation Responses*; John Wiley & Sons, Ltd.: Hoboken, NJ, USA, 2015; pp. 265–287. ISBN 978-11-1895-885-8.
114. Wang, L.; Liu, D. Functions and regulation of phosphate starvation-induced secreted acid phosphatases in higher plants. *Plant Sci.* **2018**, *271*, 108–116. [CrossRef] [PubMed]
115. Tran, H.T.; Qian, W.; Hurley, B.A.; She, Y.M.; Wang, D.; Plaxton, W.C. Biochemical and molecular characterization of AtPAP12 and AtPAP26: The predominant purple acid phosphatase isozymes secreted by phosphate-starved *Arabidopsis thaliana*. *Plant Cell Environ.* **2010**, *33*, 1789–1803. [CrossRef]
116. Wang, L.; Li, Z.; Qian, W.; Guo, W.; Gao, X.; Huang, L.; Wang, H.; Zhu, H.; Wu, J.W.; Wang, D.; et al. The *Arabidopsis* purple acid phosphatase *AtPAP10* is predominantly associated with the root surface and plays an important role in plant tolerance to phosphate limitation. *Plant Physiol.* **2011**, *157*, 1283–1299. [CrossRef] [PubMed]
117. Wang, L.; Lu, S.; Zhang, Y.; Li, Z.; Du, X.; Liu, D. Comparative genetic analysis of *Arabidopsis* purple acid phosphatases *AtPAP10*, *AtPAP12*, and *AtPAP26* provides new insights into their roles in plant adaptation to phosphate deprivation. *J. Integr. Plant Biol.* **2014**, *56*, 299–314. [CrossRef] [PubMed]
118. Robinson, W.D.; Park, J.; Tran, H.T.; Del Vecchio, H.A.; Ying, S.; Zins, J.L.; Patel, K.; McKnight, T.D.; Plaxton, W.C. The secreted purple acid phosphatase isozymes AtPAP12 and AtPAP26 play a pivotal role in extracellular phosphate-scavenging by *Arabidopsis thaliana*. *J. Exp. Bot.* **2012**, *63*, 6531–6542. [CrossRef]

119. Del Vecchio, H.A.; Ying, S.; Park, J.; Knowles, V.L.; Kanno, S.; Tanoi, K.; She, Y.M.; Plaxton, W.C. The cell wall-targeted purple acid phosphatase AtPAP25 is critical for acclimation of *Arabidopsis thaliana* to nutritional phosphorus deprivation. *Plant J.* **2014**, *80*, 569–581. [CrossRef]
120. Shane, M.W.; Stigter, K.; Fedosejevs, E.T.; Plaxton, W.C. Senescence-inducible cell wall and intracellular purple acid phosphatases: Implications for phosphorus remobilization in *Hakea prostrata* (*Proteaceae*) and *Arabidopsis thaliana* (*Brassicaceae*). *J. Exp. Bot.* **2014**, *65*, 6097–6106. [CrossRef]
121. Kaida, R.; Hayashi, T.; Kaneko, T.S. Purple acid phosphatase in the walls of tobacco cells. *Phytochemistry* **2008**, *69*, 2546–2551. [CrossRef]
122. Kaida, R.; Serada, S.; Norioka, N.; Norioka, S.; Neumetzler, L.; Pauly, M.; Sampedro, J.; Zarra, I.; Hayashi, T.; Kaneko, T.S. Potential role for purple acid phosphatase in the dephosphorylation of wall proteins in tobacco cells. *Plant Physiol.* **2010**, *153*, 603–610. [CrossRef] [PubMed]
123. Stebbins, G.L. Comparative aspects of plant morphogenesis: A cellular molecular and evolutionary approach. *Am. J. Bot.* **1992**, *79*, 589–598. [CrossRef]

International Journal of
Molecular Sciences

MDPI

Article

Novel Insights from Comparative In Silico Analysis of Green Microalgal Cellulases

Gea Guerriero [1,*], Kjell Sergeant [1], Sylvain Legay [1], Jean-Francois Hausman [1], Henry-Michel Cauchie [1], Irshad Ahmad [2] and Khawar Sohail Siddiqui [2,*]

1 Environmental Research and Innovation (ERIN) Department, Luxembourg Institute of Science and Technology (LIST), 5 Avenue des Hauts-Fourneaux, L-4362 Esch/Alzette, Luxembourg; kjell.sergeant@list.lu (K.S.); sylvain.legay@list.lu (S.L.); jean-francois.hausman@list.lu (J.-F.H.); henry-michel.cauchie@list.lu (H.-M.C.)

2 Life Sciences Department, King Fahd University of Petroleum and Minerals (KFUPM), Dhahran 31261, Saudi Arabia; irshad@kfupm.edu.sa

* Correspondence: gea.guerriero@list.lu (G.G.); ksiddiqui@kfupm.edu.sa (K.S.S.); Tel.: +352-275-888-5023 (G.G.); +966-13-860-8397 (K.S.S.)

Received: 10 May 2018; Accepted: 8 June 2018; Published: 15 June 2018

Abstract: The assumption that cellulose degradation and assimilation can only be carried out by heterotrophic organisms was shattered in 2012 when it was discovered that the unicellular green alga, *Chlamydomonas reinhardtii* (Cr), can utilize cellulose for growth under CO_2-limiting conditions. Publications of genomes/transcriptomes of the colonial microalgae, *Gonium pectorale* (Gp) and *Volvox carteri* (Vc), between 2010–2016 prompted us to look for cellulase genes in these algae and to compare them to cellulases from bacteria, fungi, lower/higher plants, and invertebrate metazoans. Interestingly, algal catalytic domains (CDs), belonging to the family GH9, clustered separately and showed the highest (33–42%) and lowest (17–36%) sequence identity with respect to cellulases from invertebrate metazoans and bacteria, respectively, whereas the identity with cellulases from plants was only 27–33%. Based on comparative multiple alignments and homology models, the domain arrangement and active-site architecture of algal cellulases are described in detail. It was found that all algal cellulases are modular, consisting of putative novel cysteine-rich carbohydrate-binding modules (CBMs) and proline/serine-(PS) rich linkers. Two genes were found to encode a protein with a putative Ig-like domain and a cellulase with an unknown domain, respectively. A feature observed in one cellulase homolog from Gp and shared by a spinach cellulase is the existence of two CDs separated by linkers and with a C-terminal CBM. Dockerin and Fn-3-like domains, typically found in bacterial cellulases, are absent in algal enzymes. The targeted gene expression analysis shows that two Gp cellulases consisting, respectively, of a single and two CDs were upregulated upon filter paper addition to the medium.

Keywords: cellulase; glycosyl hydrolase family 9; carbohydrate binding module (CBM); bioinformatics; RT-qPCR

1. Introduction

Cellulose, a linear polysaccharide of glucose linked by β-1,4-glycosidic linkages, is the most abundant biopolymer on Earth and is found in the cell walls of plants. Cellulose consists of long chains of glucose tightly packed together due to H-bonds and constitutes the chief load-bearing polysaccharide. It is embedded in a matrix of pectins and hemicelluloses, and is additionally impregnated by lignin in some instances [1]. Cellulases are grouped into endoglucanases (EC: 3.2.1.4), that randomly hydrolyse internal β-1,4-glycosidic bonds and exoglucanases (cellobiohydrolase, EC: 3.2.1.91) that processively release mainly cellobiose from the reducing or non-reducing chain extremity [2]. Processive endoglucanases that possess the properties of both endo- and exocellulases have also been described [3,4].

Based on amino acid sequence similarity, cellulases are classified into different glycosyl hydrolase (GH) families [5,6]. For example, endocellulases span the GH-families, 5–10, 12, 26, 44, 45, 48, 51, 61, 74, and 124, whereas exocellulase members are found in the GH families, 5, 6, and 9 (CAZy database, available online: http://www.cazy.org/Glycoside-Hydrolases.html). Most cellulases involved in the degradation of cellulose deriving from plant lignocellulosic biomass are produced by bacteria, archaea, fungi, and protozoa [7]. Some bacteria, oomycetes, protozoa, sea squirts, the fungus *Microdochium nivale* [8] and especially plants synthesize cellulose for growth and development, and, hence, require cellulases to degrade, modify, and remodel cellulose [9]. Some microorganisms (bacteria, fungi, protozoa) that live in a symbiotic relationship within the guts of phytophagous organisms also produce cellulases [10]. Later it was discovered that, apart from cellulolytic enzymes from symbionts, invertebrates also possess endogenous cellulases secreted by salivary glands and the gut [11]. Until recently, it was considered that cellulose catabolism was limited to heterotrophic organisms and higher plants (for remodeling cellulose). However, in 2012, it was experimentally established that the photosynthetic microalga *Chlamydomonas reinhardtii* (Cr) can utilize cellulose for growth in the absence/limitation of other C-sources by secreting endocellulases [12]. The alga combines features of both plants and animals (it is considered a "planimal" [12]), and has a genome characterized by an expansion of transporter gene families, indicative of an adaptation to life in soil environments [13].

In view of the biotechnological applications of novel cellulases in the degradation of lignocellulosic biomass to produce biofuel, here, we bioinformatically analyze, for the first time, cellulases from three microalgal species whose complete genomes have been published and compared [14,15]. We choose cellulase homologs from microalgae with increasing multicellularity (unicellular alga *C. reinhardtii*; colonial algae *Gonium pectorale*, Gp, with 16 and *Volvox carteri*, Vc, with 2000–6000 cells) and compare their sequences with different cellulases from diverse taxonomic groups. We model all the microalgal cellulase homologs and analyze in detail conserved motifs and their phylogenetic relationship, arrangement of different domains, and active-site architecture in addition to examining carbohydrate-binding modules (CBMs) and linker regions. We conclude this study by determining the expression levels of three cellulases in Gp in a control condition and after the addition of crystalline cellulose substrate (filter paper) to the growth medium.

2. Results and Discussion

The present work is based on the discovery that the photosynthetic microalga, *C. reinhardtii*, can secrete cellulases into the medium under CO_2-limiting conditions, although cellulase secretion was not detected in the closely related Chlorella kessleri [12]. Interestingly, Chlorella has cellulose, whereas Cr, Gp, and Vc do not have cellulose in their cell walls [16]. In the present paper, we discuss the sequence and structural analysis of cellulases from three members of Chlorophyceae (Cr, Gp, and Vc) with increasing cellular complexity (from single cells to colonies).

2.1. Algal Cellulases Belong to Glycosyl Hydrolase Family 9

The structurally and functionally important conserved residues show that all algal sequences of catalytic domains (CDs) belong to the inverting GH9 family of CAZymes (Carbohydrate-Active Enzymes) with $(\alpha/\alpha)_6$-barrel topology. Glucanases, belonging to the GH family 9, are considered the most conserved cellulases and are widely distributed among bacteria, fungi, amoebozoa, invertebrate metazoans, mosses, ferns, gymnosperms, and angiosperms [17]. Three conserved regions are identified in the CDs of algal cellulases (Figure 1 lower panels, Supplementary Figure S1a), consistent with the motifs/patterns of GH9 cellulases reported from across diverse taxonomic groups [17]. The variation of amino acids at each position within each region is compared between microalgal (Figure 1, lower panels) and all other GH9 cellulases described (Figure 1, upper panels) [17]. Region I of microalgal cellulases contains the characteristic DAGD motif where, in addition to

H-bonding of acidic residues with water (Figure 1, lower panel and asterisks), the C-terminal D acts as the catalytic base that extracts a proton from the nucleophilic water and the N-terminal D acts as an essential supporting residue [3,18–20]. The pattern corresponding to Region I, ([LVS]-x-[GK]-G-[WFYLM]-[YHF]-D-[ACGS]-G-[DSN]-X(2)-[KMR]-[FAILY]-X-[FWYLQTV]-[APTNS]-[MLGAQS]) has now been included in the PROSITE database (PS60032) [21]. Interestingly, in Region I of GH9 from all other organisms, D (catalytic base) is replaced by an N in the Angiosperm *Medicago* and G in few sea-squirt isoenzymes [17]; however, the activities of these enzymes have not been determined. In Region I of GH9 from all other organisms, two G and a K residues are also conserved, however, their role in catalysis has not yet been elucidated (Figure 1, upper panel).

The comparison of Region II (PROSITE pattern, PS00592) reveals that, although H and R residues (Figure 1, upper panels) are involved in substrate-binding via H-bonding [18,19], both residues are not conserved. The H is replaced by V in Vc and by S in *Panesthia cribrata* (Metazoa), whereas R is replaced by K and S in microalgae and by A or G in many GH9 cellulases [17]. An interesting finding about Region II of algal cellulases is the presence of an extra four residue sequence (PT[PTA][YSG]) (Figure 1, lower middle panel), which is missing in non-algal GH9 enzymes, with the exception being two cellulase homologs (CrCel9D and Gp KXZ44756) (Supplementary Figure S1a). The PROSITE pattern, PS00592, has now been revised to [HLY]-[AILMV]-[FIL]-G-x-[NSTW]-x(2,4)-[SCTV]-[FY]-[LIVMFY]-[SITV]-G-x(1,5)-[GSY]-x(2)-[AFPSTY]-[FLPSV]-x(2)-[AILPQVM]-[HV]-[DHLS]-[KRS] [21].

Residues in Region III are involved in substrate-binding and catalysis (Figure 1, asterisks), with fully conserved E acting as an acid that protonates the leaving group [19,22] and stabilizes the positively-charged oxocarbonium transition-state [18,23]. The fully conserved nucleophilic D forms H-bonds with the residues of the active-site loop, comprising of regions I and II, to bring it in proper alignment [18].

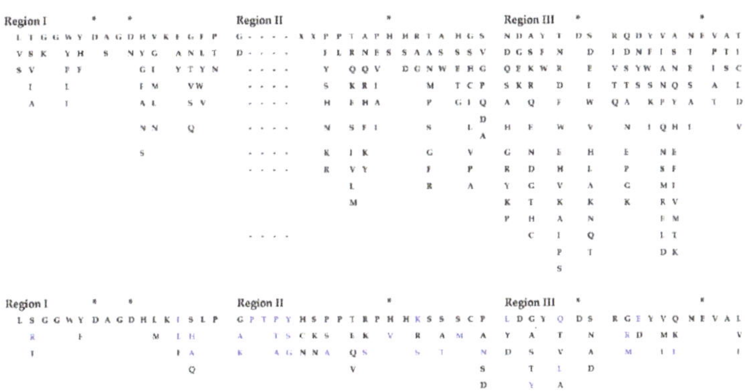

Figure 1. Variation in amino acids at each position within three conserved motifs of microalgal cellulases compared to other consensus sequences of GH9 cellulases from across the taxonomic groups. The sequences from different taxonomic groups were chosen, as mentioned in Figure 2. Upper panels, sequences from [17]; lower panels, microalgal sequences (this study). The gaps are denoted by dashes. *, catalytic, and binding residues. Blue residues, variations in algal sequences. The extra four residues in Region II is found in all algal cellulases, except CrCel9D and Gp KXZ44756. "X" refers to extra residues in Region II not shown by [17]. The pattern corresponding to Region I updates the PROSITE Database.

2.2. Algal Cellulases Are Closest to Invertebrate Metazoan GH9 Enzymes

The percentage identity matrix (Supplementary Figure S1b) and the phylogenetic analysis (Figure 2) of CD regions (such as blue highlighted regions, Supplementary Figure S1a) from GH9 cellulases reveal that algal cellulases are closer to invertebrate metazoan enzymes than to plants.

The identity matrix of GH9 family cellulases from selected groups show that algal enzymes have the highest (33–42%) and lowest (17–36%) sequence identity, with proteins from invertebrates and bacteria, respectively, whereas the identity with cellulases from plants is 27–33% (Supplementary Figure S1b).

Full length sequences of GH9 cellulases have other domains, such as CBM, immunoglobulin (Ig-like), fibronectin type III (Fn3-like), and dockerin, which can produce a bias during phylogenetic analysis [17]. To accurately determine the similarity, all cellulases were truncated to comprise only CDs (Supplementary Figure S2) and the phylogenetic tree was then constructed. The phylogenetic analysis revealed that truncated GH9 cellulases cluster together within a taxonomic group and that algal cellulases are closer to invertebrate metazoan enzymes (Figure 2), as suggested by the identity matrix (Supplementary Figure S1b). Notably, the GH9 belonging to the representatives of a specific (Sub)Kingdom (notably, Eubacteria, Fungi, Metazoa, Plantae) cluster together, except for *Ciona savignyi*, which does not form a cluster with the other representatives of Chordata *Branchiostoma floridae* and *Ciona intestinalis*, but clusters, instead, with arthropods. *C. intestinalis* forms a separate branch where the two cellulases analyzed are found together. The only representative of the Kingdom Protista (Phylum Amoebozoa), *Dictyostelium discoideum*, clusters in a sister clade to the one formed by bacteria. Interestingly, the cellulases from the green microalgae, *G. pectorale*, *C. reinhardtii*, *V. carteri*, and *Chlorella zofingiensis*, do not cluster with GH9 from Plantae, but instead form separate groups. This is consistent with the hypothesis that GH9 genes are related by vertical descent and not by horizontal gene transfer [17]. Particularly, the *G. pectoral*, GH9 KXZ44756, is sister to *C. reinhardtii* CrCel9D, which is the gene described as being strictly induced by crystalline cellulose [12].

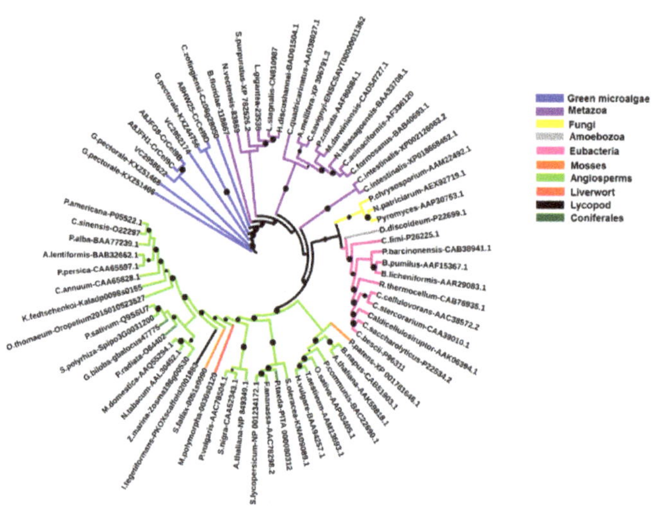

Figure 2. Maximum likelihood phylogenetic tree of GH9 cellulases (built using catalytic domains, CD in the protein sequences; see Supplementary Figure S2) from different species (100 bootstraps). The circles refer to the bootstraps (range 0.6–1; the size of the circles is proportional to the bootstrap values). The name of the species analyzed and their accession numbers are indicated in the tree. The *V. carteri* cellulases are indicated VC2958622 and VC2952174. The different colors represent the different taxonomic groups, i.e., either (Sub) Kingdoms, Phyla, Divisions, or Orders.

2.3. Algal Cellulases Are Multimodular

Sequence alignment (Supplementary Figure S1a) and homology models (Figure 3) reveal that algal GH9 cellulases consist of catalytic and non-catalytic modules. Multidomain cellulases are widespread among many taxonomic groups, however, cellulases from anaerobic bacteria, found in cellulosomes, have the most complex architecture consisting of different types of modules (Supplementary Figure S3). For example, *Clostridium cellulolyticum* produces 13 GH9 modular cellulases containing a different number and arrangement of CD (single), CBM (0–2), dockerin (0–1), and Ig-like domain (0–1) [24]. However, among templates, only the full sequences of 1JS4/4TF4 and 1KFG/1GA2, comprising CD, linker, and CBM, have been crystallized (Supplementary Figure S3) [18,25]. Multimodular cellulases are more efficient than free enzyme (with only CD) due to synergism because of the close proximity between the enzyme and the cellulosic substrate [1,2,26]. Glycosylated linkers provide flexibility to the CD for higher activity [27] and protease protection, as well as increased binding to the cellulose surface [28] (see also Section 2.6). The statistics regarding homology-based modelling are given in Supplementary Table S1, showing the top templates employed by I-TASSER, such as 1KS8 (an endocellulase from a termite), 1JS4, 1TF4 (mixed endo-/exocellulase from *Thermobifida fusca*), and 1UT9 (exocellulase from *Clostridium thermocellum*).

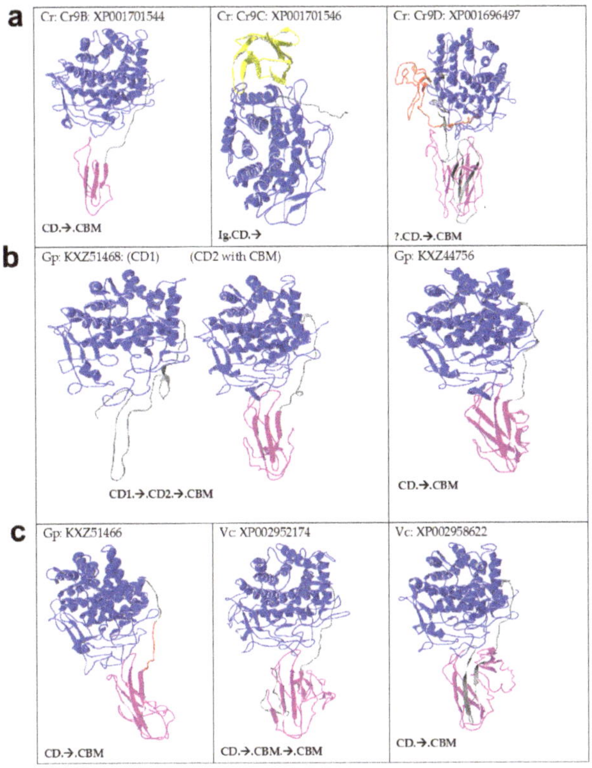

Figure 3. Homology models of selected family GH9 cellulases. Blue, CD (catalytic domain); pink, CBM (carbohydrate binding module); grey, linker; yellow, Ig-like domain; red/?, unknown. Organism names, accession/PDB codes, and cellulase types are given alongside the structures. (**a**) Cr, *Chlamydomonas reinhardtii*; (**b**) and (**c**) Gp, *Gonium pectoral* and Vc, *Volvox carteri*. The X-ray structures of templates (PDB: 1JS4/4TF4, 1KFG/1GA2, 1UT9, 1KS8, 1CLC and 2YIK) used by I-TASSER for generating homology models are given in Supplementary Figure S3. The domain arrangement is given below the structure, with a dot showing separation between two adjacent domains. CD, catalytic domain; CBM, carbohydrate-binding module; arrow, linker. The I-TASSER statistics are given in Supplementary Table S1 and the X-ray structures of templates are given in Supplementary Figure S3.

Some physicochemical properties (Supplementary Table S2) and arrangement of various domains in algal (Figure 3) and some other non-algal cellulases (Supplementary Figure S3) are given. In addition to GH9 CDs, all algal cellulases were found to have putative CBMs linked to CD by linkers, except CrCel9C (hereafter, indicated, for simplicity, as Cr9C; Figure 3a). Only Cr9C was found to have an Ig-like domain and no CBM, whereas Cr9D had an unknown sequence at the N-terminus (Figure 3a). Interestingly, Gp KXZ51468 (hereafter, referred to as Gp51468) had two consecutive CD, separated by two linkers with a single CBM (CD1-linker-CD2-linker-CBM), whereas Vc2952174 has two putative CBMs and a single CD. Spinach cellulase (Supplementary Figure S3b,c) was found to have a similar domain arrangement to Gp51468, whereas the bacterial cellulase from *Caldocellum saccharolyticum* has one family GH9 CD and another family GH48 CD, along with three CBMs [29]. In contrast to bacterial cellulases [30], dockerin and Fn3-like domains were found to be absent in algal, plant, and invertebrate metazoan cellulases (Figure 3). Multimodular microalgal cellulases were found to be closest to invertebrate metazoan homologs (Figure 2, Supplementary Figure S1b); however, in contrast to the modular cellulase from abalone, many invertebrate cellulases (such as termite) were found to be non-modular (Supplementary Figure S3).

2.4. Active-Site Architecture Shows Different Types of Cellulolytic Activities in Algal GH9 Cellulases

Cellulolytic organisms secrete different types of cellulases, in addition to β-glucosidases, xylanases and lignin-degrading enzymes [1,7]. In this work, we focus on microalgal cellulases. Various residues (white) involved in substrate binding (S-labelled), Ca^{++}-binding (M-labelled), catalytic residues (C-labelled), and loops (pink boxed) within the CD (blue highlighted region) are shown in the multiple alignment (Supplementary Figure S1a). All the catalytic acidic residues are strictly conserved in all GH9 cellulases, including microalgae (Figure 1 upper panel, Supplementary Figure S1a and Figure 4a). However, there is variation in Ca^{++}-binding residues among algal and other GH9 cellulases, probably reflecting different metal-binding affinities. For example, a D residue is substituted by an A and G in Vc2952174 and 1UT9, respectively (M-labelled, blue highlighted, Supplementary Figure S1a), that are unable to bind calcium. In glucanases, the active-site is mostly lined by aromatic residues in order to bind sugar moieties, although polar amino acids are also present (Figure 4b). These amino acids bind to cellulose via H-bonding and hydrophobic interaction, whereas aromatic amino acids interact via CH-π interaction with the sugar rings [18,31]. The substrate-binding residues around the active-site pocket are mostly conserved among GH9 sequences (Figure 4b), however, there are exceptions (red residues, Supplementary Figure S1a). In Region II (Figure 1, lower panel), highly conserved H and R are replaced by V465 and K467, respectively, in Vc2952174 (Supplementary Figure S1a).

Cellulases can be classified into endo-, exo-, and exo/endo- (also called processive endoglucanases) [3,4] (Figure 5). It has been shown that cellulases with similar sequences have different specificities, implying that exo- versus endo-versus exo/endo activities are a consequence of subtle differences in and around the active-site cleft [32]. However, in spite of this limitation, modelling of algal CDs may give valuable insight into their likely mode of action, although determination of X-ray structures and experimental data obtained via enzyme assays using various substrates (soluble, amorphous, and crystalline), as well as product analysis (cellobiose versus oligosaccharides), are more reliable options [4].

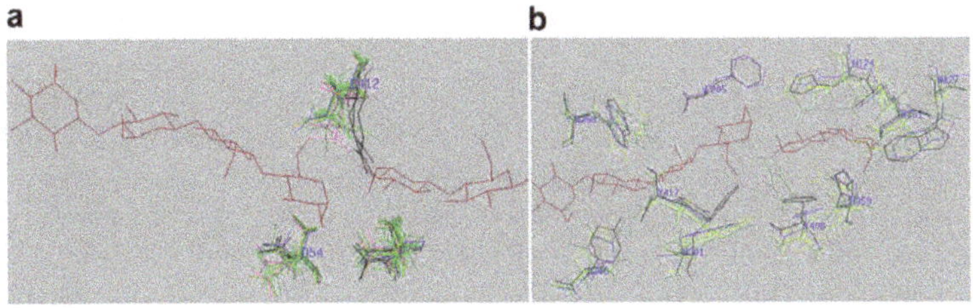

Figure 4. Active-site pocket of selected GH9 cellulases showing conserved residues around the substrate superimposed on each other. (**a**) Catalytic-residues (above, E412; below left to right, D54, D57) and (**b**) binding-residues (above from left to right: W253, F205, H124, R361, W127; below from left to right: H306, Y417, W301, Y408, H359). The residue number refers to that of termite (PDB, 1KS8). Red, substrate (C4 + C2); blue, termite; green, algae; pink, spinach; black, *T. fusca* (4TF4); orange, fungus (*Neocallimastix patriciarum*).

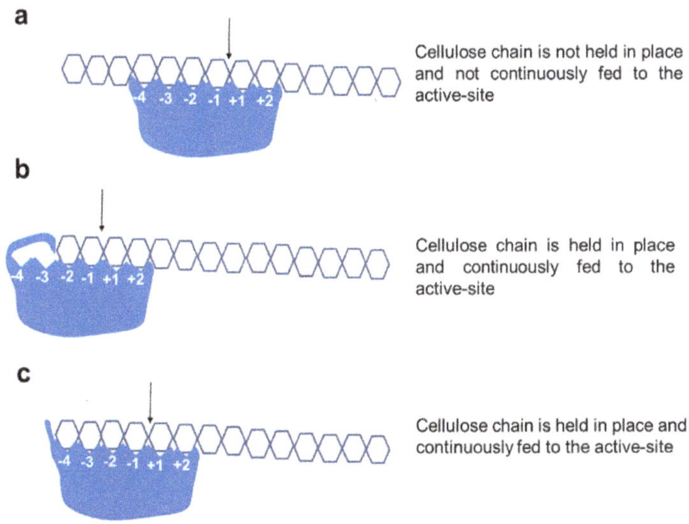

Figure 5. Various mechanisms of GH9 family cellulases found in algal enzymes. (**a**) Random cleavage of cellulose by endoglucanases to form oligosaccharides; (**b**) sequential cleavage of cellulose by GH9 exoglucanase-like enzyme due to partial blockage of the active site similar to that found in CbhA from Ruminiclostridium, 1UT9; (**c**) sequential cleavage of cellulose by processive endoglucanases (also called exo/endo cellulases) into oligosaccharides not longer than cellotetraoses due to blockage after the −4 binding site. Numbers (−4 to +2) show binding subsites (non-reducing to reducing) in the cellulase catalytic domain (blue). Arrows show the cleavage site; hexagon, glucose units.

Space-filled active-sites of algal and selected GH9 cellulases derived from homology models of CD were compared with the X-ray structures of endocellulases, an exoglucanase (both from GH6 and GH9 families), and mixed exo/endo processive endoglucanases (Figure 6). Binding and catalytic residues are shown as space-filled atoms, along with cellotetraose + cellobiose substrates (Figure 6b, top and lower panels). The X-ray structures of 4TF4 from *T. fusca* and 1KFG from *C. cellulolyticum*, with their respective oligosaccharides, were found to superimpose perfectly on each other, implying that the substrate can be modelled on algal cellulases, with the aim to locate the position of sthe ubstrate and orientation of subsites within the active-site [3,25]. To show the accessibility of the substrate, the degree of indentations and obstructions in the form of blockages and cavities around the active-site are depicted in different shades of color ranging from orange/red (humps) to dark blue (depressions) (Figure 6, middle panels).

Cellulose is a very recalcitrant crystalline polymer due to intensive H-bonding between tightly packed glucose chains [1] and the cleavage of β-glycosidic bonds is much more energy-demanding than α-linked glycosidic bonds [33]. Glucanases that act on cellulose are broadly divided into endocellulases and exocellulases [34]. Endocellulases have an open cleft or groove structure that allows the binding of many sugar units to randomly cleave internal bonds, with concomitant release of cellulose chain after every cleavage and, finally, producing oligosaccharides (Figure 5a). In contrast, exocellulases (cellobiohydrolases) from the families, GH5 and GH6, such as from *Humicola insolens* (Figure 6f), have a tunnel structure due to loops that partially cover the active-site, enabling the cellulose chain to thread through [5]. This geometry allows exoglucanase to hold on to the cellulose chain during product release without losing the chain to the surroundings. During such a processive movement, every alternate glucose unit is presented to the active site, resulting in the liberation of the cellobiose product [35].

In nature, however, the differentiation between an exo- and endo-acting cellulase is, at best, blurred, since many cellulases show characteristics of both endo- and exocellulase, depending on the active-site architecture and/or the presence of CBM close to the reducing end of the active-site [3,18,35,36]. True exocellulase with unique tunnel-like active-sites, such as in GH6 cellobiohydrolase [35] (Figure 6f, *H. insolens*), have not yet been found in GH9 family cellulases. However, one GH9 member (CbhA from *Ruminiclostridium*, 1UT9) was shown to have exocellulase activity (despite the absence of a tunnel-like active site), which was explained by the abrupt blockage of the active-site after the −2 subsite by a GEDNGLW loop, which is absent in other GH9 cellulases (Figures 5b and 6e). However, a transient tunnel formation by extended loops (DIYA-NDDY, Supplementary Figure S1a; Figure 6p–r) upon substrate binding has also been proposed [23].

The mixed exo/endo type cellulases show some type of blockage of the active-site [3,18]. A classic example showing this type of active-site architecture is a processive cellulase (E4) from Thermobifida fusca in which the non-reducing end is blocked (Figure 6c grey "tower block"). This block acts as a "measuring stick", resulting in cleavage products that are not any longer than cellotetraose that exit towards the bottom, whereas the remaining chain is held in place by the C-terminal CBM and is fed to the active-site in a processive manner (Figure 5c) [3,18]. With time, the enzyme cleaves cellulose and cellooligosaccharides (G5-G6) into cellotetraose and smaller oligosaccharides (G1-G3). Further incubation of G3-G6 cellooligosaccharides with the E4 results in the formation of a mixture of cellotetraose, cellotriose, cellobiose, and glucose products, as determined by thin-layer chromatography using purified enzymes [37]. The formation of cellotriose and cellobiose from amorphous and crystalline cellulose by cloned and purified *Clostridium thermocellum* and *Saccharophagus degradans* processive endoglucanases has also been demonstrated [4]. Other mechanisms have also been described to account for processive endoglucanases, including the presence of a CBM that binds cellulose, disrupts its crystalline structure, and feeds substrate to the active-site. Additionally, an increase in the substrate affinity for the active-site to prevent instant dissociation of the cellulose chain after initial attack has also been proposed [36]. Interestingly, a change discovered in a single amino acid around the active-site can convert a non-processive into a processive pectinase [38]. Later work extended this to *C. cellulolyticum* cellulases and proposed that the presence of a single critical aromatic residue around the active-site can influence the processive behaviour [22].

The active-site architecture of all algal enzymes, except Vc2952174, illustrates a fully open cleft (Figure 6a–n upper panels) with tower blocks towards the non-reducing end, suggesting that these may either be GH9 exo (as 1UT9) [23] or exo/endo processive enzymes [5,19,39]. The active-site architecture and the accessibility analysis of algal enzymes, such as Cr9D and Gp44756, is indicative of an open cleft and the absence of any "tower blocks" on the non-reducing end of the active-site (Figure 5a; Figure 6i,l, upper and lower panels), implying that these may simply be pure endocellulases [22].

The cavity analysis gives additional support for the presence of humps (greenish to yellowish shades) near the non-reducing end (−4 subsite) in the exo/endo-type of cellulases, whereas the active-site cleft is depicted in shades of blue depending on the depth (Figure 6j,k,m,n, middle panels). The lower panels (Figure 6) show the view from the reducing end looking down to the active-site cleft. This view confirms that, whereas pure endoglucanases (1KS8, a; 1CLC, b, Cr9D, i; Gp44756, l; Figure 6, lower panels) show low-height obstructions, GH9 exo/endo and exo-type of cellulases are characterized by tower blocks (Figure 6, lower panels; 4TF4/1JS4, c; 1KFG, d; 1UT9, e; Cr9B, g; Cr9C, h; Gp51468, j and k; Gp51466, m and both Vc cellulases, n and o). As both GH9 exo- and exo/endo processive cellulases are characterized by tower blocks (Figure 6), to unambiguously distinguish between these two types, loops responsible for purely exocellulase activity in CbhA from *Ruminoclostridium* (1UT9) were modelled for all algal enzymes (Figure 6p–r). Modeling of CbhA (1UT9) shows that these extra loops (QGY-WGS and NSPH-GCFT, Supplementary Figure S1a, pink boxed) are responsible for exo-activity by either partially covering the active-site near the non-reducing end (−4 subsite) or by running parallel along the active-site (IYAE-NDDY, Supplementary Figure S1a). This specific conformation means that they are modeled to cover the active-site upon substrate binding (Figure 6p–r, red loops) [23].

Among all algal cellulases (Figure 6g–o), only Vc2952174 has the necessary loop (CVSR-GSAR, Supplementary Figure S1a) that can block the active-site (Figure 6o,r; light pink) similar to that of CbhA exo-cellulase (red loops). In purely endo (1KS8) and exo/endo (4TF4) cellulases, these loops point away from the active-site, as seen in all algal cellulases except Vc2952174 (Figure 6p–r). The loops in all microalgal cellulases (Figure 6p–r), equivalent to the CbhA long loop (Figure 6p–r, red), running parallel to the active-site are much shorter. In the absence of X-ray structures and experimental data, it is not clear whether these shorter loops in microalgal cellulases will occlude the active-site upon substrate binding, such as in CbhA [23]. It is noteworthy that, in Vc2952174 (CVSR-GSAR) and Cr9B (THTD-GSSS), there is an extra loop that is absent in all cellulases described here (Supplementary Figure S1a). This loop covers the active-site in Vc2952174 (Figure 5b; Figure 6r, light pink loop), but is farther away from the active-site in Cr9B (Figure 6p, yellow loop).

Currently available experimental data on Chlamydomonas (Cr) cellulases can be exploited in support of our assignment of different algal cellulases as GH9 exo-, endo-, and exo/endo types. These cellulase types can be differentiated on the basis of substrates utilized and products released [12]. Cellulases with open clefts, such as exo-acting GH9, endo, and exo/endo, can hydrolyze filter paper, as well as carboxymethyl cellulose (CMC). However, whereas endo-acting enzymes form oligosaccharides preferably from amorphous cellulose, exo and exo/endo-acting enzymes can also produce cellobiose [18,34,37]. The published results showed that a mixture of all three Cr cellulases can utilize both CMC, crystalline Avicel, and filter paper, with the release of C5, C4, and C3, as well as C2 (cellobiose) as products, suggesting the presence of either an exo- or an endo- and at least one processive mixed exo/endo types of cellulase [12].

Collectively, based on the experimental data of *Cr cellulases* [12] and active-site and loop analysis described here (Figure 6), it can be deduced that all algal cellulases are likely to be exo/endo processive enzymes (presence of tower blocks with exo-loops shortened and pointing away from the active-site), except for Cr9D (Figure 6i) and Gp44756 (Figure 6l), which seem to be endoglucanases (absence of tower blocks with exo-loops shortened and pointing away from the active-site). The corresponding tower block (Figure 6e, 1UT9), due to extra loops [23] responsible for exocellulase activity in CbhA, seems to be pointing away from the active-site or is shortened in all microalgal cellulases, except in Vc2952174 (Figure 6o) where the loop (light pink) covers the active-site (Figures 5b and 6r). However, it is possible that these shortened loops in microalgal enzymes (Figure 6p–r) may close the active-site upon substrate binding. It is interesting that both the C- and N-terminal CDs in Gp51468 seem to have similar activities (processive exo/endo). It is noteworthy that Gp51468 is composed of two CDs separated by a linker, which is also found for spinach cellulase (Figure 3).

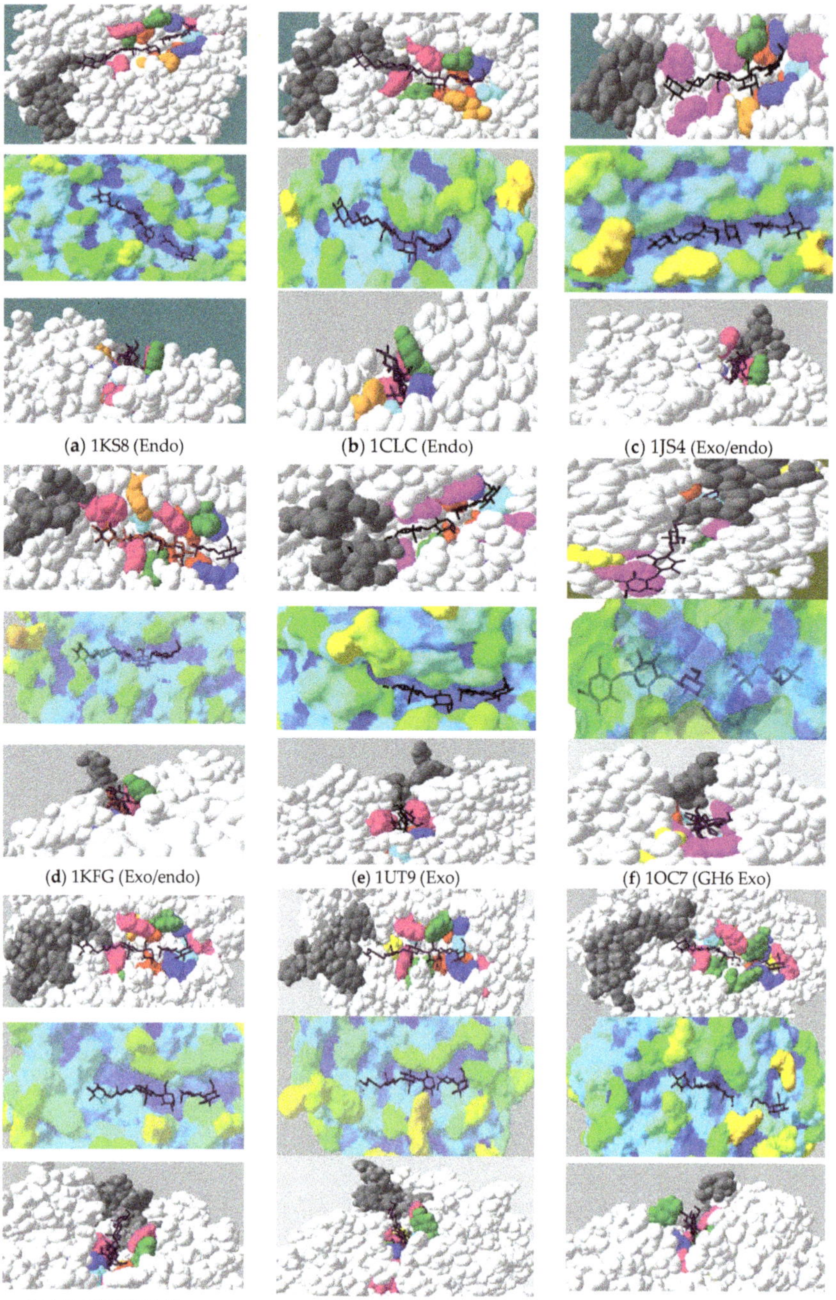

(**a**) 1KS8 (Endo) (**b**) 1CLC (Endo) (**c**) 1JS4 (Exo/endo)

(**d**) 1KFG (Exo/endo) (**e**) 1UT9 (Exo) (**f**) 1OC7 (GH6 Exo)

(**g**) XP_001701544 (Exo or Exo/Endo) (**h**) XP_001701546 (Exo or Exo/endo) (**i**) XP_001696497 (Endo or Exo)

Figure 6. *Cont.*

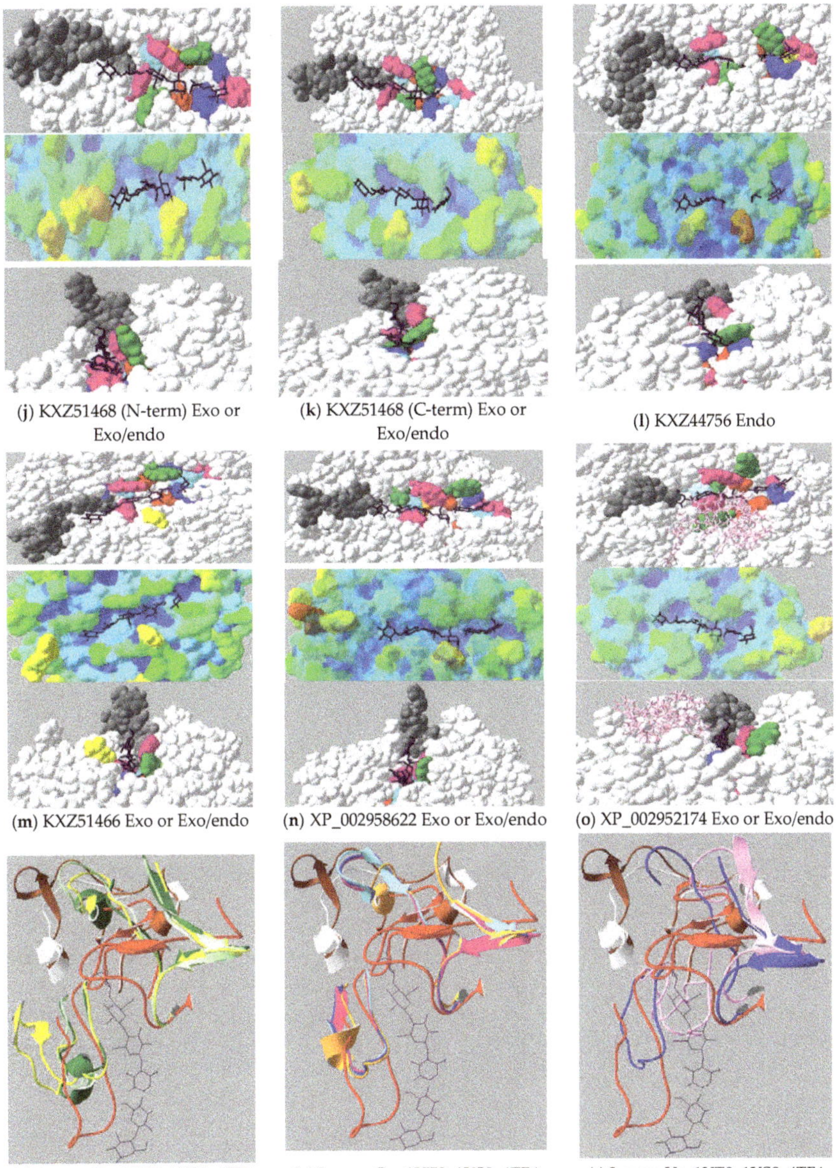

(**j**) KXZ51468 (N-term) Exo or Exo/endo
(**k**) KXZ51468 (C-term) Exo or Exo/endo
(**l**) KXZ44756 Endo

(**m**) KXZ51466 Exo or Exo/endo
(**n**) XP_002958622 Exo or Exo/endo
(**o**) XP_002952174 Exo or Exo/endo

(**p**) Loops: Cr, 1UT9, 1KS8, 4TF4
(**q**) Loops: Gp, 1UT9, 1KS8, 4TF4
(**r**) Loops: Vc, 1UT9, 1KS8, 4TF4

Figure 6. Active-site architecture of selected family GH9 cellulases determined from X-ray structures and homology models. Upper panels (**a–o**): Top view of the active-site. Critical residues surrounding the active-site. Blue, H; pink, W; turquoise, R/K; yellow, S; in lieu of substrate-binding, W; green, Y; orange, F; red, catalytic residues (E/D); grey, blocking residues/loops; Middle panels (**a–o**): Cavity analysis of the active-site pocket highlighting clefts, tunnels, and blocks in various shades. Dark blue, completely buried; orange/red, at least 75% surface accessible. Upper and middle panels showing substrate (C4 + C2) from −4 non-reducing (left) to +2 reducing end (right); Lower panels (**a–o**): View of the active-site from +2 to −4 subsite looking down the cleft/barrel highlighting the absence or presence of "tower blocks" (grey) at the non-reducing end. The extra loop in Vc2952174 (**o**) is shown as ball and stick (pink); (**p–r**): Analysis of the blocking loops/secondary structure elements in microalgal CDs compared with 1KS8, (endo-type, white), 4TF4 (exo/endo-type, brown) and 1UT9 (exo-type, red);

(p) Cr9B (XP_001701544), yellow; Cr9C (XP_001701546), light green, Cr9D (XP_001696497), dark green; (q) N-Gp (KXZ51468), turquoise; C-Gp (KXZ51468), blue; Gp (KXZ51466), magenta, Gp (KXZ44756), orange; (r) Vc (XP_002952174), light pink; Vc (XP_002958622), dark blue. Black, cleaved hexose substrate. The text description is as in Figure 3.

2.5. Novel Cysteine-Rich CBM in Algal Cellulases

The non-catalytic CBMs recognize polysaccharides and promote the association of the enzyme with its substrate, although standalone CBMs that are not linked to CDs have also been described [40]. Based on sequence similarity, CBMs are currently divided into 83 families (CAZy database available online: http://www.cazy.org/Carbohydrate-Binding-Modules.html). Three main functions of CBMs have been described [40] that include concentrating CDs of enzymes on the surface of polysaccharides for enhanced degradation, targeting distinct regions of a polysaccharide, such as crystalline cellulose [41,42], and, possibly, disrupting polysaccharide structure via replacement of H-bonds in crystalline cellulose by H-bonds from polar residues in CBM [3,43]. In addition, CBMs were proposed to help feed cellulose chain into the catalytic site, especially in the case of processive endocellulases [18].

Among the microalgal cellulases that have been described here, only Cr9C does not have a CBM, whereas all other enzymes have CBMs located on the C-terminal side, with Vc2952174 having two CBMs (Figure 3). Multiple alignment (Supplementary Figure S1a, pink highlighted), phylogenetic analysis (Supplementary Figure S4 built using the sequences in Supplementary Figure S5), and identity matrix of putative microalgal CBMs compared to CBMs across different families (1–6, 10, 11, 12, 14, 17/28, 18, 20, 41, 43–45, 47–50, 53, 81) and taxonomic groups (bacteria, fungi, microalgae, invertebrates, plants) shows low similarity between them. The identity matrix of top hits (Supplementary Figure S6) shows that, although Cr, Gp, and Vc putative CBMs have high identity (15–75%) with each other, microalgal CBMs show lower identity (19–27%) across members of known families, implying that Cr, Gp, and Vc CBMs do not belong to any of the previously described families in the CAZy database (available online: http://www.cazy.org/Carbohydrate-Binding-Modules.html). Like CBM14 and 18 family members, Cr, Gp, and Vc putative CBMs have a high percentage of cysteine residues.

To identify motifs in Cr, Gp, and Vc, CBM sequences from members belonging to different families and taxonomic groups were subjected to MEME analysis (Table 1). In addition to 2-C motifs that were found in all algal CBMs, 6-C and 4-C residue motifs were only found in Cr9B, Gp51466, Gp51468, and Vc2958622 (Table 1, motifs 1–2). It is noteworthy that none of the microalgal CBMs (Cr, Gp and Vc) have a Hevein motif characteristic of cysteine-rich CBM18 members. None of the motifs 1–3 are found in any other algal, bacterial, fungal, or plant CBMs, including Cys-rich CBM1, CBM14, and CBM18, nor in CBMs that are commonly associated with endo- (EC:3.2.1.4) and exoglucanases (EC: 3,2.1.91). Based on our results (Supplementary Figures S3 and S4, Table 1), we propose that Cys-rich algal GH9-appended CBMs are classified into a new CBM family or two separate families. One family may include Cr9B, Gp51466, Gp51468, and Vc2958622, whereas another family may include Cr9D, Gp44756, and Vc2952174.

It has been proposed that a lack of aromatic residues in the CBM binding region, along with flexible linkers, results in a decreased cellulose-CMB affinity that can promote movement and feeding of the cellulose chain to the catalytic site of processive endoglucanes [18]. In the context of microalgal GH9-appended exo/endo processive glucanase (Section 2.4, Figure 6), described here, this feature may be crucial, however, the absence of structural data precludes drawing any further conclusions. The presence of multiple C residues (10–16) in algal cellulases is also intriguing. For example, a CBM-like region on the C-terminal side of a CD/linker containing an eight cysteine-box with 4-disulfide bridges has been proposed to promote substrate binding, help in the folding of secretory proteins, maintain conformational stability, and induce a conformational change required for activity [44]. In the present study, the identification of novel CBMs in Cr, Gp, and Vc is solely based on the evidence that modular cellulases require CBM modules, along with CD and linkers (Supplementary Figure S1a). In view of the novelty of algal CBMs described here, binding data between CBMs and cellulose is vital for unequivocal confirmation.

Table 1. Motif analysis of GH9-appended microalgal CBMs (carbohydrate binding modules) by MEME.

E-Value/Cys	Motifs	Proteins
Motif 1 6-C 5.4e-30		Cr9B, Gp51466, Gp51468, Vc2958622
Motif 2 4-C 7.8e-10		Cr9B, Gp51466, Gp51468, Vc2958622
Motif 3 2-C 4.1e-7		Cr9B, Cr9D, all three Gp, Vc2958622
Hevein Motif: 7 C		General CBM18 motif
6 C	$[CG]X_{5-7}[C]X_4[CCS]X_4[C]X_6[C]X_3$ [C] $[CG]X_8[C]X_5[CCS]X_4[C]X_7[C]$	Plants (MS), Fungus (BD) Diatom (Fc)

None of the CBM1 and CBM14 motifs were found in any Cr, Gp, and Vc sequences. For a comparison, the consensus sequence of the Hevein motif is provided in the table.

2.6. *Algal Cellulases Have PS-Rich Linkers*

The linkers were, generally, thought of simply as a connecting "rope" between CD and CBMs, which, due to their flexibility, allow cellulases to nudge forward on the surface of cellulose with a caterpillar-like movement [45]. However, recent data point to many crucial functions, such as binding of glycosylated linkers to cellulose substrate [28] and the modulation of endoglucanase activity [42]. Putative linkers have been found in all algal cellulases (Figure 7). All these linkers are located between CD and CBMs, except in Cr9C, which is lacking a CBM (Figure 3). Collectively, algal linkers can be roughly classified into P/S-rich (N-Gp51468, C-Vc2952174 and Vc2958622), P/S/T-rich (Cr9D) and non-P/S/T linkers (Cr9B, Cr9C, Gp51466, C-Gp51468 and N-Vc2952174). Although it is straightforward to identify PS/PT linkers [29,46], it is not possible to confidently assign linker functions to non-P/S/T regions because of high sequence variability [27,30,46] (Figure 7). In cellulases, the average length of linkers is 20–50 residues [47]. However, linkers as small as 6–14 residues long and lacking S and/or T residues and as long as >100 residues have been reported in addition to the substitution of Ig-like or Fn3-like domains in lieu of linkers between CD and CBD modules (Figure 3) [27,30,32,46]. Interestingly, Vc2952174 has two linkers (N-Vc2952174 lacking P/S/T residues and C-Vc2952174 rich in P/S residues), one for each CBM (Figures 3 and 7). On the other hand, Gp51468 has two linkers on the C-terminal side of a single CBM (Figure 3), where N-Gp51468 is PS-rich and C-Gp51468 is lacking P, S, and T residues (Figure 7). It is intriguing to note that GH9-appended microalgal linkers have a preponderance of PS residues (Figure 7), which contrasts with PT residues found in the invertebrate metazoan abalone [48] and *Caldocellum saccharolyticum* cellulase linkers [29]). While many cellulase-appended linkers from *Pseudomonas fluorescence* are S-rich, these have very low P content [32,47], unlike microalgal linkers, which are found to be P/S-rich (Figure 7).

Glycosylation shows great diversity that depends on the sugars (type, sequence, chain length, branching point, anomeric nature) attached to various amino acid side chains that, generally, include N for N-linked modification; S, T, and Y for O-linked modification via the OH-group, and W for C-mannosylation. Whereas N-linked glycosylation requires a N-X-S/T consensus sequence (X can be any amino acid, except P), no consensus motif has been described for *O*-glycosylation. As glycans are secondary gene products, glycosylation is also cell/tissue and species specific [49]. Although *N*-glycosylation of the secreted proteins in microalgae is well documented [50], no information is available regarding *O*-glycosylation in modular algal glucanases. However, glycosylation in fungal glucanases, including *Trichoderma reesei*, has been described that displays extensive modification of linkers with di- and tri-saccharides at OH-groups of T and S residues [51]. S and T residues confer different properties to glycosylated peptides. In contrast to S, the steric repulsion between the side chain methyl group in T and the carbohydrate moiety can drastically alter the sugar to peptide backbone orientation with the possibility of altered water structure and/or H-bond formation [52]. These modifications likely lead to changes in the binding affinity of S versus T *O*-glycosylated peptides to the polysaccharide substrate, however, further experimental verification is required in the case of cellulase linker-cellulose interaction.

The presence of linkers with different amino acid sequences implies different functions. Linkers are highly divergent in lengths and sequences, but typically contain G, P, S, and T residues. P imparts extended conformation [53] and does not form H-bonds, while G provides flexibility, and S/T are often involved in *O*-glycosylation, which confers rigidity, stability, and protease-resistance [27,47]. Recent studies have found that the length (distance between the CD and CBM) and rigidity/flexibility of linkers play a critical role in the efficient functioning of cellulases; however, the precise role of a linker in the structure-function of modular cellulase is not yet fully understood [54]. For example, an increase in the number of PS/T boxes enhanced the cellulolytic activity on crystalline cellulose due to desorption of the enzyme from the substrate [55], whereas progressive shortening of linkers were shown to cause a decrease in flexibilty, with concomitant reduction in activity and enhancement in stability [56].

Cr-9D	CD→GLLQLGASTPDWNAYCGATPSPVPSPTPTPSPSPAPIPSPSPSPTPSPSPSPSATPSPSPSPSPSPSPSPSPSPSPSPSGNPSCPPT→CBM→SKAPAT
Gp44756	CD→GLLQLSASTPDWDAYCGTAPTPTTCP←CBM
Vc2952174	CD→GLIEVESLLHEGGCSWEGYCVSTSCGTPG←CBM→PSQPPPQPPAPPPPPSPPLGLPSDPSKAPAT←CBM
Cr-9B	CD→GLAAVDAAIKAAGCTWTSYCALTCTGL←CBM
Cr-9C	CD→ALSGLDAAITKAGCTWSSYCSMACT←No CBM
Vc2958622	CD→GLTDLEQGLQRSGPPPPPSPSPSLSPPPSPPPRPSSPPSQSPPPPPLASCSPT←CBM
N-Gp51468	CD→GLTALDQALRATGCNLVDYCGGKCSLAPSPSPPPAASPSLKPSPPPSSSPPPSPSPSASPSPSPSPSPSPSPG←CD
C-Gp51468	CD→GLTALDQALRSTGCNLACTNALITAQGAC←CBM
Gp51466	CD→ACTNALITAQGAC←CBM

Figure 7. Linkers in microalgal cellulases (between the arrows). Black, PS, or PST linkers; red, putative linker sequence (or may be part of C-terminal CD or N-terminal CBM).

2.7. Expression of Cellulases in Gonium Pectorale (Gp)

RT-qPCR revealed an increased expression when *Gonium* was cultivated in the presence of cellulose for two of the three cellulases analyzed in this study (Figure 8). The genes, *Gp51466* and *Gp51468*, showed a statistically-significant increase in expression in the presence of filter paper. The cellulase encoded by *Gp44756* shows a trend, which is, however, not statistically-significant. Nevertheless, if we consider the phylogenetic position of *Gp44756*, the modelling results shown in Figure 6 and the gene expression analysis, it is reasonable to assume *Gp44756* is the *G. pectorale* ortholog of *CrCel9D*.

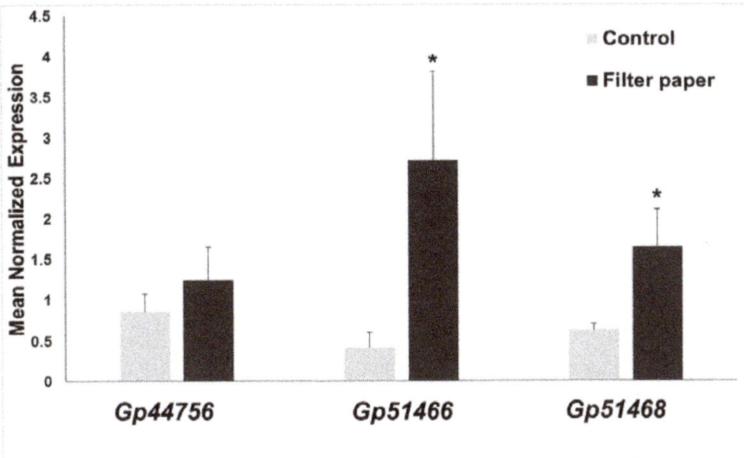

Figure 8. Gene expression analysis of the three *G. pectorale* cellulases after growth for 14 days under continuous light in the presence/absence of 0.1% (*w*/*v*) filter paper. Asterisks denote statistically-significant values after Student's *t*-test (* *p*-value < 0.05).

3. Materials and Methods

3.1. Computational Methods

The cellulase accession numbers are indicated in the phylogenetic tree. The sequences were taken from [17], manually truncated to CDs, and enriched by blasting in the Metazome database (Available online: https://metazome.jgi.doe.gov/pz/portal.html#!search?show=BLAST). The physico-chemical properties of algal cellulases were determined using the ProtParam tool (Available online: http://web.expasy.org/protparam/). Conserved domains and GH-family assignment were identified with the MotifScan (Available online: http://myhits.isb-sib.ch/cgi-bin/motif_scan) and ScanProsite [21] algorithms. The pair-wise multiple alignment of algal cellulases for identifying conserved residues and motifs were determined by using CLUSTAL-Ω (Available online: http://www.ebi.ac.uk/Tools/msa/clustalo/) [57]. The 3D homology models of the algal sequences comprising complete sequences, as well as only CD regions, were generated with I-TASSER Suite (Available online: http://zhanglab.ccmb.med.umich.edu/I-TASSER/) [58] utilizing LOMETS, SPICKER, and TM-align. The models were then refined using REMO by optimizing the backbone hydrogen-bonding networks and FG-MD by removing the steric clashes and improving the torsion angles. Separate homology models for Gp1468 and the spinach homolog were generated due to the presence of two CDs. The residues implicated in substrate binding and activity were manually annotated using the 3D structures of cellulase templates available in the PDBsum database (Available online: http://www.ebi.ac.uk/thornton-srv/databases/cgi-bin/pdbsum/GetPage.pl?pdbcode=index.html) [59], CLUSTAL-Ω, and COACH/ COFACTOR tools within the I-TASSER Suite and published literature. The final structures showing various domains, conserved regions, motifs, and active-site architecture (including surface accessibility, blocks, clefts, and tunnels) were visualized by superimposing each model on the *T. fusca* template (4TF4) in the presence of cellotetraose substrate from −1 to −4 subsites and cellobiose from +1 to +2 subsites with DeepView Swiss-PdbViewer v4.1 (Available online: http://www.expasy.org/spdbv/) [60]. For the phylogenetic analysis, truncated CD and CBM sequences (the sequences are given in Supplementary Figures S2 and S5) were aligned with Clustal-Ω and the alignment submitted to PhyML [61] (available at http://phylogeny.lirmm.fr/phylo_cgi/one_task.cgi?task_type=phyml) to obtain a maximum likelihood phylogenetic tree (100 bootstraps). The tree was visualized with iTOL-Interactive Tree of Life (Available online: http://itol.embl.de/). Putative CBMs in Cr, Gp, and Vc algae were analyzed using MotifScan and the CAZY database (Available online: http://www.cazy.org/Carbohydrate-Binding-Modules.html) by a manual search through all 83 families. For CBM analysis, the selected sequences from the CAZY database belonging to various families were truncated to only CBM parts by subjecting these sequences to MotifScan (Available online: https://myhits.isb-sib.ch/cgi-bin/motif_scan). Full sequences were used for standalone CBMs, especially from the algae and for those sequences where CBM motifs were not identified. The CBM sequences were aligned using CLUSTAL-Ω and subjected to the MEME tool (Available online: http://meme-suite.org/tools/meme) for the discovery of novel motifs [62].

3.2. Growth of G. pectorale, RNA Extraction, cDNA Synthesis and RT-qPCR

G. pectorale (strain K3-F3–4, mating type minus, NIES-2863 obtained from the Microbial Culture Collection at National Institute for Environmental Studies, Tsukuba, Japan; Available online: http://mcc.nies.go.jp/) was grown under continuous light (1300 lux) in 50 mL of modified Bold's 3N medium (UTEX, Austin, TX, USA) for 14 days in the presence/absence of autoclaved 0.1% *w*/*v* Whatman Grade 1 filter paper (Merck, Darmstadt, Germany). Algae were centrifuged for 10 min at 15,000× *g*, the pellet immediately frozen in liquid nitrogen, and cells disrupted using sterilized 5 mm stainless steel beads and a bead beater (Retsch MM400, Aartselaar, Belgium) set at 20 Hz for 2 min (the holders were previously cooled with liquid nitrogen to avoid heating of the samples during disruption). Total RNA was extracted using the Qiagen RNA extraction kit (Qiagen, Leusden, The Netherlands) coupled to the on-column DNase I digestion. The RNA purity and quality were measured with a Nanodrop ND-1000 (Thermo Scientific, Villebon-sur-Yvette, France) and a 2100 Bioanalyzer (Agilent, Santa Clara, CA, USA), respectively. Two hundred nanograms of RNA were retrotranscribed into cDNA with the ProtoScript II RTase (New England Biolabs, Leiden, The Netherlands) and random primers, according to the manufacturer's instructions.

The cDNA was diluted to 2 ng/µL and 2 µL were used for the RT-qPCR analysis (final volume of the reaction: 10 µL) in 384-wells plates. An automated liquid handling robot (epMotion 5073, Eppendorf, Hamburg, Germany) was used to prepare the plates, which were run on a Viia™ 7 System (Thermo Scientific, Villebon-sur-Yvette, France). The TaqMan Low ROX 2x Mix was used (Takyon, Eurogentec, Seraing, Belgium). To ensure robust results, the TaqMan chemistry was used to evaluate *Gonium* cellulase relative expression (fluorescent dye and quencher used are FAM-TAMRA, target-specific primers and probes are given in Supplementary Table S3). The expression of the *Gonium* cellulases was calculated with qbase$^+$ version 3.1 (Biogazelle, Zwijnaarde, Belgium; Available online: www.qbaseplus.com) after normalization using the genes, *rpl23* and *eef1*, that the program, geNorm™, identified as the most stable. Normalized relative quantities were calculated according to [63], by considering specific target PCR efficiency and multiple reference gene normalization. Here, four candidate reference genes were validated for gene expression analysis, the eukaryotic translation elongation factor, 1α *eef1*, *rpl23*, encoding the 60S ribosomal protein, L23, *tbpA* coding for the TATA-box binding protein and *tubA1* coding for α tubulin. For statistical analysis, the normalized relative quantities exported from qbase$^+$ were subjected to a Student's *t*-test, as implemented in Excel.

The primers used in this study are reported in Supplementary Table S3. Primers were designed using Primer3Plus (Available online: http://www.bioinformatics.nl/cgi-bin/primer3plus/primer3plus.cgi/) and further checked with the OligoAnalyzer 3.1 tool from Integrated DNA technologies (Available online: http://eu.idtdna.com/calc/analyzer). Primer efficiencies were checked via qPCR using 6 points of a serial five-fold dilution of cDNA starting at 20 ng.

4. Conclusions and Future Direction

This is the first report on the bioinformatics of algal family GH9 cellulases. The GH9 catalytic domains of algal cellulases form a distinct group, which is phylogenetically closer to invertebrate metazoan than plant or bacterial homologs. All algal enzymes were found to be modular and analysis of the active-site architecture of the considered CDs indicates endoglucanase and mixed exo/endo (processive endoglucanase) types of activities. It has been suggested that the lack of pure cellobiohydrolases (exo-acting) in algae are compensated by the presence of many processive endoglucanases, along with endocellulases, to produce a simple and efficient enzyme system for the degradation of cellulose [4]. Except for Cr9C, all cellulase homologs have at least one putative C-terminal novel cysteine-rich CBM. The presence of novel CBMs and PS-rich linkers, in combination with CDs, indicate that the studied cellulases may have enhanced catalytic properties suitable for the efficient degradation of cellulosic biomass. In this context, Gp51468 is of special interest as it is composed of two CDs with exo/endo activities, two different linkers, and a single CBM. Future work will involve cloning, purification, and crystallization of Gp51468 to fully understand its mode of action, as well as growing it in the presence of different cellulosic substrates for the production of valuable biochemicals.

Supplementary Materials: Supplementary materials can be found at http://www.mdpi.com/1422-0067/19/6/1782/s1.

Author Contributions: G.G. and K.S.S. conceived the idea of writing the paper and designed the experiments; G.G., K.S.S. performed the experiments; K.S., S.L. and I.A. contributed to the experimental work concerning bioinformatics and the Taqman-based RT-qPCR; G.G., I.A., K.S.S. analyzed the data; G.G. and K.S.S. wrote the paper draft; all authors discussed, contributed to and edited the paper.

Acknowledgments: The personal assistance of K.S.S. and I.A. by KFUPM is acknowledged.

Conflicts of Interest: The authors declare no conflict of interest.

References

1. Guerriero, G.; Hausman, J.F.; Strauss, J.; Ertan, H.; Siddiqui, K.S. Lignocellulosic biomass: Biosynthesis, degradation, and industrial utilization. *Eng. Life Sci.* **2016**, *16*, 1–16. [CrossRef]
2. Guerriero, G.; Siddiqui, K.S. Modular and Multifunctional Enzyme Systems for Plant Cell Wall Degradation: Diversity, Synergy, Chimeras and Magnetic-Glycosidases. *Ref. Modul. Life Sci* **2017**. [CrossRef]
3. Li, Y.; Irwin, D.C.; Wilson, D.B. Processivity, substrate binding, and mechanism of cellulose hydrolysis by *Thermobifida fusca* Cel9A. *Appl. Environ. Microbiol.* **2007**, *73*, 3165–3172. [CrossRef] [PubMed]
4. Watson, B.J.; Zhang, H.; Longmire, A.G.; Moon, Y.H.; Hutcheson, S.W. Processive endoglucanases mediate degradation of cellulose by *Saccharophagus degradans*. *J. Bacteriol.* **2009**, *191*, 5697–5705. [CrossRef] [PubMed]
5. Davies, G.; Henrissat, B. Structure and mechanism of glycosyl hydrolases. *Structure* **1995**, *3*, 853–859. [CrossRef]
6. Lombard, V.; Golaconda Ramulu, H.; Drula, E.; Coutinho, P.M.; Henrissat, B. The carbohydrate-active enzymes database (CAZy) in 2013. *Nucl. Acids Res.* **2014**, *42*, D490–D495. [CrossRef] [PubMed]
7. Guerriero, G.; Hausman, J.F.; Strauss, J.; Ertan, H.; Siddiqui, K.S. Destructuring plant biomass: Focus on fungal and extremophilic cell wall hydrolases. *Plant. Sci.* **2015**, *234*, 180–193. [CrossRef] [PubMed]
8. Schweiger-Hufnagel, U.; Ono, T.; Izumi, K.; Hufnagel, P.; Morita, N.; Kaga, H.; Morita, M.; Hoshino, T.; Yumoto, I.; Matsumoto, N.; et al. Identification of the extracellular polysaccharide produced by the snow mold fungus *Microdochium Nivale*. *Biotechnol. Lett.* **2000**, *22*, 183–187. [CrossRef]
9. Brummell, D.A.; Lashbrook, C.C.; Bennett, A.B. Plant endo-1, 4-*b*-D-glucanases. In Enzymatic Conversion of Biomass for Fuels Production; ACS Symp: Washington DC, WA, USA, 1994; Volume 566, pp. 100–129.4-b-D-glucanases. In *Enzymatic Conversion of Biomass for Fuels Production*; ACS Symp: Washington DC, WA, USA, 1994; Volume 566, pp. 100–129.
10. Weimer, P.J. Cellulose Degradation by Ruminal Microorganisms. *Crit. Rev. Biotechnol.* **1992**, *12*, 189–233. [CrossRef]
11. Watanabe, H.; Tokuda, G. Animal cellulases. *Cell. Mol. Life Sci.* **2001**, *58*, 1167–1178. [CrossRef] [PubMed]
12. Blifernez-Klassen, O.; Klassen, V.; Doebbe, A.; Kersting, K.; Grimm, P.; Wobbe, L.; Kruse, O. Cellulose degradation and assimilation by the unicellular phototrophic eukaryote *Chlamydomonas reinhardtii*. *Nat. Commun.* **2012**, *3*, 1214. [CrossRef] [PubMed]

13. Merchant, S.S.; Prochnik, S.E.; Vallon, O.; Harris, E.H.; Karpowicz, S.J.; Witman, G.B.; Terry, A.; Salamov, A.; Fritz-Laylin, L.K.; Maréchal-Drouard, L.; et al. The Chlamydomonas genome reveals the evolution of key animal and plant functions. *Science* **2007**, *318*, 245–250. [CrossRef] [PubMed]
14. Prochnik, S.E.; Umen, J.; Nedelcu, A.M.; Hallmann, A.; Miller, S.M.; Nishii, I.; Ferris, P.; Kuo, A.; Mitros, T.; Fritz-Laylin, L.K.; et al. Genomic analysis of organismal complexity in the multicellular green alga *Volvox carteri*. *Science* **2010**, *329*, 223–226. [CrossRef] [PubMed]
15. Hanschen, E.R.; Marriage, T.N.; Ferris, P.J.; Hamaji, T.; Toyoda, A.; Fujiyama, A.; Neme, R. The Gonium pectorale genome demonstrates co-option of cell cycle regulation during the evolution of multicellularity. *Nat. Commun.* **2016**, *7*, 11370. [CrossRef] [PubMed]
16. Domozych, D.S.; Ciancia, M.; Fangel, J.U.; Mikkelsen, M.D.; Ulvskov, P.; Willats, W.G. The cell walls of green algae: A journey through evolution and diversity. *Front. Plant. Sci.* **2012**, *3*, 82. [CrossRef] [PubMed]
17. Davison, A.; Blaxter, M. Ancient origin of glycosyl hydrolase family 9 cellulase genes. *Mol. Biol. Evol.* **2005**, *22*, 1273–1284. [CrossRef] [PubMed]
18. Sakon, J.; Irwin, D.; Wilson, D.B.; Karplus, P.A. Structure and mechanism of exo/endocellulase E4 from *Thermomonospora fusca*. *Nat. Struct. Biol.* **1997**, *4*, 810–818. [CrossRef] [PubMed]
19. Khademi, S.; Guarino, L.A.; Watanabe, H.; Tokuda, G.; Meyer, E.F. Structure of an endoglucanase from termite, *Nasutitermes takasagoensis*. *Acta Cryst. D Biol. Cryst.* **2002**, *58 Pt 4*, 653–659. [CrossRef]
20. Zhou, W.; Irwin, D.C.; Escovar-Kousen, J.; Wilson, D.B. Kinetic studies of *Thermobifida fusca* Cel9A active site mutant enzymes. *Biochemistry* **2004**, *43*, 9655–9663. [CrossRef] [PubMed]
21. Sigrist, C.J.A.; de Castro, E.; Cerutti, L.; Cuche, B.A.; Hulo, N.; Bridge, A.; Bougueleret, L.; Xenarios, I. New and continuing developments at PROSITE. *Nucl. Acids Res.* **2013**, *41*, D344–D347. [CrossRef] [PubMed]
22. Parsiegla, G.; Belaïch, A.; Belaïch, J.P.; Haser, R. Crystal structure of the cellulase Cel9M enlightens structure/function relationships of the variable catalytic modules in glycoside hydrolases. *Biochemistry* **2002**, *41*, 11134–11142. [CrossRef] [PubMed]
23. Schubot, F.D.; Kataeva, I.A.; Chang, J.; Shah, A.K.; Ljungdahl, L.G.; Rose, J.P.; Wang, B.C. Structural basis for the exocellulase activity of the cellobiohydrolase CbhA from *Clostridium thermocellum*. *Biochemistry* **2004**, *43*, 1163–1170. [CrossRef] [PubMed]
24. Ravachol, J.; Borne, R.; Tardif, C.; de Philip, P.; Fierobe, H.P. Characterization of all family-9 glycoside hydrolases synthesized by the cellulosome-producing bacterium *Clostridium cellulolyticum*. *J. Biol. Chem.* **2014**, *289*, 7335–7348. [CrossRef] [PubMed]
25. Mandelman, D.; Belaich, A.; Belaich, J.P.; Aghajari, N.; Driguez, H.; Haser, R. X-Ray crystal structure of the multidomain endoglucanase Cel9G from Clostridium cellulolyticum complexed with natural and synthetic cello-oligosaccharides. *J. Bacteriol.* **2003**, *185*, 4127–4135. [CrossRef] [PubMed]
26. Fontes, C.M.G.A.; Gilbert, H.J. Cellulosomes: Highly efficient nanomachines designed to deconstruct plant cell wall complex carbohydrates. *Ann. Rev. Biochem.* **2010**, *79*, 655–681. [CrossRef] [PubMed]
27. Sammond, D.W.; Payne, C.M.; Brunecky, R.; Himmel, M.E.; Crowley, M.F.; Beckham, G.T. Cellulase linkers are optimized based on domain type and function: Insights from sequence analysis, biophysical measurements, and molecular simulation. *PLoS ONE* **2012**, *7*, e48615. [CrossRef] [PubMed]
28. Payne, C.M.; Resch, M.G.; Chen, L.; Crowley, M.F.; Himmel, M.E.; Taylor, L.E.; Sandgren, M.; Ståhlberg, J.; Stals, I.; Tan, Z.; et al. Glycosylated linkers in multimodular lignocellulose-degrading enzymes dynamically bind to cellulose. *Proc. Natl. Acad. Sci. USA* **2013**, *110*, 14646–14651. [CrossRef] [PubMed]
29. Teo, V.S.; Saul, D.J.; Bergquist, P.L. Cel A, Another Gene Coding for A Multidomain Cellulase from the Extreme Thermophile *Caldocellum Saccharolyticum*. *Appl. Microbiol. Biotechnol.* **1995**, *43*, 291–296. [CrossRef]
30. Zverlov, V.V.; Velikodvorskaya, G.V.; Schwarz, W.H.; Bronnenmeier, K.; Kellermann, J.; Staudenbauer, W.L. Multidomain structure and cellulosomal localization of the *Clostridium thermocellum* cellobiohydrolase CbhA. *J. Bacteriol.* **1998**, *180*, 3091–3099. [PubMed]
31. Tian, L.; Liu, S.; Wang, S.; Wang, L. Ligand-binding specificity and promiscuity of the main lignocellulolytic enzyme families as revealed by active-site architecture analysis. *Sci. Rep.* **2016**, *6*, 23605. [CrossRef] [PubMed]
32. Gilkes, N.R.; Henrissat, B.; Kilburn, D.G.; Miller, R.C., Jr.; Warren, R.A. Domains in microbial beta-1, 4-glycanases: Sequence conservation, function, and enzyme families. *Microbiol. Rev.* **1991**, *155*, 303–315.
33. Wolfrom, M.L.; Thompson, A.; Timberlake, C.E. Comparative hydrolysis rates of the reducing disaccharides of D-glucopyranose. *Cereal. Chem.* **1963**, *40*, 82–86.

34. Behera, B.C.; Sethi, B.R.; Mishra, R.R.; Dutta, S.K.; Thatoi, H.N. Microbial cellulases-Diversity and biotechnology with reference to mangrove environment: A review. *J. Gen. Eng. Biotechnol.* **2017**, *15*, 197–210. [CrossRef]
35. Varrot, A.; Frandsen, T.P.; von Ossowski, I.; Boyer, V.; Cottaz, S.; Driguez, H.; Schülein, M.; Davies, G.J. Structural basis for ligand binding and processivity in cellobiohydrolase Cel6A from Humicola insolens. *Structure* **2003**, *11*, 855–864. [CrossRef]
36. Tomme, P.; Kwan, E.; Gilkes, N.R.; Kilburn, D.G.; Warren, R.A. Characterization of CenC, an enzyme from Cellulomonas fimi with both endo-and exoglucanase activities. *J. Bacteriol.* **1996**, *178*, 4216–4223. [CrossRef] [PubMed]
37. Irwin, D.; Shin, D.H.; Zhang, S.; Barr, B.K.; Sakon, J.; Karplus, P.A.; Wilson, D.B. Roles of the catalytic domain and two cellulose binding domains of *Thermomonospora fusca* E4 in cellulose hydrolysis. *J. Bacteriol.* **1998**, *180*, 1709–1714. [PubMed]
38. Pages, S.; Kester, H.C.M.; Visser, J.; Benen, J.A.E. Changing a single amino acid residue switches processive and non-processive behavior of *Aspergillus niger* endopolygalacturonase I and II. *J. Biol. Chem.* **2001**, *276*, 33652–562001. [CrossRef] [PubMed]
39. Guérin, D.M.; Lascombe, M.B.; Costabel, M.; Souchon, H.; Lamzin, V.; Béguin, P.; Alzari, P.M. Atomic (0.94 A) resolution structure of an inverting glycosidase in complex with substrate. *J. Mol. Biol.* **2002**, *316*, 1061–1069. [CrossRef] [PubMed]
40. Boraston, A.B.; Bolam, D.N.; Gilbert, H.J.; Davies, G.J. Carbohydrate-binding modules: Fine-tuning polysaccharide recognition. *Biochem. J.* **2004**, *382*, 769–781. [CrossRef] [PubMed]
41. Gao, P.J.; Chen, G.J.; Wang, T.H.; Zhang, Y.S.; Liu, J.; Sheng, W. Non-hydrolytic Disruption of Crystalline Structure of Cellulose by Cellulose Binding Domain and Linker Sequence of Cellobiohydrolase I from Penicillium janthinellum. *Acta Biochim. Biophys. Sin.* **2001**, *33*, 13–18. [PubMed]
42. Petkun, S.; Rozman Grinberg, I.; Lamed, R.; Jindou, S.; Burstein, T.; Yaniv, O.; Shoham, Y.; Shimon, L.J.; Bayer, E.A.; Frolow, F. Reassembly and co-crystallization of a family 9 processive endoglucanase from its component parts: Structural and functional significance of the intermodular linker. *Peer. J.* **2015**, *15*, e1126. [CrossRef] [PubMed]
43. Tormo, J.; Lamed, R.; Chirino, A.J.; Morag, E.; Bayer, E.A.; Shoham, Y.; Steitz, T.A. Crystal structure of a bacterial family-III cellulose-binding domain: A general mechanism for attachment to cellulose. *EMBO J.* **1996**, *15*, 5739–5751. [PubMed]
44. Popolo, L.; Ragni, E.; Carotti, C.; Palomares, O.; Aardema, R.; Back, J.W.; Dekker, H.L.; de Koning, L.J.; de Jong, L.; de Koster, C.G. Disulfide bond structure and domain organization of yeast beta(1, 3)-glucanosyltransferases involved in cell wall biogenesis. *J. Biol. Chem.* **2008**, *283*, 18553–18565. [CrossRef] [PubMed]
45. Receveur, V.; Czjzek, M.; Schülein, M.; Panine, P.; Henrissat, B. Dimension, shape, and conformational flexibility of a two domain fungal cellulase in solution probed by small angle X-ray scattering. *J. Biol. Chem.* **2002**, *277*, 40887–40892. [CrossRef] [PubMed]
46. Jung, E.D.; Lao, G.; Irwin, D.; Barr, B.K.; Benjamin, A.; Wilson, D.B. DNA sequences and expression in *Streptomyces lividans* of an exoglucanase gene and an endoglucanase gene from *Thermomonospora fusca*. *Appl. Environ. Microbiol.* **1993**, *59*, 3032–3043. [PubMed]
47. Rabinovich, M.L.; Melnick, M.S.; Bolobova, A.V. The structure and mechanism of action of cellulolytic enzymes. *Biochemistry* **2002**, *67*, 850–871. [CrossRef] [PubMed]
48. Suzuki, K.; Ojima, T.; Nishita, K. Purification and cDNA cloning of a cellulase from abalone Haliotis discus hannai. *Eur. J. Biochem.* **2003**, *270*, 771–778. [CrossRef] [PubMed]
49. Lis, H.; Sharon, N. Protein glycosylation. structural and functional aspects. *Eur. J. Biochem.* **1993**, *218*, 1–27. [CrossRef] [PubMed]
50. Elodie, M.-R.; Marie-Christine, K.-M.; Gaëtan, V.; Clément, O.; Carole, B.; Patrice, L.; Muriel, B. Protein *N*-glycosylation in eukaryotic microalgae and its impact on the production of nuclear expressed biopharmaceuticals. *Front. Plant. Sci.* **2014**, *5*, 359. [CrossRef]
51. Hui, J.P.; Lanthier, P.; White, T.C.; McHugh, S.G.; Yaguchi, M.; Roy, R.; Thibault, P. Characterization of cellobiohydrolase I (Cel7A) glycoforms from extracts of *Trichoderma reesei* using capillary isoelectric focusing and electrospray mass spectrometry. *J. Chromatogr. B Biomed. Sci. Appl.* **2001**, *752*, 349–368. [CrossRef]

52. Corzana, F.; Busto, J.H.; Jiménez-Osés, G.; de Luis, M.G.; Asensio, J.L.; Jiménez-Barbero, J.; Peregrina, J.M.; Avenoza, A. Serine versus threonine glycosylation: The methyl group causes a drastic alteration on the carbohydrate orientation and on the surrounding water shell. *J. Am. Chem. Soc.* **2007**, *129*, 9458–9467. [CrossRef] [PubMed]
53. Boze, H.; Marlin, T.; Durand, D.; Perez, J.; Vernhet, A.; Canon, F.; Sarni-Manchado, P.; Cheynier, V.; Cabane, B. Proline-rich salivary proteins have extended conformations. *Biophys. J.* **2010**, *99*, 656–665. [CrossRef] [PubMed]
54. Ruiz, D.M.; Turowski, V.R.; Murakami, M.T. Effects of the linker region on the structure and function of modular GH5 cellulases. *Sci. Rep.* **2016**, *6*, 28504. [CrossRef] [PubMed]
55. Li, Y.; Wang, J.; Wang, L.; Tong, H.; Bu, M.; Gao, G.; Han, W.; Zhang, Z. The PT/S-Box of modular cellulase AcCel12B plays a key role in the hydrolysis of insoluble cellulose. *Catalysts* **2018**, *8*, 123. [CrossRef]
56. Sonan, G.K.; Receveur-Brechot, V.; Duez, C.; Aghajari, N.; Czjzek, M.; Haser, R.; Gerday, C. The linker region plays a key role in the adaptation to cold of the cellulase from an Antarctic bacterium. *Biochem. J.* **2007**, *407*, 293–302. [CrossRef] [PubMed]
57. McWilliam, H.; Li, W.; Uludag, M.; Squizzato, S.; Park, Y.M.; Buso, N.; Cowley, A.P.; Lopez, R. Analysis Tool Web Services from the EMBL-EBI. *Nucl. Acids Res.* **2013**, *4*, W597–W600. [CrossRef] [PubMed]
58. Yang, J.; Yan, R.; Roy, A.; Xu, D.; Poisson, J.; Zhang, Y. The I-TASSER Suite: Protein structure and function prediction. *Nat. Meth.* **2015**, *12*, 7–8. [CrossRef] [PubMed]
59. Laskowski, R.A.; Jabłońska, J.; Pravda, L.; Vařeková, R.S.; Thornton, J.M. PDBsum: Structural summaries of PDB entries. *Prot. Sci.* **2018**, *27*, 129–134. [CrossRef] [PubMed]
60. Guex, N.; Peitsch, M.C. SWISS-MODEL and the Swiss-Pdb Viewer. An environment for comparative protein modeling. *Electrophor* **1997**, *18*, 2714–2723. [CrossRef] [PubMed]
61. Guindon, S.; Gascuel, O. A simple, fast, and accurate algorithm to estimate large phylogenies by maximum likelihood. *Syst. Biol.* **2003**, *52*, 696–704. [CrossRef] [PubMed]
62. Bailey, T.L.; Johnson, J.; Grant, C.E.; Noble, W.S. The MEME Suite. *Nucl. Acids Res.* **2015**, *43*, W39–W49. [CrossRef] [PubMed]
63. Vandesompele, J.; De Preter, K.; Pattyn, F.; Poppe, B.; Van Roy, N.; De Paepe, A.; Speleman, F. Accurate normalization of real-time quantitative RT-PCR data by geometric averaging of multiple internal control genes. *Gen. Biol.* **2002**, *3*. [CrossRef]

International Journal of
Molecular Sciences

MDPI

Review

Fascinating Fasciclins: A Surprisingly Widespread Family of Proteins that Mediate Interactions between the Cell Exterior and the Cell Surface

Georg J. Seifert

Department of Applied Genetics and Cell Biology, University of Natural Resources and Life Science, Muthgasse 18, 1190 Vienna, Austria; georg.seifert@boku.ac.at; Tel.: +431-47654-94144

Received: 9 April 2018; Accepted: 17 May 2018; Published: 31 May 2018

Abstract: The Fasciclin 1 (FAS1) domain is an ancient structural motif in extracellular proteins present in all kingdoms of life and particularly abundant in plants. The FAS1 domain accommodates multiple interaction surfaces, enabling it to bind different ligands. The frequently observed tandem FAS1 arrangement might both positively and negatively regulate ligand binding. Additional protein domains and post-translational modifications are partially conserved between different evolutionary clades. Human FAS1 family members are associated with multiple aspects of health and disease. At the cellular level, mammalian FAS1 proteins are implicated in extracellular matrix structure, cell to extracellular matrix and cell to cell adhesion, paracrine signaling, intracellular trafficking and endocytosis. Mammalian FAS1 proteins bind to the integrin family of receptors and to protein and carbohydrate components of the extracellular matrix. FAS1 protein encoding plant genes exert effects on cellulosic and non-cellulosic cell wall structure and cellular signaling but to establish the modes of action for any plant FAS1 protein still requires biochemical experimentation. In fungi, eubacteria and archaea, the differential presence of FAS1 proteins in closely related organisms and isolated biochemical data suggest functions in pathogenicity and symbiosis. The inter-kingdom comparison of FAS1 proteins suggests that molecular mechanisms mediating interactions between cells and their environment may have evolved at the earliest known stages of evolution.

Keywords: arabinogalactan proteins; cellulose; pectin; matricellular proteins; SOS5; periostin; Mpb83

---

1. A Brief History of FAS1 Domain Proteins

The first FAS1 protein was identified in an insect model for central nervous system development, the grasshopper *Schistocerca americana*. In order to identify cell surface molecules potentially involved in the formation of axon bundles (fascicles), monoclonal antibodies (mAbs) recognizing cell surface antigens on specific fascicles were characterized. One of these antibodies recognized a 70 kDa glycoprotein named Fasciclin 1 (SaFas1 (Appendix A)) [1]. The genes coding for grasshopper SaFas1 and *Drosophila melanogaster* DmFas1 were cloned soon afterwards [2] and a homologous fruit fly gene called Midline fasciclin (*DmMfas*) was identified later [3]. In the fruit fly, a *DmFas1* knockout affected neuronal branching as well as synaptic function [4] and laser ablation of the grasshopper ortholog *SaFas1* led to disrupted cell adhesion of pioneer axons [5]. The crystal structure of DmFas1 provided the prototype for the structurally novel FAS1 domain [6]. In the meantime, molecular techniques and sequence comparison tools revealed the widespread occurrence of homologous proteins defined by the FAS1 domain (IPR000782; PF02469). The *Homo sapiens* genome encodes four FAS1 domain proteins named transforming growth factor-β induced protein (HsTgfbi), Periostin (HsPn), Stabilin-1 (HsStab1) and Stabilin-2 (HsStab2). The *HsTgfbi* gene (Appendix B) was identified in human adenocarcinoma cells as a transcript that was induced 20-fold by transforming growth factor-β [7]. Likewise, *HsPn*, was cloned based on its expression in an osteoblast cell line [8] and subsequently found to be enriched in the periosteum [9]. Finally, HsStab1 and HsStab2 were identified using two different antibodies binding to a subpopulation of endothelial cells and to the hyaluronic acid (HA) clearance receptor, respectively [10]. Due to their association with a multitude of clinical conditions, human and mammalian FAS1 proteins have been the focus of numerous detailed studies that are instructive for a general understanding of FAS1 domain proteins. Unexpectedly for proteins associated with cell adhesion, FAS1 domain proteins exist not only in animals but also in plants, fungi and prokaryotes. The first plant FAS1 protein was discovered using antibody interference in the alga *Volvox carteri* [11], a simple model for multi-cellularity consisting of just two cell types. When specific mAbs raised against a crude membrane preparation were added to volvox cultures they inhibited embryo development. The cognate protein was named algal cell adhesion molecule (CAM) based on its apparent role in the formation of intercellular contacts during early embryogenesis. The existence and physiological role of algal-CAM, which contains two FAS1 domains, raised the exciting possibility of a cell adhesion mechanism conserved between animals and plants. In higher plants FAS1 domain proteins were also identified by the biochemical and bioinformatic analysis of a group of highly *O*-glycosylated hydroxyproline-rich glycoproteins called arabinogalactan-proteins (AGPs) [12,13]. The bioinformatic investigation of the *Arabidopsis thaliana* genome revealed the existence of many fasciclin-like AGPs (FLAs) in plants [12–14]. At the same time a different investigation mapped one of several *Arabidopsis thaliana* salt overly sensitive (*sos*) mutations to the Salt overly sensitive 5 (*AtSos5*) gene encoding the AtFla4 protein [15]. In a mutant screen in a crop plant, the rice locus Microspore and tapetum regulator 1 (*OsMtr1*) was found to be required for male reproductive development and to encode a tandem FAS1 glycoprotein [16]. In fungi, FAS1 proteins have been identified in transcriptomics or proteomics studies in the Shiitake mushroom *Lentinula edodes* [17] and the rice pathogen *Magnaporthe oryzae* [18], while in the fission yeast *Schizosaccharomyces pombe* the FAS1 domain protein SpFsc1 was identified in a screen for autophagy related loci [19]. Apparently, FAS1 proteins already existed before the evolution of eukaryotes. The best-known eubacterial FAS1 proteins are Mpb70 and Mpb83, which were identified in *Mycobacterium bovis* culture filtrates [8,20–24]. Database queries reveal FAS1 proteins in both eubacteria and archaea, suggesting the inception of the domain preceded the existence of last universal common ancestor (LUCA) [25]. FAS1 proteins are often implicated in the interaction between the cell and the extracellular matrix (ECM). Considering the diversity of ECM architectures and compositions FAS1 domain proteins are surprisingly widespread between different kingdoms of uni- and multicellular life. However, despite their seemingly boundless presence throughout the tree of life, FAS1 proteins are not ubiquitous, especially in microbes whose genomes rapidly adapt to differing life styles. This suggests that FAS1 domain proteins are not essential for life per se but are suited for specialized cellular interactions that for some organisms are not required. I will next describe what is known about the structure of the FAS1 domain itself and discuss diverse additional structural features of FAS1 proteins in various kingdoms. This will be followed by a review of the biological roles of mammalian and plant FAS1 domain proteins, including the relationship of structure to function, which should help elucidate the mechanisms of FAS1 proteins in plant development.

2. The Structure of the Fasciclin 1 Domain

2.1. The Fasciclin 1 Domain

The FAS1 domain extends to approximately 140 amino acids. Although sequence conservation between different FAS1 proteins can be quite low, there exist two more highly conserved sequence stretches of around 15 residues called H1 and H2 and a conserved central YH motif (Figure 1A). Therefore, to identify FAS1 domain proteins in sequence databases, domain enhanced lookup time accelerated BLAST (DELTA-BLAST) should be used [26]. Using X-ray crystallography and NMR spectroscopy, several studies have elucidated the structures of isolated FAS1 domains or of entire FAS1 proteins [27–33]. The FAS1 domain is globular and contains a central structural fold of two β-sheets oriented at an almost perpendicular angle, varyingly described as β-wedge or β-sandwich (Figure 1B).

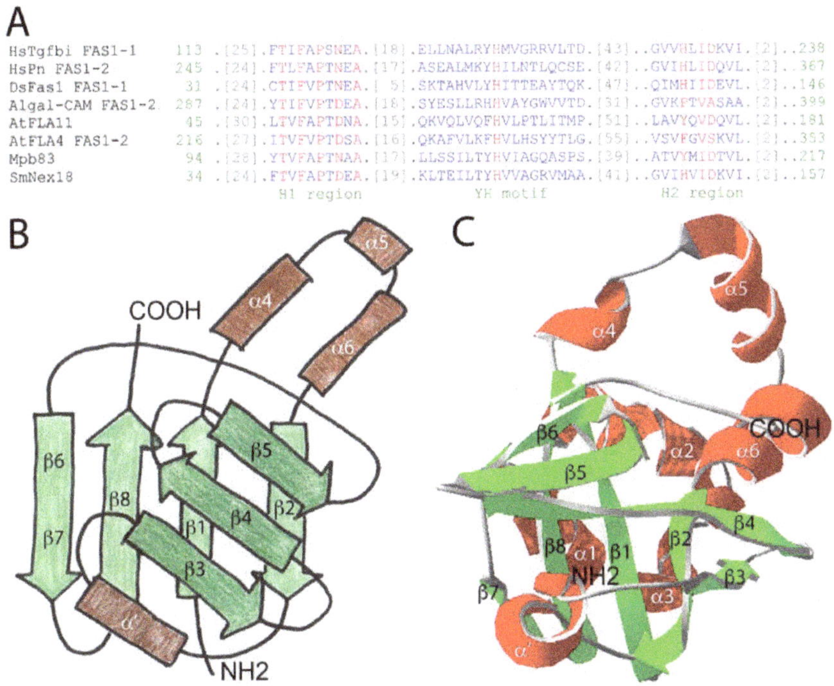

Figure 1. The FAS1 domain across kingdoms of life. (**A**) Delta-BLAST alignment of some FAS1 domains mentioned in this article. Note the conservation of the N and C-proximal H1 and the H2 region as well as the central YH motif. The sequences used by Delta-BLAST were HsTgfbi (NP_000349.1), HsPn (gi 93138709), DsFas1 FAS1-3 (1O70_A), Algal-CAM (gi 75282282), AtFLA11 (gi 116247778), AtFLA4 (gi 75206907), MbMpb83 (gi 614094354), and SmNex18 (gi 81635876). Red/blue indicates highly/not conserved residues; color bits threshold set to 2.5; (**B**) The general topology of the FAS1 domain features is reminiscent of the "thumbs-up" gesture. The secondary structure of HsTgfbi FAS1-1 is annotated omitting the three N-terminal helices for clarity; (**C**) The crystal structure of HsTgfbi FAS1-1 [33] in ribbon display.

In the human HsTgfbi structure, the first β-sheet encompasses strands β1–β2–β8–β6/7 [33] (Figure 1C). The two inner strands are oriented parallel and the two outer ones antiparallel. The second β-sheet consists of β3–β4–β5. There are three α-helices at the N-terminus (not shown in Figure 1B), three more (α4 to α6) between β1 and β2 and a less highly conserved α-helix (α′) between β2 and β3. The FAS1 domain (Figure 1B, redrawn from [25]) is a member of the "β-grasp fold" superfamily [25] and may be imagined as a "thumbs-up" gesture of the right hand with the palm representing the first β-sheet (light green in Figure 1B), the bent index, middle and ring fingers symbolizing the second β-sheet (dark green in Figure 1B) and the thumb and pinkie resembling α4 to α6 (light brown in Figure 1B) and α′ (dark brown in Figure 1B), respectively. Among all FAS1 proteins, structure to function relations have been most intensely studied for HsTgfbi (reviewed in [34]). Therefore, the elucidation of the entire HsTgfbi crystal structure [33] could be seen as the "Rosetta stone" for a better mechanistic understanding of the many biological roles of FAS1 proteins in different organisms including plants. Several studies identified individual regions and amino acid residues on the four

Tgfbi FAS1 domains that are critical for function. The main approach was to use in vitro cell adhesion as a functional assay (Figure 2).

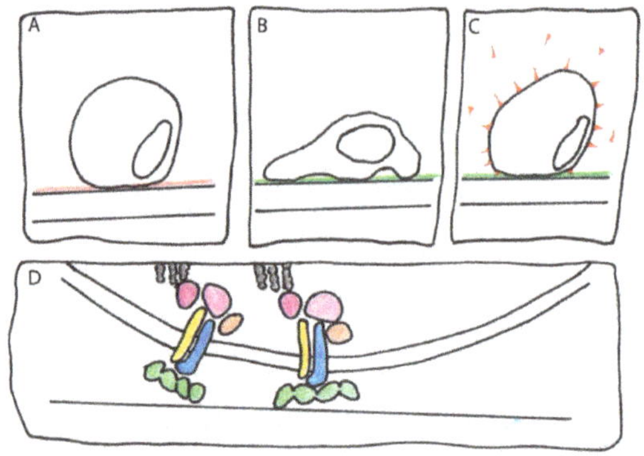

Figure 2. Cell adhesion assay. (**A**) On an uncoated or control-coated (red) plastic substrate, cell adhesion is inefficient; (**B**) cells adhere rapidly when plastic is coated with an adhesion protein such as Tgfbi or Pn (green); (**C**) to identify the receptor for the adhesion protein, integrin isotype specific antibodies (red) are co-incubated; (**D**) mammalian FAS1 proteins are thought to bind to different types of dimeric integrins (blue and yellow) that mediate mechanical contact between the cytoskeleton (grey) and the ECM as well as transduce intracellular signals using numerous associated proteins (pink and nude).

Briefly, cells more efficiently adhere to surfaces coated with adhesion proteins such as Tgfbi than to control-coated surfaces (Figure 2A,B). That this adhesion is dependent on the family of ECM receptors called integrins can be tested by adding integrin antibodies that block the Tgfbi-stimulated cell adhesion (Figure 2C). To identify sites on Tgfbi that might mediate cell adhesion, peptides corresponding to conserved Tgfbi sequence motifs were added. For instance, the NKDIL and the EPDIM peptides that are part of the FAS1-2 and the FAS1-4 (see following section) domains, respectively, both interfered with Tgfbi stimulated cell adhesion. By contrast, the KADHH peptide in the corresponding region of FAS1-1 had no effect [35]. In a study of a different cell type expressing a different integrin, the NKDIL and EPDIM peptides did not interfere with adhesion but an 18-amino acid peptide that covered the YH motif did [36]. A HsPn specific mAb identified a corresponding integrin interaction region on FAS1-2 [37]. The cell adhesion assay is too crude to demonstrate *direct* binding between FAS1 proteins and integrin; however, in combination with the crystal structure of HsTgfbi it showed that different integrins interact with different surface regions of the FAS1 domains [33]. Although integrins are not known in plants, given the complex biological roles of some FAS1 proteins of plants, this insight should be valuable for the prediction of their molecular function.

2.2. *Single and Tandem Fasciclin 1 Domains*

While prokaryotic representatives typically contain a single FAS1 domain, many eukaryotic family members contain two or more arranged in tandem (numbered FAS1-1, FAS1-2 etc. from N- to C-terminus). The two insect family members Fas1 and Mfas, and mammalian Tgfbi and Pn each contain four FAS1 domains, while there are seven in the two mammalian stabilins (Figure 3).

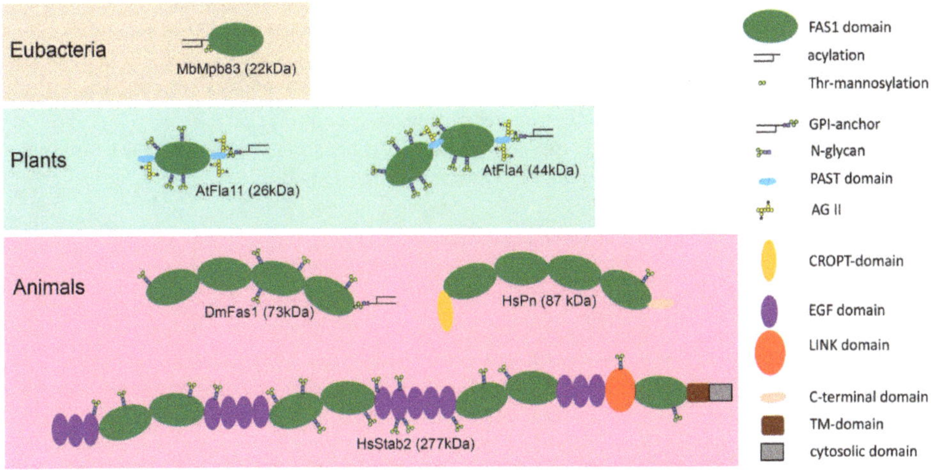

Figure 3. Domain organization and post translational modifications of six representative FAS1 proteins discussed in this article. Proteins are shown with their N-termini on the left. As opposed to the C-terminally attached GPI-anchors Mpb83 is acylated near its N-terminus. Note that the annotation of AG II glycosylation is tentative and that the annotation of *N*-glycans is based on bioinformatic predictions. Abbreviations used in this Figure, GPI: glycosylphosphoinositidyl; TM: transmembrane; AG II: arabinogalactan type II; CROPT: Cysteine-Rich domain of Pn and TGFBIp; EGF: epidermal growth factor; PAST: domain rich in Pro, Ala, Ser and Thr.

In plants, there exist both single FAS1 FLAs as well as family members with two domains in tandem. FAS1 tandem proteins also exist in fungi, where dual and five-fold tandems have been reported [18,19]. The tandem FAS1 domains structurally interact with each other, as was found in a structure of DmFas1 domains FAS1-3 an FAS1-4 [27] and of whole HsTgfbi [33]. It was suggested that the C-proximal domain in a tandem pair might bind to interaction partners such as integrins while the N-proximal domains might compete with this binding acting either as negative regulators or as safeguards against unwanted ligand binding [28]. A recent structure to function study of *AtFla4* in *Arabidopsis thaliana* showed that the C-proximal FAS1 domain was necessary and sufficient for genetic function [38]. By contrast, some studies showed cooperativity between all FAS1 domains [39]. To which degree the intramolecular interaction between neighboring FAS1 domains might affect the intermolecular interaction with other binding partners in vivo remains an open question.

2.3. Membrane Association

Most FAS1 domain proteins contain additional structural features and post-translational modifications some of which are shared between phylogenetically distinct family members. Practically all FAS1 proteins have an N-terminal secretion peptide; however, the final localization is determined by other signals. Mycobacterial Mpb83 is lipid-anchored to the plasma membrane by Cys acylation of the LAGC motif close to the N-terminus [40,41]. In plants, most FAS1 proteins are anchored to the outer sheet of the plasma membrane by a glycosylphosphoinositidyl (GPI) lipid anchor [14]. This modification is attached to proteins in the endoplasmic reticulum in exchange of a C-terminal signal peptide and exists in FAS1 proteins of plants, insects and fungi [17,18,42]. However, only one of the two *D. melanogaster* FAS1 proteins, DmFas1, is GPI-anchored [43]. The two mammalian stabilins and SpFsc1 from fission yeast are type 1 transmembrane proteins with cytosolic tails at their C-termini [10,19]. By contrast, Mpb70 from mycobacteria, mammalian Tgfbi and Pn, DmMfas and a group of FLAs are soluble proteins without any membrane anchoring motif [14,34]. Hence most FAS1 proteins are either bound to the plasma membrane or are soluble. In case of Mpb83 both surface anchored and soluble forms exist. Likewise, GPI-anchored proteins can be released from the plasma membrane by the action of phospholipases. The GPI anchor is probably needed for efficient transport of GPI-anchored proteins to the plasma membrane and potentially for their controlled release. In plants the GPI-anchor acts as a sorting signal for plasmodesmata [44]. Whether in non-animal eukaryotes such as plants and fungi that lack a cholesterol-rich plasma membrane, the GPI modification functions similar to animals to sequester proteins into membrane microdomains is under debate [45].

2.4. Glycosylation

Secreted proteins are often glycosylated. This is also the case for many FAS1 proteins, but not for all of them. For instance, although Mpb70 and Mpb83 share 75% sequence identity, but the latter is *O*-glycosylated close to its N-terminus with three threonine-linked mannose residues [46]. Another example is the *N*-linked glycosylation of four animal FAS1 proteins, mammalian Tgfbi and Pn and insect Fas1 and Mfas that are similar in size and all have four FAS1 domains, however, they carry zero, one, five and eight *N*-glycans, respectively. Likewise, in plant FAS1 proteins, *N*-glycosylation of FAS1 domains ranges from two to eight *N*-glycans [14,16]. A distinguishing feature of FAS1 proteins in plants is the presence of regions rich in proline, alanine, serine and threonine (PAST). Especially sequences containing the repetitive pattern $(P\text{-}A/S/T)_n$, whose Pro residues are post-translationally converted to Hyp and galactosylated are typical for AGPs [47]. Indeed, the first FLAs were identified in precipitates obtained with β-glucosyl Yariv reagent [12,13], which binds to $\beta(1{\rightarrow}3)$ galactose chains present on many AGPs [48,49]. However, the peptide patterns for *O*-glycosylation are not as precisely defined as the N-X-T/S *N*-glycosylation motif [50], and biochemical identification of *O*-glycosylation on defined proteins is not as straightforward as the molecular weight shift caused by enzymatic removal of *N*-glycans. Hence one should be aware of potential discrepancies as to which FAS1 proteins are indeed FLAs, e.g., Algal-CAM contains an extensive PAST-rich N-terminal region and so does OsMtr1, and after removing *N*-glycans the apparent molecular weights of both proteins were larger than expected for the apoprotein [11,16]. Therefore, both proteins might well be members of the FLA family. By contrast, other FAS1 proteins such as the wheat locus named *TaFLA9* lack any PAST-rich regions [51] and might be wrongly annotated as FLAs. However, the AtFla4 protein is decorated with AGP-like glycan epitopes despite the absence of canonical $(P\text{-}A/S/T)_n$ repeats [38]. Clearly, the simple presence or absence of PAST-rich regions is an inaccurate predictor for AGP-like *O*-glycosylation. The *O*-glycans of FLAs and other AGPs are known as type II arabinogalactans (AG II). They consist of Hyp-linked galactose that is the basis of a (1–3) β-linked galactan backbone with (1–6) β-linked galactose branches and kinks. AG II contains terminal modifications, mainly by L-arabinofuranose, L-rhamnose, L-fucose, D-glucuronic acid, 4-methyl-D-glucuronic acid and D-xylose [52–54]. Although pharmacological and genetic interference with AG II function has revealed many biological roles for these carbohydrates [55,56], models of how AGPs function remain sparse. An AGP of low abundance was shown to form a covalent link with both hemicellulose and pectin [57]. In fractionated cell walls, AGP-associated carbohydrate epitopes are abundant in pectic and hemicellulosic fractions and some are even present in the cellulose fraction [58]. Moreover, AGPs are self-adhesive in a calcium-dependent fashion [59], and can bind to pectin via carboxyl–calcium ion bridges [60]. Such results indirectly suggest that AG II-modified proteins FLAs might at least partially be linked to cell wall polysaccharides via their *O*-glycans.

2.5. Additional Domains

For human FAS1 proteins, *O*-glycosylation has not been reported [43]; however, they contain additional domains that expand their spectrum of molecular interactions. The N-terminal region of mature HsTgfbi, HsPn and DsMfas folds into a structure called Cysteine-Rich domain of Pn and TGFBIp (CROPT) [33], that mediates the interaction with various forms of collagen, fibronectin and the heterophilic binding between HsPn and HsTgfbi (reviewed in [43,61]). The human stabilins contain three dual FAS1 domain tandems and one isolated FAS1 domain, that each are separated by a varying number of Epidermal Growth Factor (EGF) repeats and an X-link domain. These domains are thought to be involved in phosphatidylserine and hyaluronic acid binding, respectively, but are not present in FAS1 proteins other than mammalian stabilins [62,63]. The C-terminal domain (CTD) of HsTgfbi contains the RGD motif that, although crucial for integrin binding of many adhesion proteins, is not required for the adhesion stimulating properties of HsTgfbi [64,65]. However, the RGD motif might modulate binding of HsTgfbi to specific integrins [39]. The CTD of HsPn, displays a high degree of intrinsic disorder and is both differentially spliced [66] and proteolytically regulated [67]. Interestingly, compared to full-length forms, HsPn forms lacking the CTD show relatively enhanced binding to ECM proteins [68]. The extreme C-terminus is highly basic and is thought to bind to heparin. Differential splicing and proteolysis might switch between forms of HsPn that more readily binds to HS and forms that more strongly interact with matrix proteins [61,67]. Compared to vertebrates, differential splicing is less frequent in plants; however, a great degree of regulatory fine-tuning might be provided by differential gene expression of the large number of plant genes encoding FAS1 proteins.

2.6. Groups of Fasciclin 1 Proteins in Plants

Compared to other organisms, plants contain the greatest number of FAS1 protein encoding genes. The majority of them are annotated as FLAs. The numbers of FLA encoding loci vary considerably between different plant genomes such as cotton (**17** and **19** loci in *Gossypium barbadense* and *G. hirsutum*, respectively [69,70]), *Eucalyptus grandis* (**18** [71]), *Arabidopsis thaliana* (**21** [14]), *Cannabis sativa* (**23** [72]), *Brassica rapa* (**33** [73]), wheat *Triticum aestivum* (**34** [51]), rice *Oryza sativa* (**27** [74]) and cottonwood *Populus trichocarpa* (**35** [75]). Apparently, diversity in this gene family varies even between closely related species (e.g., between the two *Gossypium* species or between *Arabidopsis thaliana* and *Brassica rapa*). An important step to understand the diversity of this complex gene family was sub-classification according to the presence of one or two FAS1 domains, their sequence similarity, the presence of GPI-anchors and the organization of PAST-regions [14]. Inclusion of FAS1 domain proteins that are not yet annotated as AGPs, such as OsMtr1 and Algal-CAM, might slightly enlarge the FLA family. The present FLA sub-classification has four groups: group A members are GPI-anchored and contain a single FAS1 domain flanked by two PAST-regions, group B are not GPI-anchored and sport a tandem of two FAS1 domains separated by a PAST-region. Group C contains GPI-anchored, single and tandem FAS1 proteins, with the single FAS1 domains phylogenetically closest to the N-proximal FAS1 domains of that group. Moreover, group C has PAST regions inserted close to their C-termini and between FAS1 domains. Lastly, group D contains FLAs that are not closely related to any other group and might represent separate branches of diversification.

3. Biological Aspects of FAS1 Domain Proteins across the Tree of Life

Human and Mammalian Fasciclin 1 Proteins

FAS1 proteins have been the subject of experimentation in a variety of species; however, by far the most work has been published about the four human representatives [reviewed in [34,43,76–80]]. Particularly *HsPn*, *HsTgfbi* and *HsStab2* have been implicated in a plethora of medically relevant conditions ranging from normal development [76,81] to wound healing [77], bone regeneration [82,83] and cancer [67]. A Medline search shows 624, 1229, 105 and 718 entries for Tgfbi, Pn, Stab1 and Stab2, respectively (2 March 2018, Title/Abstract, synonyms see Appendix B). A detailed medical discussion is beyond the scope of this article; however, selected highlights are presented to provide a critical viewpoint on the molecular function of FAS1 domain proteins in plants. Missense mutations in the *HsTgfbi* locus are known to cause different types of corneal dystrophy [84]. In many cases of this disease, mutated HsTgfbi forms aggregates in the corneal stroma. This is due to the thermal and pH instability of the protein and to its high abundance in the cornea [85]; however, the condition might not help us to understand HsTgfbi function. A complete *Tgfbi* gene knockout in mice caused neither corneal abnormalities nor any other major defects [86], suggesting compensation by other ECM molecules. Although *Pn* knock-out mice showed that this locus is essential for normal osteogenesis and dental development [87,88], the two closely-related *Pn* and *Tgfbi* loci function identically in many ways [43]. The most intriguing common property of mammalian FAS1 proteins is their interaction with the integrin family of ECM receptors (Table 1). Hetero-dimeric integrins are expressed in a cell-type-specific fashion and bind to many different ECM proteins such as fibronectin, laminin and fibrinogen. The cell adhesion assay described above (Figure 2) suggests that FAS1 domains might bind to integrins; however, to establish the mode of binding, alternative assays such as co-immunoprecipitation of recombinant proteins and Foerster resonance energy transfer are required (Table 1).

Table 1. Mammalian Fas1 proteins interact with integrin receptors.

Integrin Type	TGFBI	PN	STAB-2
α1β1	[64] I,A		
α2β1	[89] C,F		
α3β1	[35] A,I,P [90] A,I [91] A,I		
α4	[92] I		
α5	[92] I		
α5β1		[93] A,I,G	
α6β4	[94] A,I	[95] I	
α7β1	[96] A,I		
αvβ3	[97] A,I,P [98] A,I [99] C,F	[100] I,A [101] I [88] I [102] A,D [103] G [104] I [105] I	
αvβ5	[106] A,I,P [36] A,I,P	[100] I,A [101] I [102] A,D [104] I	[107] C,T
β3		[108] G,A	
αMβ2	[65] A,C,I	[109] A,I	[110] I,G

Assay type: G: genetic interference (siRNA, transfection, mutants & epistasis analysis etc.); A: cell adhesion (coating of plastic surfaces); I: immuno-interference (e.g., anti-integrin antibodies inhibit PN mediated responses); P: peptide interference (peptide of Fas1 domain interferes with integrin dependent binding); C: Co-IP (endogenous or added recombinant Fas1 protein); F: immunofluorescence co-localization; D: DNA-aptamer inhibition; T: FRET.

The integrin complexes are hubs of mechano-chemical signaling and interact with the cytoskeleton and with numerous signaling pathways including focal adhesion kinase (FAK), mitogen activated protein kinase (MAPK), Target of rapamycin (mTOR), and β-catenin that influence cell division and migration, differentiation and programmed cell death [76]. FAS1 proteins not only bind and influence integrins as a stationary component of the ECM but also act as mobile paracrine factors. An example for the latter is the role of HsPn in glioblastoma. In this invasive brain cancer, glioblastoma stem cells (GSCs) secrete HsPn that acts as chemoattractant for circulatory monocytes (MCs)—precursors of macrophages (Figure 4A).

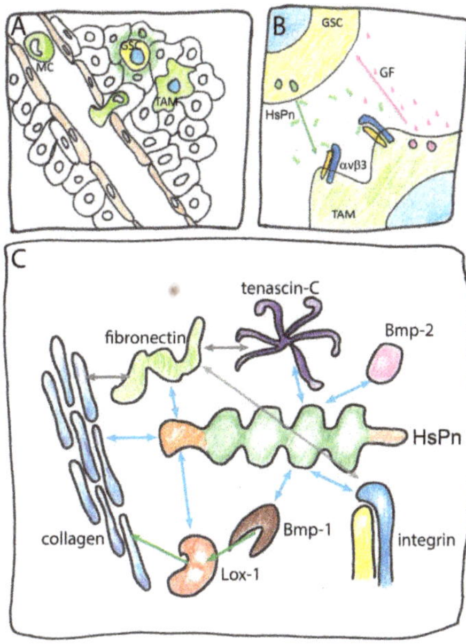

Figure 4. Multiple roles of HsPn: (**A**) Circulatory monocytes (MC) are attracted by glioblastoma stem cells (GSC) that secrete HsPn (dark green) to invade the site of the primary tumor and differentiate into tumor associated macrophages (TAM); (**B**) GSC-secreted Pn via integrin $\alpha v\beta 3$ signaling triggers the release of growth factors (GF) that establish a metastatic niche; (**C**) molecular interactions of Pn in the ECM. Blue double arrows indicate binding of HsPn. Grey double arrows indicate binding of other components. Green arrows indicate activating reactions: the metalloprotease Bone morphogenetic protein Bmp-1 proteolytically activates Lysine oxidase 1 (Lox-1) in turn enhancing collagen crosslinking.

MCs are induced to invade the site of the primary tumor and to differentiate into tumor associated macrophages (TAMs) by HsPn binding to MC-expressed integrin. In turn, TAMs secrete growth factors that further stimulate cell division in GSCs generating a metastatic niche [105] (Figure 4B). A similar model for Pn function has previously been proposed for other malignancies such as cholonangiocarcinoma and renal carcinoma [93,104]. Both HsPn and HsTgfbi directly bind to several ECM components and modulate the biosynthesis of the ECM proteins (Figure 4C). Owing to its versatile binding properties, mammalian Tgfbi was previously known as collagen-associated protein containing the RGD sequence (RGD-CAP) [111], while HsPn is known to bind to collagen and fibronectin [61]. The latter interaction affects the correct secretion and deposition of these fibrous proteins [112] and additional interactions facilitate the intramolecular crosslinking of collagen in the ECM. Via its FAS1-1 and FAS1-3 domain Pn binds to Bone morphogenetic protein-1 (Bmp-1), a metalloprotease that proteolytically activates lysine oxidase-1 (Lox-1), which, in turn, enhances collagen crosslinking [113,114]. The peptide growth factor Bone morphogenetic protein-2 (Bmp-2) that is not related to Bmp-1 also binds to Pn albeit at a different site [113]. The physiological role of this interaction remains unknown. Another matrix protein interacting with Pn is tenascin-C that forms a remarkable six-armed structure named hexabrachion and it was suggested that the link between tenascin-C and collagen formed by Pn might adapt the ECM to mechanical stress [68]. While some of the ligands of HsPn and HsTgfbi bind to the CROPT domain or the C-terminal domain, the interaction with integrin occurs via FAS1 domains. In overview, the two secreted FAS1 proteins HsTgfbi and HsPn exemplify how FAS1 domains accommodate a variety of protein–protein interactions enabling a single protein to act both in ECM integration and in intercellular signaling. By contrast, the membrane anchored stabilins mostly act as receptors for self ligands and as cell–cell interaction signals. One of their biological functions is in the removal of unwanted material from extracellular spaces and circulation by professional scavenger cells. Despite their similarity at the domain and sequence level, HsStab1 and HsStab2 are associated with distinct ligands and processes. HsStab1 acts as an intracellular sorting receptor as well as an endocytotic cell surface receptor for ligands such as acetylated low-density lipoprotein (acLDL) [115], secreted protein acidic and rich in cysteine (SPARC) [116,117], placental lactogen (PL) [118] and a chitinase-like protein called SI-CLP [119,120]. The FAS1-7 domain of HsStab1 binds to PL and SI-CLP while acLDL and SPARC interact with domains other than FAS1. HsStab1 can guide its various ligands on the secretory route from the Golgi into the lysosomes after endocytosis and through transcytosis (reviewed in [79]). HsStab2 binds ligands such as acLDL and advanced glycation products as well as various glycosaminoglycans such as HA, chondroitin sulfate and dermatan sulfate as well as the anticoagulant heparin and is required for the physiological clearance of these materials from circulation [121]. The polysaccharides bind to the X-link domain on Stab2 [122]. Both HsStab1 and HsStab2 have also been implicated in the phagocytosis of unwanted cells [123]. With their EGF domains, the stabilins recognize the "eat-me!" signal phosphatidylserine (PS) present at the surface of apoptotic cells and aged erythrocytes. HsStab2 utilizes its FAS1 domain in this process to bind to and cooperate with $\alpha v\beta 5$ integrin [107]. The interaction between HsStab2 and PS is important for the fusion of uninucleate myoblasts during myofiber formation with PS functioning as a "fuse-me!" signal [124,125]. Finally, the interaction of HsStab2 FAS1 domains with integrins mediated both homotypic and heterotypic cell to cell adhesion [110,126]. In summary, each of the four human FAS1 proteins fulfils a variety of different biological functions. While secreted HsTgfbi and HsPn act in cell–ECM adhesion, ECM structuring and intercellular signaling, the membrane-anchored HsStab1 and HsStab2 proteins act in endocytosis, intracellular trafficking and cell to cell recognition.

4. Biological Functions of Plant Fasciclin 1 Proteins

The varying gene numbers for FLAs in different plant species shows that gene duplication and gene loss are common. On the one hand, this suggests that the large repertoire of FAS1 proteins might be fine-tuned at the level of differential gene expression analogous to differential mRNA splicing in vertebrates. On the other hand, gene redundancy makes the investigation of FLA function more cumbersome. However, the non-redundant *AtFla1*, *AtFla3*, *AtFla4*, *OsMtr1* and the pair of *AtFla11* and *-12* provide a glimpse of the role of FAS1 proteins in plants.

4.1. The AtFla4 AtFei Pathway

A mutant screen in *Arabidopsis thaliana* that discovered several key regulators of salt stress response [127] also identified a mutant that carried a missense mutation in the *AtFla4* locus [15] often called *atsos5* and here referred to as *atfla4* for consistency. In *atfla4* root elongation was almost completely stalled after transfer to growth media containing moderate levels of NaCl and the mutant root developed a characteristic radially swollen tip. Likewise, elevated levels of sucrose induced this phenotype in *atfla4* [128], which is reminiscent of many conditional mutants defective in cell wall polysaccharide biosynthesis [129]. Although conditional root swelling is the most dramatic aspect of the *atfla4* phenotype, the mutant's abnormally thickened hypocotyls and inflorescence stems, its fat roots under salt-free conditions, larger leaves, shorter siliques and its abnormal seed coat mucilage indicate that *AtFla4* acts non-conditionally throughout the plant [15,128,130–134]. Cell walls in *atfla4* roots were shown to lack the pectin-rich middle lamella that is essential for intercellular adhesion, suggesting a potential function of *AtFla4* in pectin biosynthesis or structure [15]. However, double knockouts in the two similar leucine-rich receptor-like kinases (LRR-RLKs) *AtFei1* and *AtFei2* were not only phenotypically identical to *atfla4* but also non-additively interacted with it [128]. This suggested that *AtFla4* and the two *AtFei* loci act in a linear genetic pathway. The *atfei1 atfei2* double mutants were hypersensitive to cellulose biosynthesis inhibition and showed a decrease in crystalline cellulose production under restrictive conditions. As these phenotypic features were reminiscent of cellulose biosynthetic mutants it was proposed that the Fei-RLKs and by association AtFla4, might function in the biosynthesis of this important polymer [128]. The phenotype of both *atfla4* and *atfei1 atfei2* was modulated by variety of growth regulators. Firstly, the *atfei1 atfei2* root phenotype was suppressed by inhibition of the biosynthesis and oxidation of ACC but not of the perception and signaling of its product ethylene [128]. This suggested that, in line with *AtFla4*, the *AtFei* loci might negatively regulate a novel ethylene-independent ACC signaling pathway that is activated by cell wall defects [135]. Both AtFei proteins interacted with ACC synthase in yeast two hybrid assays but did not phosphorylate them in vitro [128]. Secondly, abscisic acid (ABA) signaling acted synergistically with *AtFla4* to regulate root elongation and abiotic stress response as the *atfla4* phenotype was suppressed by exogenous ABA and by ABA-oversensitive mutants and a *atfla4*-like phenotype was induced by ABA inhibition [130]. Interestingly, *AtFla4* did not interact with two NADPH respiratory burst oxidases that are required for stress responses triggered by cellulose synthase inhibition [136,137]. Finally, both the *atfei1 atfei2* double and the *atfla4* single mutants non-additively interacted with the IAA-alanine resistant 4 (*AtIar4*) locus [138]. These studies implicate the *AtFla4 AtFei* pathway with a variety of intracellular signaling pathways; however, they do not provide a model for the relation between the cell wall and AtFla4. Insights into this problem were stimulated by the characterization of the seed coat mucilage phenotype of *atfla4* and *atfei2*. During seed maturation, pectic mucilage polymers are secreted into a pocket between the plasma membrane and the primary cell wall. Upon hydration of the mature seed the mucilage polymers rapidly swell and rupture the primary walls. However, the mucilage normally adheres to the ruptured primary cell wall [139]. There exist numerous mutants affected in aspects of seed coat mucilage biosynthesis and maturation. They are instructive models for cell wall biosynthesis throughout the plant [140]. Because in *atfla4* and *atfei2* mutants and in the cellulose synthase mutant *atcesA5* seed coat mucilage adhesion was found to be defective it was initially suggested that *AtFla4* and *AtFei2* might act in cellulose biosynthesis [131,132]. However, the subsequent comparative analysis of *atfla4, atfei2, atcesA5* and other mucilage adherence mutants showed distinct roles of *AtFla4* and *AtFei2* on the one hand and *AtCesA5* on the other hand [133,134]. Rather than influencing cellulose or hemicellulose structure, *AtFla4* acted on the pectin network, which was clearly distinct from the genetic actions of *AtCesA5* and of the xylan biosynthetic locus Mucilage-modified5 (*AtMum5*) [133,141,142]. In contrast to mutants in the latter genes, *atfla4* non-additively interacted with two pectin-related loci. Firstly, Mucilage-modified2 (*AtMum2*) encodes a β-galactosidase acting on galactan side chains of pectic rhamnogalacturonan type I (RG I) and its loss of function prevents the extrusion of mucilage upon hydration [143]. The extrusion defect of *atmum2* was suppressed in the *atfla4 atmum2* double mutant [133]. Secondly, Flying saucer 1 (*AtFly1*) encodes a transmembrane E3 ubiquitin ligase thought to regulate the degree of pectin methyl esterification. Loss of *AtFly1* function led to ectopic adhesion between primary cell walls and mucilage [144] and *atfla4* suppressed this phenotype [133,134]. A possible explanation for these observations is that AtFla4 might physically interact with galactan side chains on rhamnogalacturonan type I (RGI) pectin structures that are normally remodeled by the β-galactosidase encoded by *AtMum2*, to control timely mucilage maturation and adherence. This hypothesis is consistent with the localization of *AtFla4* in the mucilage pocket during seed coat development [133,134]. How does AtFei2 fit into this scenario, and can the characterization of the seed coat mucilage phenotype be extrapolated to the whole plant? In the case of seed coat mucilage, AtFei2 might attach AtFla4 to the plasma membrane and thereby contribute to the architecture of the network—an entirely structural explanation. However, it cannot be ruled out that the interaction between the hypothetical AtFla4-pectin matrix and AtFei2 might also convey signals to the cell interior [133,134]. While AtFla4 localized to the developing mucilage pocket and the adjacent plasma membrane of seed coat epidermal cells, AtFla10 and AtFla17 but not AtFla4 were identified in a proteome of mature mucilage [145]. Whatever the biological function of these FLAs in mature mucilage, due to the *atfla4* single mutant phenotype they apparently are not functionally equivalent to AtFla4. It is unclear whether the mode of action of *AtFla4* in seed coat mucilage is identical to its mechanism throughout the plant; however, the overlapping genetic role with the *AtFei* genes in both tissues suggests a common mechanism. The AtFla4 protein might interact with various cellular components. It was hypothesized that AtFla4 might bind to the AtFei RLKs [128,146]. The localization of the majority of yellow fluorescent protein-tagged AtFla4 at the plasma membrane is compatible with this possibility [38]. However, AtFla4 was also released into the apoplast, where it might interact with cell wall carbohydrates [38,133]. The interaction of AtFla4 with cell wall polysaccharides remains to be investigated; however, both covalent and non-covalent interactions between pectic polymers and AG II glycan have previously been demonstrated (see above). In fact, it was shown that protein *O*-galactosylation is required for the function of AtFla4. Several galactosyl transferases (GALTs) mediate the initial glycosylation of hydroxyl proline residues of AGPs. Two of them called AtGalt2 and AtGalt5 together are responsible for the biosynthesis of the bulk of AGPs and their mutation caused a root and seed phenotype identical to *atfla4* [147]. Intriguingly, the phenotypes of *atgalt2 atgalt5 atfla4 atfei1 atfei2* quintuple and the *atfla4 atfei1 atfei2* triple mutants were identical [146]. It was hypothesized that AtFla4 might physically interact with the Fei RLKs via its *O*-glycans. This hypothesis was seemingly contradicted by a structure to function study where the deletion of potential *O*-glycosylation sites on AtFla4 did not interfere with genetic complementation of *atfla4* [38]. However, the AtFla4 level in the transgenic plants might have been higher than normal or cryptic *O*-glycosylation sites might still have been present. How AtFla4 and AtFei interact and if *O*-linked glycosylation plays a direct or indirect role for AtFla4 and AtFei function remains to be tested. Taken together, *AtFla4* is a unique genetic paradigm for FAS1 domain proteins in plants. A speculative model (Figure 5) combined from previous studies proposes that after its release from the plasma membrane and via its *O*-glycan AtFla4 possibly interacts with pectic polymers and contributes to biophysical properties such as swelling and interpolymer connectivity.

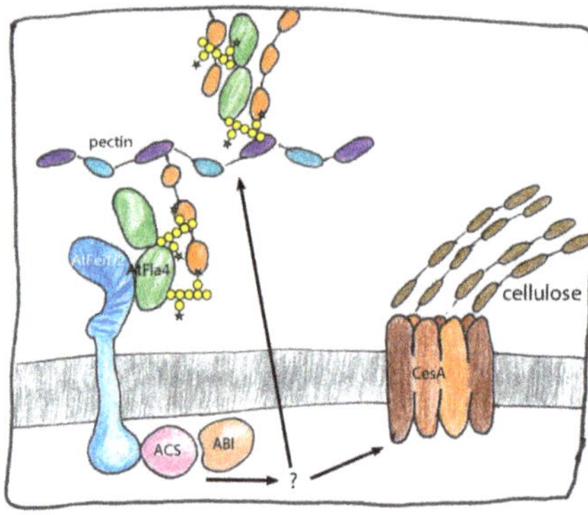

Figure 5. Hypothetical roles of AtFla4 in cell wall structure and signaling. AtFla4 might interact with the pectic network with covalent or non-covalent interactions of its glycans (yellow discs and green stars: *O*-glycans which are yet to be precisely defined, *N*-glycans are omitted). It might mechanically link pectin with AtFei1 and AtFei2 receptor kinases and the plasma membrane. The signals transduced by the receptor kinases might involve ACC synthase (ACS) and Abcisic Acid Insensitive 1 (ABI1). How signaling connects the kinases with cellulose synthase (CesA) and pectin biosyn thesis is unclear.

In parallel AtFla4 might also interact with AtFei1 and AtFei2 via its FAS1 domain to link pectin with AtFei1 and -2 and indirectly to ACC synthase. The access of the C-terminal Fas1 domain to the AtFei receptor domain might be modulated by the interaction to the cell wall. Under standard conditions relatively firm binding of AtFla4 to AtFei might suppress ACS. Under cell wall stress conditions binding of AtFla4 to AtFei might be reduced and the ACS suppression by AtFei might be relaxed leading to ACC production and decreased cellulose production and other responses. However, any postulated physical interactions of *AtFla4* remain to be demonstrated.

4.2. A Conserved Function of AtFla11 and AtFla12 in Secondary Cell Walls

One of the functionally and economically most interesting aspects of some FLAs is their role for secondary cell wall formation and structure. Secondary cell walls are by far the most abundant re-growing fiber and energy feedstock on this planet and its major constituents—cellulose, xylans and lignin—are among the most abundant biopolymers. Their sustainable and socially compatible production is thought crucial in the global transition to a carbon neutral economy. *AtFla11* and *AtFla12* (group A; [14]) and their orthologs (Appendix C) from diverse fiber crops and model plants were noted in *Arabidopsis thaliana* for their expression during the secondary thickening of stems [148], their co-expression with secondary cell–wall–specific cellulose synthase loci [149] and their expression in sclerenchyma cells [150]. A correlative and causative association of *AtFla11* and *AtFla12* orthologs with secondary cell wall biosynthesis has been observed in many plant species. The *Zinnia elegans ZeFla11* gene was found to be expressed in trans-differentiating cell cultures and in a specific subset of secondary cell wall forming vessel cells and adjacent parenchyma cells [151]. *AtFla11* and *-12* orthologs from poplar, eucalyptus and willow were highly expressed in tension wood (TW), a reaction wood forming at the upper side of branches and stems growing sideways. TW consists of almost pure cellulose deposited in the gelatinous secondary cell wall layer (G-layer) in elongated G-fiber cells and shows a characteristically low-cellulose microfibril angle (MFA, with respect to the growth axis), which entails superior tensile strength [152–154]. The three *AtFla11/12* eucalyptus orthologs *EgFla1*, *-2* and *-3* were highly expressed in stems correlating with secondary cell wall formation [71]. Also in flax and hemp *AtFla11/12* orthologs were highly expressed in fiber forming tissues [72,155]. A cotton *AtFla12* ortholog was expressed at a higher level and for a longer time period in *Gossypum barbadense* compared to *G. hirsutum* correlating with greater fiber strength and length of the former [69]. Hence, *AtFla11* orthologs might be natural genetic determinants for fiber crops. So, throughout the higher plants, *AtFla11/12* and their orthologs correlate with secondary cell wall formation, specifically the deposition of cellulose microfibrils aligned along the axis of mechanical stress. What is the role of FLAs in this process?

In *Arabidopsis thaliana* there exists a genetic redundancy between *AtFla11* and *-12*; however, the *atfla11 atfla12* double mutant was affected in the mechanical properties of its secondary cell–wall–rich stems [156]. In *atfla11 atfla12* tensile strength and stiffness were reduced and MFA was increased. Moreover, the chemical composition and cell wall structure showed a reduction of galactose, arabinose and cellulose and a concomitant increase in lignin content. The authors speculated that the FLAs via their FAS1 domains might form a heteromeric higher-order network strengthening the interaction between cellulose microfibrils [156]. The most highly expressed hybrid aspen locus called *PtFLA6* was silenced with antisense RNA. This also reduced the transcript level of many other *PtFla* genes [157,158]. Although no apparent morphological phenotype was observed, the *PtFla6* knockdown resulted in reduced flexural strength and stiffness in stems of transgenic plants. Moreover, the content of AGPs, cellulose and lignin was reduced alongside a reduction of transcript levels for numerous genes involved in cellulose and lignin formation [157]. When TW formation was stimulated in the *PtFla6* knockdown plant, the relative frequency of abnormal G-fibers increased, but no other changes were observed [158]. Taken together, the complex response to *PtFla6* interference is more compatible with a regulatory role than it is with a direct structural role in the cellulose layer. In *Eucalyptus grandis*, *EgFla1*, *-2* and *-3* were experimentally overexpressed in cambial sectors or in whole tobacco plants [71]. Despite their sequence similarity, each transgene exerted a specific effect on phenotypic parameters such as MFA, cell wall thickness, cell wall composition and xylem cell type specification. This meant that structurally even very similar *AtFla11/12* orthologs exerted different effects in the same cellular context [71]. The function of *AtFla11/12* orthologs has also been investigated during cotton fiber formation. In this system, trichome cells are initiated at the outer integument of the ovule during anthesis (2 d before to 5 d after anthesis; dpa). Then the trichomes undergo a period of rapid cell elongation (3–20 dpa), which is followed by secondary wall formation (16–40 dpa) and maturation (40–60 dpa). Hence, each cotton fiber is a single trichome cell made up of almost pure cellulose [159]. Its formation is the result of three distinct developmental processes: (1) specification and initiation, (2) expansion and (3) cell wall thickening and maturation. Two *AtFla11/12* orthologs in the cotton species *Gossypium hirsutum* named *GhAgp4* (Appendix D) and *GhFla1* were suppressed using RNAi [159,160]. *GhFla1* was specifically expressed in elongating fibers with peak expression at 10 dpa [70] and indeed, fibers of RNAi lines were much shorter than controls [159]. However, *GhFla1* RNAi suppression also resulted in delayed and reduced trichome initiation and, moreover, resulted in decreased RNA levels of other fiber-expressed FLAs and of numerous genes implicated with cell wall biosynthesis. The complex alterations at the transcript level correlated with alterations in cell wall composition including decreased cellulose formation. By contrast, the overexpression of *GhFla1* resulted in increased initiation and elongation of cotton fibers as well as upregulation of other FLAs and cell wall biosynthetic loci and in an increased cellulose content [159]. These data confirmed and extended a previous study where the *GhAGP4* locus was silenced by RNAi. Like the GhFla1 silencing, this reduced fiber initiation, elongation and quality, and affected the level of transcripts related to cell wall biosynthesis [160]. The observations made in cotton are difficult to reconcile with a biomechanical role of *GhFla1* and *GhAgp4* as an adhesive between secondary cell wall cellulose microfibrils because not only secondary cell walls were affected by the interference but also the fundamentally different processes of initiation and elongation that happen before. The apparent involvement of these genes in all three phases of fiber formation and their control of mRNA levels for an entire battery of other genes rather suggests that this class of FAS1 proteins fulfils a role in developmental control comparable to a peptide hormone such as the proteoglycan xylogen [161].

What do we know about possible molecular interactors and subcellular localization of *AtFla11/AtFla12* orthologs? Intriguingly, cotton orthologs of *AtFla7* and *AtFla11* were co-precipitated with active secondary cellulose synthase *GhCesA8* that was effectively extracted by inclusion of cellulase in the extraction buffer [162]. This supports the hypothesis that group A FLAs physically interact with the cellulose synthase complex. This hypothesis offers an attractive explanation for co-expression between group A FLAs and secondary cell wall formation related genes and for changes in MFA and mechanical properties in group A *FLA* loss of function genotypes. If one assumes that nascent cellulose fibers in combination with cellulose synthase are both bound by FLAs, this model could also explain the localization of PtFLA6 in secondary cell walls [157,158]. However, this model offers no simple explanation for the observation that the transcriptional program of secondary cell wall formation as well as cellular differentiation was heavily influenced by the experimental manipulation of FLAs [157], or for the many effects of FLA overexpression. This problem might be solved with a role of FLAs in signaling. So far, there has been one report that suggested that a RLK might physically interact with FLAs. A pull-down experiment using the *Arabidopsis thaliana* sucrose-induced receptor kinase (AtSirk1) as bait identified AtFla8 and AtFla9 among the 16 interaction partners. Interestingly, among several potential substrates of AtSirk1, there was the primary cell wall specific cellulose synthase catalytic subunit 3 (AtCesA3) [163]. At present, there is no genetic evidence linking AtFla8 and AtFla9 with primary cell wall cellulose synthesis; however, the evidence for their physical interaction with an RLK that acts on cellulose synthase offers an interesting alternative mechanism for how FLAs might influence cell wall biosynthesis. Using their potential to interact with multiple binding partners they might combine receptor kinases with substrates such as cellulose synthase. Taken together, several modes of action can be envisaged for *AtFla11/12* and its orthologs (Figure 6).

Firstly, AtFla11/12 might be directly secreted to the cell wall and bind to polymers and cross-link them. Secondly these FAS1 proteins might act as extracellular signals or be part of a receptor complex to influence transcriptional events downstream. Thirdly, they might augment cellulose deposition by acting as a physical part of the cellulose synthase machinery. Fourth, by interacting with both cellulose synthase and receptor kinase they might facilitate or modulate post-translational regulation of cellulose biosynthesis (Figure 6).

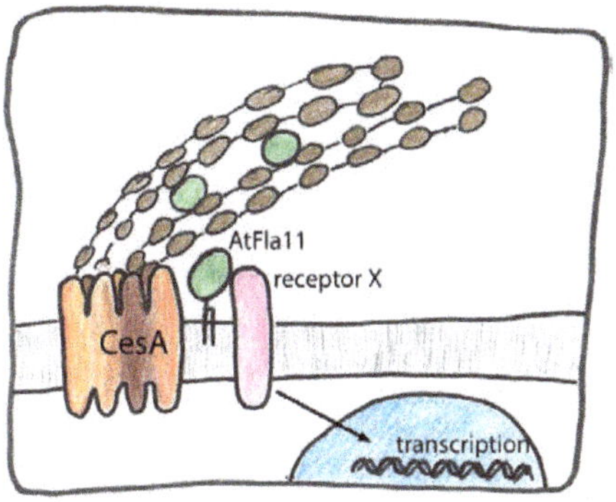

Figure 6. Putative role of *AtFla11*, *AtFla12* and their orthologs in secondary cell wall formation. Firstly they might act as secreted structural proteins in secondary cell walls binding to cellulose or other glycans. Secondly, they might be part of outside–in signaling, in connection with an unknown receptor that leads to complex changes of gene expression. Thirdly, they might bind to cellulose synthase (CesA) and directly influence or stabilize its action. Finally, they might act as adaptors between CesA and receptor kinases to modulate CesA phosphorylation.

4.3. A Cascade of FLAs Acts in Male Gametophyte Development

The FAS1 proteins encoded by the *AtFla3* and the *OsMtr1* genes, respectively, are required for male gametophyte development; however, the mRNA pattern in *Arabidopsis thaliana* suggests that several FLAs might sequentially act in this process (Figure 7).

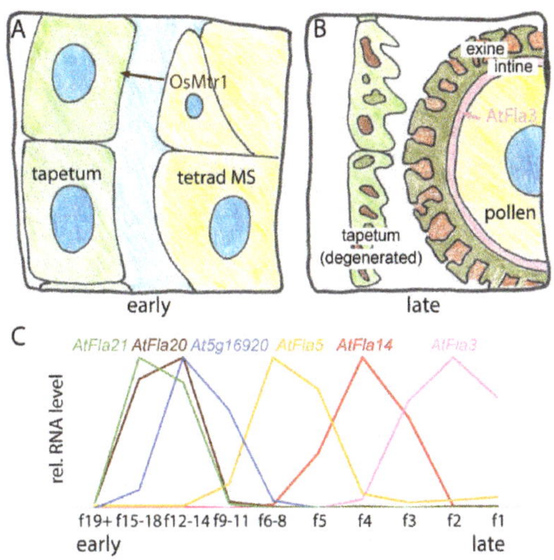

Figure 7. FAS1 proteins act sequentially in male gametophyte development: (**A**) *OsMtr1* acts in early stage of tapetum development. It is specifically expressed and secreted in microspore cells (MS) and influences the development of tapetum cells; (**B**) *AtFla3* acts at the late stage of the formation of the cellulose- and pectin-rich intine layer of pollen cell walls; (**C**) RNA-Seq data of *Arabidopsis thaliana* at different floral stages [164,165] show that six FAS1 genes are specifically expressed in anthers at different stages of floral development. f1 indicates the first open flower. The expression patterns of AtFla20 that might be the ortholog of OsMtr1 and of AtFla3 are consistent with roles in early and late pollen development, respectively.

No functional pollen was found in the rice microspore and tapetum regulator 1 (*osmtr1*) mutant [16]. *OsMtr1* expression was specifically detected in early microspore development during meiosis and tetrad formation and during the release of young microspores (Figure 7A). However, despite *OsMtr1* never being expressed in the tapetum, a nutritive maternal tissue that surrounds microspores, the *osmtr1* mutation dramatically affected tapetum development. This suggested that *OsMtr1* might act non-cell autonomously. The *osmtr1* mutant tapetum failed to undergo programmed cell death, which normally supports pollen wall formation in developing microspores. Tetrads formed normally in *osmtr1* yet, mutant microspores showed defective development as soon as they were released possibly due to the lack of cell wall material contributed by the adjacent tapetum. The localization of OsMtr1 in the microspores and the extracellular space between them and the tapetum suggested that OsMtr1 might act as a secreted signal to coordinate microspore development with tapetum degeneration (Figure 7A). Molecular interactors such as a hypothetical *OsMtr1* receptor on tapetum cells remain to be identified [16]. The expression of *AtFla3* in pollen during anthesis and pollen germination combined with the high level of expression in anthers is unique among all FLAs and *AtFla3* function was investigated using RNAi and overexpression [166]. The *AtFla3* RNAi lines were partially male-sterile due to a reduced ability to form functional microspores. Histological analysis showed that *AtFla3* RNAi affected the formation of intine, the innermost layer of the pollen wall, which consists mainly of cellulose and pectin. The earliest time point at which intine abnormalities were observed was the late uninucleate stage, suggesting that *AtFla3* might act in the formation of intine in microspore development. Abnormal *AtFla3* RNAi pollen showed defective cellulose staining, suggesting that *AtFlaA3* was required for cellulose deposition in the intine layer. More detailed analyses are required to test whether any other cell wall polymers are affected and if full loss of function of the *AtFla3* locus causes a more complete defect of pollen function. During *Arabidopsis thaliana* floral development at least six FAS1 domain loci are specifically expressed in anthers during separate developmental stages [164,165]. Like its closest rice homologue *OsMtr1, AtFla20* is expressed at an early stage, where it overlaps with *AtFLA21* and *At5g16920*—an FAS1 protein not annotated as FLA (Figure 7C). The mRNA levels of *AtFla14* and *AtFla5* sequentially peak at intermediate stages and *AtFla3* is expressed immediately before and during flower opening (anthesis) in mature and germinating pollen, as previously reported [166]. How most of these FAS1 domain proteins are involved in normal pollen wall formation remains to be investigated. Nonetheless, their expression pattern during pollen formation appears to reflect the complex development and composition of the pollen wall.

4.4. Plant FAS1 Proteins Have Many Potential Functions

Several other FLAs are known to perform at least partially non-redundant genetic roles. Firstly, the loss of function mutation of *AtFla1* caused a phenotype in the growth regulator induced formation of shoots and roots from the callus [167]. Two alleles in two different wild-type backgrounds were isolated and the wild-type background on its own not only had a dramatic influence on callus derived organ formation but it essentially inverted the genetic effect of *AtFla1* on the process. By contrast, the negative effect of the *AtFla1* locus on root length and lateral root number was constant between the two tested wild-type backgrounds. It is presently unclear how *AtFla1* might influence organ formation but initial promoter studies showed expression of *AtFla1* early during the formation of organized tissue from callus. The protein might therefore be required for pattern formation [167]. Secondly, in a gene-editing study in the non-model plant *Brassica carinata, BcFla1*, a group A FLA that is more similar to *AtFla13* than to *AtFla11*, was implicated in root hair growth, where tip-focused deposition of primary cell wall materials occurs [168]. In a third example, in maize the expression level of group A *ZmFla* genes was inversely correlated with seed abortion. Consistently, two different mutations in the group A *AtFla9* locus that showed abnormally low and high *AtFla9* RNA levels displayed increased and decreased seed abortion, respectively [169]. Lastly, a comparative study of 30 different wheat (*Triticum aestivum*) varieties found an association between the expression of a FAS1 protein and grain milling properties [170]. According to this report, the locus *TaFla8* [51] coding for a FAS1 domain that does not contain any PAST-rich domains, is specifically expressed in grain. Interestingly, its level was relatively low in varieties displaying beneficial milling properties compared to varieties scoring poorly with respect to milling efficiency. Hence, human selection of wheat varieties with improved milling properties might have inadvertently reduced the basal expression level of *TaFla8*, which might normally strengthen cell walls.

5. Fungal FAS1 Proteins

In a screen for genes specifically expressed in fruiting bodies of the Shiitake mushroom *Lentinula edodes* the *LeFlp1* locus coding for a GPI-anchored single FAS1 domain protein was identified and its specific expression during fruiting body formation suggested a developmental role [17]. In another study the *MoFlp1* locus of the rice blast fungus *Magnaporthe oryzae* was shown to encode a GPI-anchored vacuolar tandem FAS1 protein [18]. Its disruption caused defects in the formation of conidia, which, according to the authors, hinted at a potential role of *MoFlp1* in autophagy. Interestingly, the fission yeast *Schizosaccharomyces pombe* FAS1 domain protein SpFsc1 was identified in a genetic screen for autophagy-related loci [19]. SpFsc1 consists of five FAS1 domains in tandem and a transmembrane domain. The protein was detected in the vacuolar membrane and was shown to specifically act in a late step in autophagy, the fusion between the phagosomes with the vacuolar membrane. The mechanistic details including the role of the FAS1 domains in this process remain to be elucidated. In summary, studies on FAS1 domain proteins in fungi are sparse; however, they provide a stimulating variation of the hypothetical modes of action of this protein family.

6. Bacterial FAS1 Proteins

The most thoroughly studied eubacterial FAS1 proteins are the homologues Mpb70 and Mpb83 from *Mycobacterium* (reviewed in [171]). In various genomes of obligate pathogenic, opportunistic and non-pathogenic mycobacteria [172], both non-pathogenic and pathogenic species contained *Mpb70/Mpb83* homologues, while the genomes of opportunistic mycobacteria did not. This suggests a relation of *Mpb70/Mpb83* to mycobacterial lifestyle and pathogenesis. While Mpb83/Mpb70 was hypothesized to undergo "homophilic" interactions with mammalian Pn thereby interfering with cell adhesion in bones [171] this idea is not compatible with the secreted nature of Pn and is derived from cell aggregation data [8,173], which could have alternative explanations. Moreover, no intermolecular interactions between Fas1 domains have been reported. However, both host cell to ECM adhesion and host cell to mycobacterium adhesion might be modulated by binding of Mpb70/Mpb83 to molecules presented at the host cell surface. Indeed, Mpb83 was demonstrated to bind to Toll-like receptors 1 and -2 (HsTlr1 and -2), two LRR-RLKs present on human monocytes [41]. The binding of recombinant Mpb83 to HsTlr2 triggered mitogen activated protein kinase (MAPK) signaling and as a consequence induced the production of matrix metalloprotease-9 (MMP-9) and cytokines [174]. The direct interaction between Tlr2 and Mpb83, as shown by surface plasmon resonance [41], is an intriguing finding, given the presence of more than 200 LRR-RLKs in plant genomes [175]. A role of bacterial FAS1 proteins in the interaction between bacteria and their eukaryotic hosts is also observed in symbiotic communities. In the first example, the FAS1 protein SmNex18 (locus SMa1077) of the nitrogen-fixing legume symbiont *Sinorhizobium meliloti* was shown to be expressed specifically during nodulation, the initial stage of symbiosis between bacterium and plant [176,177]. Secondly, a FAS1 protein was found in *Nostonoc* cyanobacteria that form symbiotic bacterial–fungal communities called lichen. *Nostonoc punctiforme* isolates living in lichen highly expressed an unnamed FAS1 protein (locus ACC81089) that was only a minor compound in a closely related free living strain [178]. Moreover, FAS1 proteins featured prominently in the metaproteome of phyllospheric (i.e., plant leaf dwelling) bacteria [179] but were absent from rhizospheric (root-dwelling) bacteria [180]. Likewise, while present in many different genera of archaea, the FAS1 domain is not necessarily ubiquitous within the same genus. To give an example, among the 13 completely sequenced *Methanobacterium* isolates, only *M. paludis* isolated from peatland contained FAS1 domain protein sequences [181]. On the one hand, such circumstantial observations support the hypothesis that FAS1 proteins are important for the interaction between prokaryotic colonizers and eukaryotic hosts; on the other hand, they also show that FAS1 proteins are not essential.

7. Concluding Remarks

Can we define a common denominator for the function of FAS1 proteins? The characterization of HsTgfbi structure and function demonstrates the versatility of the FAS1 domain in binding multiple ligands, which might be a conserved feature. The frequent assumption that FAS1 proteins are involved in "cell adhesion," however, is more context-dependent. It is derived from insightful experiments performed with animal cell cultures; however, it might be misleading for "hard-shelled" cells such as bacteria, fungi and plants that usually adhere to their ECM as a result of turgor pressure and build up adhesive materials such as pectin. By contrast, the designation of Tgfbi and Pn as matricellular proteins, defined as non-structural ECM-components that interact with cell surface receptors as mediators between the cell and its microenvironment [77,182], seems a more productive concept for approaching the common function of FAS1 proteins in all kingdoms of life including plants. The crucial challenge for future work on plant FAS1 proteins will be the identification of molecular interactors both of the FAS1 domain and of their abundant but enigmatic glycan modifications.

Funding: This research was funded by the Austrian Science Fund FWF (Grant number I1182-B22).

Conflicts of Interest: The author declares no conflict of interest.

Appendix A

Since many genes and proteins from different organisms are discussed in this review, I adopted universal nomenclature rules according to a highly popular textbook [183]. Gene symbols are written in italics, with the first letter in uppercase indicating a wild-type allele and in lowercase indicating a mutant allele. Proteins are written in non-italics. If the text relates to genes or proteins from a certain species, a two-letter prefix (Hs for *Homo sapiens,* At for *Arabidopsis thaliana* etc.) is used.

Appendix B

Because many human proteins were independently discovered several times, and maybe also for asserting the status of discovery, there is an abundance of synonyms. Tgfbi is also known as β-ig H3, βIG, BIGH3, kerato-epithelin, RGD-CAP and MP78/70. Pn is synonymous with Osteoblast specific factor-2 (Osf-2). Stab1 has also been named Common lymphatic endothelial and vascular endothelial receptor-1 (Clever-1) and FEEL-1 is an acronym for fasciclin, EGF-like, laminin-type EGF-like, and link domain-containing scavenger receptor-1. Frequent Stab2 synonyms are lymphatic vessel endothelial hyaluronan receptor-1 (LYVE-1) and hyaluronic acid receptor (HARE).

Appendix C

The term "*AtFla11/12* ortholog" is here used to designate FLAs from other plant species that are more similar in protein sequence to *AtFla11* or *AtFla12* than to any other *Arabidopsis thaliana* protein.

Appendix D

The nomenclature in the cotton literature is somewhat confusing. Some FLAs are named AGPs and GhFla1 refers to two entirely different protein sequences. The GhFla1 that was mentioned by Li et al. (2010) [160] being most similar to *AtFla9/6/13* and the *GhFla1* silenced by Huang et al. (2013) [159] is an ortholog of *AtFla11/12*. The *GhAgp4* locus (gb EF470295) silenced by Li (2010) [160] is identical to the sequence annotated as *GhFla12*.

References

1. Bastiani, M.J.; Harrelson, A.L.; Snow, P.M.; Goodman, C.S. Expression of fasciclin I and II glycoproteins on subsets of axon pathways during neuronal development in the grasshopper. *Cell* **1987**, *48*, 745–755. [CrossRef]
2. Snow, P.M.; Zinn, K.; Harrelson, A.L.; McAllister, L.; Schilling, J.; Bastiani, M.J.; Makk, G.; Goodman, C.S. Characterization and cloning of fasciclin I and fasciclin II glycoproteins in the grasshopper. *Proc. Natl. Acad. Sci. USA* **1988**, *85*, 5291–5295. [CrossRef] [PubMed]
3. Hu, S.; Sonnenfeld, M.; Stahl, S.; Crews, S.T. Midline fasciclin: A drosophila fasciclin-I-related membrane protein localized to the CNS midline cells and trachea. *J. Neurobiol.* **1998**, *35*, 77–93. [CrossRef]
4. Zhong, Y.; Shanley, J. Altered nerve terminal arborization and synaptic transmission in drosophila mutants of cell adhesion molecule fasciclin I. *J. Neurosci.* **1995**, *15*, 6679–6687. [CrossRef] [PubMed]
5. Jay, D.G.; Keshishian, H. Laser inactivation of fasciclin I disrupts axon adhesion of grasshopper pioneer neurons. *Nature* **1990**, *348*, 548–550. [CrossRef] [PubMed]
6. Clout, N.J.; Tisi, D.; Hohenester, E. Novel fold revealed by the structure of a FAS1 domain pair from the insect cell adhesion molecule fasciclin I. *Structure* **2003**, *11*, 197–203. [CrossRef]
7. Skonier, J.; Neubauer, M.; Madisen, L.; Bennett, K.; Plowman, G.D.; Purchio, A.F. cDNA cloning and sequence analysis of βig-h3, a novel gene induced in a human adenocarcinoma cell line after treatment with transforming growth factor-β. *DNA Cell Biol.* **1992**, *11*, 511–522. [CrossRef] [PubMed]
8. Takeshita, S.; Kikuno, R.; Tezuka, K.; Amann, E. Osteoblast-specific factor 2: Cloning of a putative bone adhesion protein with homology with the insect protein fasciclin I. *Biochem. J.* **1993**, *294*, 271–278. [CrossRef] [PubMed]
9. Horiuchi, K.; Amizuka, N.; Takeshita, S.; Takamatsu, H.; Katsuura, M.; Ozawa, H.; Toyama, Y.; Bonewald, L.F.; Kudo, A. Identification and characterization of a novel protein, periostin, with restricted expression to periosteum and periodontal ligament and increased expression by transforming growth factor β. *J. Bone Miner. Res.* **1999**, *14*, 1239–1249. [CrossRef] [PubMed]
10. Politz, O.; Gratchev, A.; McCourt, P.A.; Schledzewski, K.; Guillot, P.; Johansson, S.; Svineng, G.; Franke, P.; Kannicht, C.; Kzhyshkowska, J.; et al. Stabilin-1 and -2 constitute a novel family of fasciclin-like hyaluronan receptor homologues. *Biochem. J.* **2002**, *362*, 155–164. [PubMed]
11. Huber, O.; Sumper, M. Algal-CAMS: Isoforms of a cell adhesion molecule in embryos of the alga *Volvox* with homology to *Drosophila* fasciclin I. *EMBO J.* **1994**, *13*, 4212–4222. [PubMed]
12. Schultz, C.J.; Johnson, K.L.; Currie, G.; Bacic, A. The classical arabinogalactan protein gene family of arabidopsis. *Plant Cell* **2000**, *12*, 1751–1768. [CrossRef] [PubMed]
13. Schultz, C.J.; Rumsewicz, M.P.; Johnson, K.L.; Jones, B.J.; Gaspar, Y.M.; Bacic, A. Using genomic resources to guide research directions. The arabinogalactan protein gene family as a test case. *Plant Physiol.* **2002**, *129*, 1448–1463. [CrossRef] [PubMed]
14. Johnson, K.L.; Jones, B.J.; Bacic, A.; Schultz, C.J. The fasciclin-like arabinogalactan proteins of Arabidopsis. A multigene family of putative cell adhesion molecules. *Plant Physiol.* **2003**, *133*, 1911–1925. [CrossRef] [PubMed]
15. Shi, H.; Kim, Y.; Guo, Y.; Stevenson, B.; Zhu, J.K. The Arabidopsis *Sos5* locus encodes a putative cell surface adhesion protein and is required for normal cell expansion. *Plant Cell* **2003**, *15*, 19–32. [CrossRef] [PubMed]
16. Tan, H.; Liang, W.; Hu, J.; Zhang, D. MTR1 encodes a secretory fasciclin glycoprotein required for male reproductive development in rice. *Dev. Cell* **2012**, *22*, 1127–1137. [CrossRef] [PubMed]
17. Miyazaki, Y.; Kaneko, S.; Sunagawa, M.; Shishido, K.; Yamazaki, T.; Nakamura, M.; Babasaki, K. The fruiting-specific *LeFlp1* gene, encoding a novel fungal fasciclin-like protein, of the basidiomycetous mushroom *Lentinula edodes*. *Curr. Genet.* **2007**, *51*, 367–375. [CrossRef] [PubMed]
18. Liu, T.B.; Chen, G.Q.; Min, H.; Lin, F.C. *MoFlp1*, encoding a novel fungal fasciclin-like protein, is involved in conidiation and pathogenicity in *Magnaporthe oryzae*. *J. Zhejiang Univ. Sci. B* **2009**, *10*, 434–444. [CrossRef] [PubMed]
19. Sun, L.L.; Li, M.; Suo, F.; Liu, X.M.; Shen, E.Z.; Yang, B.; Dong, M.Q.; He, W.Z.; Du, L.L. Global analysis of fission yeast mating genes reveals new autophagy factors. *PLoS Genet.* **2013**, *9*, e1003715. [CrossRef] [PubMed]

20. Nagai, S.; Matsumoto, J.; Nagasuga, T. Specific skin-reactive protein from culture filtrate of *Mycobacterium bovis* BCG. *Infect. Immun.* **1981**, *31*, 1152–1160. [PubMed]
21. Nagai, S.; Miura, K.; Tokunaga, T.; Harboe, M. Mpb70, a unique antigenic protein isolated from the culture filtrate of BCG substrain Tokyo. *Dev. Biol. Stand.* **1986**, *58*, 511–516. [PubMed]
22. Terasaka, K.; Yamaguchi, R.; Matsuo, K.; Yamazaki, A.; Nagai, S.; Yamada, T. Complete nucleotide sequence of immunogenic protein Mpb70 from *Mycobacterium bovis* BCG. *FEMS Microbiol. Lett.* **1989**, *49*, 273–276. [CrossRef] [PubMed]
23. Radford, A.J.; Wood, P.R.; Billman-Jacobe, H.; Geysen, H.M.; Mason, T.J.; Tribbick, G. Epitope mapping of the *Mycobacterium bovis* secretory protein Mpb70 using overlapping peptide analysis. *J. Gen. Microbiol.* **1990**, *136*, 265–272. [CrossRef] [PubMed]
24. Ulstrup, J.C.; Jeansson, S.; Wiker, H.G.; Harboe, M. Relationship of secretion pattern and Mpb70 homology with osteoblast-specific factor 2 to osteitis following *Mycobacterium bovis* BCG vaccination. *Infect. Immun.* **1995**, *63*, 672–675. [PubMed]
25. Burroughs, A.M.; Balaji, S.; Iyer, L.M.; Aravind, L. Small but versatile: The extraordinary functional and structural diversity of the β-grasp fold. *Biol. Dir.* **2007**, *2*, 18. [CrossRef] [PubMed]
26. Boratyn, G.M.; Schaffer, A.A.; Agarwala, R.; Altschul, S.F.; Lipman, D.J.; Madden, T.L. Domain enhanced lookup time accelerated blast. *Biol. Dir.* **2012**, *7*, 12. [CrossRef] [PubMed]
27. Clout, N.J.; Hohenester, E. A model of FAS1 domain 4 of the corneal protein βig-h3 gives a clearer view on corneal dystrophies. *Mol. Vis.* **2003**, *9*, 440–448. [PubMed]
28. Moody, R.G.; Williamson, M.P. Structure and function of a bacterial fasciclin I domain protein elucidates function of related cell adhesion proteins such as TGFBIp and periostin. *FEBS Open Biol.* **2013**, *3*, 71–77. [CrossRef] [PubMed]
29. Carr, M.D.; Bloemink, M.J.; Dentten, E.; Whelan, A.O.; Gordon, S.V.; Kelly, G.; Frenkiel, T.A.; Hewinson, R.G.; Williamson, R.A. Solution structure of the *Mycobacterium tuberculosis* complex protein Mpb70: From tuberculosis pathogenesis to inherited human corneal desease. *J. Biol. Chem.* **2003**, *278*, 43736–43743. [CrossRef] [PubMed]
30. Basaiawmoit, R.V.; Oliveira, C.L.; Runager, K.; Sorensen, C.S.; Behrens, M.A.; Jonsson, B.H.; Kristensen, T.; Klintworth, G.K.; Enghild, J.J.; Pedersen, J.S.; et al. SAXS models of TGFBIp reveal a trimeric structure and show that the overall shape is not affected by the Arg124His mutation. *J. Mol. Biol.* **2011**, *408*, 503–513. [CrossRef] [PubMed]
31. Yun, H.; Kim, E.H.; Lee, C.W. ^{1}H, ^{13}C, and ^{15}N resonance assignments of FAS1-IV domain of human periostin, a component of extracellular matrix proteins. *Biomol. NMR Assign.* **2017**, *12*, 95–98. [CrossRef] [PubMed]
32. Underhaug, J.; Koldso, H.; Runager, K.; Nielsen, J.T.; Sorensen, C.S.; Kristensen, T.; Otzen, D.E.; Karring, H.; Malmendal, A.; Schiott, B.; et al. Mutation in transforming growth factor β induced protein associated with granular corneal dystrophy type 1 reduces the proteolytic susceptibility through local structural stabilization. *Biochim. Biophys. Acta* **2013**, *1834*, 2812–2822. [CrossRef] [PubMed]
33. Garcia-Castellanos, R.; Nielsen, N.S.; Runager, K.; Thogersen, I.B.; Lukassen, M.V.; Poulsen, E.T.; Goulas, T.; Enghild, J.J.; Gomis-Ruth, F.X. Structural and functional implications of human transforming growth factor β-induced protein, TGFBIp, in corneal dystrophies. *Structure* **2017**, *25*, 1740–1750. [CrossRef] [PubMed]
34. Thapa, N.; Lee, B.H.; Kim, I.S. TGFBIp/βig-h3 protein: A versatile matrix molecule induced by TGF-β. *Intern. J. Biochem. Cell. Biol.* **2007**, *39*, 2183–2194. [CrossRef] [PubMed]
35. Kim, J.E.; Kim, S.J.; Lee, B.H.; Park, R.W.; Kim, K.S.; Kim, I.S. Identification of motifs for cell adhesion within the repeated domains of transforming growth factor-β-induced gene, βig-h3. *J. Biol. Chem.* **2000**, *275*, 30907–30915. [CrossRef] [PubMed]
36. Kim, J.E.; Jeong, H.W.; Nam, J.O.; Lee, B.H.; Choi, J.Y.; Park, R.W.; Park, J.Y.; Kim, I.S. Identification of motifs in the fasciclin domains of the transforming growth factor-β-induced matrix protein βig-h3 that interact with the αvβ5 integrin. *J. Biol. Chem.* **2002**, *277*, 46159–46165. [CrossRef] [PubMed]
37. Orecchia, P.; Conte, R.; Balza, E.; Castellani, P.; Borsi, L.; Zardi, L.; Mingari, M.C.; Carnemolla, B. Identification of a novel cell binding site of periostin involved in tumour growth. *Eur. J. Cancer* **2011**, *47*, 2221–2229. [CrossRef] [PubMed]
38. Xue, H.; Veit, C.; Abas, L.; Tryfona, T.; Maresch, D.; Ricardi, M.M.; Estevez, J.M.; Strasser, R.; Seifert, G.J. *Arabidopsis thaliana* FLA4 functions as a glycan-stabilized soluble factor via its carboxy-proximal fasciclin 1 domain. *Plant J.* **2017**, *91*, 613–630. [CrossRef] [PubMed]

39. Son, H.N.; Nam, J.O.; Kim, S.; Kim, I.S. Multiple FAS1 domains and the RGD motif of TGFBI act cooperatively to bind αvβ3 integrin, leading to anti-angiogenic and anti-tumor effects. *Biochim. Biophys. Acta* **2013**, *1833*, 2378–2388. [CrossRef] [PubMed]
40. Vosloo, W.; Tippoo, P.; Hughes, J.E.; Harriman, N.; Emms, M.; Beatty, D.W.; Zappe, H.; Steyn, L.M. Characterisation of a lipoprotein in *Mycobacterium bovis* (BCG) with sequence similarity to the secreted protein MPB70. *Gene* **1997**, *188*, 123–128. [CrossRef]
41. Chambers, M.A.; Whelan, A.O.; Spallek, R.; Singh, M.; Coddeville, B.; Guerardel, Y.; Elass, E. Non-acylated *Mycobacterium bovis* glycoprotein MPB83 binds to TLR1/2 and stimulates production of matrix metalloproteinase 9. *Biochem. Biophys. Res. Commun.* **2010**, *400*, 403–408. [CrossRef] [PubMed]
42. Rittenour, W.R.; Harris, S.D. Glycosylphosphatidylinositol-anchored proteins in *Fusarium graminearum*: Inventory, variability, and virulence. *PLoS ONE* **2013**, *8*, e81603. [CrossRef] [PubMed]
43. Mosher, D.F.; Johansson, M.W.; Gillis, M.E.; Annis, D.S. Periostin and TGF-β-induced protein: Two peas in a pod? *Crit. Rev. Biochem Mol. Biol.* **2015**, *50*, 1–13.
44. Zavaliev, R.; Dong, X.; Epel, B.L. Glycosylphosphatidylinositol (GPI) modification serves as a primary plasmodesmal sorting signal. *Plant Physiol.* **2016**, *172*, 1061–1073. [PubMed]
45. Tapken, W.; Murphy, A.S. Membrane nanodomains in plants: Capturing form, function, and movement. *J. Exp. Bot.* **2015**, *66*, 1573–1586. [CrossRef] [PubMed]
46. Michell, S.L.; Whelan, A.O.; Wheeler, P.R.; Panico, M.; Easton, R.L.; Etienne, A.T.; Haslam, S.M.; Dell, A.; Morris, H.R.; Reason, A.J.; et al. The mpb83 antigen from *Mycobacterium bovis* contains *O*-linked mannose and (1→3)-mannobiose moieties. *J. Biol. Chem.* **2003**, *278*, 16423–16432. [CrossRef] [PubMed]
47. Kieliszewski, M.J. The latest hype on Hyp-*O*-glycosylation codes. *Phytochemistry* **2001**, *57*, 319–323. [CrossRef]
48. Kitazawa, K.; Tryfona, T.; Yoshimi, Y.; Hayashi, Y.; Kawauchi, S.; Antonov, L.; Tanaka, H.; Takahashi, T.; Kaneko, S.; Dupree, P.; et al. β-galactosyl Yariv reagent binds to the β-1,3-galactan of arabinogalactan proteins. *Plant Physiol.* **2013**, *161*, 1117–1126. [CrossRef] [PubMed]
49. Paulsen, B.S.; Craik, D.J.; Dunstan, D.E.; Stone, B.A.; Bacic, A. The Yariv reagent: Behaviour in different solvents and interaction with a gum arabic arabinogalactan-protein. *Carbohydr. Pol.* **2014**, *106*, 460–468. [CrossRef] [PubMed]
50. Shimizu, M.; Igasaki, T.; Yamada, M.; Yuasa, K.; Hasegawa, J.; Kato, T.; Tsukagoshi, H.; Nakamura, K.; Fukuda, H.; Matsuoka, K. Experimental determination of proline hydroxylation and hydroxyproline arabinogalactosylation motifs in secretory proteins. *Plant J.* **2005**, *42*, 877–889. [CrossRef] [PubMed]
51. Faik, A.; Abouzouhair, J.; Sarhan, F. Putative fasciclin-like arabinogalactan-proteins (FLA) in wheat (*Triticum aestivum*) and rice (*Oryza sativa*): Identification and bioinformatic analyses. *Mol. Genet Genom.* **2006**, *276*, 478–494. [CrossRef] [PubMed]
52. Tan, L.; Qiu, F.; Lamport, D.T.; Kieliszewski, M.J. Structure of a hydroxyproline (Hyp)-arabinogalactan polysaccharide from repetitive Ala-Hyp expressed in transgenic *Nicotiana tabacum*. *J. Biol. Chem.* **2004**, *279*, 13156–13165. [CrossRef] [PubMed]
53. Tryfona, T.; Liang, H.C.; Kotake, T.; Tsumuraya, Y.; Stephens, E.; Dupree, P. Structural characterization of Arabidopsis leaf arabinogalactan polysaccharides. *Plant Physiol.* **2012**, *160*, 653–666. [CrossRef] [PubMed]
54. Tryfona, T.; Theys, T.E.; Wagner, T.; Stott, K.; Keegstra, K.; Dupree, P. Characterisation of FUT4 and FUT6 α-(1→2)-fucosyltransferases reveals that absence of root arabinogalactan fucosylation increases Arabidopsis root growth salt sensitivity. *PLoS ONE* **2014**, *9*, e93291. [CrossRef] [PubMed]
55. Seifert, G.J.; Roberts, K. The biology of arabinogalactan proteins. *Ann. Rev. Plant Biol.* **2007**, *58*, 137–161. [CrossRef] [PubMed]
56. Tan, L.; Showalter, A.M.; Egelund, J.; Hernandez-Sanchez, A.; Doblin, M.S.; Bacic, A. Arabinogalactan-proteins and the research challenges for these enigmatic plant cell surface proteoglycans. *Front. Plant Sci.* **2012**, *3*, 140. [CrossRef] [PubMed]
57. Tan, L.; Eberhard, S.; Pattathil, S.; Warder, C.; Glushka, J.; Yuan, C.; Hao, Z.; Zhu, X.; Avci, U.; Miller, J.S.; et al. An Arabidopsis cell wall proteoglycan consists of pectin and arabinoxylan covalently linked to an arabinogalactan protein. *Plant Cell* **2013**, *25*, 270–287. [CrossRef] [PubMed]
58. Moller, I.; Sorensen, I.; Bernal, A.J.; Blaukopf, C.; Lee, K.; Obro, J.; Pettolino, F.; Roberts, A.; Mikkelsen, J.D.; Knox, J.P.; et al. High-throughput mapping of cell-wall polymers within and between plants using novel microarrays. *Plant J.* **2007**, *50*, 1118–1128. [CrossRef] [PubMed]

59. Tan, L.; Tees, D.; Qian, J.; Kareem, S.; Kieliszewski, M.J. Intermolecular interactions between glycomodules of plant cell wall arabinogalactan-proteins and extensins. *Cell Surf.* **2018**, *1*, 25–33. [CrossRef]
60. Huang, Y.; Wang, Y.; Tan, L.; Sun, L.; Petrosino, J.; Cui, M.Z.; Hao, F.; Zhang, M. Nanospherical arabinogalactan proteins are a key component of the high-strength adhesive secreted by English ivy. *Proc. Natl. Acad. Sci. USA* **2016**, *113*, E3193–3202. [CrossRef] [PubMed]
61. Kii, I.; Ito, H. Periostin and its interacting proteins in the construction of extracellular architectures. *Cell Mol. Life Sci.* **2017**, *74*, 4269–4277. [CrossRef] [PubMed]
62. Park, S.Y.; Jung, M.Y.; Lee, S.J.; Kang, K.B.; Gratchev, A.; Riabov, V.; Kzhyshkowska, J.; Kim, I.S. Stabilin-1 mediates phosphatidylserine-dependent clearance of cell corpses in alternatively activated macrophages. *J. Cell Sci.* **2009**, *122*, 3365–3373. [CrossRef] [PubMed]
63. Harris, E.N.; Parry, S.; Sutton-Smith, M.; Pandey, M.S.; Panico, M.; Morris, H.R.; Haslam, S.M.; Dell, A.; Weigel, P.H. *N*-glycans on the link domain of human HARE/Stabilin-2 are needed for hyaluronan binding to purified ecto-domain, but not for cellular endocytosis of hyaluronan. *Glycobiology* **2010**, *20*, 991–1001. [CrossRef] [PubMed]
64. Ohno, S.; Noshiro, M.; Makihira, S.; Kawamoto, T.; Shen, M.; Yan, W.; Kawashima-Ohya, Y.; Fujimoto, K.; Tanne, K.; Kato, Y. RGD-CAP (βig-h3) enhances the spreading of chondrocytes and fibroblasts via integrin α1β1. *Biochim. Biophys. Acta* **1999**, *1451*, 196–205. [CrossRef]
65. Kim, H.J.; Kim, I.S. Transforming growth factor-β-induced gene product, as a novel ligand of integrin αMβ2, promotes monocytes adhesion, migration and chemotaxis. *Int. J. Biochem. Cell Biol.* **2008**, *40*, 991–1004. [CrossRef] [PubMed]
66. Hoersch, S.; Andrade-Navarro, M.A. Periostin shows increased evolutionary plasticity in its alternatively spliced region. *BMC Evol. Biol.* **2010**, *10*, 30. [CrossRef] [PubMed]
67. Kudo, A. Introductory review: Periostin-gene and protein structure. *Cell. Mol. Life Sci.* **2017**, *74*, 4259–4268. [CrossRef] [PubMed]
68. Kii, I.; Nishiyama, T.; Li, M.; Matsumoto, K.; Saito, M.; Amizuka, N.; Kudo, A. Incorporation of tenascin-C into the extracellular matrix by periostin underlies an extracellular meshwork architecture. *J. Biol. Chem.* **2010**, *285*, 2028–2039. [CrossRef] [PubMed]
69. Liu, H.; Shi, R.; Wang, X.; Pan, Y.; Li, Z.; Yang, X.; Zhang, G.; Ma, Z. Characterization and expression analysis of a fiber differentially expressed fasciclin-like arabinogalactan protein gene in sea island cotton fibers. *PLoS ONE* **2013**, *8*, e70185. [CrossRef] [PubMed]
70. Huang, G.Q.; Xu, W.L.; Gong, S.Y.; Li, B.; Wang, X.L.; Xu, D.; Li, X.B. Characterization of 19 novel cotton *FLA* genes and their expression profiling in fiber development and in response to phytohormones and salt stress. *Physiol. Plant.* **2008**, *134*, 348–359. [CrossRef] [PubMed]
71. MacMillan, C.P.; Taylor, L.; Bi, Y.; Southerton, S.G.; Evans, R.; Spokevicius, A. The fasciclin-like arabinogalactan protein family of *Eucalyptus grandis* contains members that impact wood biology and biomechanics. *New Phytol.* **2015**, *206*, 1314–1327. [CrossRef] [PubMed]
72. Guerriero, G.; Mangeot-Peter, L.; Legay, S.; Behr, M.; Lutts, S.; Siddiqui, K.S.; Hausman, J.F. Identification of fasciclin-like arabinogalactan proteins in textile hemp (*Cannabis sativa* L.): In silico analyses and gene expression patterns in different tissues. *BMC Genom.* **2017**, *18*, 741. [CrossRef] [PubMed]
73. Li, J.; Wu, W.G. Genome-wide identification, classification and expression analysis of genes encoding putative fasciclin-like arabinogalactan proteins in Chinese cabbage (*Brassica rapa* L.). *Mol. Biol. Rep.* **2012**, *39*, 1573–4978.
74. Ma, H.; Zhao, J. Genome-wide identification, classification, and expression analysis of the arabinogalactan protein gene family in rice (*Oryza sativa* L.). *J. Exp. Bot.* **2010**, *61*, 2647–2668. [CrossRef] [PubMed]
75. Zang, L.; Zheng, T.; Chu, Y.; Ding, C.; Zhang, W.; Huang, Q.; Su, X. Genome-wide analysis of the fasciclin-like arabinogalactan protein gene family reveals differential expression patterns, localization, and salt stress response in *Populus*. *Front. Plant Sci.* **2015**, *6*, 1140. [CrossRef] [PubMed]
76. Bonnet, N.; Garnero, P.; Ferrari, S. Periostin action in bone. *Mol. Cell. Endocrinol.* **2016**, *432*, 75–82. [CrossRef] [PubMed]
77. Walker, J.T.; McLeod, K.; Kim, S.; Conway, S.J.; Hamilton, D.W. Periostin as a multifunctional modulator of the wound healing response. *Cell Tissue Res.* **2016**, *365*, 453–465. [CrossRef] [PubMed]
78. Kzhyshkowska, J.; Gratchev, A.; Goerdt, S. Stabilin-1, a homeostatic scavenger receptor with multiple functions. *J. Cell. Mol. Med.* **2006**, *10*, 635–649. [CrossRef] [PubMed]

79. Kzhyshkowska, J. Multifunctional receptor stabilin-1 in homeostasis and disease. *Sci. World J.* **2010**, *10*, 2039–2053. [CrossRef] [PubMed]
80. Conway, S.J.; Izuhara, K.; Kudo, Y.; Litvin, J.; Markwald, R.; Ouyang, G.; Arron, J.R.; Holweg, C.T.; Kudo, A. The role of periostin in tissue remodeling across health and disease. *Cell. Mol. Life Sci.* **2014**, *71*, 1279–1288. [CrossRef] [PubMed]
81. Kim, J.E.; Kim, E.H.; Han, E.H.; Park, R.W.; Park, I.H.; Jun, S.H.; Kim, J.C.; Young, M.F.; Kim, I.S. A TGF-β-inducible cell adhesion molecule, βig-h3, is downregulated in melorheostosis and involved in osteogenesis. *J. Cell. Biochem.* **2000**, *77*, 169–178. [CrossRef]
82. Klamer, S.; Voermans, C. The role of novel and known extracellular matrix and adhesion molecules in the homeostatic and regenerative bone marrow microenvironment. *Cell. Adh. Migr.* **2014**, *8*, 563–577. [CrossRef] [PubMed]
83. Klamer, S.E.; Kuijk, C.G.; Hordijk, P.L.; van der Schoot, C.E.; von Lindern, M.; van Hennik, P.B.; Voermans, C. BIGH3 modulates adhesion and migration of hematopoietic stem and progenitor cells. *Cell. Adh. Migr.* **2013**, *7*, 434–449. [CrossRef] [PubMed]
84. Han, K.E.; Choi, S.I.; Kim, T.I.; Maeng, Y.S.; Stulting, R.D.; Ji, Y.W.; Kim, E.K. Pathogenesis and treatments of TGFBI corneal dystrophies. *Prog. Retin. Eye Res.* **2016**, *50*, 67–88. [CrossRef] [PubMed]
85. Murugan, E.; Venkatraman, A.; Lei, Z.; Mouvet, V.; Rui Yi Lim, R.; Muruganantham, N.; Goh, E.; Swee Lim Peh, G.; Beuerman, R.W.; Chaurasia, S.S.; et al. pH induced conformational transitions in the transforming growth factor β-induced protein (TGFβIp) associated corneal dystrophy mutants. *Sci. Rep.* **2016**, *6*, 23836. [CrossRef] [PubMed]
86. Poulsen, E.T.; Runager, K.; Nielsen, N.S.; Lukassen, M.V.; Thomsen, K.; Snider, P.; Simmons, O.; Vorum, H.; Conway, S.J.; Enghild, J.J. Proteomic profiling of TGFBI-null mouse corneas reveals only minor changes in matrix composition supportive of TGFBI knockdown as therapy against TGFBI-linked corneal dystrophies. *FEBS J.* **2017**, *285*, 101–114. [CrossRef] [PubMed]
87. Rios, H.; Koushik, S.V.; Wang, H.; Wang, J.; Zhou, H.M.; Lindsley, A.; Rogers, R.; Chen, Z.; Maeda, M.; Kruzynska-Frejtag, A.; et al. Periostin null mice exhibit dwarfism, incisor enamel defects, and an early-onset periodontal disease-like phenotype. *Mol. Cell. Biol.* **2005**, *25*, 11131–11144. [CrossRef] [PubMed]
88. Bonnet, N.; Conway, S.J.; Ferrari, S.L. Regulation of β catenin signaling and parathyroid hormone anabolic effects in bone by the matricellular protein periostin. *Proc. Natl. Acad. Sci. USA* **2012**, *109*, 15048–15053. [CrossRef] [PubMed]
89. Guo, Y.S.; Zhao, R.; Ma, J.; Cui, W.; Sun, Z.; Gao, B.; He, S.; Han, Y.H.; Fan, J.; Yang, L.; et al. βig-h3 promotes human osteosarcoma cells metastasis by interacting with integrin α2β1 and activating PI3K signaling pathway. *PLoS ONE* **2014**, *9*, e90220. [CrossRef] [PubMed]
90. Bae, J.S.; Lee, S.H.; Kim, J.E.; Choi, J.Y.; Park, R.W.; Yong Park, J.; Park, H.S.; Sohn, Y.S.; Lee, D.S.; Bae Lee, E.; et al. βig-h3 supports keratinocyte adhesion, migration, and proliferation through α3β1 integrin. *Biochem. Biophys. Res. Commun.* **2002**, *294*, 940–948. [CrossRef]
91. Park, S.W.; Bae, J.S.; Kim, K.S.; Park, S.H.; Lee, B.H.; Choi, J.Y.; Park, J.Y.; Ha, S.W.; Kim, Y.L.; Kwon, T.H.; et al. βig-h3 promotes renal proximal tubular epithelial cell adhesion, migration and proliferation through the interaction with α3β1 integrin. *Exp. Mol. Med.* **2004**, *36*, 211–219. [CrossRef] [PubMed]
92. Maeng, Y.S.; Choi, Y.J.; Kim, E.K. TGFBIp regulates differentiation of EPC ($CD133^+$ c-kit^+ lin^- cells) to EC through activation of the notch signaling pathway. *Stem Cells* **2015**, *33*, 2052–2062. [CrossRef] [PubMed]
93. Utispan, K.; Sonongbua, J.; Thuwajit, P.; Chau-In, S.; Pairojkul, C.; Wongkham, S.; Thuwajit, C. Periostin activates integrin α5β1 through a PI3K/AKT-dependent pathway in invasion of cholangiocarcinoma. *Int. J. Oncol.* **2012**, *41*, 1110–1118. [CrossRef] [PubMed]
94. Kim, M.O.; Yun, S.J.; Kim, I.S.; Sohn, S.; Lee, E.H. Transforming growth factor-β-inducible gene-h3 (βig-h3) promotes cell adhesion of human astrocytoma cells in vitro: Implication of α6β4 integrin. *Neurosci. Lett.* **2003**, *336*, 93–96. [CrossRef]
95. Lu, Y.; Liu, X.; Jiao, Y.; Xiong, X.; Wang, E.; Wang, X.; Zhang, Z.; Zhang, H.; Pan, L.; Guan, Y.; et al. Periostin promotes liver steatosis and hypertriglyceridemia through downregulation of PPARα. *J. Clin. Investig.* **2014**, *124*, 3501–3513. [CrossRef] [PubMed]
96. Ferguson, J.W.; Thoma, B.S.; Mikesh, M.F.; Kramer, R.H.; Bennett, K.L.; Purchio, A.; Bellard, B.J.; LeBaron, R.G. The extracellular matrix protein βig-h3 is expressed at myotendinous junctions and supports muscle cell adhesion. *Cell Tissue Res.* **2003**, *313*, 93–105. [CrossRef] [PubMed]

97. Nam, J.O.; Kim, J.E.; Jeong, H.W.; Lee, S.J.; Lee, B.H.; Choi, J.Y.; Park, R.W.; Park, J.Y.; Kim, I.S. Identification of the αvβ3 integrin-interacting motif of βig-h3 and its anti-angiogenic effect. *J. Biol. Chem.* **2003**, *278*, 25902–25909. [CrossRef] [PubMed]
98. Thapa, N.; Kang, K.B.; Kim, I.S. βig-h3 mediates osteoblast adhesion and inhibits differentiation. *Bone* **2005**, *36*, 232–242. [CrossRef] [PubMed]
99. Choi, S.I.; Maeng, Y.S.; Kim, T.I.; Lee, Y.; Kim, Y.S.; Kim, E.K. Lysosomal trafficking of TGFBIp via caveolae-mediated endocytosis. *PLoS ONE* **2015**, *10*, e0119561. [CrossRef] [PubMed]
100. Gillan, L.; Matei, D.; Fishman, D.A.; Gerbin, C.S.; Karlan, B.Y.; Chang, D.D. Periostin secreted by epithelial ovarian carcinoma is a ligand for αvβ3 and αvβ5 integrins and promotes cell motility. *Cancer Res.* **2002**, *62*, 5358–5364. [PubMed]
101. Li, G.; Jin, R.; Norris, R.A.; Zhang, L.; Yu, S.; Wu, F.; Markwald, R.R.; Nanda, A.; Conway, S.J.; Smyth, S.S.; et al. Periostin mediates vascular smooth muscle cell migration through the integrins αvβ3 and αvβ5 and focal adhesion kinase (FAK) pathway. *Atherosclerosis* **2010**, *208*, 358–365. [CrossRef] [PubMed]
102. Lee, Y.J.; Kim, I.S.; Park, S.A.; Kim, Y.; Lee, J.E.; Noh, D.Y.; Kim, K.T.; Ryu, S.H.; Suh, P.G. Periostin-binding DNA aptamer inhibits breast cancer growth and metastasis. *Mol. Ther.* **2013**, *21*, 1004–1013. [CrossRef] [PubMed]
103. Matsuzawa, M.; Arai, C.; Nomura, Y.; Murata, T.; Yamakoshi, Y.; Oida, S.; Hanada, N.; Nakamura, Y. Periostin of human periodontal ligament fibroblasts promotes migration of human mesenchymal stem cell through the αvβ3 integrin/FAK/PI3K/Akt pathway. *J. Period. Res.* **2015**, *50*, 855–863. [CrossRef] [PubMed]
104. Chuanyu, S.; Yuqing, Z.; Chong, X.; Guowei, X.; Xiaojun, Z. Periostin promotes migration and invasion of renal cell carcinoma through the integrin/focal adhesion kinase/c-Jun N-terminal kinase pathway. *Tumour Biol.* **2017**, *39*. [CrossRef] [PubMed]
105. Zhou, W.; Ke, S.Q.; Huang, Z.; Flavahan, W.; Fang, X.; Paul, J.; Wu, L.; Sloan, A.E.; McLendon, R.E.; Li, X.; et al. Periostin secreted by glioblastoma stem cells recruits M2 tumour-associated macrophages and promotes malignant growth. *Nat. Cell Biol.* **2015**, *17*, 170–182. [CrossRef] [PubMed]
106. Lee, B.H.; Bae, J.S.; Park, R.W.; Kim, J.E.; Park, J.Y.; Kim, I.S. βig-h3 triggers signaling pathways mediating adhesion and migration of vascular smooth muscle cells through αvβ5 integrin. *Exp. Mol. Med.* **2006**, *38*, 153–161. [CrossRef] [PubMed]
107. Kim, S.; Park, S.Y.; Kim, S.Y.; Bae, D.J.; Pyo, J.H.; Hong, M.; Kim, I.S. Cross talk between engulfment receptors stabilin-2 and integrin αvβ5 orchestrates engulfment of phosphatidylserine-exposed erythrocytes. *Mol. Cell. Biol.* **2012**, *32*, 2698–2708. [CrossRef] [PubMed]
108. Ghatak, S.; Misra, S.; Norris, R.A.; Moreno-Rodriguez, R.A.; Hoffman, S.; Levine, R.A.; Hascall, V.C.; Markwald, R.R. Periostin induces intracellular cross-talk between kinases and hyaluronan in atrioventricular valvulogenesis. *J. Biol. Chem.* **2014**, *289*, 8545–8561. [CrossRef] [PubMed]
109. Johansson, M.W.; Annis, D.S.; Mosher, D.F. $\alpha_M\beta_2$ integrin-mediated adhesion and motility of IL-5-stimulated eosinophils on periostin. *Am. J. Respir. Cel.l Mol. Biol.* **2013**, *48*, 503–510. [CrossRef] [PubMed]
110. Jung, M.Y.; Park, S.Y.; Kim, I.S. Stabilin-2 is involved in lymphocyte adhesion to the hepatic sinusoidal endothelium via the interaction with αMβ2 integrin. *J. Leukoc. Biol.* **2007**, *82*, 1156–1165. [CrossRef] [PubMed]
111. Hashimoto, K.; Noshiro, M.; Ohno, S.; Kawamoto, T.; Satakeda, H.; Akagawa, Y.; Nakashima, K.; Okimura, A.; Ishida, H.; Okamoto, T.; et al. Characterization of a cartilage-derived 66-kDa protein (RGD-CAP/βig-h3) that binds to collagen. *Biochim. Biophys. Acta* **1997**, *1355*, 303–314. [CrossRef]
112. Kii, I.; Nishiyama, T.; Kudo, A. Periostin promotes secretion of fibronectin from the endoplasmic reticulum. *Biochem. Biophys. Res. Commun.* **2016**, *470*, 888–893. [CrossRef] [PubMed]
113. Hwang, E.Y.; Jeong, M.S.; Park, E.K.; Kim, J.H.; Jang, S.B. Structural characterization and interaction of periostin and bone morphogenetic protein for regulation of collagen cross-linking. *Biochem. Biophys. Res. Commun.* **2014**, *449*, 425–431. [CrossRef] [PubMed]
114. Maruhashi, T.; Kii, I.; Saito, M.; Kudo, A. Interaction between periostin and BMP-1 promotes proteolytic activation of lysyl oxidase. *J. Biol. Chem.* **2010**, *285*, 13294–13303. [CrossRef] [PubMed]
115. Kzhyshkowska, J.; Gratchev, A.; Brundiers, H.; Mamidi, S.; Krusell, L.; Goerdt, S. Phosphatidylinositide 3-kinase activity is required for stabilin-1-mediated endosomal transport of acLDL. *Immunobiology* **2005**, *210*, 161–173. [CrossRef] [PubMed]

116. Kzhyshkowska, J.; Workman, G.; Cardo-Vila, M.; Arap, W.; Pasqualini, R.; Gratchev, A.; Krusell, L.; Goerdt, S.; Sage, E.H. Novel function of alternatively activated macrophages: Stabilin-1-mediated clearance of SPARC. *J. Immunol.* **2006**, *176*, 5825–5832. [CrossRef] [PubMed]
117. Workman, G.; Sage, E.H. Identification of a sequence in the matricellular protein SPARC that interacts with the scavenger receptor stabilin-1. *J. Cell. Biochem.* **2011**, *112*, 1003–1008. [CrossRef] [PubMed]
118. Kzhyshkowska, J.; Gratchev, A.; Schmuttermaier, C.; Brundiers, H.; Krusell, L.; Mamidi, S.; Zhang, J.; Workman, G.; Sage, E.H.; Anderle, C.; et al. Alternatively activated macrophages regulate extracellular levels of the hormone placental lactogen via receptor-mediated uptake and transcytosis. *J. Immunol.* **2008**, *180*, 3028–3037. [CrossRef] [PubMed]
119. Kzhyshkowska, J.; Mamidi, S.; Gratchev, A.; Kremmer, E.; Schmuttermaier, C.; Krusell, L.; Haus, G.; Utikal, J.; Schledzewski, K.; Scholtze, J.; et al. Novel stabilin-1 interacting chitinase-like protein (SI-CLP) is up-regulated in alternatively activated macrophages and secreted via lysosomal pathway. *Blood* **2006**, *107*, 3221–3228. [CrossRef] [PubMed]
120. Meng, G.; Zhao, Y.; Bai, X.; Liu, Y.; Green, T.J.; Luo, M.; Zheng, X. Structure of human stabilin-1 interacting chitinase-like protein (SI-CLP) reveals a saccharide-binding cleft with lower sugar-binding selectivity. *J. Biol. Chem.* **2010**, *285*, 39898–39904. [CrossRef] [PubMed]
121. Harris, E.N.; Weigel, J.A.; Weigel, P.H. The human hyaluronan receptor for endocytosis (HARE/stabilin-2) is a systemic clearance receptor for heparin. *J. Biol. Chem.* **2008**, *283*, 17341–17350. [CrossRef] [PubMed]
122. Pandey, M.S.; Weigel, P.H. A hyaluronan receptor for endocytosis (HARE) link domain *N*-glycan is required for extracellular signal-regulated kinase (ERK) and nuclear factor-κB (Nf-κB) signaling in response to the uptake of hyaluronan but not heparin, dermatan sulfate, or acetylated low density lipoprotein (LDL). *J. Biol. Chem.* **2014**, *289*, 21807–21817. [PubMed]
123. Penberthy, K.K.; Ravichandran, K.S. Apoptotic cell recognition receptors and scavenger receptors. *Immunol. Rev.* **2016**, *269*, 44–59. [CrossRef] [PubMed]
124. Park, S.Y.; Yun, Y.; Lim, J.S.; Kim, M.J.; Kim, S.Y.; Kim, J.E.; Kim, I.S. Stabilin-2 modulates the efficiency of myoblast fusion during myogenic differentiation and muscle regeneration. *Nat. Commun.* **2016**, *7*, 10871. [CrossRef] [PubMed]
125. Kim, G.W.; Park, S.Y.; Kim, I.S. Novel function of stabilin-2 in myoblast fusion: The recognition of extracellular phosphatidylserine as a "fuse-me" signal. *BMB Rep.* **2016**, *49*, 303–304. [CrossRef] [PubMed]
126. Park, S.Y.; Jung, M.Y.; Kim, I.S. Stabilin-2 mediates homophilic cell–cell interactions via its FAS1 domains. *FEBS Lett.* **2009**, *583*, 1375–1380. [CrossRef] [PubMed]
127. Zhu, J.K. Salt and drought stress signal transduction in plants. *Annu. Rev. Plant Biol.* **2002**, *53*, 247–273. [CrossRef] [PubMed]
128. Xu, S.L.; Rahman, A.; Baskin, T.I.; Kieber, J.J. Two leucine-rich repeat receptor kinases mediate signaling, linking cell wall biosynthesis and ACC synthase in Arabidopsis. *Plant Cell* **2008**, *20*, 3065–3079. [CrossRef] [PubMed]
129. Hauser, M.T.; Morikami, A.; Benfey, P.N. Conditional root expansion mutants of Arabidopsis. *Development* **1995**, *121*, 1237–1252. [PubMed]
130. Seifert, G.J.; Xue, H.; Acet, T. The *Arabidopsis thaliana* fasciclin like arabinogalactan protein 4 gene acts synergistically with abscisic acid signalling to control root growth. *Ann. Bot.* **2014**, *114*, 1125–1133. [CrossRef] [PubMed]
131. Harpaz-Saad, S.; McFarlane, H.E.; Xu, S.; Divi, U.K.; Forward, B.; Western, T.L.; Kieber, J.J. Cellulose synthesis via the FEI2 rlk/Sos5 pathway and cellulose synthase 5 is required for the structure of seed coat mucilage in Arabidopsis. *Plant J.* **2011**, *68*, 941–953. [CrossRef] [PubMed]
132. Harpaz-Saad, S.; Western, T.L.; Kieber, J.J. The FEI2-SOS5 pathway and CELLULOSE SYNTHASE 5 are required for cellulose biosynthesis in the Arabidopsis seed coat and affect pectin mucilage structure. *Plant Signal. Behav.* **2012**, *7*, 285–288. [CrossRef] [PubMed]
133. Griffiths, J.S.; Crepeau, M.J.; Ralet, M.C.; Seifert, G.J.; North, H.M. Dissecting seed mucilage adherence mediated by FEI2 and Sos5. *Front. Plant Sci.* **2016**, *7*, 1073. [CrossRef] [PubMed]
134. Griffiths, J.S.; Tsai, A.Y.; Xue, H.; Voiniciuc, C.; Sola, K.; Seifert, G.J.; Mansfield, S.D.; Haughn, G.W. Salt-overly sensitive5 mediates Arabidopsis seed coat mucilage adherence and organization through pectins. *Plant Physiol.* **2014**, *165*, 991–1004. [CrossRef] [PubMed]

135. Tsang, D.L.; Edmond, C.; Harrington, J.L.; Nuhse, T.S. Cell wall integrity controls root elongation via a general 1-aminocyclopropane-1-carboxylic acid-dependent, ethylene-independent pathway. *Plant Physiol.* **2011**, *156*, 596–604. [CrossRef] [PubMed]
136. Xue, H.; Seifert, G.J. Fasciclin like arabinogalactan protein 4 and respiratory burst oxidase homolog d and F independently modulate abscisic acid signaling. *Plant Signal. Behav.* **2015**, *10*, e989064. [CrossRef] [PubMed]
137. Wormit, A.; Butt, S.M.; Chairam, I.; McKenna, J.F.; Nunes-Nesi, A.; Kjaer, L.; O'Donnelly, K.; Fernie, A.R.; Woscholski, R.; Barter, M.C.; et al. Osmosensitive changes of carbohydrate metabolism in response to cellulose biosynthesis inhibition. *Plant Physiol.* **2012**, *159*, 105–117. [CrossRef] [PubMed]
138. Steinwand, B.J.; Xu, S.; Polko, J.K.; Doctor, S.M.; Westafer, M.; Kieber, J.J. Alterations in auxin homeostasis suppress defects in cell wall function. *PLoS ONE* **2014**, *9*, e98193. [CrossRef] [PubMed]
139. North, H.M.; Berger, A.; Saez-Aguayo, S.; Ralet, M.C. Understanding polysaccharide production and properties using seed coat mutants: Future perspectives for the exploitation of natural variants. *Ann. Bot.* **2014**, *114*, 1251–1263. [CrossRef] [PubMed]
140. Voiniciuc, C.; Yang, B.; Schmidt, M.H.; Gunl, M.; Usadel, B. Starting to gel: How Arabidopsis seed coat epidermal cells produce specialized secondary cell walls. *Int. J. Mol. Sci.* **2015**, *16*, 3452–3473. [CrossRef] [PubMed]
141. Ralet, M.C.; Crepeau, M.J.; Vigouroux, J.; Tran, J.; Berger, A.; Salle, C.; Granier, F.; Botran, L.; North, H.M. Xylans provide the structural driving force for mucilage adhesion to the Arabidopsis seed coat. *Plant Physiol.* **2016**, *171*, 165–178. [CrossRef] [PubMed]
142. Voiniciuc, C.; Gunl, M.; Schmidt, M.H.; Usadel, B. Highly branched xylan made by IRX14 and MUCI21 links mucilage to Arabidopsis seeds. *Plant Physiol.* **2015**, *169*, 2481–2495. [CrossRef] [PubMed]
143. Dean, G.H.; Zheng, H.; Tewari, J.; Huang, J.; Young, D.S.; Hwang, Y.T.; Western, T.L.; Carpita, N.C.; McCann, M.C.; Mansfield, S.D.; et al. The Arabidopsis *Mum2* gene encodes a β-galactosidase required for the production of seed coat mucilage with correct hydration properties. *Plant Cell* **2007**, *19*, 4007–4021. [CrossRef] [PubMed]
144. Voiniciuc, C.; Dean, G.H.; Griffiths, J.S.; Kirchsteiger, K.; Hwang, Y.T.; Gillett, A.; Dow, G.; Western, T.L.; Estelle, M.; Haughn, G.W. Flying saucer1 is a transmembrane RING E3 ubiquitin ligase that regulates the degree of pectin methylesterification in Arabidopsis seed mucilage. *Plant Cell* **2013**, *25*, 944–959. [CrossRef] [PubMed]
145. Tsai, A.Y.; Kunieda, T.; Rogalski, J.; Foster, L.J.; Ellis, B.E.; Haughn, G.W. Identification and characterization of Arabidopsis seed coat mucilage proteins. *Plant Physiol.* **2017**, *173*, 1059–1074. [CrossRef] [PubMed]
146. Basu, D.; Tian, L.; Debrosse, T.; Poirier, E.; Emch, K.; Herock, H.; Travers, A.; Showalter, A.M. Glycosylation of a fasciclin-like arabinogalactan-protein (SOS5) mediates root growth and seed mucilage adherence via a cell wall receptor-like kinase (FEI1/FEI2) pathway in Arabidopsis. *PLoS ONE* **2016**, *11*, e0145092. [CrossRef] [PubMed]
147. Basu, D.; Wang, W.; Ma, S.; DeBrosse, T.; Poirier, E.; Emch, K.; Soukup, E.; Tian, L.; Showalter, A.M. Two hydroxyproline galactosyltransferases, Galt5 and Galt2, function in arabinogalactan-protein glycosylation, growth and development in Arabidopsis. *PLoS ONE* **2015**, *10*, e0125624. [CrossRef] [PubMed]
148. Brown, D.M.; Zeef, L.A.; Ellis, J.; Goodacre, R.; Turner, S.R. Identification of novel genes in Arabidopsis involved in secondary cell wall formation using expression profiling and reverse genetics. *Plant Cell* **2005**, *17*, 2281–2295. [CrossRef] [PubMed]
149. Persson, S.; Wei, H.; Milne, J.; Page, G.P.; Somerville, C.R. Identification of genes required for cellulose synthesis by regression analysis of public microarray data sets. *Proc. Natl. Acad. Sci. USA* **2005**, *102*, 8633–8638. [CrossRef] [PubMed]
150. Ito, S.; Suzuki, Y.; Miyamoto, K.; Ueda, J.; Yamaguchi, I. AtFLA11, a fasciclin-like arabinogalactan-protein, specifically localized in sclerenchyma cells. *Biosci. Biotechnol. Biochem.* **2005**, *69*, 1963–1969. [CrossRef] [PubMed]
151. Dahiya, P.; Findlay, K.; Roberts, K.; McCann, M.C. A fasciclin-domain containing gene, *ZeFla11*, is expressed exclusively in xylem elements that have reticulate wall thickenings in the stem vascular system of *Zinnia elegans* cv Envy. *Planta* **2006**, *223*, 1281–1291. [CrossRef] [PubMed]
152. Lafarguette, F.; Leplé, J.-C.; Déjardin, A.; Laurans, F.; Costa, G.; Lesage-Descauses, M.-C.; Pilate, G. Poplar genes encoding fasciclin-like arabinogalactan proteins are highly expressed in tension wood. *New Phytol.* **2004**, *164*, 107–121. [CrossRef]

153. Gritsch, C.; Wan, Y.; Mitchell, R.A.; Shewry, P.R.; Hanley, S.J.; Karp, A. G-fibre cell wall development in willow stems during tension wood induction. *J. Exp. Bot.* **2015**, *66*, 6447–6459. [CrossRef] [PubMed]
154. Qiu, D.; Wilson, I.W.; Gan, S.; Washusen, R.; Moran, G.F.; Southerton, S.G. Gene expression in eucalyptus branch wood with marked variation in cellulose microfibril orientation and lacking G-layers. *New Phytol.* **2008**, *179*, 94–103. [CrossRef] [PubMed]
155. Hobson, N.; Deyholos, M.K. $LuFLA_{1PRO}$ and $LuBGAL_{1PRO}$ promote gene expression in the phloem fibres of flax (*Linum usitatissimum*). *Plant Cell Rep.* **2013**, *32*, 517–528. [CrossRef] [PubMed]
156. MacMillan, C.P.; Mansfield, S.D.; Stachurski, Z.H.; Evans, R.; Southerton, S.G. Fasciclin-like arabinogalactan proteins: Specialization for stem biomechanics and cell wall architecture in Arabidopsis and Eucalyptus. *Plant J.* **2010**, *62*, 689–703. [CrossRef] [PubMed]
157. Wang, H.; Jiang, C.; Wang, C.; Yang, Y.; Yang, L.; Gao, X.; Zhang, H. Antisense expression of the fasciclin-like arabinogalactan protein *FLA6* gene in *Populus* inhibits expression of its homologous genes and alters stem biomechanics and cell wall composition in transgenic trees. *J. Exp. Bot.* **2015**, *66*, 1291–1302. [CrossRef] [PubMed]
158. Wang, H.; Jin, Y.; Wang, C.; Li, B.; Jiang, C.; Sun, Z.; Zhang, Z.; Kong, F.; Zhang, H. Fasciclin-like arabinogalactan proteins, PtFLAs, play important roles in GA-mediated tension wood formation in *Populus*. *Sci. Rep.* **2017**, *7*, 6182. [CrossRef] [PubMed]
159. Huang, G.Q.; Gong, S.Y.; Xu, W.L.; Li, W.; Li, P.; Zhang, C.J.; Li, D.D.; Zheng, Y.; Li, F.G.; Li, X.B. A fasciclin-like arabinogalactan protein, GhFLA1, is involved in fiber initiation and elongation of cotton. *Plant Physiol.* **2013**, *161*, 1278–1290. [CrossRef] [PubMed]
160. Li, Y.; Liu, D.; Tu, L.; Zhang, X.; Wang, L.; Zhu, L.; Tan, J.; Deng, F. Suppression of GhAgp4 gene expression repressed the initiation and elongation of cotton fiber. *Plant Cell Rep.* **2010**, *29*, 193–202. [CrossRef] [PubMed]
161. Motose, H.; Sugiyama, M.; Fukuda, H. A proteoglycan mediates inductive interaction during plant vascular development. *Nature* **2004**, *429*, 873–878. [CrossRef] [PubMed]
162. Li, A.; Wang, R.; Li, X.; Liu, M.; Fan, J.; Guo, K.; Luo, B.; Chen, T.; Feng, S.; Wang, Y.; et al. Proteomic profiling of cellulase-aid-extracted membrane proteins for functional identification of cellulose synthase complexes and their potential associated- components in cotton fibers. *Sci. Rep.* **2016**, *6*, 26356. [CrossRef] [PubMed]
163. Wu, X.N.; Sanchez Rodriguez, C.; Pertl-Obermeyer, H.; Obermeyer, G.; Schulze, W.X. Sucrose-induced receptor kinase Sirk1 regulates a plasma membrane aquaporin in Arabidopsis. *Mol. Cell. Proteom.* **2013**, *12*, 2856–2873. [CrossRef] [PubMed]
164. Klepikova, A.V.; Logacheva, M.D.; Dmitriev, S.E.; Penin, A.A. RNA-seq analysis of an apical meristem time series reveals a critical point in *Arabidopsis thaliana* flower initiation. *BMC Genom.* **2015**, *16*, 466. [CrossRef] [PubMed]
165. Klepikova, A.V.; Kasianov, A.S.; Gerasimov, E.S.; Logacheva, M.D.; Penin, A.A. A high resolution map of the *Arabidopsis thaliana* developmental transcriptome based on RNA-seq profiling. *Plant J.* **2016**, *88*, 1058–1070. [CrossRef] [PubMed]
166. Li, J.; Yu, M.; Geng, L.L.; Zhao, J. The fasciclin-like arabinogalactan protein gene, *FLA3*, is involved in microspore development of Arabidopsis. *Plant J.* **2010**, *64*, 482–497. [CrossRef] [PubMed]
167. Johnson, K.L.; Kibble, N.A.; Bacic, A.; Schultz, C.J. A fasciclin-like arabinogalactan-protein (FLA) mutant of *Arabidopsis thaliana*, FLA1, shows defects in shoot regeneration. *PLoS ONE* **2011**, *6*, e25154. [CrossRef] [PubMed]
168. Kirchner, T.W.; Niehaus, M.; Debener, T.; Schenk, M.K.; Herde, M. Efficient generation of mutations mediated by CRISPR/Cas9 in the hairy root transformation system of *Brassica carinata*. *PLoS ONE* **2017**, *12*, e0185429. [CrossRef] [PubMed]
169. Cagnola, J.I.; de Chassart, G.J.D.; Ibarra, S.E.; Chimenti, C.; Ricardi, M.M.; Delzer, B.; Ghiglione, H.; Zhu, T.; Otegui, M.E.; Estevez, J.M.; et al. Reduced expression of selected fasciclin-like arabinogalactan protein genes associates with the abortion of kernels in field crops of *Zea mays* (maize) and of Arabidopsis seeds. *Plant Cell Environ.* **2018**. [CrossRef] [PubMed]
170. Nirmal, R.C.; Furtado, A.; Rangan, P.; Henry, R.J. Fasciclin-like arabinogalactan protein gene expression is associated with yield of flour in the milling of wheat. *Sci. Rep.* **2017**, *7*, 12539. [CrossRef] [PubMed]
171. Wiker, H.G. MPB70 and MPB83—Major antigens of *Mycobacterium bovis*. *Scand. J. Immunol.* **2009**, *69*, 492–499. [CrossRef] [PubMed]

172. Rahman, S.A.; Singh, Y.; Kohli, S.; Ahmad, J.; Ehtesham, N.Z.; Tyagi, A.K.; Hasnain, S.E. Comparative analyses of nonpathogenic, opportunistic, and totally pathogenic mycobacteria reveal genomic and biochemical variabilities and highlight the survival attributes of *Mycobacterium tuberculosis*. *MBio* **2014**, *5*, e02020. [CrossRef] [PubMed]
173. Elkins, T.; Hortsch, M.; Bieber, A.J.; Snow, P.M.; Goodman, C.S. *Drosophila* fasciclin I is a novel homophilic adhesion molecule that along with fasciclin III can mediate cell sorting. *J. Cell Biol.* **1990**, *110*, 1825–1832. [CrossRef] [PubMed]
174. Chen, S.T.; Li, J.Y.; Zhang, Y.; Gao, X.; Cai, H. Recombinant MPT83 derived from *Mycobacterium tuberculosis* induces cytokine production and upregulates the function of mouse macrophages through TLR2. *J. Immunol.* **2012**, *188*, 668–677. [CrossRef] [PubMed]
175. Shiu, S.H.; Bleecker, A.B. Receptor-like kinases from Arabidopsis form a monophyletic gene family related to animal receptor kinases. *Proc. Natl. Acad. Sci. USA* **2001**, *98*, 10763–10768. [CrossRef] [PubMed]
176. Oke, V.; Long, S.R. Bacteroid formation in the *Rhizobium*-legume symbiosis. *Curr. Opin. Microbiol.* **1999**, *2*, 641–646. [CrossRef]
177. Oke, V.; Long, S.R. Bacterial genes induced within the nodule during the *Rhizobium*-legume symbiosis. *Mol. Microbiol.* **1999**, *32*, 837–849. [CrossRef] [PubMed]
178. Paulsrud, P.; Lindblad, P. Fasciclin domain proteins are present in *Nostoc* symbionts of lichens. *Appl. Environ. Microb.* **2002**, *68*, 2036–2039. [CrossRef]
179. Delmotte, N.; Knief, C.; Chaffron, S.; Innerebner, G.; Roschitzki, B.; Schlapbach, R.; von Mering, C.; Vorholt, J.A. Community proteogenomics reveals insights into the physiology of phyllosphere bacteria. *Proc. Natl. Acad. Sci. USA* **2009**, *106*, 16428–16433. [CrossRef] [PubMed]
180. Knief, C.; Ramette, A.; Frances, L.; Alonso-Blanco, C.; Vorholt, J.A. Site and plant species are important determinants of the methylobacterium community composition in the plant phyllosphere. *ISME J.* **2010**, *4*, 719–728. [CrossRef] [PubMed]
181. Tejerizo, G.T.; Kim, Y.S.; Maus, I.; Wibberg, D.; Winkler, A.; Off, S.; Puhler, A.; Scherer, P.; Schluter, A. Genome sequence of *Methanobacterium congolense* strain buetzberg, a hydrogenotrophic, methanogenic archaeon, isolated from a mesophilic industrial-scale biogas plant utilizing bio-waste. *J. Biotechnol.* **2017**, *247*, 1–5. [CrossRef] [PubMed]
182. Bornstein, P. Diversity of function is inherent in matricellular proteins: An appraisal of thrombospondin 1. *J. Cell Biol.* **1995**, *130*, 503–506. [CrossRef] [PubMed]
183. Alberts, B.; Johnson, A.; Lewis, J.; Morgan, D.; Raff, M.; Roberts, K.; Walter, P. *Molecular Biology of the Cell*, 6th ed.; Garland Science: New York, NY, USA, 2015.

International Journal of
Molecular Sciences

MDPI

Article

Evolution Analysis of the Fasciclin-Like Arabinogalactan Proteins in Plants Shows Variable Fasciclin-AGP Domain Constitutions

Jiadai He [1,†], Hua Zhao [1,†], Zhilu Cheng [2], Yuwei Ke [3], Jiaxi Liu [1] and Haoli Ma [1,*]

1 College of Agronomy, Northwest A&F University, Xianyang 712100, Shaanxi, China; 2015014922@nwsuaf.edu.cn (J.H.); zhaohua362@nwsuaf.edu.cn (H.Z.); 2015014872@nwsuaf.edu.cn (J.L.)
2 College of Landscape Architecture and Arts, Northwest A&F University, Xianyang 712100, Shaanxi, China; czl980525@163.com
3 College of Life Sciences, Northwest A&F University, Xianyang 712100, Shaanxi, China; keyuwei98@163.com
* Correspondence: mahaoli@nwsuaf.edu.cn; Tel.: +86-29-87081551
† These authors contribute equally to this work.

Received: 22 February 2019; Accepted: 19 April 2019; Published: 20 April 2019

Abstract: The fasciclin-like arabinogalactan proteins (FLAs) play important roles in plant development and adaptation to the environment. FLAs contain both fasciclin domains and arabinogalactan protein (AGP) regions, which have been identified in several plants. The evolutionary history of this gene family in plants is still undiscovered. In this study, we identified the *FLA* gene family in 13 plant species covering major lineages of plants using bioinformatics methods. A total of 246 *FLA* genes are identified with gene copy numbers ranging from one (*Chondrus crispus*) to 49 (*Populus trichocarpa*). These FLAs are classified into seven groups, mainly based on the phylogenetic analysis of plant FLAs. All FLAs in land plants contain one or two fasciclin domains, while in algae, several FLAs contain four or six fasciclin domains. It has been proposed that there was a divergence event, represented by the reduced number of fasciclin domains from algae to land plants in evolutionary history. Furthermore, introns in *FLA* genes are lost during plant evolution, especially from green algae to land plants. Moreover, it is found that gene duplication events, including segmental and tandem duplications are essential for the expansion of *FLA* gene families. The duplicated gene pairs in *FLA* gene family mainly evolve under purifying selection. Our findings give insight into the origin and expansion of the *FLA* gene family and help us understand their functions during the process of evolution.

Keywords: fasciclin-like AGP; FLA; evolution; phylogeny

1. Introduction

The cell wall plays an important role in plant growth and development by providing structural support and protection, and acting as a filtering mechanism. Although cell wall proteins account for less than 10% of the cell wall mass, they are predominantly involved in the wall structure, support, signaling, and interactions with other wall components and with the plasma membrane [1,2]. Hydroxyproline-rich glycoproteins (HRGPs) are a major group of cell wall glycoproteins that play important roles in plant growth and development [3].

HRGPs are characterized by a protein backbone rich in hydroxyproline (Hyp). The HRGPs superfamily can be divided into three main subfamilies based on the varying degrees of *O*-glycosylation: Arabinogalactan proteins (AGPs), extensins (EXTs), and proline-rich proteins (PRPs) [4–6]. The protein backbones of AGPs are rich in hydroxyproline/proline (Hyp/Pro), alanine (Ala), serine (Ser), and threonine (Thr), and these amino acids are regularly arranged as Ala–Pro, Ser–Pro, and Thr–Pro, which were introduced as arabinogalactan (AG) glycomodules [7–9]. The carbohydrate side chains of

AGPs are attached to Hyp and enriched in arabinose and galactose [10]. Based on the variable protein backbones [6], AGPs can be classified into classical AGPs, chimeric AGPs, and AGP-EXT hybrids. The chimeric AGPs can be further categorized into three subclasses based on different conserved domains: Fasciclin-like AGPs (FLAs) [11–13], phytocyanin-like AGPs (PAGs) [14,15], and xylogen-like AGPs (XYLPs) [16,17]. As one subclass of the chimeric AGPs, FLAs consist of both fasciclin domains and AGP regions. In most plant species, FLAs contain one or two fasciclin domains. The fasciclin domains contain two highly conserved motifs (H1 and H2) of about 10 amino acids long each and a conserved central YH motif [18]. Proteins with fasciclin domains were first identified in grasshoppers [19] and as adhesion factors were first identified in fruit flies [20]. Since then, more and more fasciclin domains have been identified in animal, yeast, bacteria and plant proteins [18]. The majority of plant fasciclin-like proteins are FLAs and the functions of FLAs are related to many important processes in development and stress responses, such as contributing to biophysical properties (e.g., swelling and interpolymer connectivity), affecting secondary cell wall formation and structure, acting in male gametophyte development, influencing organ formation, and sensing salt stress in roots [18].

To date, FLAs have been identified in several plants, including *Arabidopsis* (*Arabidopsis thaliana*) [21], rice (*Oryza sativa*) [12,22], wheat (*Triticum aestivum*) [22], poplar (*Populus trichocarpa*) [23,24], zinnia (*Zinnia elegans*) [25], cotton (*Gossypium raimondii*) [26], sea island cotton (*Gossypium barbadense*) [27], Chinese cabbage (*Brassica rapa*) [28], eucalyptus (*Eucalyptus grandis*) [13], and textile hemp (*Cannabis sativa*) [29]. The analysis of HRGPs from 1000 plant transcriptomes has provided new insights into the evolution of HRGPs across major evolutionary milestones and reveals the origin and diversity of Glycosylphosphatidylinositol (GPI)-anchored AGPs [3]. However, the evolutionary history of the FLA family in plants is little known. In a previous study, it was proposed that a conserved group of FLAs with a single fasciclin domain was specific to the evolution of flowering plant secondary cell wall formation and properties through phylogenetic analysis of >100 FLA mature proteins [30]. In this study, we identify 246 FLAs from 13 plant species belonging to algae, liverworts, mosses, lycophytes, gymnosperms, dicots, and monocots. Moreover, bioinformatics methods are adopted to reveal the evolutionary mechanisms of the FLA family. In order to understand the functions of the FLAs, the evolutionary history of FLAs is investigated in this study. It is found that the *FLA* genes are abundant in most investigated green plants, but only in one red alga. Additionally, our study shows that there is a reduction in the number of fasciclin domains in FLAs from algae to land plants, which indicates that the reduced number of fasciclin domains plays a crucial role in land plant evolution.

2. Results and Discussions

2.1. Identification of the FLA Family in Plants

FLAs contain both fasciclin domains and AGP regions [6]. We first used the HMM profile of fasciclin downloaded from Pfam (available online: http://pfam.xfam.org/family/PF02469) to identify the proteins with fasciclin domains from 13 plant species (*C. crispus, Chlamydomonas reinhardtii, Chara braunii, Marchantia polymorpha, Physcomitrella patens, Selaginella moellendorffii, Picea abies, Amborella trichopoda, Brachypodium distachyon, O. sativa, A. thaliana, E. grandis,* and *P. trichocarpa*) [31–43]. Then, the obtained proteins were examined by using Batch CD-search tool in the NCBI conserved domain database (available online: http://www.ncbi.nlm.nih.gov/Structure/bwrpsb/bwrpsb.cgi). After that, the AGP regions were identified from these fasciclin proteins by using Finding-AGP program [7]. The proteins that contained both AGP regions and fasciclin domains were identified as FLAs. A total of 235 *FLA* genes were identified by the HMMER-Finding-AGP program method.

However, the number of *FLA* genes found in some plants was different from those described in former studies. In *A. thaliana*, *FLA20* (*AT5G40940*) and *FLA21* (*AT5G06920*) [21] were not identified, while a new putative *FLA* gene, *AT5G16920*, was identified. In *E. grandis*, *Eucgr.A01741* and *Eucgr.K02662* were missing [13], and *Eucgr.K00086* was a newly identified *FLA* gene. In *P. trichopoda*, 46 *FLA* genes were identified compared with the 50 *FLA* genes analyzed in a previous study [24]: *Potri.013G152200, Potri.T130300, Potri.001G440800, Potri.018G005100, Potri.008G127500, Potri.008G128200*, and *Potri.005G079500* were not identified, whereas *Potri.019G049600, Potri.T118500* and *Potri.012G006200* were new putative *FLA* genes identified in this study. In *O. sativa*, two *FLA* genes found in a previous study (*LOC_Os02g49420* and *LOC_Os02g26290*) [12] were not identified, while a putative new *FLA* gene (*LOC_Os12g13160*) was identified in our work. Among 13 *FLA* genes that were not identified by the HMMER-Finding-AGP program method, it was found that Potri.T130300, Potri.018G005100, LOC_Os02g49420, and LOC_Os02g26290 did not contain a fasciclin domain by using Batch CD-Search tool. Besides, because the AGP regions of Eucgr.K02662, Potri.008G127500, and Potri.008G128200 were found in the fasciclin domain, they were not identified as FLAs in this study. Then, the remaining six FLAs (AT5G40940, AT5G06920, Eucgr.A01741, Potri.013G152200, Potri.001G440800, and Potri.005G079500) were included in this study and also used as queries to perform BLAST searches to identify their homologous FLAs in other plant species: Phpat.003G041000 in *P. patens*, MA_89859g0010 and MA_10360g0010 in *P. abies*, scaffold00024.69 in *A. trichopoda*, and Eucgr.H00590.1 in *E. grandis*. As a result, 246 *FLA* genes were identified.

The number of *FLA* genes ranged from 1 to 49 across the different plant species; in most species, the number of *FLA* genes was between 11 and 26. *C. crispus* had only one *FLA* gene, while *P. trichocarpa* contained the highest number of *FLA* genes (49), almost double the number of the second one, *O. sativa* (26). It was found that the number of *FLA* genes and genome size were uncorrelated. *P. abies*, for instance, which had the largest genome size (19,600 Mb) among these 13 plant species, had only 24 *FLA* genes compared with *P. trichocarpa* which had 49 *FLA* genes with a much smaller genome size (434.29 Mb) (Table 1). The number of *FLA* genes was also uncorrelated with the number of predicted genes in plant species. For example, *E. grandis* contained more genes (45,226) than *O. sativa*, while *O. sativa* had more *FLA* genes (26) than *E. grandis* (18) (Table 1). Overall, higher plants contained the highest number of *FLA* genes and the number of *FLA* genes increased from lower plants to higher plants. For example, the number of *FLA* genes was doubled from lycophytes to gymnosperm.

Moreover, the intron-exon structures of 246 *FLA* genes were retrieved from the OrcAE website (available online: https://bioinformatics.psb.ugent.be/orcae/overview/Chbra), Phytozome website (Version 12; available online: https://phytozome.jgi.doe.gov/pz/portal.html), and ConGenIE website (available online: http://congenie.org/) and were displayed by GSDS 2.0 (available online: http://gsds.cbi.pku.edu.cn/) [44]. Green algae *FLA* genes contained a large number of introns, while most land plants *FLA* genes contained one intron or even had no intron (Table S1). It seemed that introns in FLA genes were lost during plant evolution, especially from green algae to land plants.

Table 1. Information about genome size and fasciclin-like arabinogalactan protein (FLA) gene number in the plants of interest for this study.

Lineage	Organism	Genome Size (Mb)	No. of Predicted Genes	No. of *FLA* Genes	Reference
Red algae	*Chondrus crispus*	104.98	9843	1	This study
Green algae	*Chlamydomonas reinhardtii*	120.405	14,488	11	This study
	Chara braunii	1751.21	35,424	24	This study
Liverworts	*Marchantia polymorpha*	215.739	19,287	14	This study
Mosses	*Physcomitrella patens*	472.081	23,733	12	This study
Lycophytes	*Selaginella moellendorffii*	212.315	34,782	9	This study
Gymnosperm	*Picea abies*	19,600	28,354	24	This study
Amborellales	*Amborella trichopoda*	706.495	19,354	12	This study
Eudicots	*Arabidopsis thaliana*	119.148	38,093	22	Schultz et al. [21]
	Eucalyptus grandis	691.43	45,226	19	MacMillan et al. [13]
	Populus trichocarpa	434.29	37,197	49	Showalter et al. [24]
Monocots	*Brachypodium distachyon*	218.345	34,310	23	This study
	Oryza sativa	374.423	33,185	26	Ma and Zhao [12]

2.2. Phylogenetic Analysis and Classification of FLAs

In order to understand the relationships between FLAs with different numbers of fasciclin domains, evolutionary analysis was performed based on multiple sequence alignments of FLAs. First, all the FLA protein sequences were filtered by BLAST+ [45] with a −5 expect (E) threshold. The sequences (CreFLA2, CreFLA3, CreFLA4, CreFLA5, CreFLA6, and CreFLA7 in *C. reinhardtii*, CbrFLA5, CbrFLA6, CbrFLA8, CbrFLA10, CbrFLA12, CbrFLA13, CbrFLA14, CbrFLA17, CbrFLA18, and CbrFLA21 in *C. braunii*) with low similarity to other plant species were removed, and classified into Group F (Table S1). Next, after removing sequences of signal peptides and GPI anchor addition signals, the filtered 230 FLA sequences were aligned by Clustal Omega 1.2.2, and the HMM profile of fasciclin domains was used as a guide [46,47]. Then, the fasciclin domains could be divided into two types (Type 1 and Type 2) based alignment results (Figure 1 and Figure S1). The FLA sequences with Type 1 and Type 2 fasciclin domains were further aligned, respectively (Figures S2 and S3). Interestingly, for some algae FLA sequences that contained more than two fasciclin domains, only one or two fasciclin domains had hits in other FLA sequences: The first and the fourth fasciclin domains in CreFLA11, the second fasciclin domain in CreFLA10. It was likely that the other fasciclin domains with low similarity to those in higher plants were lost in the course of evolution from algae to land plants.

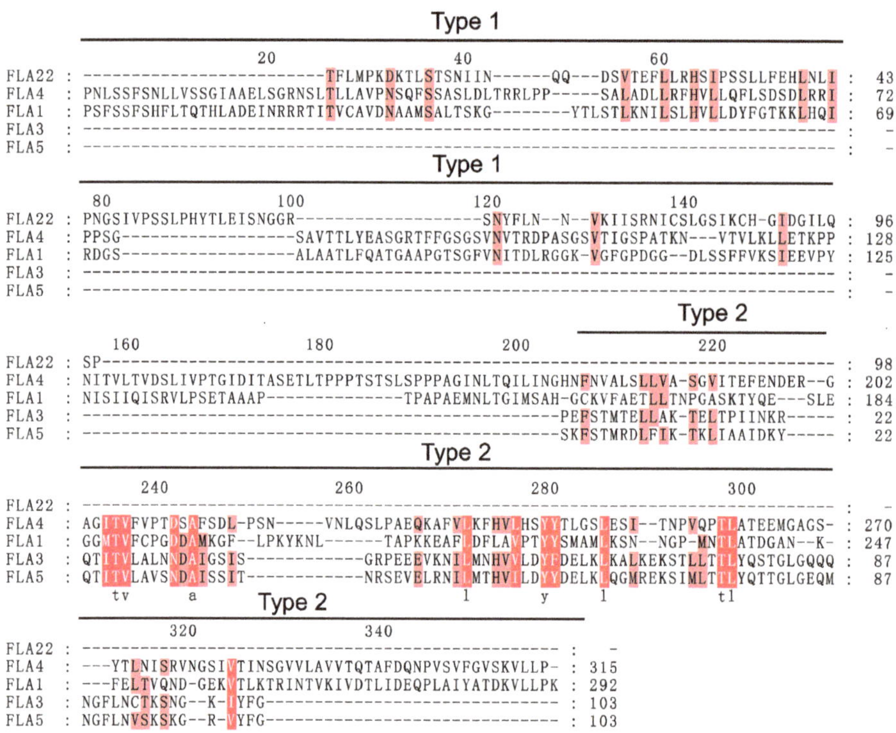

Figure 1. Multiple sequence alignment of representative FLA sequences. Fasciclin domains were divided into two types (Type 1 and Type 2). Residues with high similarity (80%, 60%) were highlighted in dark pink and light pink, respectively.

The phylogenetic tree of filtered 230 FLA sequences could not be built because the identity of alignment was very low (<30%). Once the identity was above 30%, the accuracy of alignment was acceptable [48–50]. The accuracy of the FLA alignment results was tested by computing the overall mean distance with the P-distance method in Mega 7 [49,51]. As P-distance equals 1 minus the identity of amino acids, the identities of Type 1 and Type 2 fasciclin domains were 31.7% and 30.4%, respectively. The accuracy results of Type 1 and Type 2 were 0.683 and 0.696, respectively. These indicators made it suitable for building the phylogenetic trees. The Maximum Likelihood (ML) trees for each type were built using the best models: Le_Gascuel_2008 model [52] + Gamma distribution + evolutionarily invariable (LG + G + I) for Type 1, Le_Gascuel_2008 model + Gamma distribution (LG + G) for Type 2, with 85% partial deletion by Mega 7. Bootstrap analyses with 1000 replicates were performed for support estimation. Confidence values below 50% were cut off, and confidence values higher than 70% were shown on nodes (Figures 2 and 3). Although the similarity between full-length sequences of FLAs are quite low, the fasciclin domains exhibited two highly conserved motifs (H1 and H2) and a conserved central YH motif [18]. MEME web server (available online: http://meme-suite.org/tools/meme) [53] was used to find the conserved motif (H1, H2, and YH motifs) of Type 1 and Type 2 sequences. The H1 and YH motif were similar between Type 1 and Type 2 sequences, while the H2 region was quite different. In Type 1 sequences, the H2 motif was characterized by [Gly/Ile/Val/Leu/Phe]–X–[Ile/Val/Cys]–His–Gly–[Ile/Val/Leu]–X–X–[Leu/Val/Pro/Ile]–[Leu/Met/Ile] sequence. In Type 2 sequences, the H2 motif was characterized by [Val/Ile/Met/Leu]–[Tyr/His/Phe/Gln]–X–[Val/Ile/Leu]–X–X–[Val/Leu]–Leu–[Leu/Phe/Val]–Pro sequence (X represents any amino acid) (Figures 2 and 3). Interestingly, most FLAs with single fasciclin domain was of Type 2, while only a few FLAs with single fasciclin domain was of Type 1.

Based on the sequence similarity, phylogenetic analysis, and previous study [11], we have classified FLAs into seven groups: Group A (including FLA6, FLA7, FLA 9, FLA 11–13 from *A. thaliana*), Group B (FLA 15–18 from *A. thaliana*), Group C (including FLA 1–3, FLA 5, FLA8, FLA10, FLA14 from *A. thaliana*), Group D1 (including FLA 19–22 from *A. thaliana*), Group D2 (including FLA4 from *A. thaliana*), Group E, and Group F (Table S1). Group F sequences were all algae FLAs which were not included in building phylogenetic trees. The remaining algae FLAs were all in Group D1 and Group E, which meant that Group D1 and Group E might be traced back to the origin of the FLA family in plants. Moreover, *FLA3, 5, 14, 20, 21,* and *22* were specifically expressed in anthers at different stages of floral development [18,54,55]. *FLA3* was involved in microspore development, and its knock-down plants showed reduced female fertility [56]. There was a probability that Group C and Group D1 FLAs were mainly related to male gametophyte development. Group C and Group D1 FLAs were also related to the growth regulator. For instance, FLA1 and FLA2 might play an important role in root development [57,58]. Interestingly, in Group A, all FLAs were with single fasciclin domain. A previous study proposed that Group A FLAs were specific to the evolution of flowering plant secondary cell wall formation and properties [30]. For example, *FLA11*, *FLA12*, and *ZeFLA11* are highly expressed in vascular tissue and double mutants of *FLA11* and *12* showed defects in secondary cell wall thickening [25,30]. *EgrFLA1, 2,* and 3 were also highly expressed in stems. EgrFLA2 was involved in altering fiber cellulose deposition in woody tissue, and EgrFLA3 influenced flexural strength [13]. In *Eucalyptus nitens*, EniFLA1, 2, and 3, which were closely related to FLA11 and 12, as well as highly similar to EgrFLA1 and 2, could affect stem biomechanics [30]. These Group A FLAs and their homologs in other plants (poplar, zinnia) were also involved in secondary cell wall biosynthesis [23,25]. In addition, FLA9 in Group A was also related to seed development. It had been shown that the stress-induced reductions of *FLA9* gene expression enhanced the abortion of fertilized ovaries [59].

In addition, the variable fasciclin number of FLAs had a tight relationship to the phylogenetic tree. All the FLAs with multiple fasciclin domains (>2) were in Group D1 and Group E. As these FLAs were only identified in algae, they might be the most original FLAs in the course of evolutionary history. In Group A, all the FLAs were with single fasciclin domain and belonged to seed plants. Group A FLAs were the latest FLAs generated in the course of evolutionary history. From Group E to Group A, the number of fasciclin domains reduced over the course of evolutionary history. Except for Group A FLAs, the structures of FLAs were quite diverse, especially for Group E FLAs, which included the most original FLAs. Moreover, Group E *FLA* genes contained more introns than other groups. The number of introns also reduced over the course of evolutionary history.

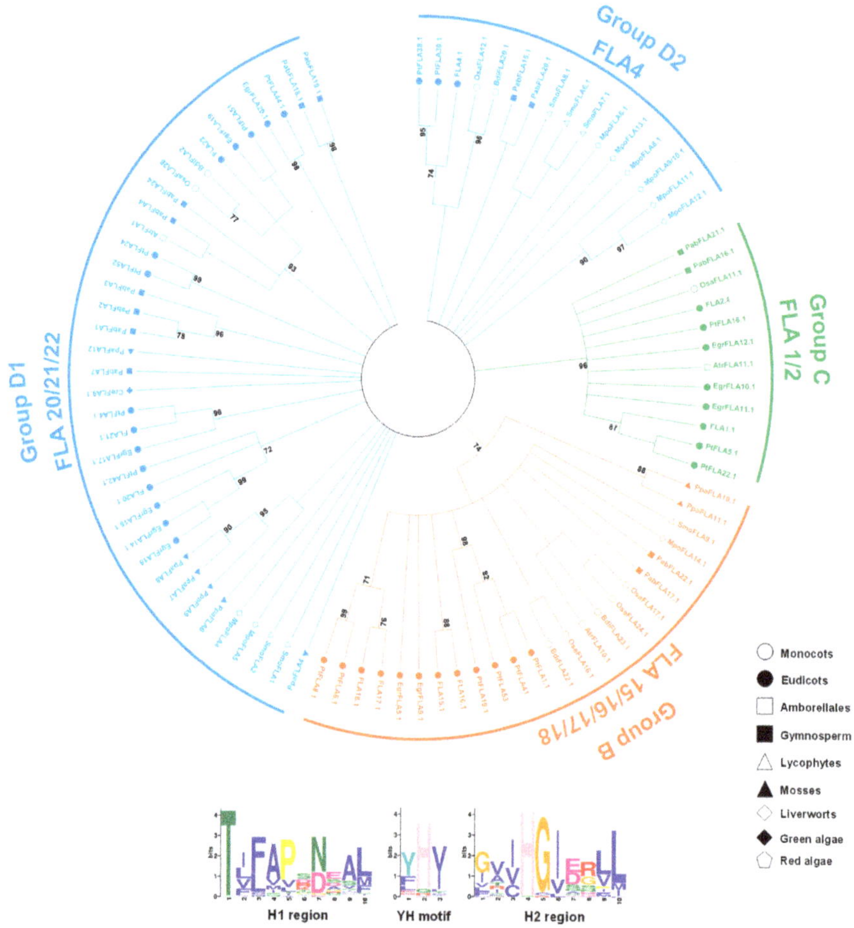

Figure 2. Phylogenetic relationships between Type 1 fasciclin domains in plant species. The amino acid sequences of fasciclin domains in FLAs were aligned by Clustal Omega 1.2.2 with the guide of HMM profile of fasciclin domains, and the phylogenetic trees were built by Mega 7 using the Maximum Likelihood (ML) method with 85% partial deletion. Bootstrap analyses with 1,000 replicates were performed for support estimation. The confidence values below 50% were cut off, and the confidence values higher than 70% are shown on nodes. The tree was divided into four major clades: Group B, Group C, Group D1, and Group D2. Plant species from different lineages are shown in different shape. FLAs from *A. thaliana* are indicated for each clade. The order of fasciclin domains was designated from the N-terminus to the C-teminus (e.g., FLA4.1, FLA4.2, and so on). The conserved motifs (H1, H2, and YH motifs) shown below the tree were found using the MEME web server.

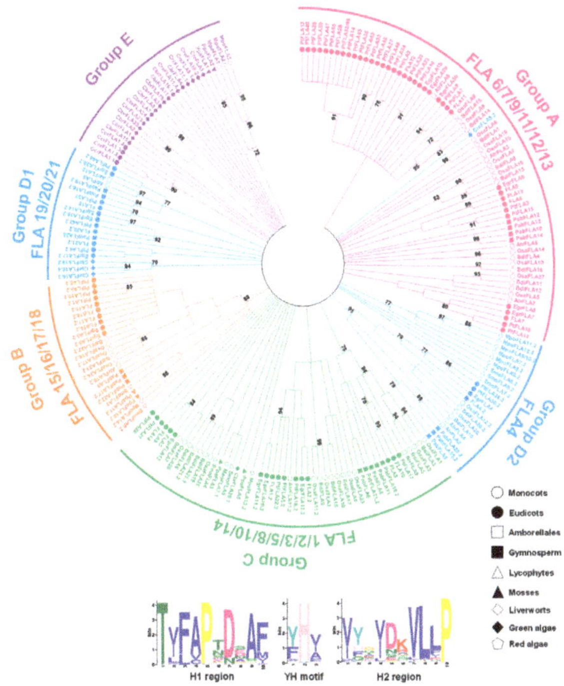

Figure 3. Phylogenetic relationships between Type 2 fasciclin domains in plant species. The amino acid sequences of fasciclin domains in FLAs were aligned by Clustal Omega 1.2.2 with the guide of HMM profile of fasciclin domains, and the phylogenetic trees were built by Mega 7 using the Maximum Likelihood (ML) method with 85% partial deletion. Bootstrap analyses with 1000 replicates were performed for support estimation. The confidence values below 50% were cut off, and the confidence values higher than 70% are shown on nodes. The tree was divided into six major clades: Group A, Group B, Group C, Group D1, Group D2, and Group E. Plant species from different lineages are shown in different shape. FLAs from *A. thaliana* are indicated for each clade. The domain closest to the N-terminus is indicated by .1 and the second by .2. The conserved motifs (H1, H2, and YH motifs) shown below the tree were found using the MEME web server.

Moreover, to understand the relationship between FLAs with single fasciclin domain, a phylogenetic tree of FLAs with single fasciclin domain from nine plant species (*C. reinhardtii*, *C. crispus*, *M. polymorpha*, *P. patens*, *S. moellendorffii*, *P. abies*, *A. trichopoda*, *B. distachyon*, and *A. thaliana*) was built by the Maximum Likelihood (ML) method under the LG + G model with 85% partial deletion. Bootstrap analyses with 1000 replicates were performed for support estimation; confidence values higher than 50% were shown on nodes. The structure displays of these FLAs were generated by GSDS 2.0 (available online: http://gsds.cbi.pku.edu.cn/) [44] (Figure 4). The structure of Group A *FLA* genes was very similar. Except for *PabFLA12*, *PabFLA14*, and *AtrFLA6*, the remaining Group A *FLA* genes did not contain introns, and most of their fasciclin domains were flanked by two AGP regions. The structures of FLAs with single fasciclin domains in Group E were quite diverse. By contrast, the phylogenetic relationship of FLAs with single fasciclin domain was similar to the phylogenetic relationships of Type 2 (Figure 3). The main type of fasciclin domain in these FLAs was Type 2 fasciclin domain. Most of Group D1 FLAs contained Type 1 fasciclin domains. It is likely that the Type 1 fasciclin domain was lost mainly in FLAs with single fasciclin domain over the course of evolutionary history. Different from phylogenetic relationships of Type 1 and Type 2 fasciclin domains (Figures 2 and 3), Group C appeared to be divergent (Figure 4). Some Group C FLAs were close to Group D2, while others were close to Group B. Moreover, the structure of these diverged Group C was different. The fasciclin domains of FLAs tailed with AGP regions belonged to Group C, which were close to Group B. For FLAs from Group C which was close to Group D2, their fasciclin domains were covered by two AGP regions.

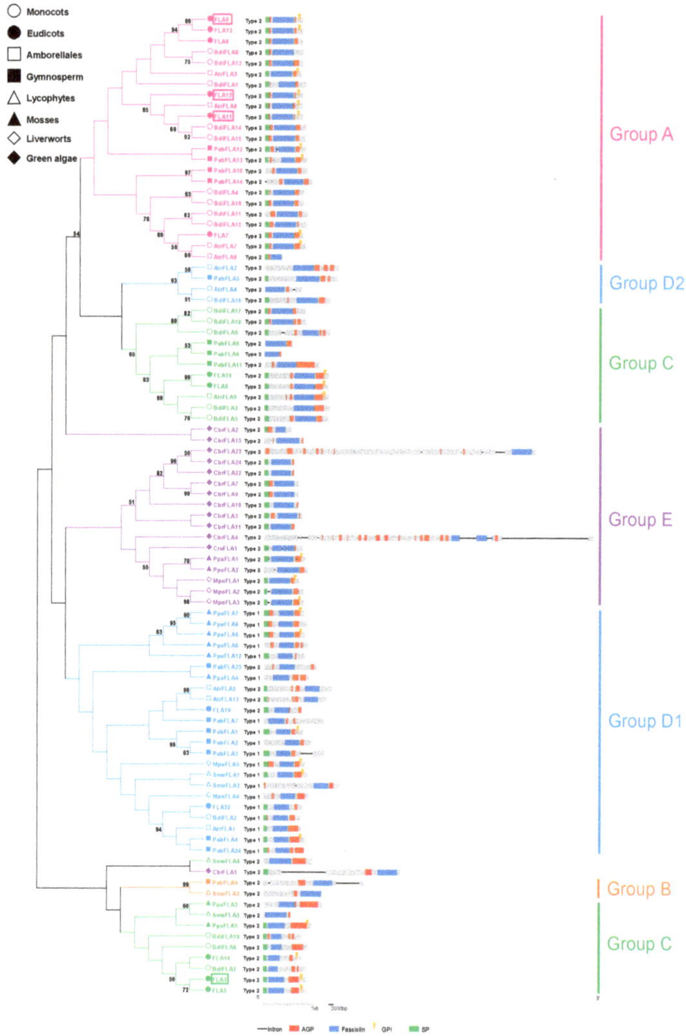

Figure 4. Phylogenetic relationships and structure display of FLAs with single fasciclin domain in nine plant species (*C. reinhardtii*, *C. crispus*, *M. polymorpha*, *P. patens*, *S. moellendorffii*, *P. abies*, *A. trichopoda*, *B. distachyon*, and *A. thaliana*). Plant species from different lineages are shown in different shapes. The phylogenetic trees were built by Mega 7 using the Maximum Likelihood (ML) method under LG+G model with 85% partial deletion. Bootstrap analyses with 1000 replicates were performed for support estimation, the confidence values higher than 50% are shown on nodes. The tree was divided into six groups according to the classifications based on two types fasciclin domains (Figures 2 and 3): Group A, Group B, Group C, Group D1, Group D2 and Group E. The structure displays were generated by GSDS 2.0. Black lines represent introns, gray rectangles the CDS regions, red rectangles the AGP regions, blue rectangles the fasciclin domains, green rectangles signal peptides, and yellow wedges GPI-anchor modification sites. The framed FLAs denote functionally characterized FLAs (FLA3, FLA9, FLA11, and FLA12).

2.3. Structural and Evolutionary Analysis of FLAs

The amino acid sequences of 246 FLAs identified in our work were shown in Figure S4. One hundred seventy-six of them contained a single fasciclin domain, and 66 of them contained two fasciclin domains. Only four FLAs with more than two fasciclin domains were found in algae, one in red algae and three in green algae. Moreover, FLAs with a single fasciclin domain, as well as with two domains first appeared in green algae (Figure 5). It was likely that divergence happened in green algae. From green algae to land plants, the number of fasciclin domains in FLAs was reduced. It had been proven that FLAs with a single fasciclin domain had conserved roles in secondary cell wall biology and properties [13]. Besides, there was an example of the functional roles of different fasciclin domains in one FLA protein. The C-proximal fasciclin domain of FLA4 was responsible for its genetic functions, while the N-proximal fasciclin domain was required for stabilization of plasma membrane localization [60,61]. It was likely that the number of fasciclin domains was related to the functions of FLAs.

FLAs were classified into seven groups based on the sequence similarity, phylogenetic analysis, and previous study [11]. Different from the previous study [11], Group D was divided into Group D1 and Group D2 because of their difference in phylogenetic analysis. Moreover, Group E and Group F present in non-seed plants are the groups newly proposed in this work. The evolutionary history of FLA family was shown in Figure 5. FLAs evolved very early during plant evolution. Group E first appeared in the plant kingdom, then Group F, Group D1, Group C, Group D2, Group B, Group A appeared successively. The Group E FLA from red algae was the most original FLA. Group F was largely dissimilar to the other groups and only existed in green algae. Group D1 and Group C evolved early during green plant evolution. The divergence of FLAs occurred in green algae; Group D1 and Group C remained, while Group F was lost after the separation between green algae and land plants. Group B and Group D2 evolved after plants conquered the land. Group A, the latest group appeared, evolved during seed plant evolution. By contrast, Group E, the earliest appeared group, was lost in seed plants.

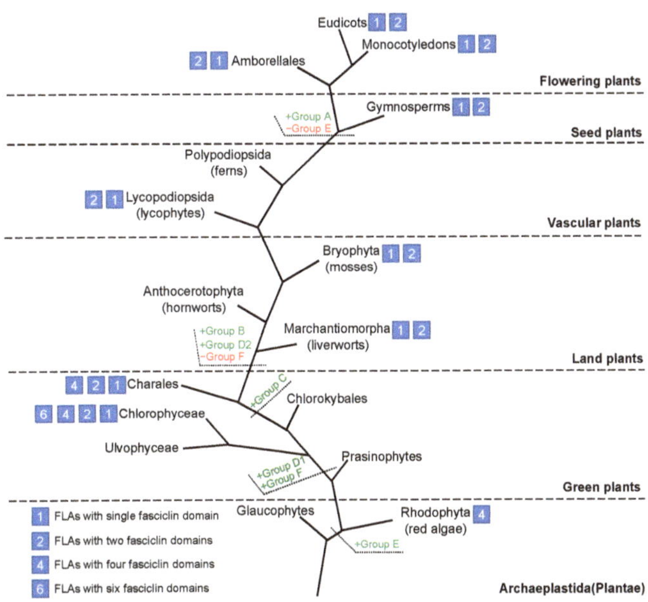

Figure 5. Evolutionary model of the FLA family in plants. The green letters display the appearance of different groups of FLAs. The red letters display the disappearance of Group E and Group F FLAs. The cubes display the number of fasciclin domains in FLAs.

2.4. Analysis of FLA Duplication Patterns during the Process of Evolution

The evolution of genomes and genetic systems is mainly driven by gene duplications [62]. The three elementary gene expansion patterns are tandem duplication, segmental duplication, and transposition events [63,64]. In the plant kingdom, tandem duplication and segmental duplication are the main processes of gene family expansion compared with transposition events [65,66]. We investigated these two duplication events to understand the *FLA* genes' expansion patterns in the plant kingdom. The paralogous genes that exist in the same chromosome within a 50 kb physical distance are examples of tandem duplication [65]. First, in order to find the chromosomal locations, the annotation information for the *FLA* genes was downloaded from OrcAE (available online: https://bioinformatics.psb.ugent.be/orcae/overview/Chbra), Phytozome (available online: https://phytozome.jgi.doe.gov/pz/portal.html) and ConGenIE (available online: http://congenie.org/). Then, the distances between *FLA* genes' locations were compared in the same chromosome. The locus search tool on PGDD (available online: http://chibba.agtec.uga.edu/duplication/index/locus) and MCSCAN were used to find the segmental duplications (Table S2). The duplications in *FLA* genes were related to whole-genome duplication events (Figure 6). The higher plants exhibited more duplications than lower plants. *P. trichocarpa* had the highest number of duplicated *FLA* genes, which made it have more *FLA* genes than other plant species. Although most duplicated pairs shared the same structure type, some duplicated genes had different structure types. For example, in *C. reinhardtii*, *Cre16.g687742* containing two fasciclin domains and *Cre16.g687854* containing single fasciclin domain most probably result from tandem duplication. It seemed that some *FLA* genes with single fasciclin domain evolved from *FLA* genes with two fasciclin domains. FLAs with single fasciclin domain evolved from FLAs with multiple fasciclin domains, and the number of fasciclin domains was reduced in evolutionary history.

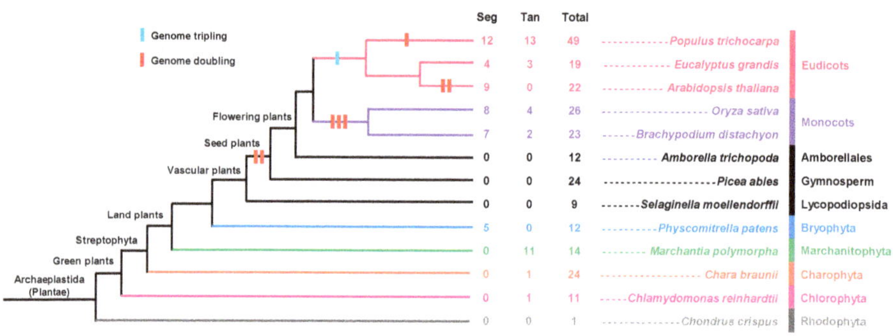

Figure 6. Duplication events of *FLA* genes in the plant kingdom. The phylogenetic tree on the left was built based on the Tree of Life Web project (available online: http://www.tolweb.org/Green_plants) and whole-genome duplication events in PGDD (available online: http://chibba.pgml.uga.edu/duplication/index/home). The number next to the tree is the number of *FLA* genes resulting from segmental duplication, tandem duplication, and total *FLA* genes in the species. Seg: Segmental duplication (pairs); Tan: Tandem duplication (pairs); Total: Total number of *FLA* genes in the species.

In order to understand the evolution processes of the *FLA* gene family in the plant kingdom, duplicated gene pairs among *FLAs* were used to estimate the molecular evolutionary rates by calculating their Ka/Ks value (Table S2). The Ka/Ks values of all the duplicated gene pairs except the *Mapoly0075s0013.1/Mapoly0075s0013.2* gene pair were lower than 1. It was assumed that FLA duplicated gene pairs evolved under purifying selection, indicating that the functions of the *FLAs* gene family were crucial to plant development and functional mutations in *FLA* genes might have negative impacts on plants. The Ka/Ks ratio of *Mapoly0075s0013.1/Mapoly0075s0013.2* gene pair was 2.3512, showing that this gene pair underwent positive selection pressure during evolution. However, plants could not escape from their environment in order to adapt to changes, so positive selection, which could lead to beneficial functional changes, was also important during plant evolution [67]. The *Mapoly0075s0013.1/Mapoly0075s0013.2* gene pair, which was found to experience positive selection, might have improved the adaptation of the plant to new environments.

3. Materials and Methods

3.1. Bioinformatics Identification of FLAs

Multiple searches were carried out in order to identify FLA genes in 13 plant species (*C. crispus, C. reinhardtii, C. crispus, M. polymorpha, P. patens, S. moellendorffii, P. abies, A. trichopoda, B. distachyon, O. sativa, A. thaliana, E. grandis*, and *P. trichocarpa*) [31–43]. The predicted proteomes of *C. crispus* was downloaded from NCBI, that of *C. braunii* were from the OrcAE website (available online: https://bioinformatics.psb.ugent.be/orcae/overview/Chbra), that of P. abies were from the ConGenIE website (available online: http://congenie.org/), and that of other species from the Phytozome website (Version 12; available online: https://phytozome.jgi.doe.gov/pz/portal.html). Except for *P. abies* [35], the statistics of genome size overall number of predicted genes were from the NCBI Genome database (available online: https://www.ncbi.nlm.nih.gov/genome).

Then, the Hidden Markov Model (HMM) profile built for fasciclin domains was downloaded from Pfam (available online: http://pfam.xfam.org/family/PF02469) [68], and HMMER 3.0 [69] was used to search proteins with fasciclin domains from the selected plants. Then the presence of fasciclin domains corresponding to the obtained proteins was examined by the NCBI conserved domain database (available online: http://www.ncbi.nlm.nih.gov/cdd). Next, the Finding-AGP program [7] was used to identify AGP regions from proteins with fasciclin domains. Finally, proteins with both fasciclin domains and AGP regions were identified as FLAs. Also, the omitted FLA sequences that were identified in former studies (AT5G40940, AT5G06920, Eucgr.A01741, Potri.013G152200, Potri.001G440800, and Potri.005G079500) were used as queries to perform BLAST searches with a −3 expect (E) threshold to find FLAs that could not be identified by HMMER 3.0.

Moreover, most FLAs have a predicted signal peptide and GPI-anchor. Therefore, SignalP 4.1 Server (available online: http://www.cbs.dtu.dk/services/SignalP/) was used to predict signal peptides [70] and big-PI Plant Predictor (available online: http://mendel.imp.ac.at/gpi/plant_server.html) was used to predict GPI modification sites [71]. The intron of red algae *FLA* was detected by the GSDS website (available online: http://gsds.cbi.pku.edu.cn/) [44], and the intron of other *FLAs* were found from the OrcAE website (available online: https://bioinformatics.psb.ugent.be/orcae/overview/Chbra), the Phytozome website (Version 12; available online: https://phytozome.jgi.doe.gov/pz/portal.html), and the ConGenIE website (available online: http://congenie.org/). The amino acid sequences and the presence of AGP regions, signal peptides, fasciclin domains, and GPI-anchor signals are given in Table S1.

3.2. Multiple Sequence Alignment and Phylogenetic Analysis

All of the FLA protein sequences were searched against each other by BLAST+ with a −5 expect (E) threshold [45]. The sequences with low similarity were removed. Then, signal peptides and GPI modification sites were removed from filtered FLA sequences. These sequences were aligned by Clustal Omega 1.2.2 with HMM of the fasciclin domain as a guide in the alignment [46,47]. The fasciclin domains were designated as Type 1 and Type 2 and were also aligned by Clustal Omega 1.2.2 with the HMM of the fasciclin domain as a guide in the alignment [46,47]. GeneDoc [72] was used to display multiple sequence alignments.

The reliability of alignment results was tested by computing overall mean distance with the P-distance method by Mega 7 [49,51]. The alignments of Type 1, Type 2, and FLAs with a single fasciclin domain was then used to build phylogenetic trees with the Maximum Likelihood (ML) method. The best models for ML trees were found by Mega 7 [51,73]. Then, ML trees were built under the best model with 85% partial deletion by Mega 7. Bootstrap analyses with 1, 000 replicates were performed for support estimation [51,52].

3.3. Motif Prediction

In order to identify the conserved domains and motifs of Type 1 and Type 2 fasciclin domains, MEME web server (available online: http://meme-suite.org/tools/meme) [53] was used to identify the conserved motifs (H1 and H2 regions, YH motif). The following parameters were used when running the MEME: (1) The motif sites in sequences were distributed by 0 or 1 occurrence per sequence; (2) the maximum of motifs was set to be 10 for the H1 and H2 regions, and 3 for the YH motif; and (3) a 0-order model of sequences was used as the background model.

3.4. Gene Duplication and Molecular Evolution

The annotation information of the *FLA* genes on the phytozome website (available online: https://phytozome.jgi.doe.gov/pz/portal.html), the OrcAE website (available online: https://bioinformatics.psb.ugent.be/orcae/overview/Chbra), and the ConGenIE website (available online: http://congenie.org/) was used to find the chromosomal locations. The paralogous genes that exit in the same chromosome within a 50-kb physical distance was defined as tandem duplication [64]. The segmental duplications of 10 plants (*C. reinhardtii*, *P. patens*, *S. moellendorffii*, *P. abies*, *A. trichopoda*, *B. distachyon*, *O. sativa*, *A. thaliana*, *E. grandis*, and *P. trichocarpa*) were found by the PGDD locus search tool (available online: http://chibba.agtec.uga.edu/duplication/index/locus). Because *M. polymorpha and C. crispus* data were absent in PGDD, Multiple Collinearity Scan (MCSCAN) [74–77] was used to find the segmental duplications in *M. polymorpha*.

To calculate the molecular evolutionary rates between *FLAs* duplicated gene pairs, pairwise alignment was performed among these gene pairs by ClustalW (codons) in MEGA7 [51]. Then, the MYN (Modified YN) model in KaKs_Calculator 2.0 was used to estimate the nonsynonymous substitution rate (Ka), the synonymous substitution rate (Ks) and the Ka/Ks value of these duplicated gene pairs [78].

4. Conclusions

FLAs play an important role in plant development and adaption to the environment. Two hundred forty-six *FLA* genes in 13 plant species were identified in this study. It was found that FLAs first appeared in algae. Based on the sequence similarity and phylogenetic analysis, FLAs could be classified into seven groups: Group A, Group B, Group C, Group D1, Group D2, Group E, and Group F. Group E FLAs were the earliest to appear in evolutionary history and disappeared in seed plants, while Group A FLAs were the latest and only existed in seed plants. FLAs with multiple fasciclin domain (>2) were possibly the first FLA type to appear in Archaeplastida because they only existed in algae. FLAs with single fasciclin domain and with two fasciclin domains were dominant in green plants. The number of fasciclin domains in FLAs varied in green algae and was reduced to one or two in land plants. In addition, introns in *FLA* genes were lost during plant evolution, especially from green algae to land plants. Moreover, tandem and segmental duplications contributed to the expansion of the *FLA* gene family, and duplicated gene pairs in *FLAs* mainly evolved under purifying selection.

Supplementary Materials: Supplementary materials can be found at http://www.mdpi.com/1422-0067/20/8/1945/s1.

Author Contributions: H.M. conceived of and designed the research plans; J.H. and H.Z. performed most of the experiments and analyzed the data; Z.C., Y.K., and J.L. provided technical assistance to J.H. and H.Z.; J.H. and H.Z. wrote the article with contributions from all the authors; H.M. supervised and supported the writing.

Acknowledgments: This research was supported by the National Natural Science Foundation of China (31500212), the Natural Science Foundation of Shaanxi Province (2015JQ3090), and the Undergraduate Innovation Foundation of Northwest A&F University (No. 1201710712099).

Conflicts of Interest: The authors declare no conflict of interest.

Abbreviations

FLA	Fasciclin-like arabinogalactan protein
AGP	Arabinogalactan protein
GPI	Glycosylphosphatidylinositol
PAST%	The percentage of Pro, Ala, Ser, and Thr residues in a protein amino-acid sequence
Ccr	*Chondrus crispus*
Cre	*Chlamydomonas reinhardtii*
Mpo	*Marchantia polymorpha*
Smo	*Selaginella moellendorffii*
Pab	*Picea abies*
Atr	*Amborella trichopoda*
Egr	*Eucalyptus grandis*
Pt	*Populus trichocarpa*
Bdi	*Brachypodium distachyon*
Os	*Oryza sativa*
Ka	Nonsynonymous substitution rate
Ks	Synonymous substitution rate
PGDD	Plant Genome Duplication Database
NCBI	National Center for Biotechnology Information
ConGenIE	Conifer Genome Integrative Explorer
HMM	Hidden Markov Model
BLASTP	Protein Basic Local Alignment Search Tool
MYN	Modified Yang-Nielsen Algorithm
MCSCAN	Multiple Collinearity Scan
MEME	Multiple Expectation maximization for Motif Elicitation
LG	Le_Gascuel_2008 model
G	Gamma distribution
I	Evolutionarily invariable
GSDS	Gene Structure Display Server
OrcAE	Online Resource for Community Annotation of Eukaryotes

References

1. Nothnagel, E.A. Proteoglycans and related components in plant cells. *Int. Rev. Cytol.* **1997**, *174*, 195–291. [PubMed]
2. Chivasa, S.; Ndimba, B.K.; Simon, W.J.; Robertson, D.; Yu, X.L.; Knox, J.P.; Bolwell, P.; Slabas, A.R. Proteomic analysis of the *Arabidopsis thaliana* cell wall. *Electrophoresis* **2002**, *23*, 1754–1765. [CrossRef]
3. Johnson, K.L.; Cassin, A.M.; Lonsdale, A.; Wong, G.K.-S.; Soltis, D.E.; Miles, N.W.; Melkonian, M.; Melkonian, B.; Deyholos, M.K.; Leebens-Mack, J.; et al. Insights into the Evolution of Hydroxyproline-rich Glycoproteins from 1000 plant Transcriptomes. *Plant Physiol.* **2017**, *174*, 904–921. [CrossRef]
4. Jamet, E.; Albenne, C.; Boudart, G.; Irshad, M.; Canut, H.; Pont-Lezica, R. Recent advances in plant cell wall proteomics. *Proteomics* **2008**, *8*, 893–908. [CrossRef] [PubMed]
5. Nothnagel, E.A.; Bacic, A.; Clarke, A.E. *Cell and Developmental Biology of Arabinogalactan-Proteins*; Springer Science Business Media: New York, NY, USA, 2000.
6. Showalter, A.M. Arabinogalactan-proteins: Structure, expression and function. *Cell. Mol. Life Sci.* **2001**, *58*, 1399–1417. [CrossRef]
7. Ma, Y.; Yan, C.; Li, H.; Wu, W.; Liu, Y.; Wang, Y.; Chen, Q.; Ma, H. Bioinformatics Prediction and Evolution Analysis of Arabinogalactan Proteins in the Plant Kingdom. *Front. Plant Sci.* **2017**, *8*, 66. [CrossRef]
8. Ellis, M.; Egelund, J.; Schultz, C.J.; Bacic, A. Arabinogalactan-proteins: Key Regulators at the Cell Surface? *Plant Physiol.* **2010**, *153*, 403–419. [CrossRef]
9. Shpak, E.; Barbar, E.; Leykam, J.F.; Kieliszewski, M.J. Contiguous Hydroxyproline Residues Direct Hydroxyproline Arabinosylation in *Nicotiana tabacum*. *J. Biol. Chem.* **2001**, *276*, 11272–11278. [CrossRef] [PubMed]

10. Showalter, A.M.; Basu, D. Extensin and Arabinogalactan-Protein Biosynthesis: Glycosyltransferases, Research Challenges, and Biosensors. *Front. Plant Sci* **2016**, *7*, 814. [CrossRef] [PubMed]
11. Johnson, K.L.; Jones, B.J.; Bacic, A.; Schultz, C.J. The fasciclin-like arabinogalactan proteins of Arabidopsis. A multigene family of putative cell adhesion molecules. *Plant Physiol.* **2003**, *133*, 1911–1925. [CrossRef] [PubMed]
12. Ma, H.; Zhao, J. Genome-wide identification, classification, and expression analysis of the arabinogalactan protein gene family in rice (*Oryza sativa* L.). *J. Exp. Bot.* **2010**, *61*, 2647–2668. [CrossRef]
13. MacMillan, C.P.; Taylor, L.; Bi, Y.; Southerton, S.G.; Evans, R.; Spokevicius, A. The fasciclin-like arabinogalactan protein family of *Eucalyptus grandis* contains members that impact wood biology and biomechanics. *New Phytol.* **2015**, *206*, 1314–1327. [CrossRef] [PubMed]
14. Mashiguchi, K.; Asami, T.; Suzuki, Y. Genome-wide identification, structure and expression studies, and mutant collection of 22 early nodulin-like protein genes in Arabidopsis. *Biosci. Biotechnol. Biochem.* **2009**, *73*, 2452–2459. [CrossRef]
15. Ma, H.; Zhao, H.; Liu, Z.; Zhao, J. The phytocyanin gene family in rice (*Oryza sativa* L.): genome-wide identification, classification and transcriptional analysis. *PLoS ONE* **2011**, *6*, e25184. [CrossRef]
16. Motose, H.; Sugiyama, M.; Fukuda, H. A proteoglycan mediates inductive interaction during plant vascular development. *Nature* **2004**, *429*, 873–878. [CrossRef] [PubMed]
17. Kobayashi, Y.; Motose, H.; Iwamoto, K.; Fukuda, H. Expression and genome-wide analysis of the xylogen-type gene family. *Plant Cell Physiol.* **2011**, *52*, 1095–1106. [CrossRef] [PubMed]
18. Seifert, G.J. Fascinating Fasciclins: A Surprisingly Widespread Family of Proteins that Mediate Interactions between the Cell Exterior and the Cell Surface. *Int. J. Mol. Sci.* **2018**, *19*, 1628. [CrossRef]
19. Bastiani, M.J.; Harrelson, A.L.; Snow, P.M.; Goodman, C.S. Expression of fasciclin I and II glycoproteins on subsets of axon pathways during neuronal development in the grasshopper. *Cell* **1987**, *48*, 745–755. [CrossRef]
20. Elkins, T.; Zinn, K.; McAllister, L.; Hoffmann, F.M.; Goodman, C.S. Genetic analysis of a Drosophila neural cell adhesion molecule: interaction of fasciclin I and Abelson tyrosine kinase mutations. *Cell* **1990**, *60*, 565–575. [CrossRef]
21. Schultz, C.J.; Rumsewicz, M.P.; Johnson, K.L.; Jones, B.J.; Gaspar, Y.M.; Bacic, A. Using Genomic Resources to Guide Research Directions. The Arabinogalactan Protein Gene Family as a Test Case. *Plant Physiol.* **2002**, *129*, 1448–1463. [CrossRef]
22. Faik, A.; Abouzouhair, J.; Sarhan, F. Putative fasciclin-like arabinogalactan-proteins (FLA) in wheat (*Triticum aestivum*) and rice (*Oryza sativa*): identification and bioinformatic analyses. *Mol. Genet. Genom.* **2006**, *276*, 478–494. [CrossRef]
23. Lafarguette, F.; Leplé, J.C.; Déjardin, A.; Laurans, F.; Costa, G.; Lesage-Descauses, M.C.; Pilate, G. Poplar genes encoding fasciclin-like arabinogalactan proteins are highly expressed in tension wood. *New Phyto.* **2004**, *164*, 107–121. [CrossRef]
24. Showalter, A.M.; Keppler, B.D.; Liu, X.; Lichtenberg, J.; Welch, L.R. Bioinformatic Identification and Analysis of Hydroxyproline-Rich Glycoproteins in *Populus trichocarpa*. *Bmc Plant Biol.* **2016**, *16*, 229. [CrossRef] [PubMed]
25. Dahiya, P.; Findlay, K.; Roberts, K.; McCann, M.C. A fasciclin-domain containing gene, *ZeFLA11*, is expressed exclusively in xylem elements that have reticulate wall thickenings in the stem vascular system of *Zinnia elegans* cv Envy. *Planta* **2006**, *223*, 1281–1291. [CrossRef] [PubMed]
26. Huang, G.Q.; Xu, W.L.; Gong, S.Y.; Li, B.; Wang, X.L.; Xu, D.; Li, X.B. Characterization of 19 novel cotton *FLA* genes and their expression profiling in fiber development and in response to phytohormones and salt stress. *Physiol. Plant* **2008**, *134*, 348–359. [CrossRef] [PubMed]
27. Liu, H.; Shi, R.; Wang, X.; Pan, Y.; Li, Z.; Yang, X.; Zhang, G.; Ma, Z. Characterization and expression analysis of a fiber differentially expressed Fasciclin-like arabinogalactan protein gene in sea island cotton fibers. *PLoS ONE* **2013**, *8*, e70185. [CrossRef]
28. Jun, L.; Xiaoming, W. Genome-wide identification, classification and expression analysis of genes encoding putative fasciclin-like arabinogalactan proteins in Chinese cabbage (*Brassica rapa* L.). *Mol. Biol. Rep.* **2012**, *39*, 10541–10555. [CrossRef]

29. Guerriero, G.; Mangeot-Peter, L.; Legay, S.; Behr, M.; Lutts, S.; Siddiqui, K.S.; Hausman, J.F. Identification of fasciclin-like arabinogalactan proteins in textile hemp (*Cannabis sativa* L.): in silico analyses and gene expression patterns in different tissues. *Bmc Genom.* **2017**, *18*, 741. [CrossRef]
30. MacMillan, C.P.; Mansfield, S.D.; Stachurski, Z.H.; Evans, R.; Southerton, S.G. Fasciclin-like arabinogalactan proteins: specialization for stem biomechanics and cell wall architecture in Arabidopsis and *Eucalyptus*. *Plant J.* **2010**, *62*, 689–703. [CrossRef] [PubMed]
31. Collen, J.; Porcel, B.; Carre, W.; Ball, S.G.; Chaparro, C.; Tonon, T.; Barbeyron, T.; Michel, G.; Noel, B.; Valentin, K.; et al. Genome structure and metabolic features in the red seaweed *Chondrus crispus* shed light on evolution of the Archaeplastida. *Proc. Natl. Acad. Sci. USA* **2013**, *110*, 5247–5252. [CrossRef] [PubMed]
32. Bowman, J.L.; Kohchi, T.; Yamato, K.T.; Jenkins, J.; Shu, S.; Ishizaki, K.; Yamaoka, S.; Nishihama, R.; Nakamura, Y.; Berger, F.; et al. Insights into Land Plant Evolution Garnered from the *Marchantia polymorpha* Genome. *Cell* **2017**, *171*, 287–304. [CrossRef] [PubMed]
33. Lang, D.; Ullrich, K.K.; Murat, F.; Fuchs, J.; Jenkins, J.; Haas, F.B.; Piednoel, M.; Gundlach, H.; Van Bel, M.; Meyberg, R.; et al. The *Physcomitrella patens* chromosome-scale assembly reveals moss genome structure and evolution. *Plant J.* **2018**, *93*, 515–533. [CrossRef] [PubMed]
34. Banks, J.A.; Nishiyama, T.; Hasebe, M.; Bowman, J.L.; Gribskov, M.; dePamphilis, C.; Albert, V.A.; Aono, N.; Aoyama, T.; Ambrose, B.A.; et al. The Selaginella Genome Identifies Genetic Changes Associated with the Evolution of Vascular Plants. *Science* **2011**, *332*, 960–963. [CrossRef] [PubMed]
35. Nystedt, B.; Street, N.R.; Wetterbom, A.; Zuccolo, A.; Lin, Y.-C.; Scofield, D.G.; Vezzi, F.; Delhomme, N.; Giacomello, S.; Alexeyenko, A.; et al. The Norway spruce genome sequence and conifer genome evolution. *Nature* **2013**, *497*, 579–584. [CrossRef] [PubMed]
36. *Ambrella* Genome Project. The *Amborella* Genome and the Evolution of Flowering Plants. *Science* **2013**, *342*, 1241089. [CrossRef] [PubMed]
37. The International Brachypodium Initiative. Genome sequencing and analysis of the model grass *Brachypodium distachyon*. *Nature* **2010**, *463*, 763–768. [CrossRef]
38. Ouyang, S.; Zhu, W.; Hamilton, J.; Lin, H.; Campbell, M.; Childs, K.; Thibaud-Nissen, F.; Malek, R.L.; Lee, Y.; Zheng, L.; et al. The TIGR rice Genome Annotation Resource: improvements and new features. *Nucleic Acids Res.* **2007**, *35*, D883–D887. [CrossRef]
39. Merchant, S.S.; Prochnik, S.E.; Vallon, O.; Harris, E.H.; Karpowicz, S.J.; Witman, G.B.; Terry, A.; Salamov, A.; Fritz-Laylin, L.K.; Marechal-Drouard, L.; et al. The *Chlamydomonas* Genome Reveals the Evolution of Key Animal and Plant Functions. *Science* **2007**, *318*, 245–250. [CrossRef] [PubMed]
40. Lamesch, P.; Berardini, T.Z.; Li, D.; Swarbreck, D.; Wilks, C.; Sasidharan, R.; Muller, R.; Dreher, K.; Alexander, D.L.; Garcia-Hernandez, M.; et al. The Arabidopsis Information Resource (TAIR): improved gene annotation and new tools. *Nucleic Acids Res.* **2012**, *40*, D1202–D1210. [CrossRef]
41. Myburg, A.A.; Grattapaglia, D.; Tuskan, G.A.; Hellsten, U.; Hayes, R.D.; Grimwood, J.; Jenkins, J.; Lindquist, E.; Tice, H.; Bauer, D.; et al. The genome of *Eucalyptus grandis*. *Nature* **2014**, *510*, 356–362. [CrossRef]
42. Tuskan, G.A.; Difazio, S.; Jansson, S.; Bohlmann, J.; Grigoriev, I.; Hellsten, U.; Putnam, N.; Ralph, S.; Rombauts, S.; Salamov, A.; et al. The Genome of Black Cottonwood, *Populus trichocarpa* (Torr. & Gray). *Science* **2006**, *313*, 1596–1604.
43. Nishiyama, T.; Sakayama, H.; de Vries, J.; Buschmann, H.; Saint-Marcoux, D.; Ullrich, K.K.; Haas, F.B.; Vanderstraeten, L.; Becker, D.; Lang, D.; et al. The *Chara* Genome: Secondary Complexity and Implications for Plant Terrestrialization. *Cell* **2018**, *174*, 448–464. [CrossRef]
44. Hu, B.; Jin, J.; Guo, A.-Y.; Zhang, H.; Luo, J.; Gao, G. GSDS 2.0: an upgraded gene feature visualization server. *Bioinformatics* **2015**, *31*, 1296–1297. [CrossRef]
45. Camacho, C.; Coulouris, G.; Avagyan, V.; Ma, N.; Papadopoulos, J.; Bealer, K.; Madden, T.L. BLAST+: architecture and applications. *Bmc Bioinform.* **2009**, *10*, 421. [CrossRef] [PubMed]
46. Sievers, F.; Wilm, A.; Dineen, D.; Gibson, T.J.; Karplus, K.; Li, W.; Lopez, R.; McWilliam, H.; Remmert, M.; Soding, J.; et al. Fast, scalable generation of high-quality protein multiple sequence alignments using Clustal Omega. *Mol. Syst. Biol.* **2011**, *7*, 539. [CrossRef]
47. Sievers, F.; Higgins, D.G. Clustal Omega for making accurate alignments of many protein sequences. *Protein Sci.* **2018**, *27*, 135–145. [CrossRef]
48. Thompson, J.D.; Plewniak, F.; Poch, O. A comprehensive comparison of multiple sequence alignment programs. *Nucleic Acids Res.* **1999**, *27*, 2682–2690. [CrossRef] [PubMed]

49. Hall, B.G. *Phylogenetic Trees Made Easy: A How-To Manual*, 4th ed.; Oxford University Press: Cary, NC, USA, 2011.
50. Ogden, T.H.; Rosenberg, M.S. Multiple Sequence Alignment Accuracy and Phylogenetic Inference. *Syst. Biol.* **2006**, *55*, 314–328. [CrossRef] [PubMed]
51. Kumar, S.; Stecher, G.; Tamura, K. MEGA7: Molecular Evolutionary Genetics Analysis Version 7.0 for Bigger Datasets. *Mol. Biol. Evol.* **2016**, *33*, 1870–1874. [CrossRef]
52. Le, S.Q.; Gascuel, O. An Improved General Amino Acid Replacement Matrix. *Mol. Biol. Evol.* **2008**, *25*, 1307–1320. [CrossRef]
53. Bailey, T.L.; Elkan, C. Fitting a mixture model by expectation maximization to discover motifs in biopolymers. *Proc. Int. Conf. Intell. Syst. Mol. Biol.* **1994**, *2*, 28–36. [PubMed]
54. Klepikova, A.V.; Logacheva, M.D.; Dmitriev, S.E.; Penin, A.A. RNA-seq analysis of an apical meristem time series reveals a critical point in *Arabidopsis thaliana* flower initiation. *Bmc Genom.* **2015**, *16*, 466. [CrossRef] [PubMed]
55. Klepikova, A.V.; Kasianov, A.S.; Gerasimov, E.S.; Logacheva, M.D.; Penin, A.A. A high resolution map of the *Arabidopsis thaliana* developmental transcriptome based on RNA-seq profiling. *Plant J.* **2016**, *88*, 1058–1070. [CrossRef]
56. Li, J.; Yu, M.; Geng, L.L.; Zhao, J. The fasciclin-like arabinogalactan protein gene, *FLA3*, is involved in microspore development of Arabidopsis. *Plant J.* **2010**, *64*, 482–497. [CrossRef] [PubMed]
57. Basu, D.; Tian, L.; Debrosse, T.; Poirier, E.; Emch, K.; Herock, H.; Travers, A.; Showalter, A.M. Glycosylation of a Fasciclin-Like Arabinogalactan-Protein (SOS5) Mediates Root Growth and Seed Mucilage Adherence Via a Cell Wall Receptor-Like Kinase (FEI1/FEI2) Pathway in Arabidopsis. *PLoS ONE* **2016**, *11*, e0145092. [CrossRef]
58. Johnson, K.L.; Kibble, N.A.; Bacic, A.; Schultz, C.J. A Fasciclin-Like Arabinogalactan-Protein (FLA) Mutant of *Arabidopsis thaliana*, FLA1, Shows Defects in Shoot Regeneration. *PLoS ONE* **2011**, *6*, e25154. [CrossRef]
59. Cagnola, J.I.; Dumont de Chassart, G.J.; Ibarra, S.E.; Chimenti, C.; Ricardi, M.M.; Delzer, B.; Ghiglione, H.; Zhu, T.; Otegui, M.E.; Estevez, J.M.; et al. Reduced expression of selected *FASCICLIN-LIKE ARABINOGALACTAN PROTEIN* genes associates with the abortion of kernels in field crops of *Zea mays* (maize) and of Arabidopsis seeds. *Plant Cell Env.* **2018**, *41*, 661–674. [CrossRef] [PubMed]
60. Turupcu, A.; Almohamed, W.; Oostenbrink, C.; Seifert, G.J. A speculation on the tandem fasciclin 1 repeat of FLA4 proteins in angiosperms. *Plant Signal. Behav.* **2018**, *13*, e1507403. [CrossRef]
61. Xue, H.; Veit, C.; Abas, L.; Tryfona, T.; Maresch, D.; Ricardi, M.M.; Estevez, J.M.; Strasser, R.; Seifert, G.J. *Arabidopsis thaliana* FLA4 functions as a glycan-stabilized soluble factor via its carboxy-proximal fasciclin 1 domain. *Plant J.* **2017**, *91*, 613–630. [CrossRef]
62. Moore, R.C.; Purugganan, M.D. The early stages of duplicate gene evolution. *Proc. Natl. Acad. Sci. USA* **2003**, *100*, 15682–15687. [CrossRef] [PubMed]
63. Cao, J.; Li, X. Identification and phylogenetic analysis of late embryogenesis abundant proteins family in tomato (*Solanum lycopersicum*). *Planta* **2015**, *241*, 757–772. [CrossRef]
64. Kong, H.; Landherr, L.L.; Frohlich, M.W.; Leebens-Mack, J.; Ma, H.; dePamphilis, C.W. Patterns of gene duplication in the plant *SKP1* gene family in angiosperms: evidence for multiple mechanisms of rapid gene birth. *Plant J.* **2007**, *50*, 873–885. [CrossRef] [PubMed]
65. Cannon, S.B.; Mitra, A.; Baumgarten, A.; Young, N.D.; May, G. The roles of segmental and tandem gene duplication in the evolution of large gene families in *Arabidopsis thaliana*. *Bmc Plant Biol.* **2004**, *4*, 10. [CrossRef] [PubMed]
66. Zhu, Y.; Wu, N.; Song, W.; Yin, G.; Qin, Y.; Yan, Y.; Hu, Y. Soybean (*Glycine max*) expansin gene superfamily origins: segmental and tandem duplication events followed by divergent selection among subfamilies. *Bmc Plant Biol.* **2014**, *14*, 93. [CrossRef] [PubMed]
67. Cao, J.; Li, X.; Lv, Y.; Ding, L. Comparative analysis of the phytocyanin gene family in 10 plant species: A focus on *Zea mays*. *Front. Plant Sci.* **2015**, *6*, 515. [CrossRef] [PubMed]
68. Finn, R.D.; Coggill, P.; Eberhardt, R.Y.; Eddy, S.R.; Mistry, J.; Mitchell, A.L.; Potter, S.C.; Punta, M.; Qureshi, M.; Sangrador-Vegas, A.; et al. The Pfam protein families database: towards a more sustainable future. *Nucleic Acids Res.* **2016**, *44*, D279–D285. [CrossRef]
69. Eddy, S.R. Accelerated profile HMM searches. *Plos Comput. Biol.* **2011**, *7*, e1002195. [CrossRef]

70. Petersen, T.N.; Brunak, S.; von Heijne, G.; Nielsen, H. SignalP 4.0: discriminating signal peptides from transmembrane regions. *Nat. Methods* **2011**, *8*, 785–786. [CrossRef] [PubMed]
71. Eisenhaber, B.; Wildpaner, M.; Schultz, C.J.; Borner, G.H.; Dupree, P.; Eisenhaber, F. Glycosylphosphatidylinositol Lipid Anchoring of Plant Proteins. Sensitive Prediction from Sequence- and Genome-wide Studies for Arabidopsis and Rice. *Plant Physiol.* **2003**, *133*, 1691–1701. [CrossRef] [PubMed]
72. Nicholas, K.B.; Nicholas, H.B., Jr.; Deerfield, D.W., II. GeneDoc: Analysis and visualization of genetic variation. *Embnew* **1997**, *4*, 1–4.
73. Nei, M.; Kumar, S. *Molecular Evolution and Phylogenetics*; Oxford University Press: New York, NY, USA, 2000.
74. Lee, T.H.; Tang, H.; Wang, X.; Paterson, A.H. PGDD: a database of gene and genome duplication in plants. *Nucleic Acids Res.* **2013**, *41*, D1152–D1158. [CrossRef]
75. Tang, H.; Bowers, J.E.; Wang, X.; Ming, R.; Alam, M.; Paterson, A.H. Synteny and Collinearity in Plant Genomes. *Science* **2008**, *320*, 486–488. [CrossRef] [PubMed]
76. Tang, H.; Wang, X.; Bowers, J.E.; Ming, R.; Alam, M.; Paterson, A.H. Unraveling ancient hexaploidy through multiply-aligned angiosperm gene maps. *Genome Res.* **2008**, *18*, 1944–1954. [CrossRef] [PubMed]
77. Tang, H.; Bowers, J.E.; Wang, X.; Paterson, A.H. Angiosperm genome comparisons reveal early polyploidy in the monocot lineage. *Proc. Natl. Acad. Sci. USA* **2010**, *107*, 472–477. [CrossRef] [PubMed]
78. Wang, D.; Zhang, Y.; Zhang, Z.; Zhu, J.; Yu, J. KaKs_Calculator 2.0: A Toolkit Incorporating Gamma-Series Methods and Sliding Window Strategies. *Genom. Proteom. Bioinform.* **2010**, *8*, 77–80. [CrossRef]

International Journal of
Molecular Sciences

MDPI

Review

Membrane-Bound Class III Peroxidases: Unexpected Enzymes with Exciting Functions

Sabine Lüthje [1,*] and Teresa Martinez-Cortes [2]

[1] Oxidative Stress and Plant Proteomics Group, Institute for Plant Science and Microbiology, University of Hamburg, Ohnhorststrasse 18, 22609 Hamburg, Germany
[2] Dpto de Biología Animal, Biología Vegetal y Ecología (Lab. Fisiología Vegetal), Facultad de Ciencias—Universidade da Coruña, A Zapateira s/n, 15071 A Coruña, Spain; teresa.mcortes@udc.es
* Correspondence: sabine.luethje@uni-hamburg.de Tel.: +49-(0)40-42816-340

Received: 6 August 2018; Accepted: 17 September 2018; Published: 21 September 2018

Abstract: Class III peroxidases are heme-containing proteins of the secretory pathway with a high redundance and versatile functions. Many soluble peroxidases have been characterized in great detail, whereas only a few studies exist on membrane-bound isoenzymes. Membrane localization of class III peroxidases has been demonstrated for tonoplast, plasma membrane and detergent resistant membrane fractions of different plant species. In silico analysis revealed transmembrane domains for about half of the class III peroxidases that are encoded by the maize (*Zea mays*) genome. Similar results have been found for other species like thale-cress (*Arabidopsis thaliana*), barrel medic (*Medicago truncatula*) and rice (*Oryza sativa*). Besides this, soluble peroxidases interact with tonoplast and plasma membranes by protein–protein interaction. The topology, spatiotemporal organization, molecular and biological functions of membrane-bound class III peroxidases are discussed. Besides a function in membrane protection and/or membrane repair, additional functions have been supported by experimental data and phylogenetics.

Keywords: *Arabidopsis thaliana*; Class III peroxidase; *Medicago truncatula*; microdomains; phylogenetics; plasma membrane; protein–protein interaction; *Oryza sativa*; tonoplast; *Zea mays*

---

1. Introduction

Classical secretory plant peroxidases (EC 1.11.1.7; class III peroxidases; donor: H_2O_2 oxidoreductases) are heme-containing enzymes that belong to the peroxidase-catalase superfamily [1,2]. According to PeroxiBase, as at August 2018 [3], at least 158 class III peroxidases have been identified in the maize (*Zea mays* L.) genome, 155 isoenzymes in the rice (*Oryza sativa* L.) genome, 103 isoenzymes in the barrel medic (*Medicago truncatula* GAERTN. (Gärtner, Joseph)) genome, and 75 isoenzymes in the thale-cress (*Arabidopsis thaliana* (L.) HEYNH (Heynhold, Gustav)) genome. Differences result partially from some unique peroxidase clusters in monocotyledonous plants that were not found in dicotyledonous plants [4]. Additional isoenzymes can be produced by post-transcriptional and post-translational modifications of peroxidase transcripts [5,6].

As high as the number of peroxidases in plant cells is, so are their implications in different functions various [7–9]: peroxidases are involved in cell wall modifications like lignification, suberisation, and cross-linking of hydroxyprolin–rich glycoproteins and polysaccharides. They are involved in phytohormone metabolism, senescence, and in several biosynthetic pathways, and fulfill important functions in stress-related processes [8,10–12]. Since peroxidases exhibit an almost 1000-fold higher affinity for hydrogen peroxide as catalases and their activities can be modified in the presence of different stress factors, these enzymes play a key role in the detoxification of reactive oxygen species (ROS) [13].

2. Membrane-Bound Class III Peroxidases

Evidence for membrane-bound class III peroxidases has been presented for different plant species and tissues on the protein level [14]. Guaiacol peroxidase activities have been detected at thylakoid and peroxisomal membranes, tonoplast and plasma membranes [15–20]. Class III peroxidases from tonoplast and plasma membrane (PM) have been partially purified and characterized in more detail [20–22]. Further membrane-bound peroxidases have been identified by genomic and proteomic approaches, but still lack biochemical characterization.

A membrane-bound class III peroxidase has been identified in tonoplast of Madagascar periwinkle (*Catharantus* roseus (L.) D.DON) leaves [23]. Green fluorescence protein (GFP)-fusion constructs verified localization of CroPrx01 (CrPrx1) at the inner surface of the tonoplast [21].

At the same time, class III peroxidases have been identified in highly enriched PM preparations of maize roots. At least four PM-bound peroxidases (ZmPrx01, ZmPrx66, ZmPrx70 and pmPOX2a) have been partially purified and characterized [22,24]. Depending on the state of development and oxidative stress, further peroxidases have been identified in PM of maize roots [25].

Proteomic and genomic approaches identified PM-bound class III peroxidases in different plant species and tissues; OsPrx95 has been identified in root PM of a salt sensitive rice cultivar [26]. PsPrx13 has been identified in PM of iron deficient pea (*Pisum sativum* L.) roots [27]. AtPrx64 has been shown to interact with Casparian strip formation at the PM [28,29]. AtPrx64 showed high sequence similarity to AtPrx66 as did AtPrx47 [30]. It was shown that AtPrx66 and AtPrx47 were associated with lignification of vessels, whereas AtPrx64 was associated with lignification of sclerenchyma [30,31]. A recent study demonstrated that aluminum tolerance of tobacco (*Nicotiana tabacum* L.) plants was improved by overexpression of a*tprx64* [32].

A localization of class III peroxidases in detergent resistant membranes (DRM) has been demonstrated for PM of barrel medic, maize and sugar beet (*Beta vulgaris* L.) roots [14,33,34]. MtPrx02 was identified in DRM of barrel medic [33]. A precursor of BvPrx12 has been identified in DRM of sugar beet under iron deficiency [34].

3. Structure

The structure of class III peroxidases is well conserved [6]. The proteins contain N-terminal signal peptides, binding-sites for heme and calcium, and four conserved disulfide bridges (Figure 1). Transmembrane domains were predicted for ZmPrx01, OsPrx95, AtPrx47 and MtPrx02, whereas for CroPrx01 and AtPrx64 transmembrane helices (TMH) appear unlikely.

A hypothetical model of CroPrx01 has been published [35]. Besides an N-terminal propeptide (34 amino acids) that directed the GFP-fusion construct of CroPrx01 to the endoplasmic reticulum (ER), a C-terminal extension of 23 to 25 amino acids directed the GFP-fusion constructs of the protein to the vacuole [35]. C-terminal extensions have been found for ZmPrx01 and OsPrx95 (Figure 1). In contrast to CroPrx01, these proteins were identified in highly enriched PM preparations [21,25]. Additionally, prediction of N-glycosylation sites (Table 1) suggested a localization of these enzymes at the apoplastic side.

In accordance with earlier predictions, structures of membrane-bound peroxidases have 13–21 α-helices and between two and 11 β-sheets [14,21]. Exceptions were AtPrx47 and AtPrx64, whose hypothetical structures contained no β-sheets (Figure 2). For AtPrx64, the second calcium binding-site was not conserved (Figure 1). The structures of these peroxidases may need further elucidation.

In comparison to horseradish peroxidase (1HCH1A, HRP) [36], active sites of AtPrx47, AtPrx64, MtPrx02 and ZmPrx01 were well conserved (Figure 1). In contrast, the distal His-42 of HRP was replaced by Val-70 in OsPrx95 (Figure 1). This His facilitates formation of the initial iron-peroxide complex by deprotonating the peroxide and subsequently promote cleavage of the oxygen-oxygen bond by protonating the distal oxygen [37]. In comparison to the wild type HRP, guaiacol peroxidase activity of the H42V mutant was much lower. This observation was in accordance with a ~106-fold lower formation of compound I by the mutant in comparison to the wildtype [38]. As a consequence, reduction of compound II has been more rapid than reduction of compound I. So far, biochemical characterization of OsPrx95 is lacking.

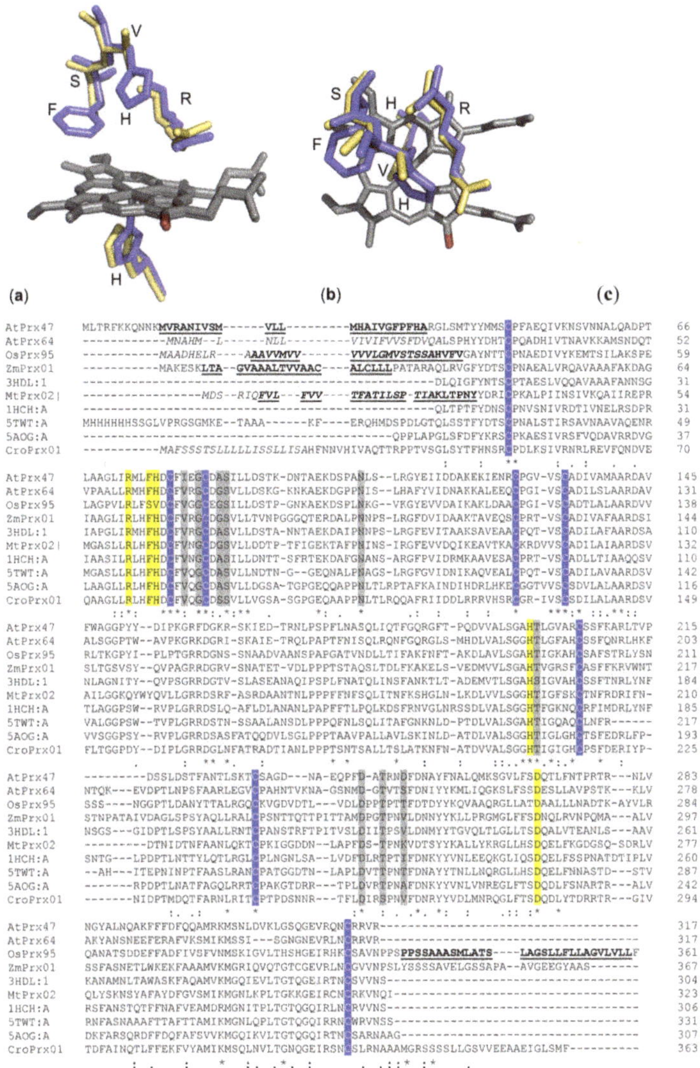

Figure 1. Active sites of horseradish peroxidase (HRP) and OsPrx95 and multiple sequence alignment of class III peroxidases. Superposition of active sites of HRP (blue) and OsPrx95 (yellow) was prepared by Phyton-enhanced molecular graphic tool (PyMOL): (**a**) site view, (**b**) top view; in HRP Arg-38 and His-42 build the distal active-site structure [39]. The heme group (grey), with δ-meso edge (red), is fixed by His-170 in the active site. (**c**) Clustal Omega (https://www.ebi.ac.uk/Tools/msa/clustalo/) was used for multiple sequence alignment of HRP [40], membrane-bound peroxidases and soluble peroxidases that were used as templates for modelling of structures shown in Figure 2 [36,41–43]. Active sites (yellow), calcium binding-sites (grey), conserved cysteine residues for formation of disulfide bridges (blue), transmembrane domains (bold and underlined), signal peptides (italic).

Table 1. Properties and physiological functions of membrane-bound class III peroxidases of arabidopsis, barrel medic, rice and maize. After identification of peroxidase sequences with trans-membrane domains [44], signal peptides [45], post translational modifications [46] and localisation [47] of proteins have been further analysed by bioinformatic tools. In case of missing experimental data on biological functions these have been predicted by a Basic Local Alignment Search Tool (BLAST) (*) search against all peroxidases (inclusive soluble isoenzymes) with highest sequence similarity and known function. ER, endoplasmatic reticulum membrane; Golgi, golgi apparatus; MoM, mitochondrial outer membrane; MS, identified by mass spectrometry [14]; MW, molecular weight; pI, point isoelectric; PM, plasma membrane; PSORT, protein subcellular localization prediction tool; Vac., vacuole. Predicted pyrrolidone carboxylic acid (PCA) modification (+).

Species	Acc. No. [1]	Protein [2]	MW [3]	pI [3]	SignalP [4]	N-Glyco [5]	NetPhos [6]	PCA [7]	PSORT [8]	Response to
A. thaliana	Q9LE15	AtPrx04	34.28	7.87	1-19	1	39	+	PM	senescence
	Q9FX85	AtPrx10	38.03	6.17	—	2	29	—	Vac	hormones and stress
	O49293	AtPrx13	36.88	5.15	—	1	23	+	ER	—
	Q9SK52	AtPrx18	35.50	5.03	1-29	2	41	—	PM	floral organ development
	O80822	AtPrx25	37.45	8.48	—	2	38	—	PM	etiolation and drought
	Q9SZB9	AtPrx47 [MS]	35.97	8.57	—	1	30	—	PM	lignification of vessels
	O81755	AtPrx48	44.79	4.99	—	2	42	—	PM	—
	P59120	AtPrx58	35.29	4.92	1-23	2	30	+	PM	—
	Q43872	AtPrx64 [MS]	34.71	9.05	1-22	3	30	—	PM	etiolation and drought
	Q96511	AtPrx69	35.68	9.47	1-23	3	29	—	ER	etiolation and cold stress
	Q9FMI7	AtPrx70	36.00	6.97	1-23	2	25	—	PM	—
M. truncatula	G7KFK6	MtPrx02 [MS]	36.02	9.20	1-23	1	27	—	PM	nitrogen starvation, wounding and pathogen
	A0A072UYJ4	MtPrx03	35.99	5.35	1-25	3	28	—	Vac	rhizobium inoculation
	G8A179	MtPrx04	35.48	8.32	1-22	5	34	—	PM	nitrogen starvation and pathogen
	A0A072URQ9	MtPrx09	35.92	6.51	1-25	6	41	+	Golgi	phosphate starvation and pathogen
	G7IC23	MtPrx13	35.26	8.01	1-25	2	22	—	PM	nodulation and pathogen
	G7KFM2	MtPrx29	35.79	9.41	1-28	1	27	+	PM	drought, wounding, nitrogen starvation, phosphate starvation, pathogen, methyl jasmonate and ultraviolet (UV) irradiation
	A0A072UJD7	MtPrx32	36.88	8.62	1-23	1	35	—	PM	nitrogen starvation and pathogen
	Q1RXM7	MtPrx34	34.38	9.11	1-22	3	32	—	Vac	drought, insect damage and pathogen
	I3S041	MtPrx37	34.90	7.15	1-22	1	30	—	Vac	drought, insect damage, phosphate starvation and pathogen
	G7IF04	MtPrx39	36.19	6.5	1-25	5	31	—	PM	pathogen
	A0A072UQ08	MtPrx40	37.48	4.73	1-28	2	33	+	PM	pathogen
	A0A072TWY1	MtPrx41	34.60	5.03	1-23	1	30	+	PM	methyl jasmonate, nodulation and pathogen
	Q1SC11	MtPrx42	36.08	9.09	1-25	1	27	+	ER	phosphate starvation and pathogen
	A0A072UHR9	MtPrx49	35.30	9.72	1-25	—	31	—	Vac	nodulation and pathogen
M. truncatula	G7I8C1	MtPrx55	34.47	9.45	1-23	2	33	—	PM	nodulation and pathogen
	G7K822	MtPrx58	36.06	6.87	1-25	2	31	+	ER	nodulation and pathogen
	G7JCW8	MtPrx66	34.64	8.21	1-19	1	27	+	PM	nitrogen starvation and pathogen
	A0A072UYL4	MtPrx74	38.49	9.28	1-29	2	28	+	PM	nodulation and pathogen
	I3T3G6	MtPrx79	36.97	8.82	—	2	48	—	PM	nodulation *
	A0A072UIE4	MtPrx81	35.06	8.14	1-25	—	30	—	Vac	pathogen
	G7JZ11	MtPrx83	41.78	5.78	1-23	—	29	—	ER	—
	G7J9S0	MtPrx84	33.82	9.17	1-26	2	44	+	PM	—
	Q1SAT8	MtPrx91	36.52	7.22	1-21	3	35	—	Vac	—
	G7LB60	MtPrx92	34.76	8.81	1-25	1	40	—	MoM	nodulation and pathogen
	G7JJA7	MtPrx93	35.84	5.79	1-23	2	36	+	ER	—
	G7JKU3	MtPrx94	46.33	9.21	—	3	49	—	Vac	methyl jasmonate
	G7JKV1	MtPrx95	38.06	6.14	—	5	36	—	PM	methyl jasmonate *
	G7J9S2	MtPrx98	33.99	8.06	1-26	3	48	+	PM	—

Table 1. *Cont.*

Species	Acc. No. [1]	Protein [2]	MW [3]	pI [3]	SignalP [4]	N-Glyco [5]	NetPhos [6]	PCA [7]	PSORT [8]	Response to
	G7INU9	MtPrx[P]113	12.74	5.56	1-24	—	10	—	ER	cadmium treatment, nodulation, nematode
	A2Q4B7	MtPrx[P]116	33.77	7.98	1-14	1	23	—	PM	infection and pathogen *
	G7JCW9	MtPrx[P]117	17.60	6.23	1-19	—	13	—	PM	nitrogen and phosphate starvation and pathogen *
O. sativa	Q5VR15	OsPrx01	34.78	6.88	1-24	1	30	—	ER	—
	Q5U1T6	OsPrx07	31.93	5.32	1-20	1	27	—	PM	—
	Q5U1T4	OsPrx09	35.50	5.67	1-11	5	31	—	PM	etiolation and drought *
	Q5U1T3	OsPrx10	36.20	5.38	—	5	34	—	Vac	—
	Q5U1T0	OsPrx13	36.18	5.16	1-21	8	18	—	PM	—
	Q6ER51	OsPrx30	34.43	7.52	1-26	1	33	+	PM	—
	Q5U1R1	OsPrx32	34.06	8.42	1-30	2	30	—	PM	pathogen *
	Q5U1Q7	OsPrx36	50.76	4.85	1-17	2	45	+	Vac	—
	Q5U1P7	OsPrx46	35.95	7.02	1-29	1	20	—	Vac	nitrogen and phosphate starvation *
	Q7XUL1	OsPrx5556	51.48	9.15	1-26	5	86	+	PM	gibberellic acid and heat *
	Q6AVZ8	OsPrx65	37.61	7.20	1-24	2	21	—	MoM	drought *
	Q8L3W2	OsPrx95 [MS]	37.58	5.63	1-28	6	38	—	PM	—
	Q6Z121	OsPrx100	32.78	5.68	1-20	2	27	—	PM	—
	Q6Z121	OsPrx101	32.78	5.68	1-20	2	27	—	PM	—
	Q8GVG6	OsPrx106-1	40.15	7.22	1-19	2	47	—	Vac	—
	A2YP47	OsPrx106-2	40.15	7.22	1-19	2	45	—	Vac	—
	Q8GVG0	OsPrx107	34.23	9.19	1-22	3	15	—	ER	nitrogen and phosphate starvation *
	Q5U1I4	OsPrx109	33.02	8.26	1-21	4	32	+	ER	anaerobic stress, brassinolide and giberellic acid treatments and nematode infection *
O. sativa	Q6Z3Y8	OsPrx116	36.05	7.69	1-19	1	29	+	MoM	pH and oxidative stress *
	Q5U1H6	OsPrx117	33.51	6.65	1-25	1	30	—	PM	pH, oxidative stress, nitrogen starvation and pathogen *
	Q6UU25	OsPrx118	36.93	4.96	1-25	2	23	—	Vac	—
	Q7XHB1	OsPrx126	35.48	4.45	1-27	1	33	+	PM	pH, oxidative stress, nematode infection and pathogen *
	Q5U1F8	OsPrx135	34.83	8.79	1-31	3	38	+	PM	—
	Q5U1F5	OsPrx138	35.66	8.40	—	2	17	—	PM	nitrogen starvation *
Z. mays	A5H8G4	ZmPrx01 [MS]	38.36	6.81	—	5	35	—	PM	cell wall modification, wounding and pathogen
	A0A1D6KUF1	ZmPrx11	35.44	5.14	1-28	3	28	+	PM	pH, oxidative stress and pathogen *
	B4FHG3	ZmPrx24 [MS]	37.82	5.9	1-25	2	36	—	PM	pH, oxidative stress, abscissic and ethylene stress, etiolation and heat shock *
	B4FD28	ZmPrx26	37.50	5.01	1-38	3	34	—	MiM	pH, drought and pathogen *
	B6T173	ZmPrx35	36.80	6.05	1-22	2	24	—	ER	—
	A0A1D6H655	ZmPrx36	33.09	4.90	1-24	2	29	+	MoM	heat and oxidative stress *
	B4FVT1	ZmPrx39	37.91	8.30	1-21	8	34	—	MoM	pathogen
	B4FJX1	ZmPrx41	35.75	4.88	—	1	30	—	PM	pH, drought and etiolation *
	A0A1D6FBJ6	ZmPrx46	38.65	6.19	1-23	1	33	—	ER	—
	Q9ZTS6	ZmPrx52	34.59	8.96	1-27	4	29	—	PM	pathogen *
	K7VQB0	ZmPrx53	46.55	4.98	1-24	1	42	+	Vac	—
	B6TU39	ZmPrx55	34.96	4.85	1-42	1	33	+	PM	cold stress
	A0A1D6IMZ0	ZmPrx56	34.57	4.35	1-25	3	36	—	PM	pathogen *
	B6T5R9	ZmPrx57	37.58	5.73	1-19	3	26	—	PM	pathogen *
	B4FH68	ZmPrx58 [MS]	36.91	6.35	1-25	1	30	—	PM	pH, oxidative stress, abscissic and ethylene stress, etiolation and heat shock *
	A5H454-1	ZmPrx66 [MS]	33.42	8.39	1-29	4	70	+	PM	drought
	A0A1D6H655	ZmPrx67	32.77	4.89	—	4	26	—	MoM	heat, drought, pathogen and oxidative stress *
	A5H452	ZmPrx70 [MS]	33.41	9.08	1-25	4	34	+	ER/PM	drought
	B4FMF8	ZmPrx71	35.71	9.64	1-28	3	30	—	PM	—
	B4F7T9	ZmPrx72	36.67	9.31	—	2	22	—	MiM	nitrogen starvation *
	B6SNF9	ZmPrx76	33.30	8.64	1-24	—	33	—	ER	drought
	B4FH35	ZmPrx77	35.71	9.64	1-32	6	20	+	ER	—
	A0A1D6IKW2	ZmPrx78	41.42	5.86	1-21	3	40	+	Vac	pH and drought *
	B4FG39	ZmPrx81	36.42	7.66	1-29	3	35	—	PM	drought
	A0A1D6E530	ZmPrx85	35.48	5.35	1-23	1	32	—	PM	pathogen

Table 1. *Cont.*

Species	Acc. No. [1]	Protein [2]	MW [3]	pI [3]	SignalP [4]	N-Glyco [5]	NetPhos [6]	PCA [7]	PSORT [8]	Response to
Z. mays	Q9ZTS8	ZmPrx86	35.51	4.37	1-21	3	34	+	PM	pathogen
	A0A1D6HQQ9	ZmPrx89	36.08	4.87	1-32	10	23	+	MiM	pH, anoxia, ethylene and gibberellic acid *
	A0A1D6HQQ8	ZmPrx94	37.24	5.82	1-29	6	30	—	PM	drought
	B6TWB1	ZmPrx102	37.11	6.00	1-26	4	24	—	Vac	pH and pathogen *
	A0A1D6IKX3	ZmPrx103	57.26	5.06	1-27	4	106	—	Vac	—
	B4FBH0	ZmPrx104	38.04	8.09	1-27	2	23	—	Vac	etiolation *
	B4FYH1	ZmPrx109	37.94	6.00	1-22	3	40	—	PM	pathogen *
	A0A1D6J1L2	ZmPrx112	41.37	7.03	1-27	6	64	—	PM	etiolation, drought and pathogen *
	A0A1D6KQI0	ZmPrx113	38.28	10.26	—	3	27	—	PM	—
	C0PPB6	ZmPrx114	32.71	5.20	1-23	5	38	+	PM	pH, drought and pathogen *
	B6U6W0	ZmPrx115	37.59	8.31	?	3	40	+	PM	etiolation, cell wall modification, wounding and pathogen *
	A0A1D6H652	ZmPrx122	34.37	6.88	1-28	2	29	+	PM	drought *
	B4FA32	ZmPrx123	34.01	6.50	1-25	1	25	+	PM	etiolation, abscisic acid and pathogen *
	A0A1D6N9N5	ZmPrx124	37.42	4.84	1-23	2	25	—	PM	pH and sulfur deficiency *
	K7TMB0	ZmPrx128	35.80	8.05	1-36	3	30	—	MoM	—
	K7U151	ZmPrx129	33.75	6.80	1-26	1	20	—	PM	—
	A0A1D6JNY2	ZmPrx132	46.64	4.99	1-22	1	40	+	Vac	—
	K7VFH6	ZmPrx135	35.67	4.98	1-31	—	22	—	ER	abscisic acid and heat *
	C4IZ20	ZmPrx136	34.48	8.45	1-22	2	21	—	PM	etiolation, drought and pathogen *
	A0A1D6JF04	ZmPrx140	37.88	8.88	1-21	3	31	—	PM	pH *

[1] http://www.uniprot.org/; [2] http://peroxibase.toulouse.inra.fr/; [3] https://web.expasy.org/protparam/; [4] http://www.cbs.dtu.dk/services/SignalP/; [5] http://www.cbs.dtu.dk/services/NetNGlyc/; [6] http://www.cbs.dtu.dk/services/NetPhos/; [7] https://prosite.expasy.org/; [8] http://psort1.hgc.jp/form.html.

In silico analyzes of putative membrane-bound peroxidases revealed at least one glycosylation-site with the exception of some endoplasmatic reticulum (ER) and vacuolar peroxidases (Table 1). These membrane-bound peroxidases were predicted to be highly phosphorylated. Glycosylation of class III peroxidases seem to be necessary for protein folding, stability and catalytic activity [48]. Glycosylation is necessary for activity of PM-bound peroxidases [22]. Phosphorylation may change properties of the enzymes and be involved in protein-protein interaction. These predictions have to be verified by experimental data.

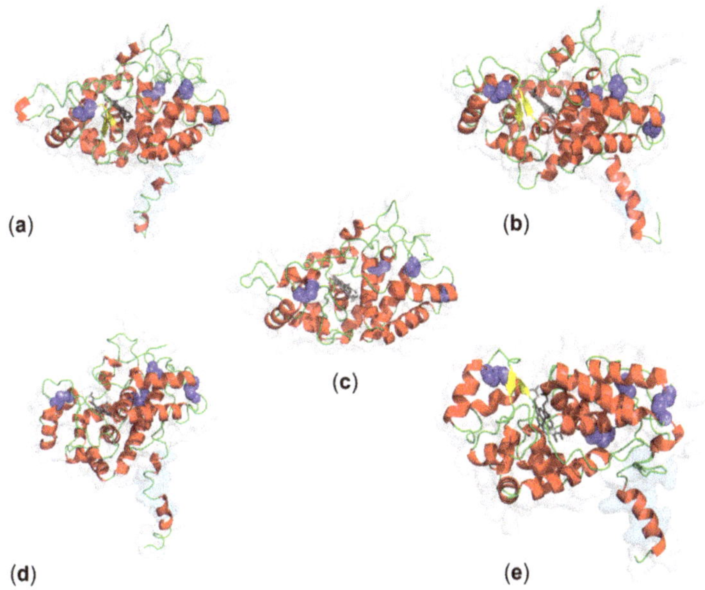

Figure 2. Putative tertiary structures of PM-bound class III peroxidases. Structures were predicted by SWISS-MODEL for (**a**) ZmPrx01 (3hdlA, 1 TMH, root PM), (**b**) OsPrx95 (3hdl1A, 3 TMH, root PM), (**c**) AtPrx64 (3hdl1A, 0 TMH, PM), (**d**) AtPrx47 (5twt, 1 TMH, PM), and (**e**) MtPrx02 (5twt1A, 1 TMH, root tip DRM). N-terminal amino acid sequences containing transmembrane domains (marine) were modeled by ab initio with Phyre2 in intensive mode [49]. Ligand binding-sites were predicted by 3DLigandSite [50]. The hypothetical models were prepared by PyMOL (http://pymol.org/) and present the structural characteristics of class III peroxidases. Colors indicate secondary structures: α-helix (red), β-sheets (yellow), loops (green), disulfides (blue), and haem group (grey).

4. Topology and Spatiotemporal Organization

All class III peroxidases appear to have a cleavable signal peptide in common. In silico analyzes of 142 class III peroxidase sequences from maize predicted putative signal peptides, transmembrane domains and the localization of these proteins. This approach revealed a ratio of 53% soluble to 47% membrane-bound isoenzymes [14]. Among the membrane proteins, signals for localization at the endoplasmatic reticulum (ER, 42%), PM (55%), Golgi (2%) and tonoplast (1%) were predicted.

A comparison between rice, barrel medic, arabidopsis and maize showed similar results about the number of isoenzymes with putative transmembrane helices, 51% in rice ($n = 78$), 49% in maize ($n = 77$), 35% in barrel medic ($n = 36$) and 44% in arabidopsis ($n = 33$). The protein subcellular localization prediction tool (PSORT) indicated a PM localization for at least eight class III peroxidases in arabidopsis, 17 in barrel medic, 13 in rice and 27 in maize (Table 1). Additionally, peroxidases were predicted for ER, Golgi, mitochondria outer (MoM) and inner (MiM) membranes. A vacuolar localization was predicted for seven peroxidases in rice and barrel medic, for six in maize, and for one in arabidopsis. A possible localization of these vacuolar peroxidases at the tonoplast will need further proof. For maize roots, peroxidases have been detected at the tonoplast [51].

The properties of PM-bound peroxidases from maize and the identification of class III peroxidases in DRM suggested a tight interaction with the membrane [14,22,24,33,34]. According to membrane-bound ascorbate peroxidases that interact with the membrane via a C-terminal membrane-spanning segment [16,52–59], PM-bound class III peroxidases appeared to be anchored by an uncleavable N-terminal signal peptide with a transmembrane helix. Beyond that, proteomic approaches suggest protein-protein interaction [60].

4.1. *Transmembrane Spanning Domains*

In general, N-terminal signal peptides are cleaved off during maturation of soluble secretory peroxidases from eukaryotic organisms. In many cases, the resulting N-terminal positioned glutamine is catalyzed to a cyclic molecule, the so-called pyrrolidone carboxylic acid (PCA) [61]. Prosite predicted such a modification for 32% of the peroxidases shown in Table 1 [62]. Four (OsPrx36, ZmPrx53, ZmPrx78, ZmPrx132) out of 21 vacuolar peroxidases are modified this way potentially. Pyrrolidone carboxylic acid appear absent in CroPrx01 [21] and vacuolar peroxidases from arabidopsis and barrel medic. One third of the putative PM-bound peroxidases showed an N-terminal glutamine that may be modified after cleavage of the signal peptide [21].

Three transmembrane helices (TMH) were predicted for OsPrx95 by different prediction tools [14]. PSORT identified a cleavable signal peptide at the N-terminus of OsPrx95 and predicted a type I membrane protein (TMH 345...361, C-terminus) with a localization at the PM (Table 1). This prediction was in accordance with the identification of the protein in the PM [26]. In contrast, the hypothetical model showed a transmembrane domain at the N-terminus (Figure 2). All other peroxidases from Table 1 have one TMH at the N-terminus that is overlapping with the ER signal peptide. Signal peptides of membrane-bound proteins can remain and, due to their hydrophobicity, function as an N-terminal transmembrane domain. Actual search algorithms may be not able to make exact predictions of this possibility yet. So far, crystal structures of membrane-bound peroxidases have not been published.

In accordance with the biochemical classification and with the identification of ZmPrx01, ZmPrx66, and ZmPrx70 at the PM, five of $n = 5$ prediction programs detected N-terminal signal peptides in their amino acid sequences [22]. PSORT even predicted a non-cleavable signal peptide in ZmPrx70 and a localization of the enzyme at the PM or ER membrane (Table 1). The hypothesis of a transmembrane-spanning domain was supported by experimental data: (i) non of the maize peroxidases could be washed off from the PM by high salt concentrations or other treatments [22,24,63]; (ii) peroxidases were found in DRM of washed PM [14]; (iii) recombinant peroxidases were localized in the yeast membrane after heterologous expression of their full length amino acid sequences [64].

4.2. *Protein–Protein Interaction*

Molecular masses of ZmPrx01 (70 kDa) and pmPOX2a (155 kDa) suggested a putative protein–protein interaction [22,24]. Analyzes of the maize PM proteome by high resolution clear native electrophoresis (hrCNE) pointed to the participation of a PM-bound guaiacol peroxidase to a putative high molecular mass protein complex of 1092 kDa [60]. Interaction of AtPrx64 with PM-bound Casparian strip domain proteins (CASPs) was demonstrated [28]. For tonoplast, interaction of CroPrx01 with arabinogalactan proteins (AGPs) was suggested [35].

Protein–protein interaction may be supported by post-translational modifications (Table 1). Phosphorylation may change the properties of peroxidases and facilitate protein–protein interaction, although this hypothesis has to be verified.

4.3. *Spatiotemporal Organization*

The proteome of PM microdomains (defined by DRM) is involved in signaling and response to biotic and abiotic stress, cellular trafficking, cell wall synthesis and degradation, and metabolism [65,66].

Evidence for a localization of class III peroxidases in DRM has been presented [14,33,34]. As shown in Figure 3, microcompartimentation allows PM-bound peroxidases to co-localize with ROS producing and detoxifying enzymes in the membrane [25]. Thus PM-bound peroxidases may probably not only detoxify H_2O_2 directly at the site of origin to ensure the optimal protection of the membrane, but could also protect specific functional regions of the PM [14,25] and fulfill specific functions in dependence on their substrates.

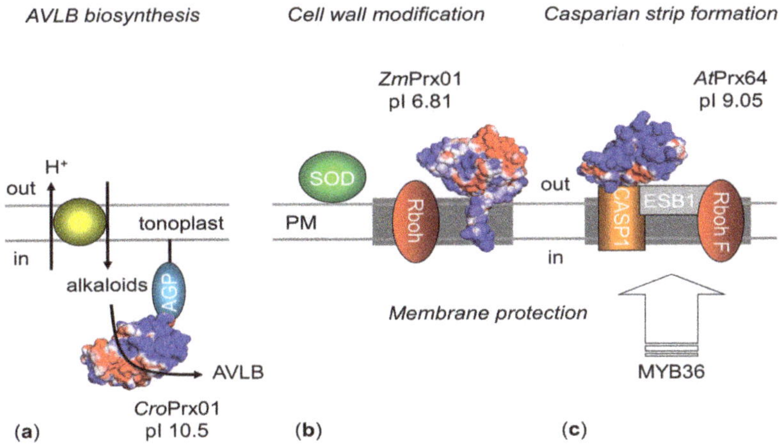

Figure 3. Membrane interaction and spatiotemporal organization of CroPrx01, ZmPrx01 and AtPrx64. (**a**) Soluble CroPrx01 interacts via an arabinogalactan protein (AGP, blue) with the inner surface of the tonoplast [35,67]. For α-3′,4′-anhydrovinblastine (AVLB) synthesis, the complex anchored next to a proton driven alkaloid antiporter (yellow ellipse) [68,69], (**b**) ZmPrx01 is anchored in the PM by a transmembrane-spanning domain with its active site at the out-side [22]. For cell wall modification, the enzyme co-localized with Rboh (ruby colored) in microdomains. Simultaneously, superoxide dismutase (SOD, green) and ZmPrx01 regulate ROS levels at the PM [14,70]. (**c**) Soluble AtPrx64 interact with a transmembrane Casparian strip protein (CASP1, orange) at the PM [28,29,71,72]. A dirigent-like protein (EBS1, grey) facilitates co-localization of the complex with RbohF (ruby colored) in sclerenchyma cells [72]. Casparian strip formation is regulated by MYB36 [29]. For peroxidases solvent accessible surface charges are shown: negative (red); neutral (white), positive (blue).

A co-localization of AtPrx64 with respiratory burst oxidase homolog F (RbohF) has been shown [28]. Besides AtPrx64, RbohF, and a dirigent-like protein (Enhanced Suberin 1, ESB1) were recruited by CASPs to assemble the lignin polymerization machinery [29]. It was shown that this process is regulated by the transcription factor MYB36.

The pattern of CroPrx01:GFP-fusion constructs pointed to localization in mircodomains. A co-localization with AGPs was postulated at the tonoplast [35]. Regulation of CroPrx01 functions were discussed in dependence on the specific localization/orientation.

5. Function

Functional analysis and biochemical characterization of membrane-bound class III peroxidases are still fragmentary. It is very difficult to identify the exact function(s) of plant peroxidases because of (i) the huge amount of similar isoenzymes, (ii) the broad substrate specificity, (iii) the high number of possible functions, and (iv) the ability of other isoenzymes to compensate the absence of an enzyme in knock-out experiments [11,73]. Progress has been made by molecular biological approaches in combination with biochemical characterization that resolve some of these problems [74].

5.1. Molecular Function

Class III peroxidases are involved in production and scavenging of ROS [7,8,75]. Various substrates are oxidized by the peroxidative cycle that play important functions in polymerization of cell wall components, auxin metabolism, and NAD(P)H oxidation via a non-catalytic reaction. Superoxide anion production via this reaction is immediately converted to hydrogen peroxide and molecular oxygen either spontaneously or through superoxide dismutase. The hydroxylic cycle can produce ROS. Both cycles control hydrogen peroxide levels of cells [25]. A function of hydrogen peroxide in redox regulation and signaling has been demonstrated [76,77].

Substrates

The PM is the first target of a stressor. Large amounts of hydrogen peroxide are produced at the PM due to cellular processes, as a response to stress factors, or to external sources in plant–pathogen interactions [78–81]. This oxidative stress causes lipid peroxidation and changes in membrane permeability [82–84]. PM-bound peroxidases were suggested to be substantially involved in the detoxification of the cell and/or repair of the PM after oxidative damage [25]. They probably play an important role in maintaining the cell functions under different stress conditions [22].

Class III peroxidases oxidize a wide variety of small phenolic compounds, including monolignols, hydoxycinnamic acids, dimeric alkaloids and others [85,86]. For the docking study shown in Table 2, several of these substrates were tested: (i) monolignols such as coniferyl, sinapyl, and *p*-coumaryl alcohols as precursors of lignin [87,88]; (ii) hydoxycinnamic acids like ferulic acid, caffeic acid and *p*-coumaric acid as precursors of suberin [89]; (iii) L-3,4-dihydroxyphenylalanine (L-DOPA) as precursor of various alkaloids, catecholamines, and melanin [90,91]; and (iv) other substrates like ascorbic acid, indol-3-acetic acid (IAA) and nicotinamide adenine dinucleotides (NAD(P)H). Auxins like IAA have a crucial role in the cell elongation process [92] and can be accumulated in vacuoles [93].

The docking analysis revealed different substrate affinities for the peroxidases in dependence on their localization and biological function. According to their classification, the peroxidases showed low affinity against ascorbic acid (Table 2).

PM-bound peroxidases of arabidopsis, barrel medic or rice (Figure 1), have not been characterized biochemically yet. For AtPrx64, the substrate affinity decreased in the order NADPH > IAA > NADH > L-DOPA > cinnamyl alcohol > sinapyl alcohol > ferulic acid (Table 2). Gene expression of AtPrx64 indicates that this peroxidase is crucial to lignin deposition in the Casparian strip of endodermis cell walls in roots [28,29]. This observation correlates with affinities of the hypothetical structure of the enzyme against alcohols such as cinnamyl and sinapyl (Table 2).

The affinity of AtPrx47 decreased from NADPH > IAA > NADH > ferulic acid > cinnamyl alcohol. The docking analyzes revealed that L-DOPA appears not to be a substrate for AtPrx47. These results supported a function of AtPrx47 in the lignification of vessels [30].

PM-bound MtPrx02 showed a high affinity for substrates related to cell wall modification, such as IAA, ferulic acid, cinnamyl and sinapyl alcohol (Table 2). This observation confirmed experimental data with this peroxidase [94]. Similar results were found for OsPrx95 that showed high affinity with substrates involved in cell wall modification such as IAA and cinnamyl alcohol (Table 2).

Both, CroPrx01 and AtPrx64, revealed similar affinities with L-DOPA, whereas the other peroxidases showed no reaction with this substrate. This observation excluded a localization of AtPrx47, MtPrx02, OsPrx95 or ZmPrx01 at the tonoplast or in the vacuole. As mentioned above, the structure of AtPrx64 may need further elucidation and results from the docking study have to be verified by biochemical characterization of the native enzyme and/or recombinant polypeptide. Biochemical differences between native proteins and recombinant polypeptides may be possible due to a distinct degree of glycosylation.

Table 2. Docking results of multiple ligand molecules by Firedock (http://bioinfo3d.cs.tau.ac.il/FireDock/). Putative structures of the membrane-bound peroxidases CroPrx01 (5aog.1.A), AtPrx64 (3hdl.1.A), AtPrx47 (5twt.1.A), MtPrx02 (5twt.1.A), OsPrx95 (3hdl.1A) and ZmPrx01 (3hdl.1.A) were modeled by SWISS-MODEL (https://swissmodel.expasy.org/). Horseradish peroxidase (HRP, 1hch) was used for comparison.

Substrate	CroPrx01	AtPrx64	AtPrx47	MtPrx02	OsPrx95	ZmPrx01	HRP
ascorbic acid	−5.47	−5.39	−5.94	−5.48	−5.56	−5.29	−5.88
L-DOPA [1]	−8.23	−8.39	0.24	12.62	7.11	12.34	−0.34
indole acetic acid	−15.41	−13.66	−12.95	−17.24	−13.05	−13.6	−11.79
NADH [2]	−9.26	−8.6	−12.82	−12.42	−13.32	−9.89	−6.29
NADPH [3]	−12.93	−13.69	−13.78	−11.69	−11.95	−12.49	−5.44
cinnamyl alcohol	−7.95	−7.98	−7.67	−8.74	−7.67	−6.39	−5.85
coniferyl alcohol	−4.56	−4.33	−3.77	−3.73	−4.85	−4.54	−3.81
sinalpyl alcohol	−7.31	−7.88	−5.23	−8.02	−6.51	−7.26	−3.35
ferulic acid	−6.8	−7.18	−8.14	−9.04	−6.11	−5.22	−6.86
caffeic acid	−5.35	−5.34	−5.06	−5.98	−5.91	−4.88	−2.07
p-coumaric acid	−4.96	−5.77	−5.32	−5.46	−6.08	−6.4	−5.54

[1] DOPA, L-3,4-Dihydroxyphenylalanin; [2] NADH, Nicotinamide adenine dinucleotide; [3] NADPH, Nicotinamide adenine dinucleotide phosphate.

Cinnamyl alcohol is easily oxidized by most of the peroxidases, but coniferyl and sinapyl alcohol seem to be poor substrates for many of these enzymes. Thus even during lignin synthesis, sinapyl alcohol dehydrogenation may be mediated by other phenolic radicals [95,96].

5.2. Biological Functions

The expression of peroxidase enzymes is regulated and induced by several stressors, such as pathogens, temperature, nutrient starvation or environmental changes [74]. Regulation of peroxidases is strongly effected by ROS levels in the plants. These ROS levels are linkups to the stressors, related besides the self-regulation of the plant [97].

Although class III peroxidases are highly conserved, the high similarity in protein structure has no simple correlation with the functions of these enzymes. Nevertheless, the structural features of class III peroxidases, like distribution of surface charges, glycosylation patterns or even the regulation of the enzymes, seem to be important in the classification by functions of the enzymes [74]. Another feature that was used to characterize peroxidases is the isoelectric point (pI) [6]. So far, little or no information can be found on the correlation between pI and biological functions.

Both, alkaline and acidic stress, generate an increase of ROS in plant tissues [98–100]. Those changes in soil pH and apoplast have a significant effect on the regulation of peroxidases that are linked to ROS generation and scavenging. On the one hand, several studies showed optima in the range between pH 4.5 to 6.5 for peroxidase activity [101,102]. On the other hand, pH is crucial for the stability of peroxidases [103].

Evaluation of Table 1 revealed several membrane peroxidases that respond to pH in rice ($n = 2$) and maize ($n = 11$), but none in arabidopsis or barrel medic. In maize, a localization of these isoenzymes was predicted for PM ($n = 7$), vacuole ($n = 2$) and mitochondria inner membrane ($n = 2$). In rice, two isoenzymes were predicted for PM. All these peroxidases were anionic with the exception of ZmPrx140 (Table 1).

A proteomic approach suggested specific functions of ZmPrx01, ZmPrx66, ZmPrx70 and pmPOX2a in response to oxidative stress [63]. Both, substrate affinity and specific activities make a function of ZmPrx01 in lignification most likely. Several putative cis-regulatory elements were identified by partial gene analysis of ZmPrx70 that suggested a regulation of its gene by wounding, methyl jasmonate, salicylic acid, and elicitors [22]. In addition, a regulation of the gene by oxidative stress was suggested by seven successive elements with sequence similarity to the consensus sequence of antioxidant-responsive elements (ARE) at the 5′-untranslated region of ZmPrx70 [63,104].

Recent studies showed that high-light conditions can increase phenol and peroxidase levels and by that the hydrogen peroxide scavenging capacity of the vacuole. In fact it was proposed that the vacuolar couple CroPrx01/secondary metabolites represent an important sink/buffer of hydrogen peroxide in green plant cells [105].

5.2.1. Orthologous

According to Welinder et al. [61], truly orthologous of peroxidases should have sequence identities greater than 90%. For the peroxidases discussed above, orthologous were not found for the species investigated. Sequence similarities above 70% might indicate related physicochemical properties and physiological functions.

Homologous of AtPrx47 appear not to exist in arabidopsis, highest sequence similarity was 47% (Table 3). For AtPrx47, a function in lignification of vessels has been shown during plant development [29,30]. This observation was supported by the docking analyzes (Table 2). Several proteins with sequence similarities above 80% to AtPrx47 were found in the Brassicaceae family, e.g., AruPrx47 (94.9%) from horseradish (*Armoracia rusticana* G.GAERTN., B.MEY (Bernhard. Meyer). and SCHERB. (Scherbius, Johannes)). Sequence similarities above 70% were indicated for MtPrx19 (G7IJT4) and CroPrx09. However, MtPrx19 was suggested to have a function in response to pathogens, whereas the function of CroPrx09 is unkown. For maize and rice, sequence similarities were below 60%. In contrast to AtPrx47, none of the putative orthologous appear to have a transmembrane-spanning domain or a localization at the PM.

Table 3. Homologous and most similar sequences of membrane-bound class III peroxidases in other species. Sequence identity is given in brackets. Protein names are given according to PeroxiBase.

Protein	*Z. mays*	*O. sativa*	*M. truncatula*	*A. thaliana*	*C. roserus*
CroPrx01	*ZmPrx16 (55%)*	*OsPrx23 (53%)*	*MtPrx48 (54%)*	*AtPrx12 (59%)*	*CroPrx03 (84%)*
AtPrx64	*ZmPrx136 (65%)*	*OsPrx32 (60%)*	*MtPrx55 (72%)*	*AtPrx66(52%)*	*CroPrx40 (72%)* *CroPrx43 (72%)*
AtPrx47	*ZmPrx15 (57%)*	*OsPrx117 (60%)*	*MtPrx19 (71%)*	*AtPrx66 (47%)* *AtPrx64 (46%)*	*CroPrx09 (75%)*
MtPrx02	*ZmPrx120 (52%)*	*OsPrx40 (59%)*	*MtPrx70 (96%)*	*AtPrx52 (46%)*	*CroPrx04 (55%)*
OsPrx95	*ZmPrx94 (73%)*	*OsPrx97 (50%)* *OsPrx134 (50%)*	*MtPrx07 (43%)*	*AtPrx56 (46%)*	*CroPrx49 (48%)* *CroPrx59 (47%)*
ZmPrx1	*ZmPrx101 (68%)*	*OsPrx12 (74%)*	*MtPrx07(52%)*	*AtPrx39 (46%)*	*CroPrx14 (48%)*

Similar results have been found for the tonoplast-bound CroPrx01. Sequence similarity of 84% was found between CroPrx01 and CroPrx03 (B2G335), but in maize, rice, barrel-medic or arabidopsis, sequence similarities were below 60%. CroPrx03 is a soluble apoplastic peroxidase that is expressed in stem and flower tissues [106]. The transcript was down-regulated under salt and dehydration stress.

For the barrel medic MtPrx02, MtPrx70 (G7KFK8) was found as homologous by sequence similarity (Table 3). Indeed this peroxidase responded to nitrogen starvation and pathogen infection. In contrast to MtPrx02, MtPrx70 respond to mycorrhiza and appears to be a soluble peroxidase. PSORT predicted localization outside the cell and the enzyme does not contain a transmembrane helix. Arabidopsis, rice and maize showed similarities below 60% (Table 3).

True orthologous of OsPrx95 were not indicated in maize, barrel medic or arabidopsis. Highest sequence similarity to OsPrx95 was found for ZmPrx94 (A0A1D6HQQ8). Both proteins were predicted at the PM (Table 1). Salt stress caused a decrease in OsPrx95 abundance [27]. In contrast, ZmPrx94 was induced by salinity. The lignin biosynthesis pathway altered significantly under salt stress. For example, cell wall thickness of vessels in sorghum [107] and thickness of Casparian strip in maize roots [108] increased during acclimation to salinity.

Orthologous of ZmPrx01 were not found in rice, Madagascar periwinkle, barrel medic or arabidopsis. In maize, ZmPx101 (B4FU88) and ZmPrx49 (B4FY83) showed highest sequence similarity to ZmPrx01. Both proteins appear to be soluble peroxidases. ZmPrx101 is induced by pathogens, whereas the function of ZmPrx49 is unknown.

Similar results were found by comparison of all ER, PM and vacuolar peroxidases from arabidopsis, barrel medic, rice and maize. Among ER-bound peroxidases, the highest sequence identity was between ZmPrx135 and MtPrx58 (47.33%), the lowest between ZmPrx46 and AtPrx69 (37.15%). In total, 45 putative PM-bound peroxidases were found. The highest sequence identities were found for ZmPrx41 and ZmPrx55 (97.95%) > ZmPrx66 and ZmPrx114 (80.89%) > OsPrx135 and ZmPrx115 (80.43%) > OsPrx30 and ZmPrx123 (79.88%), but for most of them sequence identities were below 60%.

ZmPrx41 responded to pH, drought and etiolation, whereas ZmPrx55 appears to have a function in cold stress. ZmPrx66 showed response to drought, ZmPrx114 appears to have a function in drought stress, but responded also to pathogens and pH (Table 1). ZmPrx115 was affected by etiolation, and is involved in cell wall modification whereas the function of OsPrx135 is unkown.

For vacuolar peroxidases sequence identities were below 60% with the exception of ZmPrx53 and ZmPrx132 (86.28%). Functions of these peroxidases are unknown.

5.2.2. Phylogeny

A phylogenetic tree was performed to obtain further clues to the biological functions of membrane-bound peroxidases in maize, rice, barrel medic and arabidopsis (Figure 4). The phylogenetic tree was constructed using the maximum likelihood method based on the Jones, Taylor, Thornton (JTT) matrix-based model [99]. The analysis was done with amino acid sequences of peroxidases from Table 1. In total 110 sequences were analyzed, Eleven from arabidopsis, 31 from barrel medic, 22 from rice, 45 from maize and one from the liverwort *Marchantia polymorpha* (MpPrx92). The tree was rooted to MpPrx92. The amino acid sequence reveals no signal peptide, it has a prediction for 1 TMH and for a localization of the protein at the PM. Phylogenetic analysis allows identifying the evolutionary conservation and divergence of these enzymes from the common ancestor MpPrx92.

The tree was divided into six major groups (A to F). The first group, which clusters peroxidases that respond to various stresses, is subdivided in four clusters (A1 to A4). Cluster A1 summarized peroxidases that were characterized to work against several stresses, like drought (e.g., MtPrx29, ZmPrx66 and ZmPrx70), cold (ZmPrx55), pathogens (e.g., MtPrx02, MtPrx32 and ZmPrx86) and even hormones (e.g., AtPrx10 and MtPrx94). Cluster A2 contains only two PM peroxidases that participate in oxidative stress (OsPrx117) and lignification of vessels (AtPrx47). In fact the similarity between the two peroxidases is remarkable (Table 3), but the similarities of biological functions need further proofs. In group A3, peroxidases clustered together that are involved in reactions against pathogens (e.g., AtPrx64, MtPrx55 and MtPrx92) and nodulation (e.g., MtPrx13, MtPrx55 and MtPrx92). Nodulation is characteristic for legumes like barrel medic [109]. Peroxidases from other species in this cluster work most probably in defense against pathogens. Finally, the cluster A4 is formed exclusively by PM peroxidases of maize that are involved against stresses like drought (ZmPrx81) and defense against pathogens (ZmPrx85).

Group B is subdivided into two clusters but there is not so much known about this group.

Group C is subdivided in five clusters with different responses to abiotic stress. Group C1 contains peroxidases that were not characterized experimentally. A BLAST search suggests a function in response to pH changes, such as soil acidification. This group is formed only by maize peroxidases, with the exception of OsPrx95 from rice. Maize and rice are two of the bigger crops in the world, and frequently the growing of these plants is related to soil conditions. Low pH causes less growing of the plant and even death. So it is very interesting to find peroxidases in these two species that might be involved in this kind of stress. Additionally, this stress can be related to other stresses, like nutrient starvation or drought [112]. The C2 cluster is build by two PM peroxidases that are involved in plant development. AtPrx18 is related to the development of the floral organ and OsPrx5556 could be involved in the response to hormones and heat.

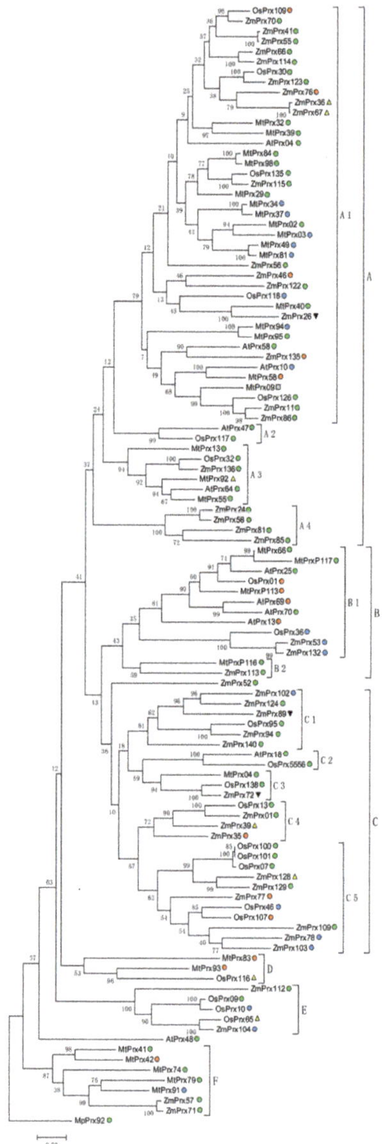

Figure 4. Phylogenetic analysis of peroxidases from Table 1 and MpPrx92 by the maximum likelihood method. The evolutionary history was inferred by using the maximum likelihood method based on the Jones, Taylor, Thornton (JTT) matrix-based model [110]. The tree with the highest log likelihood (-58984.24) is shown. The percentage of trees in which the associated taxa clustered together is shown next to the branches. Initial tree(s) for the heuristic search were obtained automatically by applying Neighbor-Join and BioNJ algorithms to a matrix of pairwise distances estimated using a JTT model, and then selecting the topology with superior log likelihood value. The tree is drawn to scale, with branch lengths measured in the number of substitutions per site. The analysis involved 110 amino acid sequences. There were a total of 841 positions in the final dataset. Evolutionary analyses were conducted by Molecular Evolutionary Genetics Analysis (MEGA7) [111]. ●, ER; ●, PM; ●, Vac; ■ Golgi; ▲, MoM; ▼, MiM.

Peroxidases from group C3 seem to be involved in nitrogen starvation response. Three peroxidases of three different species, MtPrx04, OsPrx138 and ZmPrx72, build this cluster and, despite the fact that only one of these peroxidases was studied (MtPrx04), the similarity between them is so high that they may have the same function. This could mean that these peroxidases have a common origin to protect the plant against nutrient deficiency. Cluster C4 is build by peroxidases that are involved in processes like cell wall modification (e.g., ZmPrx01). Finally, group C5 contains peroxidases that are not characterized. A BLAST search suggested a function in response to different abiotic stresses like pH and drought (ZmPrx78) or nitrogen and phosphate starvation (OsPrx46 and OsPrx107).

Group D is formed by three peroxidases from barrel medic and rice that are not characterized.

Group E include mainly peroxidases that are involved in etiolation and are regulated by light and dark (e.g., ZmPrx104 and ZmPrx112). However, these functions are only theoretical because none of these peroxidases were characterized experimentally.

Finally, group F is involved mainly in nodulation (e.g., MtPrx41, MtPrx42 and MtPrx74). In accordance with this specialization of legumes, the majority of peroxidases were from barrel medic. Two peroxidases were from maize (ZmPrx57 and ZmPrx71), and these were separated from the others. The genes showed distinct expression patterns: *zmprx57* has been found in immature leaf, whereas *zmprx71* has been expressed in primary root [113]. Both proteins have a prediction for PM localization. At least one of them may respond in defense against pathogens. This function can be easily related to the nodulation in barrel medic plants. However, biochemical characterization of the maize peroxidases is still missing.

6. Conclusions

Molecular biodiversity of membrane-bound class III peroxidases appears to be higher then expected. Although plants evolved the same molecular mechanisms, true orthologues of membrane-bound peroxidases were not identified in the four species investigated. Besides soluble class III peroxidases that interact with the membrane by protein–protein interaction, type I membrane proteins may exist. Membrane-bound peroxidases are multifunctional enzymes that fulfill essential functions in plant development and stress response. Microcompartmentation and co-localization of PM-bound peroxidases with ROS-producing and detoxifying enzymes may probably not only detoxify hydrogen peroxide directly at the site of origin, but could also protect specific functional regions of the PM and fulfill specific functions. Biochemical characterization of membrane-bound peroxidases is still fragmentary and will need further elucidation. Some of these peroxidases may have the potential for marker-assisted breeding.

Author Contributions: S.L. designed the concept and structure of the paper, she prepared Figures 1–3, Tables 1 and 3, and wrote the main text. T.M.-C. performed the docking analyzes and the phylogenetic tree, prepared Figure 4, Table 2 and wrote the interpretation of phylogenetics. S.L. and T.M.-C. discussed and corrected the manuscript.

Acknowledgments: Financial support was given to S.L. by a research grant from DFG (DFG Lu-668/4-4) and to T.M.C. by a post doctoral grant (ED481B 2016/221-0) from the Xunta de Galicia, Spain. Funds for covering the costs to publish in open access were not available.

Conflicts of Interest: The authors declare no conflict of interest.

Abbreviations

AGP	Arabinogalactan protein
ARE	Antioxidant-responsive elements
AtPrx47	Arabidopsis thaliana peroxidase 47
AtPrx64	AtPer64, Arabidopsis thaliana peroxidase 64
AVLB	-3′,4′-anhydrovinblastine
BLAST	Basic local alignment search tool
BvPrx12	Beta vulgaris peroxidase 12
CASPs	Casparian strip domain proteins

CroPrx01	CrPrx1; *Catharanthus roseus* peroxidase 1
DRM	Detergent resistant membrane
ER	Endoplasmic reticulum
ESB1	Enhanced Suberin 1
GAERTN.	Gärtner, Joseph
GFP	Green fluorescence protein
hrCNE	High resolution clear native electrophoresis
HEYNH.	Heynhold, Gustav,
HRP	Horseradish peroxidase
IAA	Indol-3-acetic acid
L-DOPA	L-3,4-dihydroxyphenylalanine
MEGA7	Molecular Evolutionary Genetics Analysis Version 7.0
B.MAY.	Bernhard. Meyer
MiM	Mitochondrial inner membrane
MoM	Mitochondrial outer membrane
MS	Mass spectrometry
MtPrx02	Medicago truncatula peroxidase 2
MW	Molecular weight
MYB36	MYB domain protein 36
NADH	Nicotinamide adenine dinucleotide
NADPH	Nicotinamide adenine dinucleotide phosphate
OsPrx95	*Oryza sativa* peroxidase 95
PCA	Pyrrolidone carboxylic acid
pI	Point isoelectric
PM	Plasma membrane
pmPOX2a	Plasma membrane peroxidase 2a
PSORT	protein subcellular localization prediction tool
PsPrx13	*Pisum sativum* peroxidase 13
PyMOL	Python-enhanced molecular graphics tool
Rboh	Respiratory burst oxidase homolouge
ROS	Reactive oxygen species
SCHERB.	Scherbius, Johannes
SOD	Superoxide dismutase
TMH	Transmembrane helix
TMHMM	Trans Membrane Helix Markov Model
Vac	Vacuole
ZmPrx01	*Zea mays* peroxidase 1

References

1. Welinder, K.G. Superfamily of plant, fungal and bacterial peroxidases. *Curr. Opin. Struct. Biol.* **1992**, *2*, 388–393. [CrossRef]
2. Zámocký, M.; Hofbauer, S.; Schaffner, I.; Gasselhuber, B.; Nicolussi, A.; Soudi, M.; Pirker, K.F.; Furtmüller, P.G.; Obinger, C. Independent evolution of four heme peroxidase superfamilies. *Arch. Biochem. Biophys.* **2015**, *574*, 108–119. [CrossRef] [PubMed]
3. Fawal, N.; Li, Q.; Savelli, B.; Brette, M.; Passaia, G.; Fabre, M.; Mathé, C.; Dunand, C. PeroxiBase: A database for large-scale evolutionary analysis of peroxidases. *Nucleic Acids Res.* **2013**, *41*, D441–D444. [CrossRef] [PubMed]
4. Duroux, L.; Welinder, K.G. The peroxidase gene family in plants: A phylogenetic overview. *J. Mol. Evol.* **2003**, *57*, 397–407. [CrossRef] [PubMed]
5. Tognolli, M.; Penel, C.; Greppin, H.; Simon, P. Analysis and expression of the class III peroxidase large gene family in *Arabidopsis thaliana*. *Gene* **2002**, *288*, 129–138. [CrossRef]
6. Welinder, K.G.; Justesen, A.F.; Kjaersgard, I.V.; Jensen, R.B.; Rasmussen, S.K.; Jespersen, H.M.; Duroux, L. Structural diversity and transcription of class III peroxidases from *Arabidopsis thaliana*. *Eur. J. Biochem.* **2002**, *269*, 6063–6081. [CrossRef] [PubMed]

7. Passardi, F.; Cosio, C.; Penel, C.; Dunand, C. Peroxidases have more functions than a swiss army knife. *Plant Cell Rep.* **2005**, *24*, 255–265. [CrossRef] [PubMed]
8. Shigeto, J.; Tsutsumi, Y. Diverse functions and reactions of class III peroxidases. *New Phytol.* **2016**, *209*, 1395–1402. [CrossRef] [PubMed]
9. Pandey, V.P.; Awasthi, M.; Singh, S.; Tiwari, S.; Dwivedi, U.N. A comprehensive review on function and application of plant peroxidases. *Biochem. Anal. Biochem.* **2017**, *6*. [CrossRef]
10. Díaz, J.; Pomar, F.; Bernal, Á.; Merino, F. Peroxidases and the metabolism of capsaicin in *Capsicum annuum* L. *Phytochem. Rev.* **2004**, *3*, 141–157. [CrossRef]
11. Hiraga, S.; Sasaki, K.; Ito, H.; Ohashi, Y.; Matsui, H. A large family of class III plant peroxidases. *Plant Cell Physiol.* **2001**, *42*, 462–468. [CrossRef] [PubMed]
12. Wang, Y.; Wang, Q.; Zhao, Y.; Han, G.; Zhu, S. Systematic analysis of maize class III peroxidase gene family reveals a conserved subfamily involved in abiotic stress response. *Gene* **2015**, *566*, 95–108. [CrossRef] [PubMed]
13. De Gara, L. Class III peroxidases and ascorbate metabolism in plants. *Phytochem. Rev.* **2004**, *3*, 195–205. [CrossRef]
14. Lüthje, S.; Meisrimler, C.N.; Hopff, D.; Möller, B. Phylogeny, topology, structure and functions of membrane-bound class III peroxidases in vascular plants. *Phytochemistry* **2011**, *72*, 1124–1135. [CrossRef] [PubMed]
15. Narendra, S.; Venkataramani, S.; Shen, G.; Wang, J.; Pasapula, V.; Lin, Y.; Kornyeyev, D.; Holaday, A.S.; Zhang, H. The arabidopsis ascorbate peroxidase 3 is a peroxisomal membrane-bound antioxidant enzyme and is dispensable for arabidopsis growth and development. *J. Exp. Bot.* **2006**, *57*, 3033–3042. [CrossRef] [PubMed]
16. Laloue, H.; Weber-Lofti, F.; Lucau-Danila, A.; Guillemaut, P. Identification of ascorbate and guaiacol peroxidase in needle chloroplasts of spruce trees. *Plant Physiol. Biochem.* **1997**, *35*, 341–346.
17. Ros Barceló, A.; Ferrer, M.A.; Florenciano, G.E.; Muñoz, R. The tonoplast localization of two basic isoperoxidases of high pI in Lupinus. *Bot. Acta* **1991**, *104*, 272–278. [CrossRef]
18. Andrews, J.; Adams, S.R.; Burton, K.S.; Evered, C.E. Subcellular localization of peroxidase in tomato fruit skin and possible implications for the regulation of fruit growth. *J. Exp. Bot.* **2002**, *53*, 2185–2191. [CrossRef] [PubMed]
19. Andrews, J.; Adams, S.R.; Burton, K.S.; Edmondson, R.N. Partial purification of tomato fruit peroxidase and its effect on the mechanical properties of tomato fruit skin. *J. Exp. Bot.* **2002**, *53*, 2393–2399. [CrossRef] [PubMed]
20. Fernandez-Calvino, L.; Faulkner, C.; Walshaw, J.; Saalbach, G.; Bayer, E.; Benitez-Alfonso, Y.; Maule, A. Arabidopsis plasmodesmal proteome. *PLoS ONE* **2011**, *6*, e18880. [CrossRef] [PubMed]
21. Costa, M.M.R.; Hilliou, F.; Duarte, P.; Pereira, L.G.; Almeida, I.; Leech, M.; Memelink, J.; Barcelo, A.R.; Sottomayor, M. Molecular cloning and characterization of a vacuolar class III peroxidases involved in the metabolism of anticancer alkaloids in *Catharanthus roseus*. *Plant Physiol.* **2008**, *146*, 403–417. [CrossRef] [PubMed]
22. Mika, A.; Buck, F.; Lüthje, S. Membrane-bound class III peroxidases: Identification, biochemical properties and sequence analysis of isoenzymes purified from maize (*Zea mays* L.) roots. *J. Proteom.* **2008**, *71*, 412–424. [CrossRef] [PubMed]
23. Sottomayor, M.; Ros Barceló, A. Peroxidase from *Catharanthus roseus* (L.) G. Don and the biosynthesis of a-30-40-anhydrovinlastine: A specific role for a multifunctional enzyme. *Protoplasma* **2003**, *222*, 97–105. [CrossRef] [PubMed]
24. Mika, A.; Lüthje, S. Properties of guaiacol peroxidase activities isolated from corn root plasma membranes. *Plant Physiol.* **2003**, *132*, 1489–1498. [CrossRef] [PubMed]
25. Lüthje, S.; Möller, B.; Perrineau, F.C.; Wöltje, K. Plasma membrane electron pathways and oxidative stress. *Antioxid. Redox Signal.* **2013**, *18*, 2163–2183. [CrossRef] [PubMed]
26. Cheng, Y.; Qi, Y.; Zhu, Q.; Chen, X.; Wang, N.; Zhao Chen, H.; Cui, X.; Xu, L.; Zhang, W. New changes in the plasma-membrane-associated proteome of rice roots under salt stress. *Proteomics* **2009**, *9*, 3100–3114. [CrossRef] [PubMed]
27. Meisrimler, C.N.; Planchon, S.; Renaut, J.; Sergeant, K.; Lüthje, S. Alteration of plasma membrane-bound redox systems of iron deficient pea roots by chitosan. *J. Proteom.* **2011**, *74*, 1437–1449. [CrossRef] [PubMed]

28. Lee, Y.; Rubio, M.C.; Alassimone, J.; Geldner, N. A mechanism for localized lignin deposition in the endodermis. *Cell* **2013**, *153*, 402–412. [CrossRef] [PubMed]
29. Kamiya, T.; Borghi, M.; Wang, P.; Danku, J.M.; Kalmbach, L.; Hosmani, P.S.; Naseer, S.; Fujiwara, T.; Geldner, N.; Salt, D.E. The MYB36 transcription factor orchestrates Casparian strip formation. *Proc. Natl. Acad. Sci. USA* **2015**, *112*, 10533–10538. [CrossRef] [PubMed]
30. Tokunaga, N.; Kaneta, T.; Sato, S.; Sato, Y. Analysis of expression profiles of three peroxidase genes associated with lignification in *Arabidopsis thaliana*. *Physiol. Plantarum* **2009**, *136*, 237–249. [CrossRef] [PubMed]
31. Zhu, T.; Budworth, P.; Han, B.; Brown, D.; Chang, H.S.; Zou, G.; Wang, X. Toward elucidating the global gene expression patterns of developing Arabidopsis: Parallel analysis of 8300 genes by a high-density oligonucleotide probe array. *Plant Physiol. Biochem.* **2001**, *39*, 221–242. [CrossRef]
32. Wu, Y.; Yang, Z.; How, J.; Xu, H.; Chen, L.; Li, K. Overexpression of a peroxidase gene (AtPrx64) of *Arabidopsis thaliana* in tobacco improves plant's tolerance to aluminum stress. *Plant Mol. Biol.* **2017**, *95*, 157–168. [CrossRef] [PubMed]
33. Lefebvre, B.; Furt, F.; Hartmann, M.A.; Michaelson, L.V.; Carde, J.P.; Sargueil-Boiron, F.; Rossignol, M.; Napier, J.A.; Cullimore, J.; Bessoule, J.J.; et al. Characterization of lipid rafts from *Medicago truncatula* root plasma membranes: A proteomic study reveals the presence of a raft-associated redox system. *Plant Physiol.* **2007**, *144*, 408–418. [CrossRef] [PubMed]
34. Gutierrez-Carbonell, E.; Takahashi, D.; Lüthje, S.; González-Reyes, J.A.; Mongrand, S.; Contreras-Moreira, B.; Abadía, A.; Uemura, M.; Abadía, J.; López-Millán, A.F. A shotgun proteomic approach reveals that Fe deficiency causes marked changes in the protein profiles of plasma membrane and detergent-resistant microdomain preparations from *Beta vulgaris* roots. *J. Proteome Res.* **2016**, *15*, 2510–2524. [CrossRef] [PubMed]
35. Sottomayor, M.; Duarte, P.; Figueiredo, R.; Ros Barceló, A. A vacuolar class III peroxidase and the metabolism of anticancer indole alkaloids in *Catharanthus roseus*. *Plant Signal. Behav.* **2008**, *3*, 899–901. [CrossRef] [PubMed]
36. Berglund, G.I.; Carlsson, G.H.; Smith, A.T.; Szöke, H.; Henriksen, A.; Hajdu, J. The catalytic pathway of horseradish peroxidase at high resolution. *Nature* **2003**, *417*, 463–468. [CrossRef] [PubMed]
37. Poulos, T.L.; Edwards, S.L.; Wariishi, H.; Gold, M.H. Crystallographic refinement of lignin peroxidase at 2 A. *J. Biol. Chem.* **1993**, *268*, 4429–4440. [PubMed]
38. Newmyer, S.L.; Ortiz de Montellano, P.R. Horseradish peroxidase His-42->Ala, His-42->Val, and Phe-41->Ala mutants. Histidine catalysis and control of substrate access to the heme iron. *J. Biol. Chem.* **1995**, *270*, 19430–19438. [CrossRef] [PubMed]
39. Finkelstein, I.J.; Ishikaea, H.; Kim, S.; Massari, A.M.; Fayer, M.D. Substrate binding and protein conformational dynamics measured by 2D-IR vibrational echo spectroscopy. *Proc. Natl. Acad. Sci. USA* **2007**, *104*, 2637–2642. [CrossRef] [PubMed]
40. Sievers, F.; Wilm, A.; Dineen, D.; Gibson, T.J.; Karplus, K.; Li, W.; Lopez, R.; McWilliam, H.; Remmert, M.; Söding, J.; et al. Fast, scalable generation of high-quality protein multiple sequence alignments using Clustal Omega. *Mol. Syst. Biol.* **2011**, *7*. [CrossRef] [PubMed]
41. Watanabe, L.; de Moura, P.R.; Bleicher, L.; Nascimento, A.S.; Zamorano, L.S.; Calvete, J.J.; Sanz, L.; Perez, A.; Bursakov, S.; Roig, M.G.; et al. Crystal structure of highly glycosylated peroxidase from Royal Palm Tree. *J. Struct. Biol.* **2010**, *169*, 226–242. [CrossRef] [PubMed]
42. Moural, T.W.; Lewis, K.M.; Barnaba, C.; Zhu, F.; Palmer, N.A.; Sarath, G.; Scully, E.D.; Jones, J.P.; Sattler, S.E.; Kang, C. Characterization of Class III Pperoxidases from switchgrass. *Plant Physiol.* **2017**, *173*, 417–433. [CrossRef] [PubMed]
43. Nnamchi, C.I.; Parkin, G.; Efimov, I.; Basran, J.; Kwon, H.; Svistunenko, D.A.; Agirre, J.; Okolo, B.N.; Moneke, A.; Nwanguma, B.C.; et al. Structural and spectroscopic characterisation of a heme peroxidase from Sorghum. *J. Biol. Inorg. Chem.* **2016**, *21*, 63–70. [CrossRef] [PubMed]
44. Krogh, A.; Larsson, B.; von Heijne, G.; Sonnhammer, E.L.L. Predicting transmembrane protein topology with a hidden Markov model: Application to complete genomes. *J. Mol. Biol.* **2001**, *305*, 567–580. [CrossRef] [PubMed]
45. Nielsen, H.; Krogh, A. Prediction of signal peptides and signal anchors by a hidden Markov model. *Proc. Int. Conf. Intell. Syst. Mol. Biol.* **1998**, *6*, 122–130. [PubMed]

46. Blom, N.; Sicheritz-Ponten, T.; Gupta, R.; Gammeltoft, S.; Brunak, S. Prediction of post-translational glycosylation and phosphorylation of proteins from the amino acid sequence. *Proteomics* **2004**, *4*, 1633–1649. [CrossRef] [PubMed]
47. Nakai, K.; Kanehisa, M. A knowledge base for predicting protein localization sites in eukaryotic cells. *Genomics* **1992**, *14*, 897–911. [CrossRef]
48. Newman, T.; De Bruijn, F.J.; Green, P.; Keegstra, K.; Kende, H.; McIntosh, L.; Ohlrogge, J.; Raikhel, N.; Somerville, S.; Thomashow, M. Genes galore: A summary of methods for accessing results from large-scale partial sequencing of anonymous Arabidopsis cDNA clones. *Plant Physiol.* **1994**, *106*, 1241–1255. [CrossRef] [PubMed]
49. Kelley, L.A.; Mezulis, S.; Yates, C.M.; Stemberg, M. The Phyre2 web portal for protein modeling, prediction and analysis. *Nat. Protoc.* **2015**, *10*, 845–858. [CrossRef] [PubMed]
50. Wass, M.N.; Kelley, L.A.; Sternberg, M.J. 3DLigandSite: Predicting ligand-binding sites using similar structures. *NAR* **2010**, *38*, 469–473. [CrossRef] [PubMed]
51. Lüthje, S.; Meisrimler, C.N.; Hopff, D.; Schütze, T.; Köppe, J.; Heino, K. Class III peroxidases. *Meth. Mol. Biol.* **2014**, *1072*, 687–706. [CrossRef]
52. Asada, K.; Miyake, C.; Ogawa, K.; Hossain, M.A. Microcompartmentation of ascorbate peroxidase and regeneration of ascorbate from ascorbate radical: Its dual role in chloroplasts. In Proceedings of the IV. International Symposium on Plant Peroxidases: Biochemistry and Physiology, Vienna, Austria, 6–10 July 1996; Obinger, C., Burner, U., Ebermann, R., Penel, C., Greppin, H., Eds.; pp. 163–167, ISBN 2881640087.
53. Jespersen, H.M.; Kjærsgard, V.H.; Østergaard, L.; Welinder, K.G. From sequence analysis of three novel ascorbate peroxidases from *Arabidopsis thaliana* to structure, function and evolution of seven types of ascorbate peroxidase. *Biochem. J.* **1997**, *326*, 305–310. [CrossRef] [PubMed]
54. Kieselbach, T.; Bystedt, M.; Hynds, P.; Robinson, C.; Schroder, W.P. A peroxidase homologue and novel plastocyanin located by proteomics to the Arabidopsis chloroplast thylakoid lumen. *FEBS Lett.* **2000**, *480*, 271–276. [CrossRef]
55. Yabuta, Y.; Motoki, T.; Yoshimura, K.; Takeda, T.; Ishikawa, T.; Shigeoka, S. Thylakoid membrane-bound ascorbate peroxidase is a limiting factor of antioxidative systems under photo-oxidative stress. *Plant J.* **2002**, *32*, 915–925. [CrossRef] [PubMed]
56. Bunkelmann, J.; Trelease, R.N. Ascorbate peroxidase. A prominent membrane protein in oilseed glyoxysomes. *Plant Physiol.* **1996**, *110*, 589–598. [CrossRef] [PubMed]
57. Ishikawa, T.; Yoshimura, K.; Sakai, K.; Tamoi, M.; Takeda, T.; Shigeoka, S. Molecular characterization and physiological role of a glyoxysome-bound ascorbate peroxidase from spinach. *Plant Cell Physiol.* **1998**, *39*, 23–34. [CrossRef] [PubMed]
58. Nito, K.; Yamaguchi, K.; Kondo, M.; Hayashi, M.; Nishimura, M. Pumpkin peroxisomal ascorbate peroxidase is localized on peroxisomal membranes and unknown membranous structures. *Plant Cell Physiol.* **2001**, *42*, 20–27. [CrossRef] [PubMed]
59. Yamaguchi, K.; Mori, H.; Nishimura, M. A novel isoenzyme of ascorbate peroxidase localized on glyoxysomal and leaf peroxisomal membranes in pumpkin. *Plant Cell Physiol.* **1995**, *36*, 1157–1162. [CrossRef] [PubMed]
60. Lüthje, S.; Hopff, D.; Schmitt, A.; Meisrimler, C.N.; Menckhoff, L. Hunting for low abundant redox proteins in plant plasma membranes. *J. Proteom.* **2009**, *72*, 475–483. [CrossRef] [PubMed]
61. Welinder, K.G.; Larsen, Y.B. Covalent structure of soybean seed coat peroxidase. *Biochim. Biophys. Acta* **2004**, *1698*, 121–126. [CrossRef] [PubMed]
62. Sigrist, C.J.A.; de Castro, E.; Cerutti, L.; Cuche, B.A.; Hulo, N.; Bridge, A.; Bougueleret, L.; Xenarios. New and continuing developments at PROSITE. *Nucleic Acids Res.* **2012**, *21*, D344–D347. [CrossRef]
63. Mika, A.; Boenisch, M.J.; Hopff, D.; Lüthje, S. Membrane-bound guaiacol peroxidases are regulated by methyl jasmonate, salicylic acid, and pathogen elicitors. *J. Exp. Bot.* **2010**, *61*, 831–841. [CrossRef] [PubMed]
64. Mika, A.; (University of Hamburg, Hamburg, Germany); Lüthje, S.; (University of Hamburg, Hamburg, Germany). Personal communication, 2007.
65. Morel, J.; Claverol, S.; Mongrand, S.; Furt, F.; Fromentin, J.; Bessoule, J.J.; Blein, J.P.; Simon-Plas, F. Proteomics of plant detergent-resistant membranes. *Mol. Cell. Proteom.* **2006**, *5*, 1396–1411. [CrossRef] [PubMed]
66. Simon-Plas, F.; Perraki, A.; Bayer, E.; Gerbeau-Pissot, P.; Mongrand, S. An update on plant membrane rafts. *Curr. Opin. Plant Biol.* **2011**, *14*, 642–649. [CrossRef] [PubMed]

67. Showalter, A.M. Arabinogalactan-proteins: Structure, expression and function. *Cell Mol. Life Sci.* **2001**, *58*, 1399–1417. [CrossRef] [PubMed]
68. Wink, M. The plant vacuole: A multifunctional compartment. *J. Exp. Bot.* **1993**, *44*, 231–246.
69. Carqueijeiro, I.; Noronha, H.; Duarte, P.; Gerós, H.; Sottomayor, M. Vacuolar transport of the medicinal alkaloids from *Catharanthus roseus* is mediated by a proton-driven antiport. *Plant Physiol.* **2013**, *162*, 1486–1496. [CrossRef] [PubMed]
70. Lüthje, S. Plasma membrane redox systems: Lipid rafts and protein assemblies. *Prog. Bot.* **2008**, *69*, 169–200. [CrossRef]
71. Roppolo, D.; De Rybel, B.; Tendon, V.D.; Pfister, A.; Alassimone, J.; Vermeer, J.E.; Yamazaki, M.; Stierhof, Y.D.; Beeckman, T.; Geldner, N. A novel protein family mediates Casparian strip formation in the endodermis. *Nature* **2011**, *473*, 380–383. [CrossRef] [PubMed]
72. Halpin, C. Cell Biology: Up Against the Wall. *Curr. Biol.* **2013**, *23*, R1050. [CrossRef] [PubMed]
73. Christensen, J.H.; Bauw, G.; van Montagu, M.; Boerjan, W. Towards the identification of lignin specific peroxidases in poplar. In Proceedings of the IV. International Symposium on Plant Peroxidases: Biochemistry and Physiology, Vienna, Austria, 6–10 July 1996; Obinger, C., Burner, U., Ebermann, R., Penel, C., Greppin, H., Eds.; pp. 113–117, ISBN 2881640087.
74. Cosio, C.; Dunand, C. Specific functions of individual class III peroxidase genes. *J. Exp. Bot.* **2009**, *60*, 391–409. [CrossRef] [PubMed]
75. Mika, A.; Minibayeva, F.; Beckett, R.; Lüthje, S. Possible functions of extracellular peroxidases in stress-induced generation and detoxification of active oxygen species. *Phytochem. Rev.* **2004**, *3*, 173–193. [CrossRef]
76. Sies, H. Hydrogen peroxide as a central redox signaling molecule in physiological oxidative stress: Oxidative eustress. *Redox Biol.* **2017**, *11*, 613–619. [CrossRef] [PubMed]
77. Noctor, G.; Reichheld, J.P.; Foyer, C.H. ROS-related redox regulation and signaling in plants. *Semin. Cell. Dev. Biol.* **2018**, *80*, 3–12. [CrossRef] [PubMed]
78. Schraudner, M.; Langebartels, C.; Sandermann, H., Jr. Plant defence systems and ozone. *Biochem. Soc. Trans.* **1996**, *24*, 456–461. [CrossRef] [PubMed]
79. Bolwell, G.P.; Bindschedler, L.V.; Blee, K.A.; Butt, V.S.; Davies, D.R.; Gardner, S.L.; Gerrish, C.; Minbayeva, F. The apoplastic oxidative burst in response to biotic stress in plants: A three-component system. *J. Exp. Bot.* **1996**, *53*, 1367–1376. [CrossRef]
80. Schützendübel, A.; Polle, A. Plant responses to abiotic stresses: Heavy metal-induced oxidative stress and protection by mycorrhization. *J. Exp. Bot.* **2002**, *53*, 1351–1365. [CrossRef] [PubMed]
81. Minibayeva, F.; Lüthje, S.; Kolesnikov, O.; Chasov, A.; Beckett, R.P.; Vylegzhanina, N.; Buck, F.; Böttger, M. Wound-induced apoplastic peroxidase activities: Their roles in the production and detoxification of reactive oxygen species. *Plant Cell Environ.* **2009**, *32*, 497–508. [CrossRef] [PubMed]
82. Cakmak, I.; van de Wetering, D.A.M.; Marschner, H.; Bienfait, H.F. Involvement of superoxide radical in extracellular ferric reduction by iron-deficient bean roots. *Plant Physiol.* **1987**, *85*, 310–314. [CrossRef] [PubMed]
83. Qiu, Q.S.; Liang, H.G. Lipid peroxidation caused by the redox system of plasma membranes from wheat roots. *J. Plant Physiol.* **1995**, *145*, 261–265. [CrossRef]
84. Rawyler, A.; Arpagaus, S.; Braendle, R. Impact of oxygen stress and energy availability on membrane stability of plant cells. *Ann. Bot.* **2002**, *90*, 499–507. [CrossRef] [PubMed]
85. Marjamaa, K.; Kukkola, E.M.; Fagerstedt, K. The role of xylem class III peroxidases in lignification. *J. Exp. Bot.* **2009**, *60*, 367–376. [CrossRef] [PubMed]
86. Vanholme, R.; Demedts, B.; Morreel, K.; Ralph, J.; Boerjan, W. Lignin biosynthesis and structure. *Plant Physiol.* **2010**, *153*, 895–905. [CrossRef] [PubMed]
87. Boerjan, W.; Ralph, J.; Baucher, M. Lignin biosynthesis. *Ann. Rev. Plant Biol.* **2003**, *54*, 519–546. [CrossRef] [PubMed]
88. Novo-Uzal, E.; Fernández-Pérez, F.; Herrero, J.; Gutiérrez, J.; Gómez-Ros, L.V.; Bernal, M.Á.; Díaz, J.; Cuello, J.; Pomar, F.; Pedreño, M.A. From Zinnia to Arabidopsis: Approaching the involvement of peroxidases in lignification. *J. Exp. Bot.* **2013**, *64*, 3499–3518. [CrossRef] [PubMed]
89. Vishwanath, S.J.; Delude, C.; Domergue, F.; Rowland, O. Suberin: Biosynthesis, regulation, and polymer assembly of a protective extracellular barrier. *Plant Cell Rep.* **2015**, *34*, 573–586. [CrossRef] [PubMed]

90. Takahama, U. Hydrogen peroxide scavenging systems in vacuoles of mesophyll cells of *Vicia faba*. *Phytochemistry* **1992**, *31*, 1127–1133. [CrossRef]
91. Soares, A.R.; Marchiosi, R.; Siqueira-Soares Rde, C.; Barbosa de Lima, R.; Dantas dos Santos, W.; Ferrarese-Filho, O. The role of L-DOPA in plants. *Plant Signal. Behav.* **2014**, *9*, e28275. [CrossRef] [PubMed]
92. Kawano, T. Roles of the reactive oxygen species-generating peroxidase reactions in plant defense and growth induction. *Plant Cell Rep.* **2003**, *21*, 829–837. [CrossRef] [PubMed]
93. Ranocha, P.; Dima, O.; Nagy, R.; Felten, J.; Corratgé-Faillie, C.; Novák, O.; Morreel, K.; Lacombe, B.; Martinez, Y.; Pfrunder, S.; et al. Arabidopsis WAT1 is a vacuolar auxin transport facilitator required for auxin homoeostasis. *Nat. Commun.* **2013**, *4*, 2625. [CrossRef] [PubMed]
94. Journet, E.P. MtPrx02, Unpublished, PeroxiBase.
95. Takahama, U.; Oniki, T. Effects of ascorbate on oxidation of hydroxycinnamic acid derivatives and the mechanism of oxidation of sinapic acid by cell wall-bound peroxidases. *Plant Cell Phys.* **1994**, *35*, 593–600. [CrossRef]
96. Pomar, F.; Merino, F.; Ros Barceló, A. *O*-4-linked coniferyl and sinapyl aldehyde in lignifying cell walls are the main targets of the Wiesner (phloroglucinol-HCl) reaction. *Protoplasma* **2002**, *220*, 17–28. [CrossRef] [PubMed]
97. Lamb, C.; Dixon, R.A. The oxidative burst in plant disease resistance. *Annu. Rev. Plant Physiol. Plant Mol. Biol.* **1997**, *48*, 251–275. [CrossRef] [PubMed]
98. Liu, T.T.; Wu, P.; Wang, H.; Zhou, Q. Response of soybean seed germination to cadmium and acid rain. *Biol. Trace Elem. Res.* **2011**, *144*, 1186–1196. [CrossRef] [PubMed]
99. Steffens, B. The role of ethylene and ROS in salinity, heavy metal, and flooding responses in rice. *Front. Plant Sci.* **2014**, *5*, 685. [CrossRef] [PubMed]
100. Zhang, H.; Liu, X.L.; Zhang, R.X.; Yuan, H.Y.; Wang, M.M.; Yang, H.Y.; Ma, H.Y.; Liu, D.; Jiang, C.J.; Liang, Z.W. Root damage under alkaline stress is associated with reactive oxygen species accumulation in rice (*Oryza sativa* L.). *Front. Plant Sci.* **2017**, *8*, 1580. [CrossRef] [PubMed]
101. Sergio, L.; Cardinali, A.; De Paola, A.; Di Venere, D. Biochemical properties of soluble and bound peroxidases from artichoke heads and leaves. *Food Technol. Biotechnol.* **2009**, *47*, 32–38.
102. Martínez-Cortés, T.; Pomar, F.; Espiñeira, J.M.; Merino, F.; Novo-Uzal, E. Purification and kinetic characterization of two peroxidases of Selaginella martensii Spring. involved in lignification. *Plant Physiol. Biochem.* **2012**, *52*, 130–139. [CrossRef] [PubMed]
103. Zhang, Y.K.; Zhu, D.F.; Zhang, Y.P.; Chen, H.Z.; Xiang, J.; Lin, X.Q. Low pH-Induced changes of antioxidant enzyme and ATPase activities in the roots of rice (*Oryza sativa* L.) seedlings. *PLoS ONE* **2015**, *10*, e0116971. [CrossRef] [PubMed]
104. Haslekås, C.; Grini, P.E.; Nordgard, S.H.; Thorstensen, T.; Viken, M.K.; Nygaard, V. ABI3 mediates expression of the peroxiredoxin antioxidant AtPER1 gene and induction by oxidative stress. *Plant Mol. Biol.* **2003**, *53*, 313–326. [CrossRef] [PubMed]
105. Ferreres, F.; Figueiredo, R.; Bettencourt, S.; Carqueijeiro, I.; Oliveira, J.; Gil-Izquierdo, A.; Pereira, D.M.; Valentao, P.; Andrade, P.B.; Duarte, P.; et al. Identification of phenolic compounds in isolated vacuoles of the medicinal plant *Catharanthus roseus* and their interaction with vacuolar class III peroxidase: An H_2O_2 affair? *J. Exp. Bot.* **2011**, *62*, 2841–2854. [CrossRef] [PubMed]
106. Kumar, S.; Jaggi, M.; Taneja, J.; Sinha, A.K. Cloning and characterization of two new Class III peroxidase genes from *Catharanthus roseus*. *Plant Physiol. Biochem.* **2011**, *49*, 404–412. [CrossRef] [PubMed]
107. Baum, S.F.; Tran, P.N.; Silk, W.K. Effects of salinity on xylem structure and water use in growing leaves of sorghum. *New Phytol.* **2000**, *146*, 119–127. [CrossRef]
108. Karahara, I.; Ikeda, A.; Kodo, T.; Uetake, Y. Development of the Casparian strip in primary roots of maize under salt stress. *Planta* **2004**, *219*, 41–47. [CrossRef] [PubMed]
109. Downie, J.A. Legume nodulation. *Curr. Biol.* **2014**, *24*, 184–190. [CrossRef] [PubMed]
110. Jones, D.T.; Taylor, W.R.; Thornton, J.M. The rapid generation of mutation data matrices from protein sequences. *Comput. Appl. Biosci.* **1992**, *8*, 275–282. [CrossRef] [PubMed]
111. Kumar, S.; Stecher, G.; Tamura, K. MEGA7: Molecular evolutionary genetics analysis version 7.0 for bigger datasets. *Mol. Biol. Evol.* **2016**, *33*, 1870–1874. [CrossRef] [PubMed]

112. Ngoune Tandzi, L.; Shelton Mutengwa, C.; Mangaptche Ngonkeu, E.L.; Gracen, V. Breeding maize for tolerance to acidic soils: A review. *Agronomy* **2018**, *8*, 84. [CrossRef]
113. Sekhon, R.S.; Lin, H.; Childs, K.L.; Hansey, C.N.; Buell, C.R.; de Leon, N.; Kaeppler, S.M. Genome-wide atlas of transcription during maize development. *Plant J.* **2011**, *66*, 553–563. [CrossRef] [PubMed]

International Journal of
Molecular Sciences

MDPI

Review

The Multifaceted Role of Pectin Methylesterase Inhibitors (PMEIs)

Alexandra Wormit [1,2,*] and Björn Usadel [1,2,3]

1 Institute of Biology 1, Botany and Molecular Genetics, RWTH Aachen University, 52074 Aachen, Germany; usadel@bio1.rwth-aachen.de
2 Bioeconomy Science Center (BioSC), Forschungszentrum Jülich, Wilhelm Johnen Straße, 52425 Jülich, Germany
3 Institute for Bio- and Geosciences (IBG-2: Plant Sciences), Forschungszentrum Jülich, 52428 Jülich, Germany
* Correspondence: awormit@bio1.rwth-aachen.de; Tel.: +49-241-80-26766

Received: 5 August 2018; Accepted: 5 September 2018; Published: 21 September 2018

Abstract: Plant cell walls are complex and dynamic structures that play important roles in growth and development, as well as in response to stresses. Pectin is a major polysaccharide of cell walls rich in galacturonic acid (GalA). Homogalacturonan (HG) is considered the most abundant pectic polymer in plant cell walls and is partially methylesterified at the C6 atom of galacturonic acid. Its degree (and pattern) of methylation (DM) has been shown to affect biomechanical properties of the cell wall by making pectin susceptible for enzymatic de-polymerization and enabling gel formation. Pectin methylesterases (PMEs) catalyze the removal of methyl-groups from the HG backbone and their activity is modulated by a family of proteinaceous inhibitors known as pectin methylesterase inhibitors (PMEIs). As such, the interplay between PME and PMEI can be considered as a determinant of cell adhesion, cell wall porosity and elasticity, as well as a source of signaling molecules released upon cell wall stress. This review aims to highlight recent updates in our understanding of the PMEI gene family, their regulation and structure, interaction with PMEs, as well as their function in response to stress and during development.

Keywords: pectin methylesterase inhibitor (PMEI), pectin; homogalacturonan (HG); cell wall properties; degree of methylesterification (DM); stress; development; applications

1. Introduction

Plant cells are surrounded by a wall composed of interacting networks of polysaccharides, highly glycosylated proteins and other polymers. Plant cell walls are complex and highly dynamic structures, responding and adapting to normal processes of growth and development as well as to biotic and abiotic stresses. They have to fulfil several different functions: on one hand, they need to be flexible to allow fast and directional cell elongation (e.g., pollen tubes) and on the other hand they have to be rigid enough to resist the internal turgor pressure, to protect against pathogens and to provide structural and mechanical support for the upright growth of the whole organism.

The major carbohydrates of primary cell walls are cellulose, hemicellulose, and pectins. Cellulose and hemicellulose form a hydrogen-bonded network that is embedded in a gel-like pectic matrix. Pectin is a galacturonan-based polysaccharide containing five distinct subclasses that can be distinguished based on the structure of their backbones and the diversity of their side chains: homogalacturonan (HG), rhamnogalacturonan I and rhamnogalacturonan II (RG-II), xylogalacturonan and apiogalacturonan [1]. In dicots and non-graminaceous plants, pectins constitute about 35% of primary cell walls, whereas in grasses only 2–10% of primary walls are pectic polysaccharides [2]. Pectin has been shown to play roles controlling cell wall porosity [3], cell elongation [4], and cell adhesion [5] and constitutes an important factor in plant development (for reviews see [6,7]).

HG constitutes ~65% of pectin and is the most abundant pectic polysaccharide in primary cell walls [1]. It consists of an α-1,4-linked D-galacturonic acid (GalA) backbone, which is synthesized by HG galacturonosyl-transferases (GAUTs) from the nucleotide sugar uridine diphosphate (UDP)-GalA in the Golgi apparatus. So far, the galacturonosyl-transferase GAUT1 has been biochemically characterized in *Arabidopsis* [8,9] and a hetero-complex formation of two GAUT1 with one GAUT7 molecule has been shown responsible for retaining GAUT1 in the Golgi apparatus [10]. In addition, GAUT4 from *Arabidopsis thaliana*, switchgrass and poplar was shown to synthesize HG and downregulate of the gene reduced HG and RG-II in the cell wall [11]. When HG is secreted into the cell wall, it is highly methyl-esterified at the C6 atom of the GalA residue (~80% [2]) and a putative pectin methyltransferase conferring this methyl-esterification has been identified in the Golgi [12,13]. In addition, some GalA residues of HG carry acetyl groups at O2 or O3, which affect the physiochemical properties of HG in vitro [14].

Once incorporated into the cell wall, HG is further selectively modified and the pattern and degree of these modifications affect pectin hydrolysis and properties such as pH, charge and crosslinking. Pectin acetyl esterases catalyze de-acetylation and the release of acetate from the cell wall, causing changes in cell wall mechanics and developmental aberrations [15,16]. Pectin methyl esterases (PMEs) catalyze the specific de-methylesterification of HG, releasing methanol and protons, and creating negatively charged carboxyl groups in the process (Figure 1). The degree and pattern of methylesterification of HG determine the biomechanical properties of the cell wall. When several consecutive GalA residues are de-methylesterified (block-wise de-methylesterification), the negatively charged carboxyl groups can form calcium bonds with other HG molecules, leading to so-called 'egg-box' structures that underlie the formation of pectin gels [17]. De-methylesterified, calcium cross-linked HG increased the amount of bound water maintaining wall hydration [18] and the hydration state was shown to affect biomechanical properties of the cell wall, such as its rigidity [19]. In addition, the strength of pectin gels is highly dependent on the amount of free calcium ions in the apoplast, as stiffness of the gel is reduced by disassociation of calcium crosslinks [20]. On the other hand, partially de-methylesterified HG (random or block-wise de-methylesterification) can become a target for pectin-degrading enzymes such as polygalacturonases and pectate/pectin lyases.

The degree of methylesterification (DM) of HG, which is controlled by the activity of large PME families, has vast consequences on the mechanical properties of the cell wall [21], affecting developmental processes such as stomata opening [22,23], cell adhesion [5], organ initiation [24] and anisotropic cell growth [25]. In addition, hydrolysis of partially de-methylesterified HG can lead to the formation of signaling molecules (oligogalacturonides), for instance, during plant-pathogen interactions [26,27]. Consequently, PME activity is tightly regulated at: (a) the transcriptional level [28], (b) by protein processing [1] and degradation [29], (c) by the pH of the cell wall environment [6,30], and (d) by endogenous inhibitor proteins called pectin methylesterase inhibitors (PMEI, Figure 1) [31,32]. The first PMEI was identified in kiwi fruit (*Actinidia deliciosa* [33]) and to date several PMEIs have been investigated in different plant species. This review therefore aims to give a comprehensive overview on the current knowledge about PMEIs and their various roles in plant development and stress response.

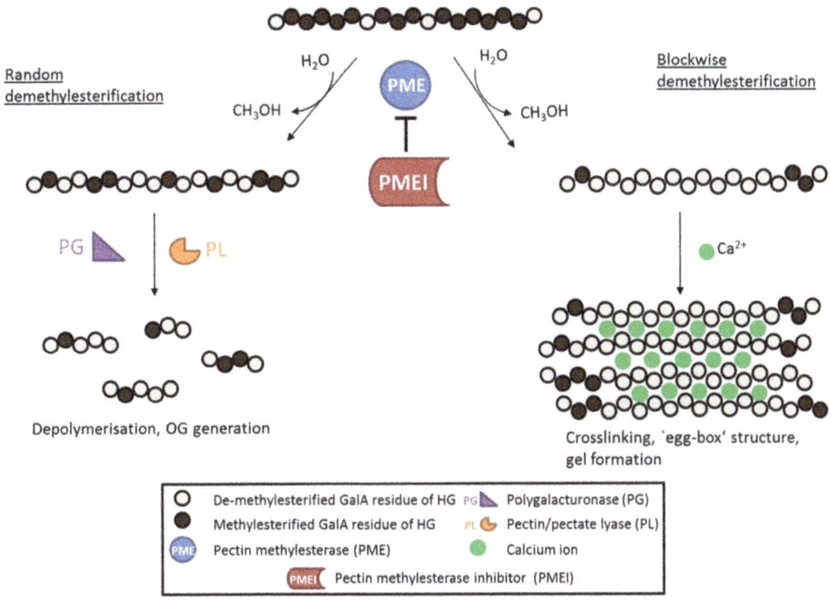

Figure 1. Schematic diagram showing the de-methylesterification of HG and the effects on its structure. HG is highly methylesterified when deposited into the cell wall. PMEs can de-methylesterify HG in a block-wise fashion, leading to several consecutive GalA residues without methylester groups. These HG backbones are negatively charged and can therefore form crosslinks with cations like calcium ions, leading to so called 'egg-box' structures responsible for gel formation. On the other hand, PMEs' can de-methylesterify single GalA residues leading to a random methylesterification pattern. Low-methylesterified HG is depolymerized by pectin-degrading enzymes such as polygalacturonases (PG) and pectin/pectate lyases (PL), which leads to the formation of oligogalacturonides (OG). PME activity is inhibited by its proteinaceous inhibitor PMEI.

2. PMEI Occurrence and Regulation

PMEIs belong to large multigene families, containing almost as many members as *PME* genes in several plant species [34]. The PMEI proteins first appeared in mosses (*Physcomitrella patens*), which coincided with the appearance of pectin in cell walls [34]. In *Arabidopsis*, 71 putative *PMEI* genes have been identified in silico (not including *proPME* genes containing PMEI domains) [34], compared to 100 open reading frames (ORFs) in *Brassica campestris* [35] and 97 putative *PMEI* genes in *Brassica rapa* [36]. Pinzón-Latorre and Deyholos [37] listed 95 PMEI ORFs in flax (*Linum usitatissimum*), whereas in poplar (*Populus trichocarpa*), only 54 genes have been annotated [34]. In monocots, the PMEI families generally contain fewer isoforms. For example, in rice (*Oryza sativa*), 49 ORFs were annotated as putative PMEIs [38] compared to 37 genes in *Sorghum bicolor* [34] and 38 putative PMEI members in *Brachypodium distachyon* [1]. The smaller family size in grasses is likely due to differences in the structure of cell wall polysaccharides, with pectins being less abundant and less methylesterified in graminaceous species [2,28]. Gene family synteny and phylogenetic analyses suggest that the expansion of the PMEI families in several angiosperm species is due to whole genome duplication and tandem duplication events during evolution [34,35,37].

PMEIs are crucial factors in regulating the DM of HG, which has tremendous effects on cell wall mechanics and affects many biological processes (Figure 2).

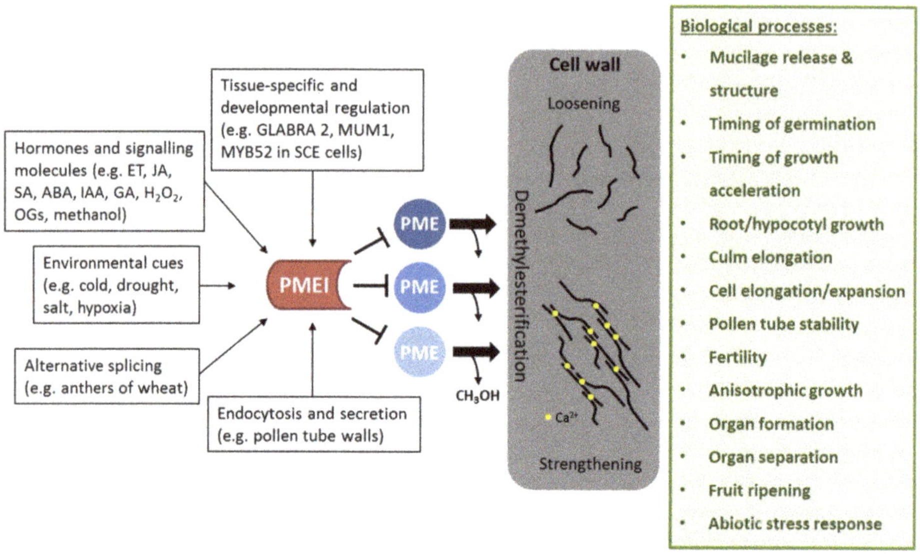

Figure 2. Effect of PMEI regulation on cell wall properties and biological processes affected by PMEI manipulation. *PMEIs* are transcriptionally regulated in a tissue-specific and development-dependent manner. Several plant hormones and signaling molecules as well as environmental stresses can activate *PMEI* gene expression. Alternative splicing and directed endocytosis and secretion regulate the level of active PMEIs in the cell wall. PMEI can inhibit several PME enzymes, thereby regulating de-methylesterification of HG. This in turn modulates cell wall properties such as loosening or strengthening, which is required for several biological processes.

Therefore, it is not surprising that their expression is highly coordinated with other pectin modifying enzymes, such as PMEs [28]. Transcriptional analyses (using microarrays, RNAseq, qRT-PCR) showed that *PMEI* expression is spatially and developmentally controlled in various species (*Brassica rapa* [36], *Brassica campestris* [39], *Arabidopsis* [40], grapevine [41], wheat [42], rice [38], and flax [37,43]). Cluster analysis for a fraction of *PME* and *PMEI* genes already identified expression clusters specific for e.g., seed coat, endosperm, pollen, shoot apex in *Arabidopsis* [1]. For example, *AtPMEI6* and *AtPMEI14* are specifically expressed in seed coat epidermal cells and their expression is modulated by the transcriptional regulators GLABRA2 and MUCILAGE-MODIFIED 1 (MUM1), known to be involved in seed coat differentiation and mucilage production [44]. In addition, the R2R3-MYB (myeloblastosis) transcription factor MYB52 has been identified as a negative regulator of pectin de-methylesterification in seed coat mucilage due to its control of *AtPMEI6* and *AtPMEI14* expression [45].

Hormones are also involved in regulation of *PMEI* expression in several species [38,46–49]. For example, transcriptional activation of *PMEI* was shown to be ethylene-dependent during the ripening process of banana fruits [46], whereas a wheat *PMEI* was inducible by salicylic and jasmonic acid as well as hydrogen peroxide, indicating a role in defense responses [48]. Moreover, post-transcriptional regulation has also been indicated for PMEIs. Rocchi et al. [42] showed that two *PMEI* genes from durum wheat undergo intron retention. Mature transcripts were predominantly found in floral organs, indicating a role in flower development and in particular, anther and pollen development.

In pollen tubes, dynamic pectin metabolism is the key to polar growth and concurrently, different PME and PMEI isoforms are highly expressed in this cell type. By immuno-labelling, it was shown that HG with a low DM is found at the flanks of the pollen tube, whereas methylesterified pectin is predominantly present at the growing tip region where wall extensibility is required for cell growth [50,51]. This transition from apical to distal pectin epitopes seems to be correlated with an increase in cell wall rigidity and decrease in visco-elasticity. The maintenance of the distribution of esterified and de-esterified HG is ensured by a dedicated mechanism involving localized secretion of specific PMEI proteins [52]. The pectin methylesterase AtPPME1 is distributed evenly in the cell wall of the pollen tube. In contrast, the pollen-specific AtPMEI2, which interacts with AtPPME1, is exclusively localized at the cell wall of the pollen tube tip. This specific localization was shown to be dependent on endocytic internalization of AtPMEI2 at the flanks of the pollen tube. Therefore, protein secretion and endocytosis seems to be an efficient mechanism for regulating DM of pectin in a temporally and spatially controlled manner in pollen tubes. It remains to be seen if other PMEIs in other plant tissues are also regulated by such a mechanism.

In general, the large number of PMEI family members with distinct expression and regulation patterns could suggest the existence of dedicated interaction pairs required for localized modifications of HG methylesterification during development. Additionally, this could indicate different inhibitory activities of PMEIs depending on the cell wall environment and/or different specificities for target PMEs to guarantee a development- and stress-dependent adjustment of cell wall mechanics.

3. PME Inhibitor Structure and Interaction with Pectin Esterases

PME inhibitors and invertase inhibitors belong to the same protein family (PF04043, http://pfam.xfam.org/family/PF04043), although their respective target proteins are part of different metabolite pathways. The overall sequence identity between PMEIs and invertase inhibitors is only moderate (20–30%), but their three-dimensional structure is highly similar [53]. As shown by crystallographic approaches, PMEIs consist of four long α-helices arranged in an up-down-up-down topology, forming a four-helix bundle. They contain an N-terminal signal peptide responsible for extracellular targeting and four highly conserved cysteine residues, which are involved in the formation of two disulfide (S-S) bridges essential for the maintenance of the secondary structure of the protein [54,55]. One S-S bridge connects helices $\alpha 2$ and $\alpha 3$; the second S-S bridge is located in a fifth, N-terminal anchor region consisting of short distorted helices [56].

Interaction studies performed in vitro at different pH values (pH 5.5, 7.5 & 8.5; pH 7.5; pH 5.5 & 7.5; pH 6.5, 7.5 & 8.5) indicated formation of a stoichiometric 1:1 complex of PMEIs with endogenous PMEs but also with enzymes from other species [41,53,56,57]. For example, Jolie et al. [58] showed that purified PMEI from kiwi can bind to PMEs from carrot and broccoli (at pH 6.5), whereas tomato and *Arabidopsis* PMEIs can interact and inhibit PMEs from orange (at pH 7.5 and pH 7, respectively) [30,59]. On the other hand, plant PMEIs do not seem capable of inhibiting fungal or bacterial PMEs, as was shown for PMEI1 and PMEI2 from *Arabidopsis*, which did not affect PME activity of *Botrytis cinerea* [26]. Similarly, Reca et al. [59] showed that purified tomato PMEI only inhibited plant PMEs but was not active against PME from *Erwinia chrysantemi*. An explanation for this lack of inhibition might be that residues critical for PME-PMEI interaction are not conserved in fungal and bacterial PMEs [56].

The three-dimensional structure of *Arabidopsis* and kiwi PMEI interacting with purified PME from tomato revealed that the inhibitor covers the putative active site of the enzyme, thus preventing substrate binding [53]. The N-terminal anchor region interacts with the C-terminus of PME, potentially supporting the structural stability. The complex formation of PMEI with PME has been shown to be pH sensitive [60]. A detailed study on kiwi PMEI indicated that PMEI-PME complex formation in vitro was reduced at pH values above 6.0, due to reversible conformation changes in the 3D structure of the PMEI protein. At pH levels above 7.5, irreversible changes in PMEI secondary structure occur, such as cleavage of the S-S bridges, inducing helix instability. Once the PMEI-PME complex is formed, it is rather stable and only dissociates under extreme pH conditions (such as pH 10), indicating that interaction with the endogenous PME enzyme protects the inhibitor from conformational changes. Different PMEIs have been shown to function at different pH values; Nguyen et al. [61] analyzed a PMEI of rice (OsPMEI28), which had a highest inhibitor activity against commercial PME at pH 8.5, whereas complex formation between PME3 and PMEI7 of *Arabidopsis* was highly sensitive to pH variation between 5 and 7 [62]. The observed sensitivity was the result of changes in the protonation of amino acid residues at the interface of the two proteins, which caused residues to shift between inter- and intra-molecular interactions. These results indicate that the prevalent cell wall environment, i.e., the pH of the cell wall, can affect the activity of PMEIs, thereby fine-tuning the DM of pectic polysaccharides.

It was also shown that one PMEI protein can bind and inhibit several endogenous PME enzymes [39,56], indicating a complex mechanism for regulation of PME activity. For example, key residues involved in the interaction between the inhibitors AtPMEI4 or AtPMEI9 and AtPME3 were identified by computational approaches and indicated that functional diversity between the two PMEIs leads to distinct consequences on pollen tube elongation [63]. However, the existence of strictly specific PMEI-PME pairs cannot be excluded at this time.

4. Role of PMEIs in Stress

The cell wall is the first barrier against pathogens and maintenance of cell wall integrity plays an important role in pathogen defense [64]. For many bacteria and fungi, secretion of cell wall degrading enzymes that hydrolyze pectin is an important component for successfully infecting a plant [65–68]. Conversely, inactivation of pectin degrading enzymes in pathogens can reduce their virulence on host plants [69–72]. Plant PME activity and the level of pectin methylesterification are highly regulated by pathogens during an infection process and an increase in PME activity is correlated with reduced DM of pectin in *Arabidopsis* during infection with different pathogens [67,68,73]. In addition, degradation of HG by pathogenic polygalacturonases releases de-esterified pectin fragments or oligogalacturonides (OGs) which in turn act as damage-associated molecular patterns (DAMPS [74]). Several studies demonstrated that OGs are sensed by specific plant receptors (wall-associated kinase, WAK) and can activate a stress response by the plant [75,76]. It has been shown that the DM of OGs influences the strength of the elicited response [27,77]. Furthermore, PME activity on HG results in release of methanol, a signaling molecule that is involved in priming of (neighboring) plants and able to retard bacterial growth [78].

Constitutive expression of the *PMEI* genes *AtPMEI-1* and *AtPMEI-2* in *Arabidopsis* reduced the PME activity and resulted in an increase in DM of HG [26]. The transgenic mutants were more resistant to the necrotrophic fungus *Botrytis cinerea*, which was not due to inhibition of fungal PME but rather to an impaired fungal growth on methylesterified pectin. Further *PMEI* genes, *AtPMEI10*, *AtPMEI11* and *AtPMEI12*, were identified as upregulated in response to *B. cinerea* infection [47]. Genetic studies indicated that AtPMEI expression is dependent on jasmonic acid and ethylene signaling but only AtPMEI11 can be induced by OGs [47]. The *pmei10*, *pmei11* and *pmei12* mutants showed increased PME activity, decreased DM of pectin and increased lesion formation upon *B. cinerea* infection. This indicates that plants modulate PME activity by expressing PMEIs in response to pathogen attack.

Similarly, transgenic wheat lines expressing the *PMEI* from kiwi fruit also showed a reduction of disease symptoms caused by two fungal pathogens, *Fusarium gramineum* and *Bipolaris sorokiniana* [79]. This was related to reduced fungal growth on pectin with high DM and the activity of fungal PG to hydrolyze the plant pectin was impaired. In addition, the transgenic wheat plants showed less disease symptoms upon infection with the biotrophic fungal pathogen *Claviceps purpurea* [80]. This also shows the importance of pectin (and its methylesterification status) in grass species, despite its low abundance [2]. In barley (*Hordeum vulgare*), a family of putative pectin esterase inhibitors (PEIs) has been associated with a resistance gene (*Rrs2* locus), which is involved in defense against *Rhynchosporium commune* which causes leaf blotch [81]. Overexpression of *HvPEI4* led to an improvement of the resistance, but none of the candidate genes alone caused a high increase in resistance level. In a recent study, a PME inhibitor from cotton (*Gossypium hirsutum*) was identified as being involved in the defense response to the fungus *Verticilium dahliae* [39]. *GhPMEI3* expression in *Arabidopsis* led to reduced disease symptoms upon infection with *V. dahlia* and silencing of *GhPMEI3* in cotton resulted in an increased susceptibility to the fungus.

Based on current evidence, the proposed role of PMEI in the immune response is the dynamic modulation of the PME activity (Figure 3). The degree of methylesterification determines the susceptibility of the plant cell wall to the pectin-degrading enzymes of the pathogen [82]. Plant PME activity generates methanol as an alarm signal of the damaged self that can regulate the transcription of pathogen-related *PMEI* genes either directly (e.g., *AtPMEI11* [47]) or via modulation of DAMP or defense hormone signaling (*AtPMEI10, AtPMEI11, AtPMEI12* [47]). OGs are released by partial hydrolysis of HG by fungal/bacterial polygalacturonases. They serve as DAMPs and bind to WAKs at the plasmamembrane, leading to activation of plant immune signaling, which induces classic defense responses. The DM of OGs can influence the intensity of the defense response. PMEIs (e.g., *AtPMEI10, AtPMEI11, AtPMEI12* [47]) are transcriptionally upregulated by defense-related signaling pathways including jasmonic acid and ethylene, and inhibit plant PMEs leading to a higher DM of pectin, which makes it less susceptible to being broken down by fungal cell wall degrading enzymes.

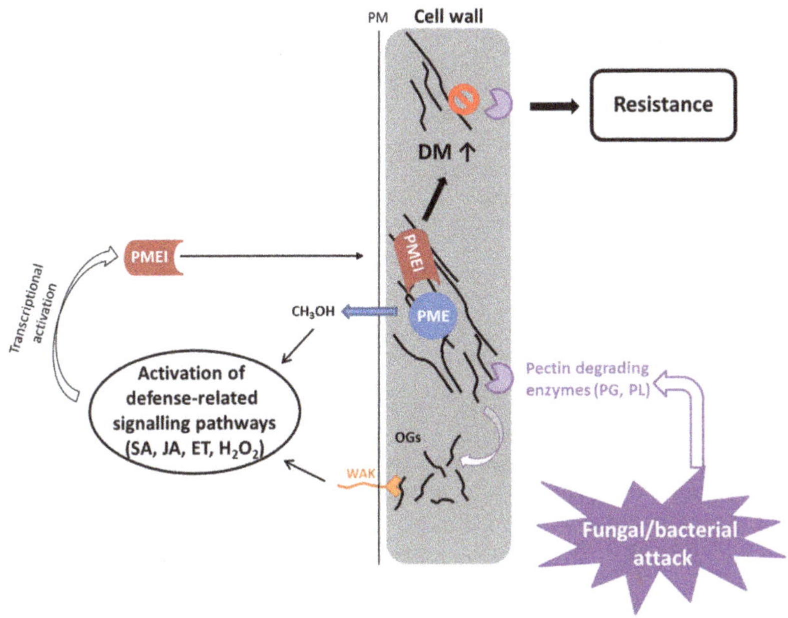

Figure 3. PMEIs modulate PME activity and DM of pectin in response to fungal and bacterial attack. Bacteria and fungi secrete pectin degrading enzymes, like polygalacturonases and pectate/pectin lyases, upon infection. PME activity leads to release of methanol, which serves as an alarm signal, activating the expression of pathogen-related PMEIs. Similarly, OGs, which are generated by degradation of pectin by fungal/bacterial PGs and PLs, serve as DAMPs, leading also to activation of defense related signaling pathways, which have been shown to activate *PMEI* gene expression. PMEIs inhibit PME activity leading to a higher degree of methylesterification of pectin, which contributes to resistance to fungal/bacterial enzymes resulting in less disease symptoms.

PMEIs have also been implicated to play a role in defense against bacteria. *CaPMEI1* from pepper (*Capsicum annuum*) was identified in pepper leaves infected with *Xanthomonas campestris* pv. *vesicatoria* (*Xcv*) [83]. *Xcv* secretes plant cell wall degrading enzymes like pectate lyases and polygalacturonases upon infection. *CaPMEI1* was transcriptionally upregulated by different biotic stress signals (e.g., salicylic acid, jasmonic acid, ethylene, H_2O_2) and consequently, silenced pepper plants exhibited an increased susceptibility to *Xcv*. Conversely, overexpression of *CaPMEI1* in *Arabidopsis* conferred an enhanced resistance to *Pseudomonas syringae* but not to the biotrophic fungus *Hyaloperonospora parasitica*.

Little is known about the role of PMEIs in viral infections. In tobacco, plant PMEs have been shown to be involved in tobacco mosaic virus movement between host cells via interaction with the viral movement protein [84]. Accordingly, overexpression of *PMEIs* in tobacco and *Arabidopsis* counteracted the action of PMEs, leading to increased resistance to tobacco mosaic virus and turnip vein clearing virus infection respectively [85]. This was characterized by the reduced or delayed movement of the viruses and decreased disease symptoms. It is suggested that *PMEI* reduces viral spreading via limiting viral-induced PME activity and by hindering the enlargement of plasmodesmata.

Plants respond to abiotic stresses, e.g., water deficit, salt stress, and temperature extremes, with changes in cell wall architecture, although available literature is more focused on transcriptomic data rather than biochemical experiments [86,87]. PMEIs are also involved in the plants' response to changing environmental conditions, however; only little is known about their molecular role. *CaPMEI1* from pepper is transcriptionally induced by cold treatment, drought stress, as well as the plant stress hormone abscisic acid and hydrogen peroxide [83]. *Arabidopsis* lines overexpressing *CaPMEI1* exhibited an increased tolerance to water stress, as shown by improved germination rate and seedling root growth in the presence of high mannitol concentrations as compared to control plants. In addition, a reduced transpiration rate was accompanied by less withering upon drought stress, and improved tolerance to oxidative stress caused by methyl viologen [83].

Several studies using expression analysis have revealed the functional importance of PMEIs in response to abiotic stresses. RT-PCR analysis of the wheat *TaPMEI* showed the responsiveness of the gene to H_2O_2 stimulus, salinity, water stress (caused by polyethylene glycol) and abscisic acid treatment [48]. Time course analysis further indicated a specificity of the transcriptional response based on the investigated tissue (leaf, stem and root). In silico meta expression analysis using Affymetrics array data for rice [38] revealed the importance for several *PMEI* family members in diverse stress responses. Drought stress, salt stress, cold and anaerobic conditions (during germination) were analyzed, and showed specific transcriptional regulations of *PMEI* genes [38]. Similarly, in *Arabidopsis,* the effect of hormone and stress treatments on the expression of *PMEI* family members was analyzed in silico. [35]. The microarray data analysis of the *PMEI* family showed that several *PMEI* are differentially expressed in response to abscisic acid, gibberellic acid, auxin and methyl jasmonate treatment, and many are involved in cold, drought, heat, oxidation, salt, and wounding regulation in *Arabidopsis* [35]. In the same study, a *cis*-element analysis of promotor regions of the *Brassica campestris* ssp. *chinensis PMEI* gene family also indicated an involvement of several *PMEI* genes in environmental stress responses. For example, of the 100 *PMEI* genes in *Brassica campestris*, 38 *BcPMEI* promotors contained both abscisic acid and auxin (indole-3-acetic acid) related *cis*-elements, 63 and 35 promotors contained heat stress elements and low-temperature response elements, respectively. Additionally, MYB binding sites involved in drought stress were found in 58 *PMEI* promotors [35]. However, these in silico results will have to be validated by experimental data.

An *Arabidopsis* mutant with reduced expression of a *PMEI* gene (At1g62760) exhibited reduced sensitivity to salt stress [88]. Enhanced root growth, higher fresh weight and less chlorosis and necrosis upon NaCl treatment were accompanied with reduced salt stress signaling in the mutant, suggesting this PMEI as a negative regulator of salinity tolerance. In contrast, overexpression of *AtPMEI13* (At5g62360) in *Arabidopsis* also led to increased salt tolerance, as transgenic lines showed higher rates of germination, root growth, and survival under salinity conditions as compared to wild type plants [89]. This phenotype was also observed in plants overexpressing the closest homolog *CbPMEI1* from *Chorispora bungeana*, an alpine plant highly tolerant to chilling and freezing stress. Both genes were repressed by salt stress and abscisic acid but were induced by cold, suggesting distinct roles of the genes in freezing and salinity tolerance. The transgenic overexpression plants showed decreased freezing tolerance; however, they exhibited increased root growth under cold conditions. Plants increase pectin levels in the cell wall under low temperatures [90] probably to reduce cell wall porosity and increase cell adhesion. It was previously reported that DM of pectin increases during cold acclimation in pea [91]. In contrast, loss of AtPME41 in *Arabidopsis* by a transfer DNA insertion caused an increase in freezing sensitivity [92]. Since pectins are important under low temperature conditions, PMEI activity could play a role in balancing the trade-off between freezing tolerance and growth by modifying the mechanical properties of cell walls.

5. Role of PMEIs in Development

PME-mediated changes in methyl-esterification of HG have been shown to affect various vegetative and reproductive stages of plant development (for reviews, see [6,7]). Therefore, we aim to give an overview of the latest results of molecular studies on PMEIs in plant development.

5.1. PMEI Function in Seeds

Upon hydration, mature *Arabidopsis* seeds rapidly extrude a gelatinous substance from seed coat epidermal (SCE) cells called mucilage, which is mainly composed of cell wall polysaccharides [93]. This layer has adhesive properties and possibly serves as a water-reservoir, which might facilitate seed germination, although it has been shown that mucilage production is not necessary for germination or plant fitness in the laboratory [94,95]. The mucilage is mainly composed of pectin, RG-I and small amounts of HG, but hemicellulose and cellulose have also been identified as minor components, as well as arabinogalactan proteins [96]. Several studies provided evidence that the DM of HG determines mucilage properties. For example, seven *PME* genes are specifically expressed in *Arabidopsis* seeds and the *pme58* mutant caused an altered pectin distribution in the mucilage [97]. Additionally, an E3 ubiquitin ligase, FLYING SAUCER1, was proposed to regulate the DM and a mutation lead to reduced mucilage extrusion and increased mucilage adherence [29]. Finally, the subtilase *atsbt1.7* mutant didn't show significant sugar compositional changes in its mucilage but a decrease in mucilage methylation, which was correlated with an increased PME activity [98].

More direct evidence was obtained by investigating the role of *PMEI6*, which was shown to be essential for proper mucilage release [44]. In three *pmei6* mutants, the outer primary cell wall of the SCE cells did not rupture into fragments but remained attached, which caused a delay in mucilage extrusion. This was accompanied by a lack of methylesterified HG (detected by antibodies) and reduced methanol release in the mucilage, as well as increased PME activity in seeds. The reduction in methylesterification in the outer primary cell wall of the three *pmei6* mutants could possibly lead to more abundant Ca^{2+} cross-linking of HG, which causes stronger attachment between adjoining cell walls and therefore impairs release of the mucilage. Further evidence indicates that PMEI14 is also involved in regulating pectin methyl-esterification in seed coat mucilage and that *PMEI14* together with *PMEI6* and *ATSBT1.7* are likely transcriptionally controlled by MYB52 [45]. Knock out mutants of *PMEI14* had an increased PME activity in seed mucilage but not in demucilated seeds. A reduced level of methylesterification of the mucilage was accompanied by an increased amount of calcium cross-linked HG detected by immuno-labelling. These studies show that PMEI6 function affects both seed coat mucilage and the primary SCE cell wall, and is necessary for proper mucilage release. In contrast, PMEI14 function is confined to inhibiting PME activity, specifically in the seed coat mucilage, potentially influencing pectin structure.

PME activity has also been shown to be tightly controlled during germination in *Arabidopsis* and garden cress (*Lepidium sativum*) [99,100], probably leading to cell wall weakening to allow radicle emergence. A small number of *PME* and *PMEIs* are strongly up-regulated upon germination, also in the presence of abscisic acid, which delays endosperm weakening and rupture [100]. Furthermore, overexpression of the PME inhibitor *PMEI5* resulted in a higher DM of seeds and reduced PME activity, which was accompanied by an earlier and faster germination process compared to wildtype [99]. This suggests that PMEI activity is essential for proper timing of seed germination but the precise molecular events are only beginning to be understood, based on modern techniques such as Chip-Seq [45] and potentially based on novel probes to dynamically monitor cell walls [101].

5.2. PMEI Function in Growth Processes

Plant cell growth is regulated by the interplay between the intracellular turgor pressure and the flexibility of the cell wall. Recent reviews have highlighted the importance of pectin, and especially HG, for mediating changes in the mechanical properties for control of growth processes [4,6,102]. In addition, studies using atomic force microscopy provided evidence that increased pectin de-methylesterification correlates with a decrease in cell wall stiffness [21,25].

A PMEI from *Arabidopsis*, AtPMEI4, has been shown to be involved in regulating hypocotyl growth. Growth acceleration of dark-grown hypocotyls requires significant cell wall remodeling, which is associated with transcriptional upregulation of pectin-modifying genes [103]. *AtPMEI4* was upregulated during this developmental transition and its expression was shown to be specific for rapidly elongating epidermis cells in hypocotyls and roots. In an overexpression line, the onset of hypocotyl growth acceleration was delayed, suggesting that the PME inhibitor controls pectin de-methylesterification essential for the timing of the growth acceleration, but not the growth process itself [103]. On the other hand, *OsPMEI28* overexpression in rice had an effect on the growth process, namely inhibition of culm elongation and decreased cell wall thickness of culms, which resulted in a dwarfed phenotype [61]. Conversely, a *pmei4* mutant with elevated PME activity in *Arabidopsis* root cell walls showed an increase in root length [62], suggesting a correlation between increased de-methylesterification and elevated growth.

Controversial results have been reported for purified kiwi PMEI applied on *Arabidopsis* root tips, where an increase of root growth rate was stated upon inhibition of PME activity [104]. Similarly, *Arabidopsis* plants overexpressing *AtPMEI2* show enhanced growth and increased biomass due to enhanced cell expansion [105] as well as increased root length due to stronger cell elongation [26]. Increased root length and area was also observed in *Arabidopsis* plants expressing a PMEI from cotton [39]. Therefore, higher levels of DM of pectin seem also to be able to promote growth and cell expansion.

This correlation was also observed in pollen tube growth. Pollen tubes undergo a fast and unidirectional growth at the apex dependent on ion (e.g., calcium) dynamics and PMEs have been shown to play a role in this process [106,107]. A pollen-specific PME inhibitor, AtPMEI2, was shown to inhibit PME activity in *Arabidopsis* [52]. Although the pollen-specific pectin methylesterase *AtPPME1* was expressed along the whole pollen tube, the localization of AtPMEI2 was exclusively restricted to the apex, the site of elongation. This supports the model that elongation requires higher DM of the cell wall, which is caused by inhibition of PME activity, specifically at the tip. Conversely, along the pollen tube, PME activity is less inhibited, resulting in de-methylesterification, which in turn would lead to formation of calcium cross-links due to the availability of free calcium ions, causing increased cell wall stiffness. Corroborative, expression of *AtPME1* in tobacco pollen tubes inhibited elongation, whereas *AtPMEI2* expression led to an increase in elongation rate [52]. Antisense expression of a pollen-specific *PMEI* from broccoli (*Brassica oleracea*) in *Arabidopsis* triggered silencing of the orthologous *Arabidopsis* gene At1g10770 and resulted in male sterility. This suggests that disturbance of PME activity imbalances the pectin distribution and affects cell wall dynamics in the pollen cell wall with detrimental effect on fertility [108]. In addition, wheat *Tdpmei2.1* and *Tdpmei2.2* transcripts have been shown to undergo complete mRNA processing, specifically in anthers [42]. The occurrence of this mechanism suggests a role of the PME inhibitors during pollen development and reproductive processes, but requires further investigations.

The present data indicate a complex relationship between changes of DM of pectin (based on PMEI activity) and growth processes. Further studies are necessary to elucidate in a tissue-specific manner how the interplay of PMEs and PMEIs regulate pectin methylesterification and affect mechanical properties of cell wall necessary for cell elongation.

5.3. PMEI Function in Organ Formation

The mechanical properties of cell walls have been proposed to play a key role in plant morphogenesis and the control of organ outgrowth [7,109,110]. Organ morphogenesis requires a transition from isotropic growth of undifferentiated cells to anisotropic growth pattern, which is dependent on local cell wall loosening and stiffening by e.g., modification of cell wall components.

AtPMEI3, which is expressed in the apical meristem, was inducible expressed in *Arabidopsis*, which led to hyper-methylesterification of HG throughout the meristem dome upon induction [24]. As a result, the formation of both floral and flower organ primordia was inhibited. Using atomic force microscopy, it was shown that the reduced de-methylesterification of HG was accompanied by increased stiffness of cell walls in the shoot apical meristem in the *AtPMEI3* overexpressing lines [25]. This further resulted in shorter cells and loss of growth asymmetry, showing that anisotropic growth requires an asymmetric loosening of longitudinal walls, which is caused by fine-tuned de-methylesterification of pectin.

Constitutive overexpression of *AtPMEI5* in *Arabidopsis* also caused strong morphological phenotypes [111,112]. The stems of the transgenic plants showed an aberrant, twisted growth pattern, especially at branching points and branching organs (e.g., leaf or inflorescence), which failed to separate properly from the main stem [111]. Additionally, root growth was disturbed in the mutant and curled leaves and misshapen siliques were observed, which was proposed to be a result of activated BR signaling [112]. This indicates that PMEIs are involved in the fine-tuning of cell wall biomechanical properties via HG methylesterification that is required for proper organ formation and separation.

PMEIs might also play a role in organ senescence as changes in pectin methylesterification have been described, e.g., for petal senescence [113,114]. However, so far, only transcriptional changes of putative invertase/pectin methylesterase inhibitor genes in petals during senescence have been reported and further, more detailed studies will be necessary to investigate their putative function.

5.4. PMEIs in Fruit Development

Fruit ripening is a highly controlled developmental event during which the fruit undergoes various changes, e.g., in sugar contents, color, flavor, aroma, texture. Softening is a very (economically) important physiological change and several cell wall-modifying enzymes are involved in this process [115]. The first inhibitor of PME was identified in ripe kiwi fruit by Balisteri et al. [33] and was found active only in the mature stage, whereas it was undetectable in unripe fruits [116]. Furthermore, the expression levels of the *PMEI* genes in kiwi fruit increased with the progression of fruit maturation [117]. In banana (*Musa acuminata*), a PME inhibitor, *MaPMEI1*, was shown to be specifically expressed in the fruit and up-regulated during the ethylene-dependent ripening process [46]. Similarly, a PMEI has been identified in the tomato fruit [59]. *SolyPMEI* was strongly expressed in red fruits and, in lower levels, also in flowers and pollen, but not in green fruits. Interaction with the tomato PME-1 enzyme in the fruit and increased *SolyPMEI* expression levels with progression of ripening stages implicates the inhibitor in modulating PME activity and degree of methyl-esterification during fruit ripening in tomato. In grape (*Vitis vinifera*), *VvPMEI1* was found to be expressed only during the early phases of berry development, characterized by a rapid berry growth both through cell division and expansion [41]. *VvPMEI1* could be involved in regulating PME activity required for rapid cell growth and to maintain firmness, or it could prevent early softening of grape berries related to pectin degradation. This indicates a role for PMEIs in modulating PME activity and pectin methylesterification in different stages of fruit development.

6. Potential of PMEIs for Applications

Lignocellulosic biomass constitutes a sustainable renewable source of chemicals and fuels, but a major bottleneck for an industrial utilization is the natural recalcitrance of the cell wall due to its complex and heterogeneous structure. Modification of cell wall polysaccharides can improve the degradation into soluble sugars, which can be used by microorganisms for fermentation, a process called saccharification [118]. Improved saccharification can make an industrial process more efficient and reduce cost for pre-treatments of the biomass. Lionetti et al. [105] showed that overexpression of *AtPMEI2* in *Arabidopsis* leads to a higher enzymatic saccharification efficiency, indicating a reduced cell wall recalcitrance likely due to the higher degree of HG methylesterification. The same was observed for wheat plants expressing the kiwi *PMEI*, suggesting that both cellulose and hemicellulose are better digestible during enzymatic hydrolysis in the transgenic plants. As a side effect, transgenic *Arabidopsis* plants had a higher biomass yield due to enhanced cell expansion. Further, a negative correlation between the level of de-methylesterified HG and cellulose degradability was found for 24 *Arabidopsis* accessions [119]. The reduction of egg-box structures or other crosslinks of methylesterified HG might be responsible for the accessibility of cellulose fibers for enzymatic digestion, which in turn results in increased saccharification efficiency. This indicates the potential of PMEI as a HG modifying enzyme to improve plant tissues for saccharification by targeted mutation or in wide-scale selection and breeding of plants for further utilization (e.g., as biofuels or for production of chemicals).

PMEIs have been shown to play a role in ripening processes (see above). Therefore, a potential application of this inhibitor would be in food (fruit and vegetable) processing. For food technology, endogenous PME activity is highly relevant because it determines the physiochemical and functional properties of pectin, thereby affecting the quality of plant-derived food products. PME activity in fruit juice causes undesired phase separation because de-methylesterified HG forms calcium crosslinks, and the resulting calcium pectate is insoluble and precipitates [120]. Addition of PMEI to orange juice was shown to control PME activity even over long storage, thereby avoiding phase separation. Application of PMEI would reduce required thermal treatments during juice processing and should result in better flavor and product quality [120]. For development or optimization of food-processing steps, it would be beneficial to determine PME activity in plant tissues. A PMEI-based molecular probe was shown to successfully label plant PMEs in various plant tissues and moreover, to differentiate between active and inactivated PME molecules [121]. This application of PMEIs would give important information about presence and localization of active PMEs after different food processing steps (e.g., thermal or high pressure treatment, microwaving). PMEIs have also been proposed for application in wine and marc production [41]. Pectin de-methylesterification releases high levels of methanol during different stages of grape processing, greatly affecting the composition of the final product [122,123]. Here, PMEIs could be used to reduce methanol formation in grape must and marc, as well as in products derived by fermentation and distillation [124]. Apart from applications in food industry, manipulation of PMEIs in planta could also be of interest. Overexpression of *PMEIs*, in banana fruits for example, would not only control over-softening but also likely extend banana shelf-life [46]. In conclusion, PMEIs could be good tools to control endogenous PME activity during food production processes, thereby positively influencing product quality.

As mentioned above, PMEIs play an important role in the response to biotic and environmental stresses. Therefore, the inhibitors could be potential targets for genetic engineering, aimed to improve plant fitness. In cold as well as salt stress conditions, PMEIs might be beneficial for balancing growth maintenance with stress tolerance and overexpression of *PMEIs* would be a promising way to improve salt tolerance of crops [89]. In addition, PMEI activity has been shown to be essential for resistance against fungal and bacterial pathogens [26,79,83]. Therefore, it could be useful as a potential molecular marker in plant breeding programs targeted to the selection of crop varieties with durable resistance.

Author Contributions: Conceptualization, A.W.; Writing-Original Draft Preparation, A.W.; Writing-Review & Editing, A.W. & B.U.; Visualization, A.W.; Funding Acquisition, B.U. & A.W.

Funding: This research was supported by DFG (grant number: US98/13-1), the German Federal Ministry of Education and Research (grant number: 031B0193B) as well as the Ministry of Innovation, Science and Research of North-Rhine Westphalia within the framework of the North-Rhine Westphalia Strategieprojekt BioEconomy Science Center (grant number 313/323-400-00213).

Conflicts of Interest: The authors declare no conflict of interest

Abbreviations

PMEI	Pectin methylesterase inhibitor
PME	Pectin methyl esterase
DM	Degree of methylation
HG	Homogalacturonan
GalA	D-galacturonic acid
GAUT	Galacturonosyltransferases
OGs	Oligogalacturonides
DAMP	Damage associated molecular pattern
H_2O_2	Hydrogen peroxide
SCE	Seed coat epidermis
MUM1	Mucilage-modified mutant 1

References

1. Wolf, S.; Mouille, G.; Pelloux, J. Homogalacturonan methyl-esterification and plant development. *Mol. Plant* **2009**, *2*, 851–860. [CrossRef]
2. Mohnen, D. Pectin structure and biosynthesis. *Curr. Opin. Plant Biol.* **2008**, *11*, 266–277. [CrossRef]
3. Braybrook, S.A.; Hofte, H.; Peaucelle, A. Probing the mechanical contributions of the pectin matrix. *Plant Signal. Behav.* **2012**, *7*, 1037–1041. [CrossRef]
4. Palin, R.; Geitmann, A. The role of pectin in plant morphogenesis. *Biosystems* **2012**, *109*, 397–402. [CrossRef]
5. Daher, F.B.; Braybrook, S.A. How to let go: Pectin and plant cell adhesion. *Front. Plant Sci.* **2015**, *6*, 523. [CrossRef]
6. Levesque-Tremblay, G.; Pelloux, J.; Braybrook, S.A.; Müller, K. Tuning of pectin methylesterification: Consequences for cell wall biomechanics and development. *Planta* **2015**, *242*, 791–811. [CrossRef]
7. Saffer, A.M. Expanding roles for pectins in plant development. *Integr. Plant Biol.* **2018**. [CrossRef]
8. Sterling, J.D.; Quigley, H.F.; Orellana, A.; Mohnen, D. The catalytic site of the pectin biosynthetic enzyme alpha-1,4-galacturonosyltransferase is located in the lumen of the Golgi. *Plant Physiol.* **2001**, *127*, 360–371.
9. Sterling, J.D.; Atmodjo, M.A.; Inwood, S.E.; Kumar Kolli, V.S.; Quigley, H.F.; Hahn, M.G.; Mohnen, D. Functional identification of an *Arabidopsis* pectin biosynthetic homogalacturonan galacturonosyltransferase. *Proc. Natl. Acad. Sci. USA* **2006**, *103*, 5236–5241. [CrossRef]
10. Atmodjo, M.A.; Sakuragi, Y.; Zhu, X.; Burrell, A.J.; Mohanty, S.S.; Atwood, J.A.; Orlando, R.; Scheller, H.V.; Mohnen, D. Galacturonosyltransferase (GAUT)1 and GAUT7 are the core of a plant cell wall pectin biosynthetic homogalacturonan:galacturonosyltransferase complex. *Proc. Natl. Acad. Sci. USA* **2011**, *108*, 20225–20230. [CrossRef]
11. Biswal, A.K.; Atmodjo, M.A.; Li, M.; Baxter, H.L.; Yoo, C.G.; Pu, Y.; Lee, Y.-C.; Mazarei, M.; Black, I.M.; Zhang, J.-Y.; et al. Sugar release and growth of biofuel crops are improved by downregulation of pectin biosynthesis. *Nat. Biotechnol.* **2018**, *36*, 249–257. [CrossRef]
12. Mouille, G.; Ralet, M.-C.; Cavelier, C.; Eland, C.; Effroy, D.; Hématy, K.; McCartney, L.; Truong, H.N.; Gaudon, V.; Thibault, J.-F.; et al. Homogalacturonan synthesis in *Arabidopsis thaliana* requires a Golgi-localized protein with a putative methyltransferase domain. *Plant J.* **2007**, *50*, 605–614. [CrossRef]
13. Krupková, E.; Immerzeel, P.; Pauly, M.; Schmülling, T. The *TUMOROUS SHOOT DEVELOPMENT2* gene of *Arabidopsis* encoding a putative methyltransferase is required for cell adhesion and co-ordinated plant development. *Plant J.* **2007**, *50*, 735–750. [CrossRef]
14. Ralet, M.-C.; Crépeau, M.-J.; Buchholt, H.-C.; Thibault, J.-F. Polyelectrolyte behaviour and calcium binding properties of sugar beet pectins differing in their degrees of methylation and acetylation. *Biochem. Eng. J.* **2003**, *16*, 191–201. [CrossRef]
15. De Souza, A.J.; Pauly, M. Comparative genomics of pectinacetylesterases: Insight on function and biology. *Plant Signal. Behav.* **2015**, *10*, e1055434. [CrossRef]
16. Gou, J.-Y.; Miller, L.M.; Hou, G.; Yu, X.-H.; Chen, X.-Y.; Liu, C.-J. Acetylesterase-mediated deacetylation of pectin impairs cell elongation, pollen germination, and plant reproduction. *Plant Cell* **2012**, *24*, 50–65. [CrossRef]

17. Braccini, I.; Pérez, S. Molecular basis of Ca^{2+}-induced gelation in alginates and pectins: The egg-box model revisited. *Biomacromolecules* **2001**, *2*, 1089–1096. [CrossRef]
18. White, P.B.; Wang, T.; Park, Y.B.; Cosgrove, D.J.; Hong, M. Water–polysaccharide interactions in the primary cell wall of *Arabidopsis thaliana* from polarization transfer solid-state NMR. *J. Am. Chem. Soc.* **2014**, *136*, 10399–10409. [CrossRef]
19. Ha, M.-A.; Apperley, D.C.; Jarvis, M.C. Molecular rigidity in dry and hydrated onion cell walls. *Plant Physiol.* **1997**, *115*, 593–598.
20. Tibbits, C.W.; MacDougall, A.J.; Ring, S.G. Calcium binding and swelling behaviour of a high methoxyl pectin gel. *Carbohydr. Res.* **1998**, *310*, 101–107. [CrossRef]
21. Peaucelle, A.; Braybrook, S.A.; Le Guillou, L.; Bron, E.; Kuhlemeier, C.; Höfte, H. Pectin-induced changes in cell wall mechanics underlie organ initiation in *Arabidopsis*. *Curr. Biol.* **2011**, *21*, 1720–1726. [CrossRef]
22. Amsbury, S.; Hunt, L.; Elhaddad, N.; Baillie, A.; Lundgren, M.; Verhertbruggen, Y.; Scheller, H.V.; Knox, J.P.; Fleming, A.J.; Gray, J.E. Stomatal function requires pectin de-methyl-esterification of the guard cell wall. *Curr. Biol.* **2016**, *26*, 2899–2906. [CrossRef]
23. Carter, R.; Woolfenden, H.; Baillie, A.; Amsbury, S.; Carroll, S.; Healicon, E.; Sovatzoglou, S.; Braybrook, S.; Gray, J.E.; Hobbs, J.; et al. Stomatal opening involves polar, not radial, stiffening of guard cells. *Curr. Biol.* **2017**, *27*, 2974.e2–2983.e2. [CrossRef]
24. Peaucelle, A.; Louvet, R.; Johansen, J.N.; Höfte, H.; Laufs, P.; Pelloux, J.; Mouille, G. *Arabidopsis* phyllotaxis is controlled by the methyl-esterification status of cell-wall pectins. *Curr. Biol.* **2008**, *18*, 1943–1948. [CrossRef]
25. Peaucelle, A.; Wightman, R.; Höfte, H. The control of growth symmetry breaking in the Arabidopsis hypocotyl. *Curr. Biol.* **2015**, *25*, 1746–1752. [CrossRef]
26. Lionetti, V.; Raiola, A.; Camardella, L.; Giovane, A.; Obel, N.; Pauly, M.; Favaron, F.; Cervone, F.; Bellincampi, D. Overexpression of pectin methylesterase inhibitors in *Arabidopsis* restricts fungal infection by *Botrytis cinerea*. *Plant Physiol.* **2007**, *143*, 1871–1880. [CrossRef]
27. Osorio, S.; Castillejo, C.; Quesada, M.A.; Medina-Escobar, N.; Brownsey, G.J.; Suau, R.; Heredia, A.; Botella, M.A.; Valpuesta, V. Partial demethylation of oligogalacturonides by pectin methyl esterase 1 is required for eliciting defence responses in wild strawberry (*Fragaria vesca*). *Plant J.* **2008**, *54*, 43–55. [CrossRef]
28. Sénéchal, F.; Wattier, C.; Rustérucci, C.; Pelloux, J. Homogalacturonan-modifying enzymes: Structure, expression, and roles in plants. *J. Exp. Bot.* **2014**, *65*, 5125–5160. [CrossRef]
29. Voiniciuc, C.; Dean, G.H.; Griffiths, J.S.; Kirchsteiger, K.; Hwang, Y.T.; Gillett, A.; Dow, G.; Western, T.L.; Estelle, M.; Haughn, G.W. FLYING SAUCER1 Is a transmembrane RING E3 ubiquitin ligase that regulates the degree of pectin methylesterification in Arabidopsis seed mucilage. *Plant Cell* **2013**, *25*, 944–959. [CrossRef]
30. Sénéchal, F.; L'Enfant, M.; Domon, J.-M.; Rosiau, E.; Crépeau, M.-J.; Surcouf, O.; Esquivel-Rodriguez, J.; Marcelo, P.; Mareck, A.; Guérineau, F.; et al. Tuning of pectin methylesterification. *J. Biol. Chem.* **2015**, *290*, 23320–23335. [CrossRef]
31. Juge, N. Plant protein inhibitors of cell wall degrading enzymes. *Trends Plant Sci.* **2006**, *11*, 359–367. [CrossRef]
32. Pelloux, J.; Rustérucci, C.; Mellerowicz, E.J. New insights into pectin methylesterase structure and function. *Trends Plant Sci.* **2007**, *12*, 267–277. [CrossRef]
33. Balestrieri, C.; Castaldo, D.; Giovane, A.; Quagliuolo, L.; Servillo, L. A glycoprotein inhibitor of pectin methylesterase in kiwi fruit (*Actinidia chinensis*). *Eur. J. Biochem.* **1990**, *193*, 183–187.
34. Wang, M.; Yuan, D.; Gao, W.; Li, Y.; Tan, J.; Zhang, X. A comparative genome analysis of PME and PMEI families reveals the evolution of pectin metabolism in plant cell walls. *PLoS ONE* **2013**, *8*, e72082. [CrossRef]
35. Liu, T.; Yu, H.; Xiong, X.; Yue, X.; Yu, Y.; Huang, L.; Cao, J. Genome-wide identification, molecular evolution, and expression profiling analysis of pectin methylesterase inhibitor genes in *Brassica campestris* ssp. *chinensis*. *Int. J. Mol. Sci.* **2018**, *19*, 1338. [CrossRef]
36. Tan, C.; Liu, Z.; Huang, S.; Li, C.; Ren, J.; Tang, X.; Liu, W.; Peng, S.; Feng, H. Pectin methylesterase inhibitor (PMEI) family can be related to male sterility in Chinese cabbage (*Brassica rapa* ssp. *pekinensis*). *Mol. Genet. Genom.* **2018**, *293*, 343–357. [CrossRef]
37. Pinzón-Latorre, D.; Deyholos, M.K. Characterization and transcript profiling of the pectin methylesterase (PME) and pectin methylesterase inhibitor (PMEI) gene families in flax (*Linum usitatissimum*). *BMC Genom.* **2013**, *14*, 742. [CrossRef]

38. Nguyen, H.P.; Jeong, H.Y.; Kim, H.; Kim, Y.C.; Lee, C. Molecular and biochemical characterization of rice pectin methylesterase inhibitors (OsPMEIs). *Plant Physiol. Biochem.* **2016**, *101*, 105–112. [CrossRef]
39. Liu, N.; Sun, Y.; Pei, Y.; Zhang, X.; Wang, P.; Li, X.; Li, F.; Hou, Y. A pectin methylesterase inhibitor enhances resistance to *Verticillium* wilt. *Plant Physiol.* **2018**, *176*, 2202–2220. [CrossRef]
40. Winter, D.; Vinegar, B.; Nahal, H.; Ammar, R.; Wilson, G.V.; Provart, N.J. An "Electronic Fluorescent Pictograph" browser for exploring and analyzing large-scale biological data sets. *PLoS ONE* **2007**, *2*, e718. [CrossRef]
41. Lionetti, V.; Raiola, A.; Mattei, B.; Bellincampi, D. The grapevine VvPMEI1 gene encodes a novel functional pectin methylesterase inhibitor associated to grape berry development. *PLoS ONE* **2015**, *10*, e0133810. [CrossRef]
42. Rocchi, V.; Janni, M.; Bellincampi, D.; Giardina, T.; D'Ovidio, R. Intron retention regulates the expression of pectin methyl esterase inhibitor (PMEI) genes during wheat growth and development. *Plant Biol.* **2012**, *14*, 365–373. [CrossRef]
43. Pinzon-Latorre, D.; Deyholos, M.K. Pectinmethylesterases (PME) and pectinmethylesterase inhibitors (PMEI) enriched during phloem fiber development in flax (*Linum usitatissimum*). *PLoS ONE* **2014**, *9*, e105386. [CrossRef]
44. Saez-Aguayo, S.; Ralet, M.-C.; Berger, A.; Botran, L.; Ropartz, D.; Marion-Poll, A.; North, H.M. PECTIN METHYLESTERASE INHIBITOR6 promotes *Arabidopsis* mucilage release by limiting methylesterification of homogalacturonan in seed coat epidermal cells. *Plant Cell* **2013**, *25*, 308–323. [CrossRef]
45. Shi, D.; Ren, A.; Tang, X.; Qi, G.; Xu, Z.; Chai, G.; Hu, R.; Zhou, G.; Kong, Y. *MYB52* negatively regulates pectin demethylesterification in seed coat mucilage. *Plant Physiol.* **2018**, *176*, 2737–2749. [CrossRef]
46. Srivastava, S.; Gupta, S.M.; Sane, A.P.; Nath, P. Isolation and characterization of ripening related pectin methylesterase inhibitor gene from banana fruit. *Physiol. Mol. Biol. Plants* **2012**, *18*, 191–195. [CrossRef]
47. Lionetti, V.; Fabri, E.; De Caroli, M.; Hansen, A.R.; Willats, W.G.T.; Piro, G.; Bellincampi, D. Three pectin methylesterase inhibitors protect cell wall integrity for Arabidopsis immunity to *Botrytis*. *Plant Physiol.* **2017**, *173*, 1844–1863. [CrossRef]
48. Hong, M.J.; Kim, D.Y.; Lee, T.G.; Jeon, W.B.; Seo, Y.W. Functional characterization of pectin methylesterase inhibitor (PMEI) in wheat. *Genes Genet. Syst.* **2010**, *85*, 97–106.
49. An, S.H.; Choi, H.W.; Hong, J.K.; Hwang, B.K. Regulation and function of the pepper pectin methylesterase inhibitor (CaPMEI1) gene promoter in defense and ethylene and methyl jasmonate signaling in plants. *Planta* **2009**, *230*, 1223–1237. [CrossRef]
50. Bosch, M.; Hepler, P.K. Pectin methylesterases and pectin dynamics in pollen tubes. *Plant Cell* **2005**, *17*, 3219–3226. [CrossRef]
51. Parre, E.; Geitmann, A. Pectin and the role of the physical properties of the cell wall in pollen tube growth of *Solanum chacoense*. *Planta* **2005**, *220*, 582–592. [CrossRef]
52. Röckel, N.; Wolf, S.; Kost, B.; Rausch, T.; Greiner, S. Elaborate spatial patterning of cell-wall PME and PMEI at the pollen tube tip involves PMEI endocytosis, and reflects the distribution of esterified and de-esterified pectins. *Plant J.* **2008**, *53*, 133–143. [CrossRef]
53. Hothorn, M.; Wolf, S.; Aloy, P.; Greiner, S.; Scheffzek, K. Structural insights into the target specificity of plant invertase and pectin methylesterase inhibitory proteins. *Plant Cell* **2004**, *16*, 3437–3447. [CrossRef]
54. Giovane, A.; Servillo, L.; Balestrieri, C.; Raiola, A.; D'Avino, R.; Tamburrini, M.; Ciardiello, M.A.; Camardella, L. Pectin methylesterase inhibitor. *Biochim. Biophys. Acta* **2004**, *1696*, 245–252. [CrossRef]
55. Camardella, L.; Carratore, V.; Ciardiello, M.A.; Servillo, L.; Balestrieri, C.; Giovane, A. Kiwi protein inhibitor of pectin methylesterase amino-acid sequence and structural importance of two disulfide bridges. *Eur. J. Biochem.* **2000**, *267*, 4561–4565.
56. Di Matteo, A.; Giovane, A.; Raiola, A.; Camardella, L.; Bonivento, D.; De Lorenzo, G.; Cervone, F.; Bellincampi, D.; Tsernoglou, D. Structural basis for the interaction between pectin methylesterase and a specific inhibitor protein. *Plant Cell* **2005**, *17*, 849–858. [CrossRef]
57. Raiola, A.; Camardella, L.; Giovane, A.; Mattei, B.; De Lorenzo, G.; Cervone, F.; Bellincampi, D. Two *Arabidopsis thaliana* genes encode functional pectin methylesterase inhibitors. *FEBS Lett.* **2004**, *557*, 199–203.

58. Jolie, R.P.; Duvetter, T.; Van Loey, A.M.; Hendrickx, M.E. Pectin methylesterase and its proteinaceous inhibitor: A review. *Carbohydr. Res.* **2010**, *345*, 2583–2595. [CrossRef]
59. Reca, I.B.; Lionetti, V.; Camardella, L.; D'Avino, R.; Giardina, T.; Cervone, F.; Bellincampi, D. A functional pectin methylesterase inhibitor protein (SolyPMEI) is expressed during tomato fruit ripening and interacts with PME-1. *Plant Mol. Biol.* **2012**, *79*, 429–442. [CrossRef]
60. Bonavita, A.; Carratore, V.; Ciardiello, M.A.; Giovane, A.; Servillo, L.; D'Avino, R. Influence of pH on the structure and function of kiwi pectin methylesterase inhibitor. *J. Agric. Food Chem.* **2016**, *64*, 5866–5876. [CrossRef]
61. Nguyen, H.P.; Jeong, H.Y.; Jeon, S.H.; Kim, D.; Lee, C. Rice pectin methylesterase inhibitor28 (OsPMEI28) encodes a functional PMEI and its overexpression results in a dwarf phenotype through increased pectin methylesterification levels. *J. Plant Physiol.* **2017**, *208*, 17–25. [CrossRef]
62. Sénéchal, F.; Mareck, A.; Marcelo, P.; Lerouge, P.; Pelloux, J. Arabidopsis PME17 activity can be controlled by pectin methylesterase inhibitor 4. *Plant Signal. Behav.* **2015**, *10*, e983351. [CrossRef]
63. Hocq, L.; Sénéchal, F.; Lefebvre, V.; Lehner, A.; Domon, J.-M.; Mollet, J.-C.; Dehors, J.; Pageau, K.; Marcelo, P.; Guérineau, F.; et al. Combined experimental and computational approaches reveal distinct pH dependence of pectin methylesterase inhibitors. *Plant Physiol.* **2017**, *173*, 1075–1093. [CrossRef]
64. Hamann, T. The plant cell wall integrity maintenance mechanism—A case study of a cell wall plasma membrane signaling network. *Phytochemistry* **2015**, *112*, 100–109. [CrossRef]
65. Espino, J.J.; Gutiérrez-Sánchez, G.; Brito, N.; Shah, P.; Orlando, R.; González, C. The *Botrytis cinerea* early secretome. *Proteomics* **2010**, *10*, 3020–3034. [CrossRef]
66. D'Ovidio, R.; Mattei, B.; Roberti, S.; Bellincampi, D. Polygalacturonases, polygalacturonase-inhibiting proteins and pectic oligomers in plant–pathogen interactions. *Biochim. Biophys. Acta* **2004**, *1696*, 237–244. [CrossRef]
67. Lionetti, V. PECTOPLATE: The simultaneous phenotyping of pectin methylesterases, pectinases, and oligogalacturonides in plants during biotic stresses. *Front. Plant Sci.* **2015**, *6*, 331. [CrossRef]
68. Raiola, A.; Lionetti, V.; Elmaghraby, I.; Immerzeel, P.; Mellerowicz, E.J.; Salvi, G.; Cervone, F.; Bellincampi, D. Pectin methylesterase is induced in Arabidopsis upon infection and is necessary for a successful colonization by necrotrophic pathogens. *Mol. Plant Microbe Interact.* **2011**, *24*, 432–440. [CrossRef]
69. Liu, N.; Zhang, X.; Sun, Y.; Wang, P.; Li, X.; Pei, Y.; Li, F.; Hou, Y. Molecular evidence for the involvement of a polygalacturonase-inhibiting protein, GhPGIP1, in enhanced resistance to *Verticillium* and *Fusarium* wilts in cotton. *Sci. Rep.* **2017**, *7*, 39840. [CrossRef]
70. Liu, C.-Q.; Hu, K.-D.; Li, T.-T.; Yang, Y.; Yang, F.; Li, Y.-H.; Liu, H.-P.; Chen, X.-Y.; Zhang, H. Polygalacturonase gene pgxB in *Aspergillus niger* is a virulence factor in apple fruit. *PLoS ONE* **2017**, *12*, e0173277. [CrossRef]
71. Bravo Ruiz, G.; Di Pietro, A.; Roncero, M.I.G. Combined action of the major secreted exo- and endopolygalacturonases is required for full virulence of *Fusarium oxysporum*. *Mol. Plant Pathol.* **2016**, *17*, 339–353. [CrossRef]
72. Cho, Y.; Jang, M.; Srivastava, A.; Jang, J.-H.; Soung, N.-K.; Ko, S.-K.; Kang, D.-O.; Ahn, J.S.; Kim, B.Y. A pectate lyase-coding gene abundantly expressed during early stages of infection is required for full virulence in *Alternaria brassicicola*. *PLoS ONE* **2015**, *10*, e0127140. [CrossRef]
73. Bethke, G.; Grundman, R.E.; Sreekanta, S.; Truman, W.; Katagiri, F.; Glazebrook, J. Arabidopsis PECTIN METHYLESTERASEs contribute to immunity against *Pseudomonas syringae*. *Plant Physiol.* **2014**, *164*, 1093–1107. [CrossRef]
74. Ferrari, S.; Savatin, D.V.; Sicilia, F.; Gramegna, G.; Cervone, F.; De Lorenzo, G. Oligogalacturonides: Plant damage-associated molecular patterns and regulators of growth and development. *Front. Plant Sci.* **2013**, *4*, 49. [CrossRef]
75. Kohorn, B.D.; Kohorn, S.L.; Saba, N.J.; Martinez, V.M. Requirement for pectin methyl esterase and preference for fragmented over native pectins for wall-associated kinase-activated, EDS1/PAD4-dependent stress response in Arabidopsis. *J. Biol. Chem.* **2014**, *289*, 18978–18986. [CrossRef]
76. Brutus, A.; Sicilia, F.; Macone, A.; Cervone, F.; De Lorenzo, G. A domain swap approach reveals a role of the plant wall-associated kinase 1 (WAK1) as a receptor of oligogalacturonides. *Proc. Natl. Acad. Sci. USA* **2010**, *107*, 9452–9457.

77. Osorio, S.; Bombarely, A.; Giavalisco, P.; Usadel, B.; Stephens, C.; Aragüez, I.; Medina-Escobar, N.; Botella, M.A.; Fernie, A.R.; Valpuesta, V. Demethylation of oligogalacturonides by FaPE1 in the fruits of the wild strawberry *Fragaria vesca* triggers metabolic and transcriptional changes associated with defence and development of the fruit. *J. Exp. Bot.* **2011**, *62*, 2855–2873. [CrossRef]
78. Komarova, T.V.; Sheshukova, E.V.; Dorokhov, Y.L. Cell wall methanol as a signal in plant immunity. *Front. Plant Sci.* **2014**, *5*, 101. [CrossRef]
79. Volpi, C.; Janni, M.; Lionetti, V.; Bellincampi, D.; Favaron, F.; D'Ovidio, R. The ectopic expression of a pectin methyl esterase inhibitor increases pectin methyl esterification and limits fungal diseases in wheat. *Mol. Plant Microbe Interact.* **2011**, *24*, 1012–1019. [CrossRef]
80. Tundo, S.; Kalunke, R.; Janni, M.; Volpi, C.; Lionetti, V.; Bellincampi, D.; Favaron, F.; D'Ovidio, R. Pyramiding PvPGIP2 and TAXI-III but not PvPGIP2 and PMEI enhances resistance against *Fusarium graminearum*. *Mol. Plant-Microbe Interact.* **2016**, *29*, 629–639. [CrossRef]
81. Marzin, S.; Hanemann, A.; Sharma, S.; Hensel, G.; Kumlehn, J.; Schweizer, G.; Röder, M.S. Are *PECTIN ESTERASE INHIBITOR* genes involved in mediating resistance to *Rhynchosporium commune* in barley? *PLoS ONE* **2016**, *11*, e0150485. [CrossRef]
82. Lionetti, V.; Cervone, F.; Bellincampi, D. Methyl esterification of pectin plays a role during plant–pathogen interactions and affects plant resistance to diseases. *J. Plant Physiol.* **2012**, *169*, 1623–1630. [CrossRef]
83. An, S.H.; Sohn, K.H.; Choi, H.W.; Hwang, I.S.; Lee, S.C.; Hwang, B.K. Pepper pectin methylesterase inhibitor protein CaPMEI1 is required for antifungal activity, basal disease resistance and abiotic stress tolerance. *Planta* **2008**, *228*, 61–78. [CrossRef]
84. Chen, M.-H.; Citovsky, V. Systemic movement of a tobamovirus requires host cell pectin methylesterase. *Plant J.* **2003**, *35*, 386–392.
85. Lionetti, V.; Raiola, A.; Cervone, F.; Bellincampi, D. Transgenic expression of pectin methylesterase inhibitors limits tobamovirus spread in tobacco and Arabidopsis. *Mol. Plant Pathol.* **2014**, *15*, 265–274. [CrossRef]
86. Le Gall, H.; Philippe, F.; Domon, J.-M.; Gillet, F.; Pelloux, J.; Rayon, C. Cell wall metabolism in response to abiotic stress. *Plants* **2015**, *4*, 112–166. [CrossRef]
87. Tenhaken, R. Cell wall remodeling under abiotic stress. *Front. Plant Sci.* **2014**, *5*, 771. [CrossRef]
88. Jithesh, M.N.; Wally, O.S.D.; Manfield, I.; Critchley, A.T.; Hiltz, D.; Prithiviraj, B. Analysis of seaweed extract-induced transcriptome leads to identification of a negative regulator of salt tolerance in Arabidopsis. *HortScience* **2012**, *47*, 704–709.
89. Chen, J.; Chen, X.; Zhang, Q.; Zhang, Y.; Ou, X.; An, L.; Feng, H.; Zhao, Z. A cold-induced pectin methyl-esterase inhibitor gene contributes negatively to freezing tolerance but positively to salt tolerance in Arabidopsis. *J. Plant Physiol.* **2018**, *222*, 67–78. [CrossRef]
90. Solecka, D.; Zebrowski, J.; Kacperska, A. Are pectins involved in cold acclimation and de-acclimation of winter oil-seed rape plants? *Ann. Bot.* **2008**, *101*, 521–530. [CrossRef]
91. Baldwin, L.; Domon, J.-M.; Klimek, J.F.; Fournet, F.; Sellier, H.; Gillet, F.; Pelloux, J.; Lejeune-Hénaut, I.; Carpita, N.C.; Rayon, C. Structural alteration of cell wall pectins accompanies pea development in response to cold. *Phytochemistry* **2014**, *104*, 37–47. [CrossRef]
92. Qu, T.; Liu, R.; Wang, W.; An, L.; Chen, T.; Liu, G.; Zhao, Z. Brassinosteroids regulate pectin methylesterase activity and AtPME41 expression in Arabidopsis under chilling stress. *Cryobiology* **2011**, *63*, 111–117. [CrossRef]
93. North, H.M.; Berger, A.; Saez-Aguayo, S.; Ralet, M.-C. Understanding polysaccharide production and properties using seed coat mutants: Future perspectives for the exploitation of natural variants. *Ann. Bot.* **2014**, *114*, 1251–1263. [CrossRef]
94. Western, T.L.; Skinner, D.J.; Haughn, G.W. Differentiation of mucilage secretory cells of the Arabidopsis seed coat. *Plant Physiol.* **2000**, *122*, 345–356.
95. Francoz, E.; Ranocha, P.; Burlat, V.; Dunand, C. Arabidopsis seed mucilage secretory cells: Regulation and dynamics. *Trends Plant Sci.* **2015**, *20*, 515–524. [CrossRef]
96. Voiniciuc, C.; Yang, B.; Schmidt, M.H.-W.; Günl, M.; Usadel, B. Starting to gel: How Arabidopsis seed coat epidermal cells produce specialized secondary cell walls. *Int. J. Mol. Sci.* **2015**, *16*, 3452–3473. [CrossRef]
97. Turbant, A.; Fournet, F.; Lequart, M.; Zabijak, L.; Pageau, K.; Bouton, S.; Van Wuytswinkel, O. PME58 plays a role in pectin distribution during seed coat mucilage extrusion through homogalacturonan modification. *J. Exp. Bot.* **2016**, *67*, 2177–2190. [CrossRef]

98. Rautengarten, C.; Usadel, B.; Neumetzler, L.; Hartmann, J.; Büssis, D.; Altmann, T. A subtilisin-like serine protease essential for mucilage release from Arabidopsis seed coats. *Plant J.* **2008**, *54*, 466–480. [CrossRef]
99. Müller, K.; Levesque-Tremblay, G.; Bartels, S.; Weitbrecht, K.; Wormit, A.; Usadel, B.; Haughn, G.; Kermode, A.R. Demethylesterification of cell wall pectins in Arabidopsis plays a role in seed germination. *Plant Physiol.* **2013**, *161*, 305–316. [CrossRef]
100. Scheler, C.; Weitbrecht, K.; Pearce, S.P.; Hampstead, A.; Büttner-Mainik, A.; Lee, K.J.D.; Voegele, A.; Oracz, K.; Dekkers, B.J.W.; Wang, X.; et al. Promotion of testa rupture during garden cress germination involves seed compartment-specific expression and activity of pectin methylesterases. *Plant Physiol.* **2015**, *167*, 200–215. [CrossRef]
101. Voiniciuc, C.; Pauly, M.; Usadel, B. Monitoring polysaccharide dynamics in the plant cell wall. *Plant Physiol.* **2018**, *176*, 2590–2600. [CrossRef]
102. Wolf, S.; Greiner, S. Growth control by cell wall pectins. *Protoplasma* **2012**, *249*, 169–175. [CrossRef]
103. Pelletier, S.; van Orden, J.; Wolf, S.; Vissenberg, K.; Ndong, Y.A.; Pelloux, J.; Bischoff, V.; Urbain, A.; Lemonnier, G.; Renou, J.; et al. A role for pectin de-methylesterification in a developmentally regulated growth acceleration in dark-grown Arabidopsis hypocotyls. *New Phytol.* **2010**, *188*, 726–739. [CrossRef]
104. Paynel, F.; Leroux, C.; Surcouf, O.; Schaumann, A.; Pelloux, J.; Driouich, A.; Mollet, J.C.; Lerouge, P.; Lehner, A.; Mareck, A. Kiwi fruit PMEI inhibits PME activity, modulates root elongation and induces pollen tube burst in *Arabidopsis thaliana*. *Plant Growth Regul.* **2014**, *74*, 285–297. [CrossRef]
105. Lionetti, V.; Francocci, F.; Ferrari, S.; Volpi, C.; Bellincampi, D.; Galletti, R.; D'Ovidio, R.; De Lorenzo, G.; Cervone, F. Engineering the cell wall by reducing de-methyl-esterified homogalacturonan improves saccharification of plant tissues for bioconversion. *Proc. Natl. Acad. Sci. USA* **2010**, *107*, 616–621. [CrossRef]
106. Jiang, L.; Yang, S.-L.; Xie, L.-F.; Puah, C.S.; Zhang, X.-Q.; Yang, W.-C.; Sundaresan, V.; Ye, D. VANGUARD1 encodes a pectin methylesterase that enhances pollen tube growth in the Arabidopsis style and transmitting tract. *Plant Cell* **2005**, *17*, 584–596. [CrossRef]
107. Tian, G.-W.; Chen, M.-H.; Zaltsman, A.; Citovsky, V. Pollen-specific pectin methylesterase involved in pollen tube growth. *Dev. Biol.* **2006**, *294*, 83–91. [CrossRef]
108. Zhang, G.Y.; Feng, J.; Wu, J.; Wang, X.W. BoPMEI1, a pollen-specific pectin methylesterase inhibitor, has an essential role in pollen tube growth. *Planta* **2010**, *231*, 1323–1334. [CrossRef]
109. Chebli, Y.; Geitmann, A. Cellular growth in plants requires regulation of cell wall biochemistry. *Curr. Opin. Cell Biol.* **2017**, *44*, 28–35. [CrossRef]
110. Zhao, F.; Chen, W.; Traas, J. Mechanical signaling in plant morphogenesis. *Curr. Opin. Genet. Dev.* **2018**, *51*, 26–30. [CrossRef]
111. Müller, K.; Levesque-Tremblay, G.; Fernandes, A.; Wormit, A.; Bartels, S.; Usadel, B.; Kermode, A. Overexpression of a pectin methylesterase inhibitor in *Arabidopsis thaliana* leads to altered growth morphology of the stem and defective organ separation. *Plant Signal. Behav.* **2013**, *8*, e26464. [CrossRef]
112. Wolf, S.; Mravec, J.; Greiner, S.; Mouille, G.; Höfte, H. Plant cell wall homeostasis is mediated by brassinosteroid feedback signaling. *Curr. Biol.* **2012**, *22*, 1732–1737. [CrossRef]
113. Hoeberichts, F.; van Doorn, W.G.; Vorst, O.; Hall, R.D.; van Wordragen, M.F. Sucrose prevents up-regulation of senescence-associated genes in carnation petals. *J. Exp. Bot.* **2007**, *58*, 2873–2885.
114. Wang, H.; Stier, G.; Lin, J.; Liu, G.; Zhang, Z.; Chang, Y.; Reid, M.; Jiang, C. Transcriptome Changes Associated with Delayed Flower Senescence on Transgenic Petunia by Inducing Expression of *etr1-1*, a Mutant Ethylene Receptor. *PLoS ONE* **2013**, *8*, e65800. [CrossRef]
115. Paniagua, C.; Posé, S.; Morris, V.J.; Kirby, A.R.; Quesada, M.A.; Mercado, J.A. Fruit softening and pectin disassembly: An overview of nanostructural pectin modifications assessed by atomic force microscopy. *Ann. Bot.* **2014**, *114*, 1375–1383. [CrossRef]
116. Giovane, A.; Balestrieri, C.; Quagliuolo, L.; Castaldo, D.; Servillo, L. A glycoprotein inhibitor of pectin methylesterase in kiwi fruit. Purification by affinity chromatography and evidence of a ripening-related precursor. *Eur. J. Biochem.* **1995**, *233*, 926–929. [CrossRef]
117. Irifune, K.; Nishida, T.; Egawa, H.; Nagatani, A. Pectin methylesterase inhibitor cDNA from kiwi fruit. *Plant Cell Rep.* **2004**, *23*, 333–338. [CrossRef]
118. Damm, T.; Commandeur, U.; Fischer, R.; Usadel, B.; Klose, H. Improving the utilization of lignocellulosic biomass by polysaccharide modification. *Process Biochem.* **2016**, *51*, 288–296. [CrossRef]

119. Francocci, F.; Bastianelli, E.; Lionetti, V.; Ferrari, S.; De Lorenzo, G.; Bellincampi, D.; Cervone, F. Analysis of pectin mutants and natural accessions of Arabidopsis highlights the impact of de-methyl-esterified homogalacturonan on tissue saccharification. *Biotechnol. Biofuels* **2013**, *6*, 4–11. [CrossRef]
120. Castaldo, D.; Lovoi, A.; Quagliuolo, L.; Servillo, L.; Balestrieri, C.; Giovane, A. Orange juices and concentrates stabilization by a proteic inhibitor of pectin methylesterase. *J. Food Sci.* **1991**, *56*, 1632–1634. [CrossRef]
121. Jolie, R.P.; Duvetter, T.; Vandevenne, E.; Van Buggenhout, S.; Van Loey, A.M.; Hendrickx, M.E. A pectin-methylesterase-inhibitor-based molecular probe for in Situ detection of plant pectin methylesterase activity. *J. Agric. Food Chem.* **2010**, *58*, 5449–5456. [CrossRef]
122. Zocca, F.; Lomolino, G.; Curioni, A.; Spettoli, P.; Lante, A. Detection of pectinmethylesterase activity in presence of methanol during grape pomace storage. *Food Chem.* **2007**, *102*, 59–65. [CrossRef]
123. Zocca, F.; Lomolino, G.; Spettoli, P.; Lante, A. A Study on the relationship between the volatile composition of moscato and prosecco grappa and enzymatic activities involved in its production. *J. Inst. Brew.* **2008**, *114*, 262–269. [CrossRef]
124. Lante, A.; Zocca, F.; Spettoli, P.; Lomolino, G.; Raiola, A.; Bellincampi, D.; Lionetti, V.; Giovane, A.; Camardella, L. Use of a Protein Inhibitor of Pectin Methylesterase for Reducing Methanol Formation in Grape Must and Marc, and Process Therefor. It. Patent WO2008104555A1, 4 September 2008.

International Journal of
Molecular Sciences

MDPI

Review

Overview of the Role of Cell Wall DUF642 Proteins in Plant Development

José Erik Cruz-Valderrama [1], Ximena Gómez-Maqueo [1], Alexis Salazar-Iribe [1], Esther Zúñiga-Sánchez [1], Alejandra Hernández-Barrera [1], Elsa Quezada-Rodríguez [1] and Alicia Gamboa-deBuen [2,*]

1 Instituto de Ecología, Universidad Nacional Autónoma de México. Mexico City 04510, Mexico
2 Universidad Nacional Autónoma de México, Mexico City 04510, Mexico
* Correspondence: agamboa@ecologia.unam.mx

Received: 30 May 2019; Accepted: 4 July 2019; Published: 6 July 2019

Abstract: The DUF642 protein family is found exclusively in spermatophytes and is represented by 10 genes in Arabidopsis and in most of the 24 plant species analyzed to date. Even though the primary structure of DUF642 proteins is highly conserved in different spermatophyte species, studies of their expression patterns in Arabidopsis have shown that the spatial-temporal expression pattern for each gene is specific and consistent with the phenotypes of the mutant plants studied so far. Additionally, the regulation of DUF642 gene expression by hormones and environmental stimuli was specific for each gene, showing both up- and down-regulation depending of the analyzed tissue and the intensity or duration of the stimuli. These expression patterns suggest that the DUF642 genes are involved throughout the development and growth of plants. In general, changes in the expression patterns of DUF642 genes can be related to changes in pectin methyl esterase activity and/or to changes in the degree of methyl-esterified homogalacturonans during plant development in different cell types. Thus, the regulation of pectin methyl esterases mediated by DUF642 genes could contribute to the regulation of the cell wall properties during plant growth.

Keywords: DUF642 family; homogalacturonans; plant cell wall; plant development; pectins

1. Introduction

Plant cell walls are dynamic compartments whose composition and chemical structure vary during plant development in response to environmental signals. Cell walls are composed of three layers: the middle lamella, and the primary and secondary cell walls. Primary cell wall deposition continues through cell growth and expansion. It is a very complex and dynamic structure composed mainly of three types of polysaccharides: cellulose, hemicelluloses, and pectins. Cellulose microfibrils and hemicelluloses constitute a network with structural proteins that are embedded in a gel-like matrix of pectins. Localization of different classes of polysaccharides within the cell wall appears to depend on species, organ, tissue, and cell type, and the pectin network is temporally and spatially regulated during plant development. Alterations of the cell wall structure and changes in properties during plant development are mainly the result of protein activities. Enzymatic activity and protein interactions respond to developmental and environmental signals and tightly regulate the complex and dynamic structure of plant cell walls.

In terrestrial plants, the chemical structure of homogalacturonans (HGs), a type of pectin that is comprised of chains of α-1,4-galacturonic acid, is modified by de-methyl-esterification processes during plant development by pectin methyl esterases, and pectin methyl esterase inhibitory proteins regulate the degree of HGs' esterification of the cell wall in different cell types. Consequently, cell or tissue-specific function could be influenced by the HGs' status [1]. An increase in complexity in the regulation of HGs' esterification status through plant evolution has been described including a diversification of the gene families involved either in pectin synthesis or in pectin modification [2,3]. The shift from an aquatic to a terrestrial environment constitutes a milestone in the complexity of pectin regulation. In the algae *Chara*, cell growth is a result of a calcium pectate cycle that does not involve enzymatic activities [4]. Charophytes are the closest related group to land plants, thus, this enzyme-less mechanism could be still present in terrestrial plants [5].

The DUF642 protein family is found exclusively in spermatophytes and is involved in regulating HGs esterification. The DUF642 family was first described in apoplast proteomes of rosette leaves in *Arabidopsis thaliana* (Arabidopsis) by Boudart et al. [6] and subsequently in cell wall proteomes from different tissues and plant species from both dicotyledons and monocotyledons [7,8]. This multigene family is represented by 10 genes in Arabidopsis as well as in other plant species [9]. DUF642 proteins contain two DUF642 domains (PF04642) that present two conserved motifs NGXFEXXP and CGPVXD, determined as the family signature. A phylogenetic analysis revealed four clades of orthologous genes for all spermatophytes (Clades A, B, C, and D). After the divergence of gymnosperms and angiosperms, a duplication of Clade A occurred, producing two subclades (A1 and A2), and a subsequent duplication occurred after the divergence of monocotyledons and eudicotyledons [9].

The DUF642 proteins are encoded by genes that contain three exons. The first exon encodes a signal peptide, and the other two exons encode the DUF642 domains. Proteomic and cellular localization studies of the Clade A proteins indicated they were located primarily in the cell wall [6,10]. However, the protein encoded by *At3g08030* was located in the nuclear pore, probably because this gene has an alternative first exon edition so one of the possible encoded protein lacks the signal peptide. The proteins in Clades B, C, and D have a GPI-anchor site at the carboxyl terminus and were detected in cell membrane proteomes [11].

The interactions of the DUF642 proteins with polysaccharides and cell wall enzymes have been demonstrated in vitro. The protein encoded by *At3g08030* (Clade A2) interacted specifically with cellulose [11]. The proteins encoded by *At4g32460/BDX* and *At5g11420* (Clade A1) interacted with the catalytic domain of a pectin methyl esterase (PME) [12]. In two transgenic Arabidopsis lines overexpressing *BDX* the PME activity increased in the seeds, seedlings, and the apical meristems, but no changes in PME activity were detected in the leaves [13]. PME activity was higher in the seeds of both overexpressing lines than it was in the control seeds during germination [13,14]. PME activity was also higher in seedlings, stems, leaves, and fruits in Arabidopsis plants overexpressing the *VqDUF642* gene (Clade A1) of *Vitis quinquangularis* [15]. The leaves of Arabidopsis plants overexpressing the *AhDGR2* gene (Clade A1) of *Amaranthus hypochondriacus* showed a decrease in PME activity, whereas an increase in PME activity was detected in the roots [16]. The phenotypes of all the Clade A1 protein mutants in Arabidopsis have been established. The *At5g25460/dgr2* mutant had a shorter root and smaller rosette leaves than the wild-type [17]. No alterations in the phenotypes of the *At5g11420* and *At1g80240/dgr1* mutants compared with the wild-type have been described. The *At4g32460/bdx* mutants had shorter siliques, seeds with altered development, and longer hypocotyl than the wild-type [18]. Among Arabidopsis Clade A2 mutants, the *At2g41800/teb* mutant is the only one for which a phenotype (longer hypocotyls) has been established [19]. No phenotypes have been established for the two Clade B genes *At1g29980* and *At2g34510*, or the Clade D gene *At5g14150*.

2. Expression Patterns of *DUF642* Genes

The expression patterns of the Clade A *DUF642* genes have been studied in Arabidopsis plants transformed with constructs containing the putative promoter region fused to the reporter genes *ER-GFP* and *GUS* (Figure 1). Differential expression patterns were detected at different stages of plant development (Table 1).

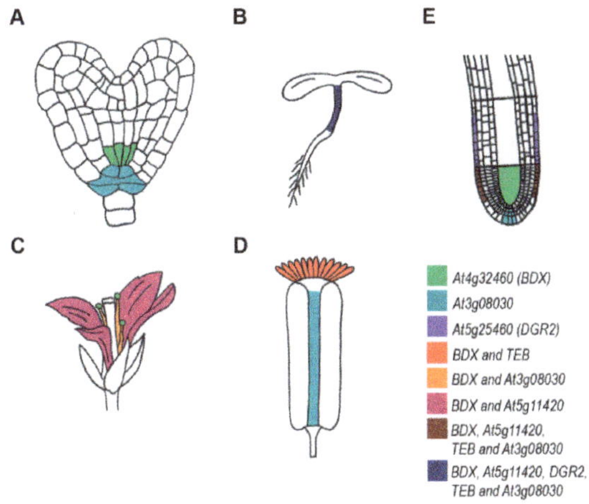

Figure 1. Expression patterns of the *DUF642* genes described to date using reporter genes during Arabidopsis plant development.

In Arabidopsis during embryonic development, *BDX* and *At3g08030* were expressed from the heart stage onwards. In the late heart stage, *At3g08030* was expressed in the quiescent center and in the columella cells, while *BDX* was expressed in the vascular tissue (Figure S1) [13].

During seedling development in Arabidopsis, *BDX, At5g11420, DGR2, TEB*, and *At3g08030* were expressed in hypocotyls and *BDX* and *DGR2* were also expressed in cotyledons, but no expression was detected for *At5g11420, TEB*, and *At3g08030* [13,17,20]. In hypocotyls, specific expression of *BDX, At5g11420, At3g08030*, and *TEB* in the epidermal cells was determined by confocal microscopy. *DGR1* and *At2g34510* expression was detected specifically in hypocotyls grown under shade conditions [21]. Transcriptomic studies in *Brassica rapa* indicated that three orthologous genes of *BDX, At5g11420*, and *DGR2* were expressed in hypocotyls [20].

The expression patterns of only *DGR1, BDX*, and *DGR2* have been studied in leaves of Arabidopsis. *BDX* expression was localized in the vascular tissue, *DGR2* expression was located in the leaf primordium, and *DGR1* expression was not detected in the leaves [13,17].

In flowers of Arabidopsis, *BDX* was expressed in petals, vascular tissue of the filament, anther, and stigmatic papilla. *At5g11420* was expressed in petals and abscission zones of floral whorls, *TEB* was expressed only in stigmatic papilla, and *At3g08030* was expressed in the replum of androecium and epidermis of the filaments of stamens [13,18] (Figure S1).

In Arabidopsis, *DGR1, BDX, At5g11420, DGR2, TEB*, and *At3g08030* expression was detected in the primary root. *BDX, At5g11420, DGR2, TEB*, and *At3g08030* were expressed in epidermal cells of the meristematic region. *BDX* also was expressed in vascular tissue and *At3g08030* was expressed in the quiescent center and in columella [13,17,18] (Figure S1). These expression patterns were similar to those found during embryonic development.

During development of the lateral roots, *BDX* and *At3g08030* expression was detected from stage II onwards, when cell proliferation begins (Figure 2). Further, at the moment of radicle protrusion, *At5g11420* and *TEB* expression was detected in the epidermal cells that surround the radicle protrusion zone. The expression patterns of *DGR1* and *DGR2* during this process have not been determined.

The expression patterns of the Clade A *DUF642* genes were altered by plant hormones (Table 1 and Table S1). For example, auxins and gibberellins increased the expression levels of *BDX* and *TEB* during Arabidopsis seed germination [22]. Exogenous auxins and gibberellins also altered the expression pattern of *BDX* by promoting its expression in cortex cells in addition to vascular tissue of Arabidopsis. The expression pattern of *At3g08030* was not altered by the addition of auxins [13].

Table 1. Expression profiles of the *DUF642* genes detected in distinct tissues/organs and species. *DUF642* genes are grouped by clades as reported by Vázquez-Lobo et al. (2012) [9] and using the locus tag of Arabidopsis as a reference for gene grouping.

Clade	Gene Name	Species	Expressed in Tissue/Organ	Reference
A1	*At1g80240*	Arabidopsis	Petal	[23]
A1	*GRMZM2G027683*	*Zea mays*	Embryo	[24]
A1	*OSNPB_040494600*	*Oryza sativa*	Floral buds, flowers, mature seed, milk grains, roots before flowering	[25]
A1	Orthologous to *At4g32460*	*Brassica rapa*	Hypocotyl epidermal cells	[21]
A1	Orthologous to *At4g32460*	*Cannabis sativa*	Stem (apical section)	[26]
A1	*OSNPB_020205200*	*Oryza sativa*	Floral buds, flowers, mature seed, milk grains	[25]
A1	Orthologous to *At5g11420*	*Brassica rapa*	Hypocotyl epidermal cells	[21]
A1	*At5g11420*	Arabidopsis	Petal	[23]
A1	*At5g11420*	Arabidopsis	Root (cortex)	[27]
A1	*OSNPB_010611000*	*Oryza sativa*	Floral buds, flowers, mature seed, milk grains, roots before flowering	[25]
A1	*OSNPB_040494800*	*Oryza sativa*	Floral buds, milk grains, roots before flowering	[25]
A1	Orthologous to *At5g25460*	*Brassica rapa*	Hypocotyl epidermal cells	[21]
A1	*AhDGR2*	*Amaranthus hypochondriacus* var. Revancha	Panicle, axillary bud, young stem, and young leaf	[16]
A1	*GRMZM2G051571*	*Zea mays*	Seed	[24]
A1	*GRMZM2G034985*	*Zea mays*	Seed, shoot, root, SAM, ear, tassel, cob, silk, anthers, ovule	[24]
A1	Orthologous to *At5g25460*	*Cannabis sativa*	Stem (apical section)	[26]
A1	*At5g25460*	Arabidopsis	Root (cortex)	[27]
A2	similar to *At3g08030*	*Brassica rapa*	Hypocotyl inner tissue cells	[21]
A2	*At3g08030*	Arabidopsis	Root (cortex/hair cell)	[27]
A2	*At2g41800*	Arabidopsis	Carpel	[23]
A2	*OSNPB_030807700*	*Oryza sativa*	Floral buds, flowers, mature seed, milk grains, roots before flowering	[25]
A2	*At2g41810*	*Brassica rapa*	Hypocotyl epidermal cells	[21]
B	Orthologous to *At1g29980*	*Brassica rapa*	Hypocotyl inner tissue cells	[21]
B	*OSNPB_010756600*	*Oryza sativa*	Floral buds, flowers, leaves before flowering, mature seed, milk grains, roots before flowering	[25]
B	*GRMZM2G324705*	*Zea mays*	Seed, shoot, root, SAM, ear, tassel, cob, silk, anthers, ovule	[24]

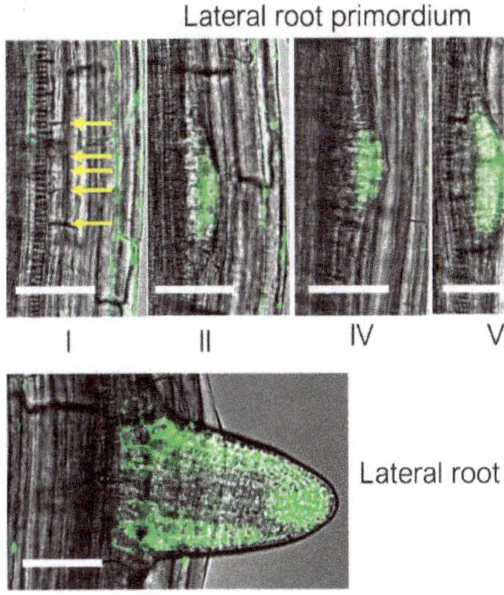

Figure 2. Expression patterns of *At3g08030* during lateral root development. The Arabidopsis *pAt3g08030::ER-GFP*-transgenic plants were obtained as described in Salazar-Iribe et al. [19]. Roman numerals indicate the developmental stages of the lateral root initiation process. The images are all section images. n = 8–10 roots for each developmental stage. Scale bars = 20 μm.

3. Subcellular Localization of DUF642 Proteins

DUF642 proteins are very abundant in cell wall proteomes from different tissues and plant species [7,8]. The At3g08030 protein has been found in all Arabidopsis cell wall proteomes reported so far. TEB and DGR1 were found in the proteomes of cell suspension cultures, whereas DGR2 was found in the proteomes of cell suspension cultures, hypocotyls, and mature stems. The At5g11420 protein was present in the proteomes of apoplast, hypocotyls, and mature stems, whereas BDX was found only in the mature stem cell wall proteome [28]. The localization of BDX and TEB in the Arabidopsis cell wall was confirmed by confocal microscopy in hypocotyl cells [18,19]. The VqDUF642 protein was localized in the cell wall of tobacco epidermal cells [15] and peach PpDU642 (TEB) was localized in the cell wall and extracellular space during a transient expression in tomato fruit [29].

Studies carried out to determine the subcellular localization of BDX in epidermal root cells of Arabidopsis suggested that it was localized intracellularly, probably in Golgi, and then relocated to the cell wall when the plants were subjected to abiotic (NaCl) or biotic (nematodes) stress stimuli [30] (Figure S2). *TEB* is highly expressed during cell division prior to cytokinesis. In Arabidopsis, the synchronization of mitosis in primary root cells treated with hydroxyurea [31] confirmed that TEB was localized intracellularly during mitosis, and was located in the cell wall until the end of the process (Figure 3). These results suggest that BDX and TEB are located transiently in the cell wall in response to endogenous or exogenous stimuli. Finally, polar localization of TEB in the cell wall of radicular hair was determined during its elongation, suggesting the formation of domains with an accumulation of these proteins within the cell walls [19].

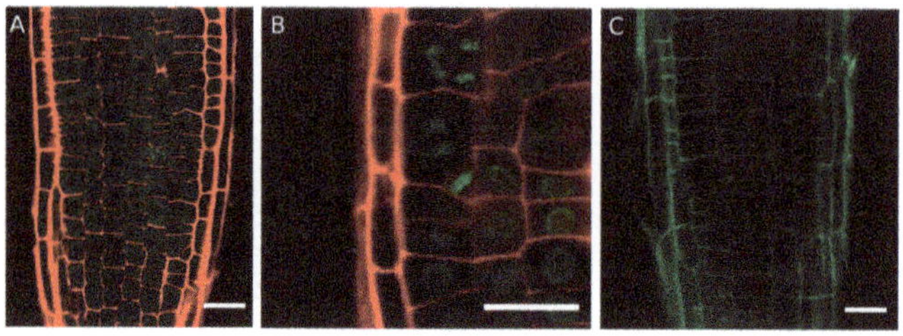

Figure 3. Subcellular localization of the TEB protein in meristematic root cells synchronized with hydroxyurea. (**A**) Arabidopsis *pTEB::TEB-GFP* seedlings were treated with hydroxyurea for 16 h, and a diffuse pattern of GFP in meristematic cells of the cortex was observed. The cell wall was stained with propidium iodide (red). (**B**) After 18 h of hydroxyurea treatment, GFP was detected as a ring surrounding the nucleus in the cortex cells. (**C**) TEB was located in the cell wall after 19 h of hydroxyurea treatment. Hydroxyurea stops DNA synthesis and after 16–17 h most of the cells are at the G2/M transition and at 18–20 h cytokinesis occurs according to Cools et al. [31]. Seedlings were sown in MS medium and transferred on day 6 to MS medium with hydroxyurea for 16–20 h. The images are all section images. Scale bars = 20 μm. The Arabidopsis transgenic plants were obtained as described in Salazar-Iribe et al. [19].

4. Function of DUF642 Proteins in Plant Development

The expression patterns of *DUF642* genes suggest they are involved throughout the development and growth of plants. In general, changes in the expression patterns of *DUF642* genes can be related directly to PME activity [13,15,16,19]. The leaves of Arabidopsis lines overexpressing *AHDGR2* had low PME activity, and this was reflected in alterations in the pattern of de-methyl-esterified HGs in the cell walls [16,19,32].

4.1. Seed Development

In Arabidopsis, *BDX*, *At5g11420*, *TEB*, and *At3g08030* were expressed during embryo development (Table 1 and Table S1), and the localized expression during development indicated differential expression in different cell types (Figure 1). However, the *bdx-1*, *At2g11420*, and *teb-1* mutants did not show alterations in gene expression during embryo development.

During endosperm development, *BDX* was the only *DUF642* gene reported at four days after pollination [33]. Of the *dgr1*, *bdx-1*, *At5g11420*, *dgr2*, and *teb-1* mutants studied during seed development, only the *bdx-1* mutant showed alterations, producing misshapen seeds with changes in the folding of the embryo. The endosperm and seed coat exert physical restrictions during elongation and folding of the embryo [34], with the walls of the endosperm adjacent to the embryo showing a decrease in the degree of HGs esterification. In the *bdx-1* mutant, a lower signal of de-methyl-esterified HGs compared with the wild-type was detected in this region. Alterations in the degree of esterification can cause greater rigidity of the cell wall of endosperm cells, thereby generating a physical restriction in the development of the embryo and compromising the subsequent survival of the seedling [32].

The accumulation of *DUF642* gene transcripts has also been reported in mature seeds. Germination pre-treatments such as priming, which improve germination performance, can induce the expression of these genes. An important increase in the expression of the orthologous gene of *At3g08030* was reported in seeds from *Brassica oleracea* subjected to an osmopriming treatment [35]. In Arabidopsis, *Ceiba aesculifolia*, and *Wigandia urens* seeds, the expression of *At3g08030* increased in response to osmopriming treatment and decreased when the seeds were subjected to controlled deterioration [36]. In a study carried out with Arabidopsis *vps29* mutant seeds (VPS29 rearrangement complex), which had a lower germination index, the abundance of *At3g08030* and *At1g29980* increased while that of *At5g11420* and *DGR2* decreased [37]. Functional studies of *At3g08030* loss of function mutants could reveal if this gene has a role in promoting seed longevity.

After seed maturation, the expression of *At5g11420*, *DGR2*, *At3g08030*, and *At1g29980* increased in after-ripened seeds (Table S1). These changes in expression may be related to changes in the biochemical machinery that are required to modify the architecture of the cell wall of the embryo, the endosperm, and the seed coat during germination.

4.2. Germination

During the germination process of barley, Brassica, and Arabidopsis seeds, progressive differential expression of several *DUF642* genes took place (Table S1). The expression levels of Clade A1 genes (*BDX*, *At5g11420* and *DGR2*) and *At3g14310*, which codes for PME3, increased after 6 h of seed imbibition. The expression of the Clade A2 gene, *TEB*, increased between 8 and 12 h, and *At3g08030* expression was detected in the mature seeds and during the entire imbibition process (Figure S3).

The lines overexpressing *BDX*, *At5g11420*, and *DGR2* showed a decrease in the onset of germination, especially in the timing of the testa rupture, compared with the wild-type seeds [13] (Figure S3). The increase in total PME activity during seed germination in the overexpressing plants may be related to the early testa rupture described previously [14]. Early testa rupture also was observed in tomato seeds from tomato plants overexpressing *VqDUF642* [15]. These results indicate that the Clade A1 genes have the same function during germination.

4.3. Hypocotyls

Under control growth conditions (long photoperiod), the Arabidopsis *bdx-1* and *teb-1* mutant lines had longer hypocotyls because of longer cell length compared with the wild-type, whereas the overexpressing lines produced shorter hypocotyls. The hypocotyls of the *bdx-1* and *teb-1* mutants had a higher signal for methyl-esterified HGs. Additionally, the *bdx-1* mutant accumulated more auxins in the epidermal cells than the wild-type [18]. During cell elongation in the hypocotyl, auxin signaling and cell wall modifications of epidermal cells were actively involved [38,39]. The accumulation of auxins in hypocotyl epidermal cells during cell elongation was facilitated by PIN transporters. In the *bdx-1* mutant, a change in the localization of PIN1 transporters was detected, possibly because of the degree of HGs esterification, as has been described for other tissues [40]. The *bdx* and *teb* mutants showed an altered hypocotyl phenotype, suggesting *DUF642* genes may be involved in regulating hypocotyl elongation during seedling development [41].

4.4. Leaves

The expression of *DGR1*, *BDX*, *At5g11420*, *DGR2*, *TEB*, *At2g41810*, *At3g08030*, *At2g34510*, and *At5g14150* was detected in rosette leaves of mature Arabidopsis plants close to flowering (Table S1). The *dgr1*, *bdx-1*, and *teb-1* mutants showed no alterations in the development of rosette leaves. The *dgr2* mutant had a smaller rosette, explained by its expression in leaf primordia [17]. The lines overexpressing *BDX*, *At5g11420*, and *TEB* showed no visible alterations in the leaves or in the size of the rosette [13,19]. No differences were found in the leaves of Arabidopsis plants overexpressing *AhDGR2* or tomato plants overexpressing *VqDUF642* [15,16]. The mechanisms that control leaf origin and growth are complex, because of the phenotypic plasticity that is present in leaves as a mechanism for adaptation to different environments [42]. The complexity of the regulation of leaf size and morphology, and the lack of alterations in the leaves of mutant lines suggest the *DUF642* genes are functionally redundant in leaves.

4.5. Reproductive Structures

During flower development in Arabidopsis, the *DUF642* genes studied to date showed differential expression patterns, although the expression of only two genes have been detected in carpel, stamens, and petals. *BDX* is the only *DUF642* gene that has been detected in the anthers of stamens. The *bdx-1* mutant line showed morphological alterations of the pollen grain that altered its viability. The in vitro germination of the pollen tube was lower in the *bdx-1* mutant pollen grains that in the wild-type.

The structural and functional alterations of the pollen grains in the *bdx-1* mutant could be caused by changes in the accumulation of auxins during the late stages of anther development (Figure 4). Similar alterations in pollen morphology have been described in mutants related to the auxin signaling pathway, including auxin synthesis and transport [43].

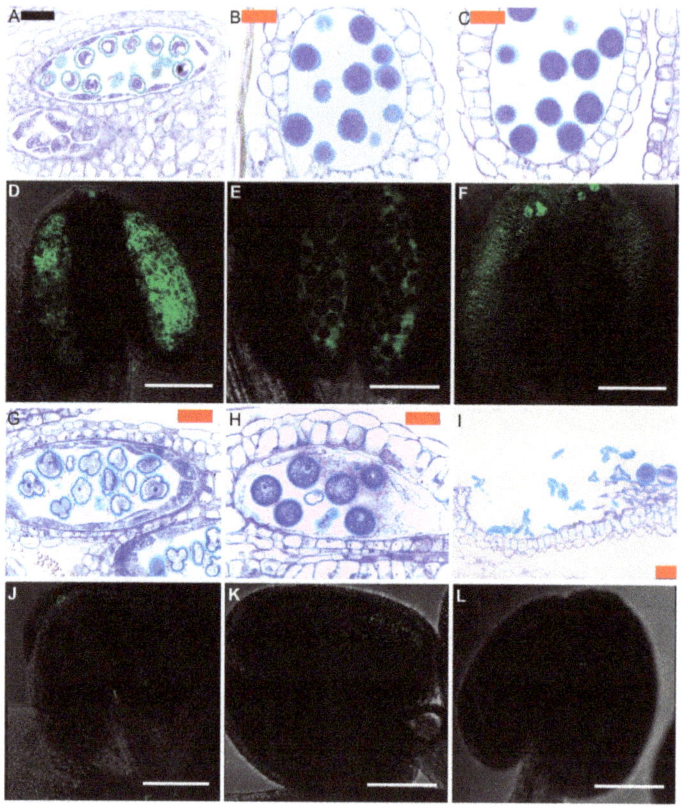

Figure 4. Pollen grain development and auxin accumulation in anthers of flowers from wild-type and *bdx-1* mutant Arabidopsis plants. Normal pollen grain development in wild-type (wt) flowers. (**A**) Anthers from 7–8 stage flower. (**B**) Anthers from 11–12 stage flower. (**C**) Anthers from 13 stage flower. GFP detection of auxins in anthers from wt/*DR5* flowers. (**D**) Anthers from 7–8 stage flower. (**E**) Anthers from 11–12 stage flower. (**F**) Anthers from 13 stage flower. Throughout the development of the anther there is an important accumulation of auxins. Pollen grain development in *bdx-1/DR5* flowers. (**G**) Anthers from 7–8 stage flower. (**H**) Anthers from 11–12 stage flower. (**I**) Anthers from 13 stage flower. The development of pollen grains is altered; at the 11–12 stage, diffuse cytoplasm is observed and, at stage 13, some pollen grains have collapsed. GFP detection of auxins in anthers from *bdx-1/DR5* flowers. (**J**) Anthers from 7–8 stage flower. (**K**) Anthers from 11–12 stage flower. (**L**) Fertilized flowers at stage 13. Throughout the development of the anther there is almost no accumulation of auxins. The Arabidopsis transgenic plants were obtained as described in Cruz-Valderrama et al. [32]. Sections: 1–2 µm. Staining was with toluidine blue. Red bar = 20 µm, Black bar = 10 µm. n = 8–10 plants for each line. White bar = 100 µm. n = 5–9 anthers from flowers from different plants.

4.6. Fruits

Studies of *DUF642* gene expression and gene function during fruit development are very scarce. Of the Arabidopsis mutants studied, only *bdx-1* showed a fruit phenotype with shorter siliques and lower seed production than wild-type plants. In the genus *Brassica*, *DGR2* is the only gene that has been detected in transcriptomes of siliques, specifically in the shells of the pods [44].

4.7. Roots

The expression of *DGR1*, *BDX*, *At5g11420*, *DGR2*, and *TEB* has been detected in primary roots of Arabidopsis. All these genes were expressed in the epidermal cells of the meristematic zone, except *DGR2*, which was expressed in the elongation zone. The *dgr2* mutant was the only one that presented a short root phenotype [17]. Overexpression of *AhDGR2* in Arabidopsis produced longer roots [16], which is understandable given that *DGR2* was expressed in the meristematic zone where the cells begin to expand and the rate of cell division is reduced.

The *bdx-1*, *At5g11420*, and *teb-1* mutants showed no alterations in the development of lateral roots. *BDX*, *At3g08030*, *At5g11420*, and *TEB* all had similar expression patterns throughout the development of lateral roots, which suggested there could be a functional redundancy that would explain the lack of altered phenotypes for these mutants.

Roots interact with microorganisms in the soil, and are particularly susceptible to pathogen attacks. Roots also are central in the perception of nutrient concentrations and other elements like heavy metals in the substrate where plants develop. The modification and restructuring of the cell wall in roots actively participate in the response to different types of stresses. Therefore, the participation of *DUF642* genes in the remodeling of the cell wall can be relevant in all these processes. Changes in the expression of *DUF642* genes in response to biotic and abiotic stresses have been described in various plant species (Table S2). However, very few functional studies are available related to the particular role of *DUF642* genes in response to stress.

5. Biotic Factors

DGR2 is negatively regulated in Arabidopsis plants during infection by the bacterium *Ralstonia solanacerum* [45]. The expression of *At1g29980* and *At3g08030* is induced by infection with *Rhodococcus fascians* [46]. *At3g08030* also is highly expressed in response to infection by *Agrobacterium tumefaciens* [47].

The abundance of the DUF642 protein encoded by *At3g08030*, which has two isoforms with different isoelectric points, was reduced significantly in the cell wall proteome of Arabidopsis suspension cells treated with chitosan, an elicitor that mimics a fungal infection [48]. Infection with the *Penicillium* strain Pc4 in post-harvest grapes increased the abundance of two isoforms of the At5g11420 orthologous protein, probably because of post-translational modification [49]. In apoplastic proteomes of maize roots (*Zea mays*) infected with the symbiont species *Trichoderma virens*, a DUF642 protein was detected 5 days after inoculation. Modification and degradation of root cell walls are essential for colonization to take place [50]. Infection of Arabidopsis plants with the phytopathogenic fungus *Botrytis cinerea* reduced the expression of *At5g11420* and *At5g25460* [51].

The overexpression of *VqDUF642* in tomato plants reduced the susceptibility to *B. cinerea* infection in both mature and immature fruits. Overexpression of this gene modified the expression of some pathogen response genes such as *SlPR1*, *SlPR2*, *SlPR3*, and *SlPR4* whose expression increased at 48 h post-infection. In transgenic Thompson grape plants, overexpression of *VqDUF642* promoted increased resistance to *B. cinerea* and induced resistance to the fungus *Erysiphe necator*, which causes oidium of vines. The leaves of the overexpressing plants infected with *E. necator* showed a less severe infection than the wild-type at 48 h post-infection. In addition, the expression levels of the pathogen-response genes *VvPR1*, *VvPR2*, *VvPR3*, and *VvPR4* increased drastically at 48- and 96-h post-infection in transgenic seedlings compared with in the wild-type [15].

In Arabidopsis, transcriptomic induction during early inoculation with the knot-nematode *Meloidogyne incognita* revealed up-regulation of *At1g29980* [52]. *BDX* and *TEB* expression were also highly induced by *M. incognita* early inoculation. The cell wall localization of BDX and TEB in the epidermal cells of primary roots was induced by *M. incognita*. Early inoculation with *Nacobus aberrans*, a nematode that cannot infect Arabidopsis, did not alter the expression of these two genes [30].

A comparative analysis of available Arabidopsis flower transcriptomes showed that changes in the expression patterns of flower-specific defense genes were critical in pathogen resistance. According to this study, the expression of *BDX* and *DGR2* was positively regulated in petals compared with in senescent leaves, in stage 15 of the flower. At this stage, an increase in the expression of the biotic stress response genes also occurred [53].

6. Abiotic Factors

The regulation of the DUF642 genes by transcription factors in response to aluminum stress has been reported. Aluminum stress has been studied extensively in aluminum-resistant plants. In particular, *Oryza sativa* (rice) is a cultivated crop plant that is resistant to aluminum, and the signaling pathway associated with its resistance has been studied widely. The Al resistance transcription factor 1 gene (*ART1*) is central to the aluminum response, and it regulates an increase in the expression of the *DUF642* gene *Os04g049490* (orthologous to *At5g11420*) [54]. *Os04g049490* was also regulated by the *SENSITIVE TO ALUMINUM RHIZOTOXICITY* gene (*STAR1*), which plays a fundamental role in aluminum resistance in rice roots [55]. In the *O. sativa indica* IR64 cultivar treated with aluminum, *Os04g41750* (orthologous to *At5g11420*) expression was up-regulated, whereas no change in its expression was detected in *O. sativa* cv. Azucena, which is more sensitive to aluminum [56]. Plants in the genus *Stylosanthes* have high tolerance to toxicity by the aluminum ion, and expression of the orthologous *DUF642* gene *Os04g0494900* increased in the roots of the Reyan 2 genotype in response to aluminum treatment [57]. A comparison between the transcriptomes of two *Citrus* species with different aluminum tolerances suggested that the process of HGs de-esterification in the cell wall of the root cells played an important role in resistance to aluminum. In *C. sinensis*, which is resistant to aluminum, four genes related to this process and a *DUF642* gene, an ortholog of *DGR2*, were up-regulated [58]. In NtSTOP-1-RNAi tobacco plants that showed a decrease in the expression of the *SENSITIVE TO PROTON RHIZOTOXICITY* gene (*STOP1*), the expression of *Nt6860* (orthologous to *At5g11420)* decreased, as has been described for *At5g11420* in the Arabidopsis *art-1* mutant [59].

In *Medicago sativa* plants exposed to cadmium, the amount of the protein orthologous to the Arabidopsis protein, DGR2, decreased [60]. The exposure of *Populus* × *canadensis* plants to high doses of zinc promoted the decrease in the expression of *POPTR_0001s27110* (orthologous to *At3g08030*) [61].

Ultraviolet radiation (UV-B) generates an imbalance in the production of reactive oxygen species. Thus, it is used to study the effects of oxidative stress on plants. In the mutant *ggt1* of the protein GAMMA-GLUTAMYL TRANSFERASE ISOFORM 1(GGT1), involved in the redox balance, the proteins DGR2 and PECTIN METHYL ESTERASE3 (At3g14310) are very abundant in both the mutant and in the wt of Arabidopsis plants exposed to UV-B [62].

In the salinity-resistant plant *Manihot esculenta* Crantz (Cassava), *RknMes02_00171* and *RknMes02_00443*, which are orthologs of *BDX* and *At5g11420* respectively, were among the 40 most highly expressed genes during NaCl treatment [63]. *AhDGR2* was significantly induced in *Amaranthus hypochondriacus* (grain amaranth), which is also resistant to high doses of salt. However, when *AhDGR2* was overexpressed in Arabidopsis, the plants showed hypersensitivity to increasing NaCl concentrations, as shown by shorter root length, smaller and slightly chlorotic rosettes, as well as considerably reduced germination rates [16].

The expression levels of *At2g34510* and *At5g11420* increased in Arabidopsis plants subjected to drought stress, although their expression levels were not altered by treatment with abscisic acid [64]. The increase in *At2g34510* expression induced by drought was corroborated in a study that tested the stress response in natural populations of Arabidopsis [65]. In *Eucalyptus calmadulensis*, the expression of *Eucgr.C02812* (orthologous to *At5g11420*) was also found to decrease in plants exposed to drought [66].

7. Conclusions

The DUF642 proteins are highly conserved, which may be related to their specific interactions with cell wall polysaccharides and proteins in different cell types. The *At3g08030*-encoded protein interacts in vitro with cellulose, a polysaccharide that has an important role in plant development [9]. However, there are still no functional studies that confirm this interaction. BDX and At5g11420 interact in vitro with a PME [12]. Studies in different plant species suggest that their function could be related to the regulation of HGs modification throughout plant development. The functional studies carried out to date are consistent with this hypothesis. For instance, in almost all plant tissues where *DUF642* gene overexpression has been induced, there was an increase in the total PME activity. Furthermore, the increase in methyl-esterified HGs in the endosperm during embryo folding and in hypocotyl epidermal cells in the *DUF642* mutants is consistent with the corresponding phenotype. However, a direct interaction of DUF642 proteins with PMEs needs to be established using different experimental approaches. Several proteomic studies have indicated the presence of two isoforms of the same DUF642 protein, suggesting these proteins undergo post-translational modifications that may be related to differential interactions within the cell wall [29,60,61].

The different functions of *DUF642* genes during development can be determined by the specific spatial-temporal expression pattern for each gene. The differential response to hormones also may participate in this process. *BDX* is the only *DUF642* gene expressed in the endosperm during seed development. This specific expression may explain the misshapen-seeds phenotype in the *bdx-1* mutant. Similarly, the expression of *DGR2* in the root elongation zone may explain the short root phenotype exclusive of the *dgr2* mutant. It is possible that there is functional redundancy during embryo development as well as during the development and emergence of the lateral root, processes in which the same temporal-spatial pattern of at least two *DUF642* genes was detected in Arabidopsis.

BDX, At5g11420, TEB, and *At3g08030* are expressed in the meristematic zone of the primary root, but the *bdx* and *teb* mutants do not show an altered root phenotype. BDX was located intracellularly in the epidermal cells of the primary root and relocated to the cell wall in response to biotic and abiotic stresses. Roots interact with different organisms in the soil and are subjected to multiple perturbations in the environment, such as decreases in the availability of water and increases in toxic ions. These interactions can alter the expression of some DUF642 genes. In the response to fungal infection, the overexpression of *VqDUF642* increased resistance to the pathogen, although the expression other *DUF642* genes decreased in response to infection. These results suggest that the *DUF642* genes participate broadly in plant responses to environmental factors and not solely in the root developmental process.

The primary structure of *DUF642* proteins is highly conserved in different spermatophyte species. However, studies of their expression patterns in Arabidopsis showed that the spatial-temporal expression pattern for each gene was specific and consistent with the phenotypes of the mutant plants studied so far. The regulation of *DUF642* gene expression by hormones and environmental stimuli also was specific for each gene. Functional studies of the *DUF642* genes in different plant species are needed to determine the relevance of the DUF642 family in the evolution of terrestrial plants.

Supplementary Materials: Supplementary materials can be found at http://www.mdpi.com/1422-0067/20/13/3333/s1. Figure S1. *At3g08030* expression during *Arabidopsis thaliana* development using *pAt3g08030::ER-GFP* plants; Figure S2. *At5g11420* expression during *Arabidopsis thaliana* development using *pAt5g11420::ER-GFP* plants; Figure S3. *At4g32460/BDX, At5g25460/DGR2, At5g11420, At3g14310/PME3, At2g41800/TEB* and *At3g08030* expression during seed germination; Figure S4. Subcellular localization of BDX in *Arabidopsis thaliana* primary roots under salinity stress conditions using *pBDX::BDX-GFP* plants. Table S1. Transcriptomic evidence of *DUF642* genes expression during specific developmental stages and/or in response to diverse environmental stimuli. Table S2. Transcriptomic evidence of *DUF642* genes expression in response to diverse environmental stimuli.

Author Contributions: J.E.C.V., X.G.-M., and A.G.-d.; participated in the conceptualization and in the review and edition of the manuscript. J.E.C.V.; A.S.-I., E.Z.-S., A.H.-B., and E.Q.-R. carried out the molecular genetic studies. J.E.C.-V., X.G.-M., A.S.-I., E.Z.-S., and A.G.-d. participated in the original draft preparation. Funding was provided to A.G.-d. All authors read and approved the final manuscript.

Funding: This research was funded by Programa de Apoyo a Proyectos de Investigación e Innovación Tecnológica, Universidad Nacional Autónoma de México grant number IN203218- X.G.-M received a fellowship from Dirección General de Asuntos del Personal Académico, Universidad Nacional Autónoma de México.

Acknowledgments: We thank K. Jiménez for assistance on Confocal studies and M. Guemez for the artwork (Figure 1). We thank M. Biswas, from Edanz Group (www.edanzediting.com/ac) for editing a draft of this manuscript.

Conflicts of Interest: No conflict of interest is declared.

References

1. De Lorenzo, G.; Ferrari, S.; Giovannoni, M.; Mattei, B.; Cervone, F. Cell wall traits that influence plant development, immunity, and bioconversion. *Plant J.* **2019**, *97*, 134–147. [CrossRef] [PubMed]
2. Yang, Z.L.; Liu, H.J.; Wang, X.R.; Zeng, Q.Y. Molecular evolution and expression divergence of the Populus polygalacturonase supergene family shed light on the evolution of increasingly complex organs in plants. *New Phytol.* **2013**, *197*, 1353–1365. [CrossRef] [PubMed]
3. McCarthy, T.W.; Der, J.P.; Honaas, L.A.; Claude, W.D.; Anderson, C.T. Phylogenetic analysis of pectin-related gene families in *Physcomitrella patens* and nine other plant species yields evolutionary insights into cell walls. *BMC Plant Biol.* **2014**, *14*, 79. [CrossRef] [PubMed]
4. Proseus, T.E.; Boyer, J.S. Calcium deprivation disrupts enlargement of *Chara corallina* cells: Further evidence for the calcium pectate cycle. *J. Exp. Bot.* **2012**, *63*, 3953–3958. [CrossRef] [PubMed]
5. Boyer, J.S. Enzyme-less growth in Chara and terrestrial plants. *Front. Plant Sci.* **2016**, *7*, 866. [CrossRef] [PubMed]
6. Boudart, G.; Jamet, E.; Rossignol, M.; Lafitte, C.; Borderies, G.; Jauneau, A.; Esquerré-Tugayé, M.-T.; Pont-Lezica, R. Cell wall proteins in apoplastic fluids of *Arabidopsis thaliana* rosettes: Identification by mass spectrometry and bioinformatics. *Proteomics* **2005**, *5*, 212–221. [CrossRef] [PubMed]
7. Jamet, E.; Canut, H.; Boudart, G.; Pont-Lezica, R.F. Cell wall proteins: A new insight through proteomics. *Trends Plant Sci.* **2006**, *11*, 33–39. [CrossRef]
8. Calderan-Rodrigues, M.J.; Guimarães Fonseca, J.; de Moraes, F.E.; Vaz Setem, L.; Carmanhanis Begossi, A.; Labate, C.A. Plant cell wall proteomics: A focus on monocot species, *Brachypodium distachyon*, *Saccharum* spp. and *Oryza sativa*. *Int. J. Mol. Sci.* **2019**, *20*, 1975. [CrossRef]
9. Vázquez-Lobo, A.; Roujol, D.; Zuñiga-Sánchez, E.; Albenne, C.; Piñero, D.; Gamboa de Buen, A.; Jamet, E. The highly conserved spermatophyte cell wall DUF642 protein family: Phylogeny and first evidence of interaction with cell wall polysaccharides in vitro. *Mol. Phylogenet. Evol.* **2012**, *63*, 510–520. [CrossRef]
10. Feiz, L.; Irshad, M.; Pont-Lezica, R.F.; Canut, H.; Jamet, E. Evaluation of cell wall preparations for proteomics: A new procedure for purifying cell walls from Arabidopsis hypocotyls. *Plant Methods* **2006**, *2*, 10. [CrossRef]
11. Borner, G.H.H.; Lilley, K.S.; Stevens, T.J.; Dupree, P. Identification of glycosylphosphatidylinositol-anchored proteins in Arabidopsis. A proteomic and genomic analysis. *Plant Physiol.* **2003**, *132*, 568–577. [CrossRef]
12. Zúñiga-Sánchez, E.; Gamboa-de Buen, A. The two DUF642 At5g11420 and At4g32460-encoded proteins interact in vitro with the AtPME3 catalytic domain. In *Protein Interactions*; Cai, J., Wang, R., Eds.; IntechOpen: London, UK, 2012.
13. Zúñiga-Sánchez, E.; Soriano, D.; Martínez-Barajas, E.; Orozco-Segovia, A.; Gamboa-deBuen, A. *BIIDXI*, the *At4g32460* DUF642 gene, is involved in pectin methyl esterase regulation during *Arabidopsis thaliana* seed germination and plant development. *BMC Plant Biol.* **2014**, *14*, 338. [CrossRef] [PubMed]
14. Müller, K.; Levesque-Tremblay, G.; Bartels, S.; Weitbrecht, K.; Wormit, A.; Usadel, B.; Haughn, G.; Kermode, A.R. Demethylesterification of cell wall pectins in Arabidopsis plays a role in seed germination. *Plant Physiol.* **2013**, *161*, 305–316. [CrossRef] [PubMed]
15. Xie, X.; Wang, Y. *VqDUF642*, a gene isolated from the Chinese grape *Vitis quinquangularis*, is involved in berry development and pathogen resistance. *Planta* **2016**, *244*, 1075–1094. [CrossRef] [PubMed]
16. Palmeros-Suárez, P.A.; Massange-Sánchez, J.A.; Sánchez-Segura, L.; Martínez-Gallardo, N.A.; Rangel, E.E.; Gómez-Leyva, J.F.; Délano-Frier, J.P. *AhDGR2*, an amaranth abiotic stress-induced DUF642 protein gene, modifies cell wall structure and composition and causes salt and ABA hyper-sensibility in transgenic Arabidopsis. *Planta* **2017**, *245*, 623–640. [CrossRef] [PubMed]
17. Gao, Y.; Badejo, A.A.; Sawa, Y.; Ishikawa, T. Analysis of two L-galactono-1, 4-lactone-responsive genes with complementary expression during the development of *Arabidopsis thaliana*. *Plant Cell Physiol.* **2012**, *53*, 592–601. [CrossRef] [PubMed]

18. Salazar-Iribe, A.; Cruz-Valderrama, J.E.; Jímenez-Durán, K.; Gómez-Maqueo, X.; Gamboa-deBuen, A. BIIDXI, a DUF642 cell wall protein, is involved in hypocotyl growth via auxin efflux. *J. Plant Physiol.* **2018**, *231*, 105–109. [CrossRef]
19. Salazar-Iribe, A.; Agredano-Moreno, L.T.; Zúñiga-Sánchez, E.; Jiménez-Garcia, L.F.; Gamboa-deBuen, A. The cell wall DUF642 At2g41800 (TEB) protein is involved in hypocotyl cell elongation. *Plant Sci.* **2016**, *253*, 206–214. [CrossRef]
20. Kohnen, M.V.; Schmid-Siegert, E.; Trevisan, M.; Petrolati, L.A.; Sénéchal, F.; Müller-Moulé, P.; Fankhauser, C. Neighbor detection induces organ-specific transcriptomes, revealing patterns underlying hypocotyl-specific growth. *Plant Cell* **2016**, *28*, 2889–2904. [CrossRef]
21. Procko, C.; Burko, Y.; Jaillais, Y.; Ljung, K.; Long, J.A.; Chory, J. The epidermis coordinates auxin-induced stem growth in response to shade. *Genes Dev.* **2016**, *30*, 1529–1541. [CrossRef]
22. Ogawa, M.; Hanada, A.; Yamauchi, Y.; Kuwahara, A.; Kamiya, Y.; Yamaguchi, S. Gibberellin biosynthesis and response during Arabidopsis seed germination. *Plant Cell* **2003**, *15*, 1591–1604. [CrossRef] [PubMed]
23. Wellmer, F.; Riechmann, J.L.; Alves-Ferreira, M.; Meyerowitz, E.M. Genome-wide analysis of spatial gene expression in Arabidopsis flowers. *Plant Cell* **2004**, *16*, 1314–1326. [CrossRef] [PubMed]
24. Chen, J.; Zeng, B.; Zhang, M.; Xie, S.; Wang, G.; Hauck, A.; Lai, J. Dynamic transcriptome landscape of maize embryo and endosperm development. *Plant Physiol.* **2014**, *166*, 252–264. [CrossRef] [PubMed]
25. Wang, H.; Niu, Q.W.; Wu, H.W.; Liu, J.; Ye, J.; Yu, N.; Chua, N.H. Analysis of non-coding transcriptome in rice and maize uncovers roles of conserved lnc RNA s associated with agriculture traits. *Plant J.* **2015**, *84*, 404–416. [CrossRef]
26. De Pauw, M.A.; Vidmar, J.J.; Collins, J.; Bennett, R.A.; Deyholos, M.K. Microarray analysis of bast fibre producing tissues of *Cannabis sativa* identifies transcripts associated with conserved and specialised processes of secondary wall development. *Funct. Plant Biol.* **2007**, *34*, 737–749. [CrossRef]
27. Shulse, C.N.; Cole, B.J.; Ciobanu, D.; Lin, J.; Yoshinaga, Y.; Gouran, M.; Turco, G.M.; Zhu, Y.; O'Malley, R.C.; Brady, S.M.; et al. High-throughput single-cell transcriptome profiling of plant cell types. *Cell Rep.* **2019**, *27*, 2241–2247. [CrossRef] [PubMed]
28. Minic, Z.; Jamet, E.; Négroni, L.; Arsene der Garabedian, P.; Zivy, M.; Jouanin, L. A sub-proteome of *Arabidopsis thaliana* mature stems trapped on Concanavalin A is enriched in cell wall glycoside hydrolases. *J. Exp. Bot.* **2007**, *58*, 2503–2512. [CrossRef]
29. Bustamante, C.A.; Budde, C.O.; Borsani, J.; Lombardo, V.A.; Lauxmann, M.A.; Andreo, C.S.; Lara, M.V.; Drincovich, M.F. Heat treatment of peach fruit: Modifications in the extracellular compartment and identification of novel extracellular proteins. *Plant Physiol. Biochem.* **2012**, *60*, 35–45. [CrossRef]
30. Salazar-Iribe, A.; Zúñiga-Sánchez, E.; Mejía, E.Z.; Gamboa-deBuen, A. Cell wall localization of two DUF642 proteins, BIIDXI and TEEBE, during *Meloidogyne incognita* early inoculation. *Plant Pathol. J.* **2017**, *33*, 614.
31. Cools, T.; Iantcheva, A.; Maes, S.; Van den Daele, H.; De Veylder, L. A replication stress-induced synchronization method for *Arabidopsis thaliana* root meristems. *Plant J.* **2010**, *64*, 705–714. [CrossRef]
32. Cruz-Valderrama, J.E.; Jiménez-Durán, K.; Zúñiga-Sánchez, E.; Salazar-Iribe, A.; Márquez-Guzmán, J.; Gamboa-deBuen, A. Degree of pectin methyl esterification in endosperm cell walls is involved in embryo bending in *Arabidopsis thaliana*. *Biochem. Biophys. Res. Commun.* **2018**, *495*, 639–645. [CrossRef] [PubMed]
33. Day, R.C.; Herridge, R.P.; Ambrose, B.A.; Macknight, R.C. Transcriptome analysis of proliferating Arabidopsis endosperm reveals biological implications for the control of syncytial division, cytokinin signaling, and gene expression regulation. *Plant Physiol.* **2008**, *148*, 1964–1984. [CrossRef] [PubMed]
34. Fourquin, C.; Beauzamy, L.; Chamot, S.; Creff, A.; Goodrich, J.; Boudaoud, A.; Ingram, G. Mechanical stress mediated by both endosperm softening and embryo growth underlies endosperm elimination in Arabidopsis seeds. *Development* **2016**, *143*, 3300–3305. [CrossRef] [PubMed]
35. Soeda, Y.; Konings, M.C.; Vorst, O.; van Houwelingen, A.M.; Stoopen, G.M.; Maliepaard, C.A.; van der Geest, A.H. Gene expression programs during *Brassica oleracea* seed maturation, osmopriming, and germination are indicators of progression of the germination process and the stress tolerance level. *Plant Physiol.* **2005**, *137*, 354–368. [CrossRef] [PubMed]
36. Garza-Caligaris, L.E.; Avendaño-Vázquez, A.O.; Alvarado-López, S.; Zúñiga-Sánchez, E.; Orozco-Segovia, A.; Pérez-Ruíz, R.V.; Gamboa-deBuen, A. *At3g08030* transcript: A molecular marker of seed ageing. *Ann. Bot.* **2012**, *110*, 1253–1260. [CrossRef] [PubMed]

37. Durand, T.C.; Cueff, G.; Godin, B.; Valot, B.; Clément, G.; Gaude, T.; Rajjou, L. Combined proteomic and metabolomic profiling of the *Arabidopsis thaliana* vps29 mutant reveals pleiotropic functions of the retromer in seed development. *Int. J. Mol. Sci.* **2019**, *20*, 362. [CrossRef] [PubMed]
38. Zheng, Z.; Guo, Y.; Novák, O.; Chen, W.; Ljung, K.; Noel, J.P.; Chory, J. Local auxin metabolism regulates environment-induced hypocotyl elongation. *Nat. Plants* **2006**, *2*, 16025. [CrossRef] [PubMed]
39. Derbyshire, P.; McCann, M.C.; Roberts, K. Restricted cell elongation in Arabidopsis hypocotyls is associated with a reduced average pectin esterification level. *BMC Plant Biol.* **2007**, *7*, 31. [CrossRef] [PubMed]
40. Braybrook, S.A.; Peaucelle, A. Mechano-chemical aspects of organ formation in *Arabidopsis thaliana*: The relationship between auxin and pectin. *PLoS ONE* **2013**, *8*, e57813. [CrossRef]
41. Irshad, M.; Canut, H.; Borderies, G.; Pont-Lezica, R.; Jamet, E. A new picture of cell wall protein dynamics in elongating cells of *Arabidopsis thaliana*: Confirmed actors and newcomers. *BMC Plant Biol.* **2008**, *8*, 94. [CrossRef]
42. Horiguchi, G.; Ferjani, A.; Fujikura, U.; Tsukaya, H. Coordination of cell proliferation and cell expansion in the control of leaf size in *Arabidopsis thaliana*. *J. Plant Res.* **2006**, *119*, 37–42. [CrossRef] [PubMed]
43. Cardarelli, M.; Cecchetti, V. Auxin polar transport in stamen formation and development: How many actors? *Front. Plant Sci.* **2014**, *5*, 333. [CrossRef] [PubMed]
44. Jaradat, M.R.; Ruegger, M.; Bowling, A.; Butler, H.; Cutler, A.J. A comprehensive transcriptome analysis of silique development and dehiscence in Arabidopsis and Brassica integrating genotypic, interspecies and developmental comparisons. *GM Crops Food* **2014**, *5*, 302–320. [CrossRef] [PubMed]
45. Hu, J.; Barlet, X.; Deslandes, L.; Hirsch, J.; Feng, D.X.; Somssich, I.; Marco, Y. Transcriptional responses of *Arabidopsis thaliana* during wilt disease caused by the soil-borne phytopathogenic bacterium, *Ralstonia solanacearum*. *PLoS ONE* **2008**, *3*, e2589. [CrossRef] [PubMed]
46. Depuydt, S.; Trenkamp, S.; Fernie, A.R.; Elftieh, S.; Renou, J.P.; Vuylsteke, M.; Holsters, M.; Vereecke, D. An integrated genomics approach to define niche establishment by *Rhodococcus fascians*. *Plant Physiol.* **2009**, *149*, 1366–1386. [CrossRef] [PubMed]
47. Lee, C.W.; Efetova, M.; Engelmann, J.C.; Kramell, R.; Wasternack, C.; Ludwig-Müller, J.; Hedrich, R.; Deeken, R. *Agrobacterium tumefaciens* promotes tumor induction by modulating pathogen defense in *Arabidopsis thaliana*. *Plant Cell* **2009**, *21*, 2948–2962. [CrossRef] [PubMed]
48. Ndimba, B.K.; Chivasa, S.; Hamilton, J.M.; Simon, W.J.; Slabas, A.R. Proteomic analysis of changes in the extracellular matrix of Arabidopsis cell suspension cultures induced by fungal elicitors. *Proteomics* **2003**, *3*, 1047–1059. [CrossRef]
49. Lorenzini, M.; Mainente, F.; Zapparoli, G.; Cecconi, D.; Simonato, B. Post-harvest proteomics of grapes infected by Penicillium during withering to produce Amarone wine. *Food Chem.* **2016**, *199*, 639–647. [CrossRef]
50. Nogueira-Lopez, G.; Greenwood, D.R.; Middleditch, M.; Winefield, C.; Eaton, C.; Steyaert, J.M.; Mendoza-Mendoza, A. The apoplastic secretome of *Trichoderma virens* during interaction with maize roots shows an inhibition of plant defense and scavenging oxidative stress secreted proteins. *Front. Plant Sci.* **2018**, *9*, 409. [CrossRef]
51. Coolen, S.; Proietti, S.; Hickman, R.; Davila Olivas, N.H.; Huang, P.P.; Van Verk, M.C.; Van Loon, J.J. Transcriptome dynamics of Arabidopsis during sequential biotic and abiotic stresses. *Plant J.* **2016**, *86*, 249–267. [CrossRef]
52. Barcalá, M.; García, A.; Cabrera, J.; Casson, S.; Lindsey, K.; Favery, B.; García-Casado, G.; Solano, R.; Fenoll, C.; Escobar, C. Early transcriptomic events in microdissected Arabidopsis nematode-induced giant cells. *Plant J.* **2010**, *61*, 698–712. [CrossRef] [PubMed]
53. Ederli, L.; Dawe, A.; Pasqualini, S.; Quaglia, M.; Xiong, L.; Gehring, C. Arabidopsis flower specific defense gene expression patterns affect resistance to pathogens. *Front. Plant Sci.* **2015**, *6*, 79. [CrossRef] [PubMed]
54. Tsutsui, T.; Yamaji, N.; Huang, C.F.; Motoyama, R.; Nagamura, Y.; Ma, J.J. Comparative genome-wide transcriptional analysis of Al-responsive genes reveals novel Al tolerance mechanisms in rice. *PLoS ONE* **2012**, *7*, e48197. [CrossRef]
55. Tsutsui, T.; Yamaji, N.; Feng, J.M. Identification of a cis-acting element of ART1, a C2H2-type zinc-finger transcription factor for aluminum tolerance in rice. *Plant Physiol.* **2011**, *156*, 925–993. [CrossRef] [PubMed]

56. Arbelaez, J.D.; Maron, L.G.; Jobe, T.O.; Pineros, M.A.; Famoso, A.N.; Rebelo, A.R.; Singh, N.; Ma, Q.; Fei, Z.; Kochian, L.V.; et al. ALUMINUM RESISTANCE TRANSCRIPTION FACTOR 1 (ART 1) contributes to natural variation in aluminum resistance in diverse genetic backgrounds of rice (*O. sativa*). *Plant Direct.* **2007**, *1*, e00014. [CrossRef] [PubMed]
57. Jiang, C.; Liu, L.; Li, X.; Han, R.; Wei, Y.; Yu, Y. Insights into aluminum-tolerance pathways in *Stylosanthes* as revealed by RNA-Seq analysis. *Sci. Rep.* **2018**, *8*, 6072. [CrossRef]
58. Guo, P.; Qi, Y.P.; Yang, L.T.; Lai, N.W.; Ye, X.; Yang, Y.; Chen, L.S. Root adaptive responses to aluminum-treatment revealed by RNA-Seq in two Citrus species with different aluminum-tolerance. *Front. Plant Sci.* **2017**, *8*, 330. [CrossRef]
59. Ito, H.; Kobayashi, Y.; Yamamoto, Y.Y.; Koyama, H. Characterization of *NtSTOP1*-regulating genes in tobacco under aluminum stress. *Soil Sci. Plant Nutr.* **2019**, *65*, 1–8. [CrossRef]
60. Gutsch, A.; Zouaghi, S.; Renaut, J.; Cuypers, A.; Hausman, J.F.; Sergean, K. Changes in the proteome of *Medicago sativa* leaves in response to long-term cadmium exposure using a cell-wall targeted approach. *Int. J. Mol. Sci.* **2018**, *19*, 2498. [CrossRef]
61. Ariani, A.; Di Baccio, D.; Romeo, S.; Lombardi, L.; Andreucci, A.; Lux, A. RNA Sequencing of *Populus x canadensis* roots identifies key molecular mechanisms underlying physiological adaption to excess zinc. *PLoS ONE* **2015**, *10*, e0117571. [CrossRef]
62. Trentin, A.R.; Pivato, M.; Mehdi, S.M.M.; Barnabas, L.E.; Giaretta, S.; Fabrega-Prats, M.; Prasad, D.; Arrigoni, G.; Masi, A. Proteome readjustments in the apoplastic space of *Arabidopsis thaliana ggt1* mutant leaves exposed to UV-B radiation. *Front. Plant Sci.* **2015**, *6*, 128. [CrossRef] [PubMed]
63. Patanun, O.; Ueda, M.; Itouga, M.; Kato, Y.; Utsumi, Y.; Matsui, A.; Tanaka, M.; Utsumi, C.; Sakakibara, H.; Yoshida, M.; et al. The histone deacetylase inhibitor suberoylanilide hydroxamic acid alleviates salinity stress in Cassava. *Front. Plant Sci.* **2017**, *7*, 2039. [CrossRef] [PubMed]
64. Huang, D.; Wu, W.; Abrams, S.R.; Cutler, A.J. The relationship of drought-related gene expression in *Arabidopsis thaliana* to hormonal and environmental factors. *J. Exp. Bot.* **2008**, *59*, 2991–3007. [CrossRef] [PubMed]
65. Des Marais, D.L.; McKay, J.K.; Richards, J.H.; Sen, S.; Wayne, T.; Juenger, T.E. Physiological genomics of response to soil drying in diverse Arabidopsis accessions. *Plant Cell* **2012**, *24*, 893–914. [CrossRef] [PubMed]
66. Thumma, B.R.; Sharma, N.; Southerton, S.G. Transcriptome sequencing of *Eucalyptus camaldulensis* seedlings subjected to water stress reveals functional single nucleotide polymorphisms and genes under selection. *BMC Genom.* **2012**, *13*, 364. [CrossRef] [PubMed]

International Journal of
Molecular Sciences

MDPI

Article

The Cell Wall PAC (Proline-Rich, Arabinogalactan Proteins, Conserved Cysteines) Domain-Proteins Are Conserved in the Green Lineage

Huan Nguyen-Kim [1,†], Hélène San Clemente [1,†], Josef Laimer [2], Peter Lackner [2], Gabriele Gadermaier [2], Christophe Dunand [1] and Elisabeth Jamet [1,*]

1 Laboratoire de Recherche en Sciences Végétales, Université de Toulouse, CNRS, UPS, 31320 Auzeville Tolosane, France; huanqnu@gmail.com (H.N-K.); sancle@lrsv.ups-tlse.fr (H.S.C.); dunand@lrsv.ups-tlse.fr (C.D.)

2 Paris-Lodron-University of Salzburg, Department of Biosciences, 5020 Salzburg, Austria; josef.laimer@sbg.ac.at (J.L.); peter.lackner@sbg.ac.at (P.L.); gabriele.gadermaier@sbg.ac.at (G.G.)

* Correspondence: jamet@lrsv.ups-tlse.fr; Tel.: +33-(0)5-3432-3830

† These two authors equally contributed to the work.

Received: 17 January 2020; Accepted: 1 April 2020; Published: 3 April 2020

Abstract: Plant cell wall proteins play major roles during plant development and in response to environmental cues. A bioinformatic search for functional domains has allowed identifying the PAC domain (Proline-rich, Arabinogalactan proteins, conserved Cysteines) in several proteins (PDPs) identified in cell wall proteomes. This domain is assumed to interact with pectic polysaccharides and *O*-glycans and to contribute to non-covalent molecular scaffolds facilitating the remodeling of polysaccharidic networks during rapid cell expansion. In this work, the characteristics of the PAC domain are described in detail, including six conserved Cys residues, their spacing, and the predicted secondary structures. Modeling has been performed based on the crystal structure of a *Plantago lanceolata* PAC domain. The presence of β-sheets is assumed to ensure the correct folding of the PAC domain as a β-barrel with loop regions. We show that PDPs are present in early divergent organisms from the green lineage and in all land plants. PAC domains are associated with other types of domains: Histidine-rich, extensin, Proline-rich, or yet uncharacterized. The earliest divergent organisms having PDPs are Bryophytes. Like the complexity of the cell walls, the number and complexity of PDPs steadily increase during the evolution of the green lineage. The association of PAC domains with other domains suggests a neo-functionalization and different types of interactions with cell wall polymers

Keywords: cell wall; evolution; green lineage; modeling; PAC domain; phylogeny; plant

1. Introduction

Plant cell walls are composite structures mainly made of polysaccharides and proteins. Cellulose microfibrils and hemicelluloses form intricate networks, which are embedded in a pectin matrix [1]. Although present in minor amounts, the cell wall proteins (CWPs) play critical roles in polysaccharides organization and remodeling processes during growth and upon environmental stresses [2,3]. Cell wall proteomics has revealed the great diversity of CWPs and allowed the discovery of unexpected CWP families [4]. The combination of genetics and biochemistry approaches has allowed demonstrating the roles of CWPs in polysaccharide metabolism, biosynthesis of lipid-rich cell wall layers, lignin monomer polymerization, but also in signaling and ROS homeostasis maintenance [5–8].

Among the newly described CWPs families, the importance of the PAC (Proline-rich Arabinogalactan protein and Conserved Cysteines) domain containing-protein (PDP) family could be stressed because of their presence in many cell wall proteomes (see *WallProtDB*, www.polebio.lrsv.ups-tlse.fr/WallProtDB/, query with "Ole e1 allergen domain" as a keyword). The name of the PDP family was initially proposed by Baldwin et al. [9], who described them as a sub-family of non-classical arabinogalactan proteins (AGPs) containing both an AGP domain and a C-terminal domain containing six Cysteines residues (named Cys 1 to Cys 6 herein). Later on, a domain partly describing the PAC domain has been proposed in the Pfam database (PF01190, http://pfam.xfam.org/). The firstly described member of this family was a protein from *Nicotiana alata* named AGPNa3 [10]. Then, several proteins very close to AGPNa3 were studied, for a review, see [11]. As examples, the following ones can be mentioned: *Daucus carota* DcAGP1 [12]; *Arabidopsis thaliana* AtAGP30 [13], and AtAGP31 (At1g28290) [14]; *Capsicum annuum* CaPRP1 [15]; *Gossypium hirsutum* GhAGP31 [16]; and *Petunia hybrida* PhPRP1 [17]. More recently, it appeared that the PAC domain could also be found alone, located at the N-terminus of the mature protein or associated with different types of domains, such as a Histidine-rich region, an *O*-glycosylated Proline/Hydroxyproline-rich domain, or an extensin domain [18,19].

Functional studies on several of the *A. thaliana* PDPs have shown their diverse roles during plant development. *PRPL1* (*Proline-Rich Protein-Like, At5g05500*) has a trichoblast-specific expression and plays roles in root hair elongation, as shown by the reduction in length of root hairs in the *prpl1* mutant [20]. Plants lacking *FOCL1* (*Fused Outer Cutiular Ledge 1, At2g16630*) produce stomata without a cuticular ledge, and thus, focl1 mutants display drought tolerance [21]. *AtAGP30* (*At2g33790*) is involved in root regeneration in vitro and in the timing of seed germination [13]. *AtAGP30* is expressed in root atrichoblasts under the control of ABA signaling [22]. *AtAGP31* is expressed in vascular tissues and repressed by methyl jasmonate at the transcriptional level [14]. AtAGP31 has also been shown to accumulate in actively growing etiolated hypocotyls [23]. In vitro interactions have been demonstrated between its PAC domain and galactans or the Gal-Ara-rich *O*-glycans of its Proline/Hydroxyproline rich domain [11]. These studies have led to the assumption that AtAGP31 could be involved in cell wall non-covalent protein/polysaccharide networks playing roles during quick cell elongation [11].

Recently, the crystal structure of the PAC domain of an allergenic protein from *Plantago lanceolata* containing an N-terminal PAC domain (Pla l 1 as a member of the Ole e 1–like protein family, PDP code 4Z8W) has been determined, highlighting the importance of β-sheets in its secondary structure [24]. In particular, the structure revealed a seven-stranded β -barrel with four loop regions. Three intramolecular disulfide bonds were found between (i) β 1b and β 6 strands (Cys 1-Cys 5), (ii) β 2 and β 5 strands (Cys 3-Cys 4), and the (iii) C-terminus and loop C-terminal of β 2 strand (Cys 2-Cys 6), thus forming a closed branched loop. A detailed characterization of allergens of the same protein family allowed proposing that they share the same core structure, whereas loop regions can be heterogeneous.

In this article, we aim at giving an evolutive overview of the PDPs throughout the green lineage, from Bryophytes to late divergent plants, such as monocots and dicots. We first define more precisely the PAC domain characteristics in order to retrieve PAC domain sequences from available genomic or RNA-seq databases using a tailor-made bioinformatic script. Since the conservation of the primary amino acid sequences of PAC domains was rather low, and since the presence of β -sheets seemed to be essential for domain folding, bona fide PAC domains were selected according to their secondary structure conservation, and protein alignment was done using a software taking into account secondary structures. Modeling of tertiary structures was done based on the available crystal structure of the Pla l 1 PAC domain. Finally, we could draw a phylogenic tree and sort the PAC domains according to their association with other domains. We could also investigate the occurrence of PAC domains in ancestor organisms.

2. Results and Discussion

2.1. Characteristics of the PAC Domain and Search for New PDP Candidates

The overall strategy used for this study is summarized in Figure 1. As a first step and in order to obtain a better definition of a PAC domain, orthologous sequences have been identified in the *A. thaliana* genome using that of the AtAGP31 PAC domain. Altogether, 14 candidate sequences were identified and manually checked for the presence of the six conserved Cys residues: At1g29140, At1g78040, At3g09925, At4g08685, At4g18596, At5g45880, At5g54855, AtAGP31, At5g05500 (PRPL1), At5g15790, At2g34790 (AtAGP30), At2g34700, At4g18596, and At2g16630 (FOCL1). These sequences were then used to identify additional PDPs by sequence similarity in eight other angiosperm genomes: *Amborella trichopoda*, *Brachypodium distachyon*, *Oryza sativa*, *Sorghum bicolor*, *Populus trichocarpa*, *Eucalyptus grandis*, *Linum usitatissimum*, and *Gossypium raimondii*. About 50 putative PDPs were collected and manually checked for the presence of the six conserved Cys residues. From this first data mining step, it appeared that the level of conservation of the amino acid sequences of the PAC domains could be low. In particular, except between the two first conserved Cys residues (Cys 1 and Cys 2), the spacing between Cys residues could be variable. Thus, the usual homology-based mining was not sufficient, and an alternative strategy was necessary to obtain exhaustive results for each plant. The alignment of angiosperms PAC domains has allowed calculating the range of spacing between the conserved Cys residues. Then, a tailor-made script based on several points detailed in Table 1 has been set up to search for additional PDPs in the same genomes or in other genomics or transcriptomics databases. However, the prediction of a signal peptide for protein secretion could not be made systematically for the proteins translated from transcriptomics data because the sequences could be incomplete. Furthermore, when genomic sequences were available, the presence of an intron between the sequences encoding, on the one hand, Cys 1 and Cys 2, and on the other hand, Cys 3 to Cys 6 was searched for to support the PAC domain identification.

Table 1. Five features of a bona fide PAC domain.

1. Present in a protein with a predicted signal peptide
2. Presence of six Cys residues downstream a Glycine residue and with a defined spacing [1]: **Gly** (3) **Cys 1** (2) **Cys 2** (10,30) **Cys 3** (20,50) **Cys 4** (8,20) **Cys 5** (25,60) **Cys 6**
3. Prediction of β-sheets according to the crystal structure of the *Plantago lanceolata* (www.rcsb.org/structure/4Z8W) PAC domain protein
4. Possibly associated to AGP, extensin, X($Proline_{n \geq 2}$) X-rich, Histidine-rich, or W-W domains
5. No prediction of additional functional domains

[1] The number of amino acids between two successive Cys residues is indicated between brackets.

Using this script, sequences encoding PAC domains have been searched for in 78 plant species belonging to the green lineage from Bryophytes (*Bryophyta*, *Marchantiophyta* and *Anthocerotophyta*) to late divergent plants. Altogether, about 450 putative PAC domain sequences were collected (S1–S4).

Three additional criteria have then been used to select bona fide PAC domain proteins. The first one was the number of conserved Cys residues. Indeed, we have found putative PAC domains showing the expected characteristics, but containing only five Cys residues, or containing more Cys residues, up to nine (S1,S5). Although some of them had sequences very similar to those of six Cys-containing PAC domains (S5), we have decided to dismiss them in case of a lack or an excess of Cys residues, which would modify the folding of the domain by generating disulfide bridges different from the expected ones. The second exclusion criterion was the absence of predicted β-sheets. Indeed, the crystal structure of the Pla I 1 PAC domain has allowed highlighting the importance of these β-sheets in its secondary structure [24]. Some proteins with large predicted α-helices and/or no predicted β-sheets have been dismissed with regard to this criterion, especially in Bryophytes, Equisetales, and Alismatales (S1,S6). The third criterion was the presence of associated predicted functional domains suggesting intracellular functions like aldehyde dehydrogenase domain (PF00171, *Tetraphis pellucida* HVBQ_2004216) or JmjC and JmjN domains of transcription factors (PF02373 and PF02375, *Pallavicinia lyelli* YFGP_2007785) (S3). In most of these latter cases, it was not possible to predict the sub-cellular localization of the proteins because they resulted from the translation of incomplete contigs obtained from RNA-seq data.

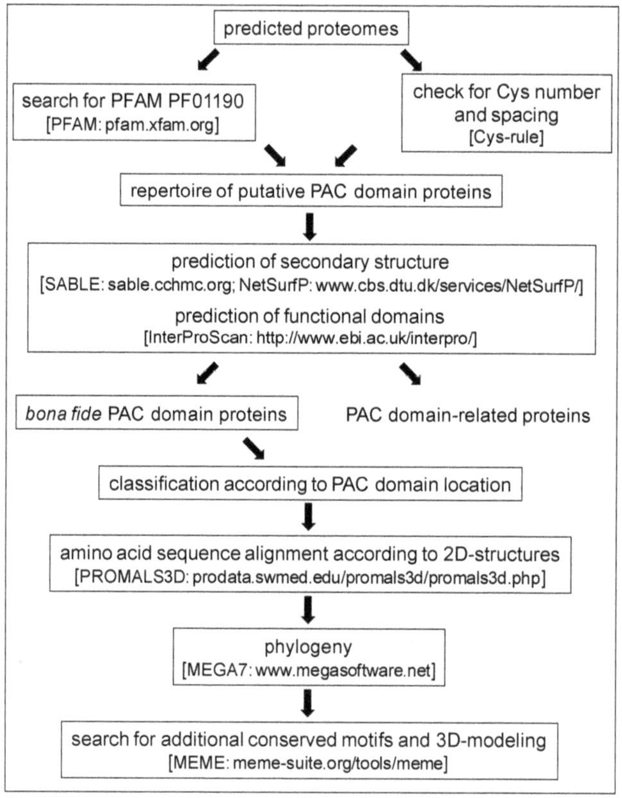

Figure 1. Pipeline for Proline-rich Arabinogalactan protein and Conserved Cys (PAC) domain protein identification and phylogeny. The name of the bioinformatics programs and resources used at each step are indicated in brackets.

2.2. The Number and the Diversity of PAC Domain Proteins Increase Along the Green Lineage

The PDPs have been classified according to the domains associated with the PAC domain. Four types were distinguished (Figure 2). Type 1 corresponds to proteins only containing a PAC domain. The corresponding genes could exhibit either no intron or one intron between the sequences encoding Cys 1 and Cys 2 and those encoding Cys 3 to Cys 6. Type 2 includes proteins with an N-terminal PAC domain, which could be associated to (i) a Proline-rich domain or (ii) a well-conserved domain of unknown function usually encoded by a specific exon and starting with the following amino acid motif: Tryptophane-X8-Tryptophane (W-W domain) (S7). As an example, At2g16630 (FOCL1) is a type 2-PAC domain protein with a W-W domain at the C-terminus. Type 3 encompasses proteins with a C-terminal PAC domain. The PAC domain could be associated with a Histidine stretch, a Proline-rich domain, and/or an AGP domain. For example, AtAGP30 and AtAGP31 are type 3-PAC domain-proteins. Finally, type 4 corresponds to proteins containing central PAC domains flanked by two extensin domains. Although a few proteins with Serine-(Proline)$_4$ motifs typical of extensins at their C-terminus were found in *Anthocerophyta* and Lycopodiales, the first bona fide type 4-PDP was found in Psilotales. There is no such PDP in *A. thaliana*.

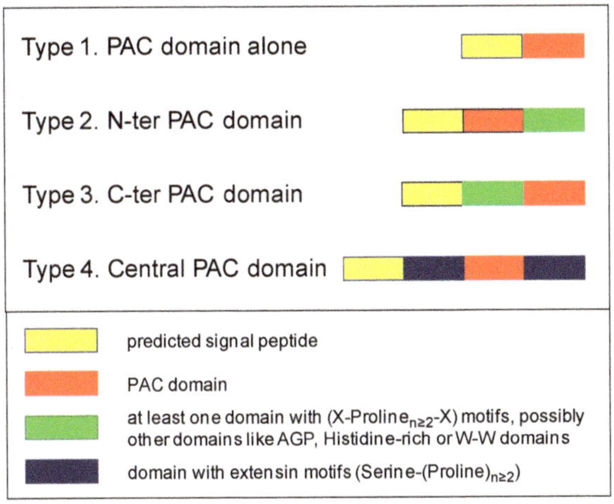

Figure 2. Classification of PAC domain proteins according to the location of the PAC domain and associated domains. **Type 1.** Proteins containing the PAC domain alone. **Type 2.** Proteins containing the PAC domain at their N-terminus together with another (several other) domain(s). **Type 3.** Proteins containing the PAC domain at their C-terminus together with another (several other) domain(s). **Type 4.** Proteins containing the PAC domain in a central position flanked by two domains with extensin motifs (Serine-(Proline)$_{n\geq2}$).

In Bryophytes and Anthocerotophyta, only one to three PAC domain proteins were found for each species (S1). The number of PDPs was higher in Psilotales and Equisetales as well as in all the plant families, which have appeared later in the green lineage. Eleven PDPs are present in *Amborella trichopoda*, which is considered as an ancestor common to angiosperms [25]. The highest numbers of PDPs, i.e., between 17 and 23, were found in Poales, *Brachypodium distachyon*, *Sorghum bicolor*, *Zea mays*, and *Oryza sativa*, as well as in *Linum usitatissimum*, *Populus trichocarpa*, and *Gossypium raimondii*. In Poales like *B. distachyon* and *O. sativa*, the genes encoding PDPs could be found in tandem (Figure 3). The PAC domains of these genes could show a high degree of identity (more than 85%), supporting the recent tandem duplication events [26]. In addition, PAC domains with various numbers of Cys residues were also found in Poales (S1). The functionality of those PAC domains has not yet been established.

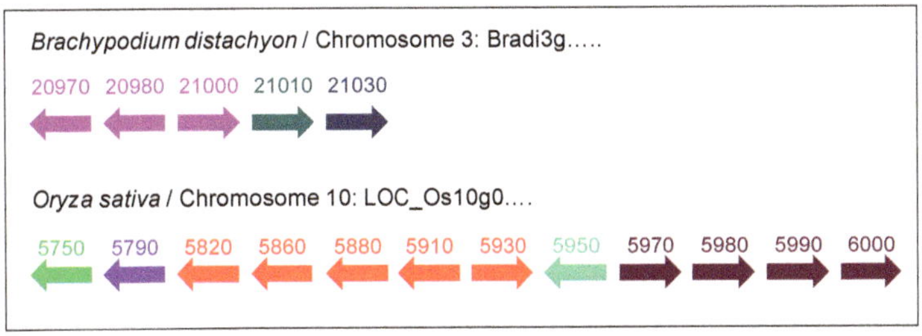

Figure 3. Examples of domain containing-protein (PDP) genes organized in tandem in the *B. distachyon* and *O. sativa* genomes. The orientation of the genes is indicated by arrows. The names of the genes are abbreviated, e.g., 20970 stands for *Bradi3g20970*, and 5750 for *O. sativa LOC_Os10g5750*. Genes sharing more than 85% identity in their PAC domain coding sequences at the amino acid level are represented with arrows of the same color.

The different types of PDPs are unevenly distributed within the different plant species (Figure 4). Only type 1- and type 2-PDPs were found in all plant families. Among the type 1-PDPs, one sub-type should be distinguished. It corresponds to highly conserved sequences throughout the green lineage since Lycopodiales with an overall percentage of identity ranging from 60% to 88% and a percentage of similarity from 69% to 92%. For comparison, the percentage of identity and of similarity between two PAC domain sequences can be rather low (15.4% and 20.7%, respectively). Among the type 2-PDPs, those including a C-terminal W-W domain are present in nearly all plant families from Bryophytes to Brassicales. They could appear as ancestors of PDPs. Type 3- and type 4-PDPs seem to have appeared more recently in the evolution of the green lineage since the most ancient type 3- and type 4-proteins were found in *A. trichopoda* and in Psilotales, respectively. Of course, one cannot exclude that some PDPs are missing in this collection since only a few complete genomes are available for plants from Psilotales to Amborellales.

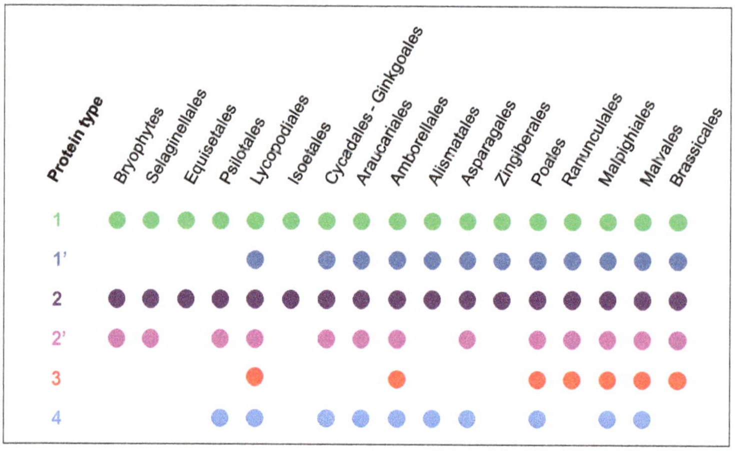

Figure 4. Distribution of the different types of PAC domains within the plant families. The different types of PAC domains are represented in Figure 2. Among type 1-PAC domains, those having a highly conserved amino acid sequence are distinguished (1′). Among type 2-PAC domains, those that are associated to a C-terminal W-W domain are highlighted (2′).

2.3. A Possible Origin for the PAC Domain

We have performed an extensive search of PAC domain sequences in the available databases dedicated to ancestors of the green lineage using both the script described above and BLAST queries using several PAC domains in case the spacing between Cys residues would be slightly different. Mining was done in the following families: Stramenopiles (*Synura petersenii*), Cryptophyta (*Chroomonas sp*), Chlorophyta (*Asteromonas gracilis*, *Chlamydomonas rheinardtii*, *Nephroselmis olivacea*, *Volvox carteri*, *Scenedesmus dimorphus*, *Scherffelia dubia*), Streptophyta (*Chara braunii*, *Coleochaete orbicularis*, *Klebsormidium flaccidum*, *Mesotaenium caldariorum*, *Penium margaritaceum*) (S4). In many cases, the proteins were incomplete either at their N-termini and it was not possible to predict a signal peptide, or at their C-termini, and they could not be classified. Whenever possible, the presence of predicted functional domains associated to the putative PAC domains was checked, and the proteins comprising functional domains associated to intracellular functions were not retained.

We could only find PAC domain-related sequences in *Chlorophyta*: 10 proteins were found in *C. rheinardtii* and one in *V. carteri* ,which both belong to Chlamydomonales. The Glycine residue located upstream the first Cys residue was always missing, and the PAC domains were associated with Proline-rich motifs of two types: either Serine-(Proline)$_n$ or (Proline)$_n$ and up to three of them could be found in a given protein. However, the secondary structures of these domains were predicted to be α-helices. In *C. rheinardtii*, the GP1 and GP2 proteins, which both have Serine-(Proline)$_n$ motifs, were described as proteins rich in Hydroxyproline resides forming the insoluble glycoprotein framework of the cell wall [27,28]. Furthermore, in *C. orbicularis*, we could find another interesting PAC domain candidate, which was associated to Proline-rich motifs but contained seven Cys residues. The highest level of identity/similarity was found with two PAC domains of *Musa acuminata*: GSMUA_Achr4T17330 (45%/51%) and GSMUA_Achr7T01790.1 (39%/50%). The highest level of identity/similarity with a *Marchantiophyta* PAC domain was found with the *Conocephalum conicum* PAC domain ILBQ_2004952 (30%/46%) and the *M. polymorpha* Mapoly0014s0128 PAC domain (33%/45%). Altogether, the sequence showing the highest level of identity to bona fide PAC domains was found in *C. orbicularis*. This is consistent with the assumption that the Coleochaetales could be one of the ancestors of the green lineage [29].

2.4. Three-Dimensional-Modeling of PAC Domain Proteins

Three-dimensional-models were calculated for 41 bona fide and 9 putative PAC domains, based on the crystal structure of the *P. lanceolata* PAC domain [24]. The sequence identities between the template and the PAC domains varied between 9.6% and 30.4% (median 15.9%). A sequence identity of 30% is generally seen as a lower limit for reliable models predicted by homology modeling algorithms, but the assumption of disulfide bridges somewhat lowers this limit. However, the low sequence similarities were still an issue. In addition, in 6 out of the 50 PAC domains, the 3D-modeling software I-Tasser was not able to find conformations enabling the formation of the three disulfide bridges between the predefined Cys residues (S8). In all these cases, either the proteins were predicted to have α-helices, or they were missing the Glycine residue upstream Cys 1.

For the bona fide PAC domains, it was possible to propose relevant 3D-models fitting with the typical structure experimentally demonstrated for the *P. lanceolata* PAC domain [24]. Four selected PAC domains from different plants are shown in Figure 5: an *Anthocerophyta* (*Anthoceros formosa*), chosen as an ancestral plant, *A. trichopoda* as the common ancestor to flowering plants, and two higher plants, *Oropetium thomaeum* and *A. thaliana*. All four 3-D models show the expected parallel β-sheets forming a β-barrel and the three disulfide bridges. They also contain loop regions as the *P. lanceolata* PAC domains. The 3D-structure of bona fide PAC domains seems to have been conserved through the evolution of the green lineage. However, the *C. orbicularis* protein, which was assumed to be an ancestor of the PDPs in the green lineage, only had three β-sheets, but the three disulfide bridges were at the predefined positions (S8).

The PAC domains that have been considered apart because of the prediction of α-helices showed completely different 3D-structures (S8). They exhibited less β-sheets or only α-helices, and as mentioned above, the three disulfide bridges were not at the expected positions. The 3-D modeling, thus, brought an additional criterion to confirm bona fide PAC domains. Interestingly, such a β-barrel structure has already been described for a mannose-binding lectin family of red algae, the *Oscillatoria Agardhii* Agglutinin-Homolog (OAAH) mannose-binding lectin family [30]. In this case, two β-barrels associate perpendicularly to build up the complete 3D-structure of the molecule, and the interaction with cell wall polymers occurs at two crevices symmetrically located at its two ends [31]. This role would be consistent with the finding that the PAC domain of AtAGP31 can interact with cell wall polysaccharides and *O*-glycans in vitro [11].

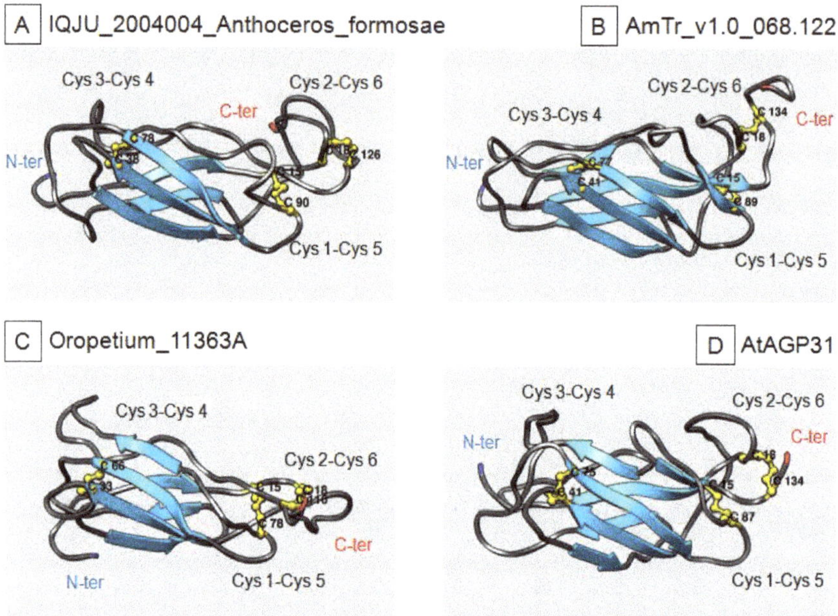

Figure 5. 3D-modeling of four PAC domains. (**A**) A representative PAC domain of Bryophytes: IQJU_2004004_Anthoceros_formosae. (**B**) A PAC domain of *A. trichopoda*: AmTr_v1.0_068.122. (**C**) A representative PAC domain of the *O. thomaeum* monocot: Oropetium_11363A. (**D**) A representative PAC domain of the *A. thaliana* dicot: At1g28290. The N-terminus (N-ter) and the C-terminus (C-ter) of the proteins are indicated in blue and red, respectively. Blue ribbons represent β-sheets. The three disulfide bridges are drawn in yellow, and the names of the Cys residues involved are indicated.

To test the role of the conserved Cys residues and, therefore, that of disulfide bridges in 3D-structure stability, in silico mutation experiments have been performed. Possible 5 Cys-PAC domain variants have been tested for the *P. lanceolata* PAC domain, and for each of the eight A. trichopoda PAC domains, which were considered as representative of the eight phylogenic clades (see below). Each Cys residue has been replaced by a Ser residue, and the change in stability was determined by MAESTRO (S11). In all cases, positive values of the ddG parameter indicating changes in unfolding free energy were found, indicating destabilization of the 3D-structure. Altogether, it seems that the conserved Cys residues are critical for the stability of the β-barrel. This could indicate that the domains lacking one Cys residue could be impaired in their biological activity or more sensitive to changes in their physiological environment. The presence of a seventh or even an eighth Cys residue could have different consequences depending on the position(s) of the additional Cys residue(s). Such residue(s) could be involved in different disulfide bridges or not. Only experimental work could allow showing any change in the biological activity of the PAC domain.

2.5. Phylogenetic Analyses Reveal the Presence of a Few Clades Grouping the PAC Domain Proteins According to Their Associated Domains

Based on all the criteria described above, 300 PAC domains have been selected for the building of phylogenetic trees (S1,S2). They have been chosen from plant families representative of the green lineage from Bryophytes to Brassicales based on a phylogenetic tree established using plastid gene sequences [32]. When several species were available for a given plant family, only one or a few of them were selected to represent it. For each plant family, the PAC domains sequences were analyzed for their percentage of identity, and the most representative plant species was retained. When the sets of PAC domain sequences were too different between plants of the same family, several species could be maintained. In addition, only PAC domains showing less than 85% of identity inside a given plant species were conserved. As a first step, the sequences were aligned according to their predicted secondary structure. Such a strategy was used in previous studies where the conservation of the primary sequences of the proteins was not sufficient to ensure relevant alignments [33–35]. The PROMALS3D software was used, and the resulting alignment was introduced in the MEGA7 software to build up a maximum likelihood tree using 500 bootstraps. Due to the low level of conservation between amino acid sequences and especially between the PAC domain sequences of the older lineages, we have decided to build up two independent trees to avoid bias due to long-branch attraction: the first one (Tree I) including plants from Bryophytes to *A. trichopoda*, and the second one (Tree II) from *A. trichopoda* to Brassicales.

Regarding Tree I, it is difficult to define clades grouping all the PAC domain sequences because most of the bootstrap values were low (S9). We only considered clades corresponding to bootstrap values higher than 30. We could define seven clades grouping 71% of the retrieved PAC domains, six of them containing one *A. trichopoda* PAC domain: clade A (AmTr.v1.0.061.7, mostly type 1-PAC domains); clade B (AmTr.v1.0.066.9, type 4-PAC domains); clade C (AmTr.v1.0.062.88, type 1-PAC domains, highly conserved sequences); clade D (AmTr.v1.0.041.161, type 2 W-W domains); clade E (AmTr.v1.000047, type 2-PAC domains); clade J (AmTr.v1.0.041.169, type 2-PAC domains); and clade K (*Equisetum sp* PAC domains). The distribution of the PAC domain sequences of the other species was not clear. PAC domains of Bryophytes were represented in clades A, D, and E, whereas a *Tmesipteris parva* (Psilotale) PAC domain was found in clade B, and a *Phylloglossum drummondii* (Lycopodiale) PAC domain in clades C, and J. Of course, one cannot exclude that PAC domains of plants, which have divergent earlier than Amborellales are still missing since only a limited number of fully sequenced genomes are available. Despite the presence of the key Cys residues and of conserved 3D-structure, the large evolutive distance existing between Bryophytes and *A. trichopoda* together with a relaxed selective pressure could explain the low sequence identity observed between sequences of Tree I. Indeed, whereas terrestrialization is assumed to have occurred 450 MYA [36], the age of angiosperms emergence was estimated to be between 169-199 MYA [37]. Based on the putative interaction with cell wall polysaccharides and *O*-glycans, the PAC domain sequence variability could be correlated with the variability of the cell wall composition from Bryophytes to angiosperms [38].

In Tree II, the PAC domains were distributed into 10 clades with high confidence bootstrap values (from 72 to 100) with the exception of clade H (28) (Figure 6, S9). An *A. trichopoda* PAC domain was found in each of them. Four clades were specific to higher plants, each of them, respectively, comprised the following *A. trichopoda* PAC domains: AmTr.v1.0.047.45 (clade F); AmTr.v1.0.068.122 (clade G); AmTr.v1.0.153.4 (clade H); and AmTr.v1.0.019.72 (clade I). Monocot and dicots were represented in all the clades, but clade J comprised a high number of grass PAC domains originating from gene duplication (see above). Interestingly, although the tree has been built up with PAC domains only, they grouped according to their association to other domains: type 1-PAC domains were found in clades A, C, F, H, and I; type 2-PAC domains were grouped in clades D, E, and J, with type 2 W-W domains in clade D; type 3-PAC domains were found in clade G with the exception of three of them in clade H with short Proline-rich motifs at their N-terminus; and type 4-PAC domains were only found in clade D. Thus, it seems that there is a link between the amino acid composition of PAC domains, their secondary structure, and the associated domains. Finally, it seems that all the PAC domains of higher plants have a counterpart in *A. trichopoda*, meaning that the modern multi-domain structures of the PDPs found in the ten angiosperm clades preceded the emergence of angiosperms.

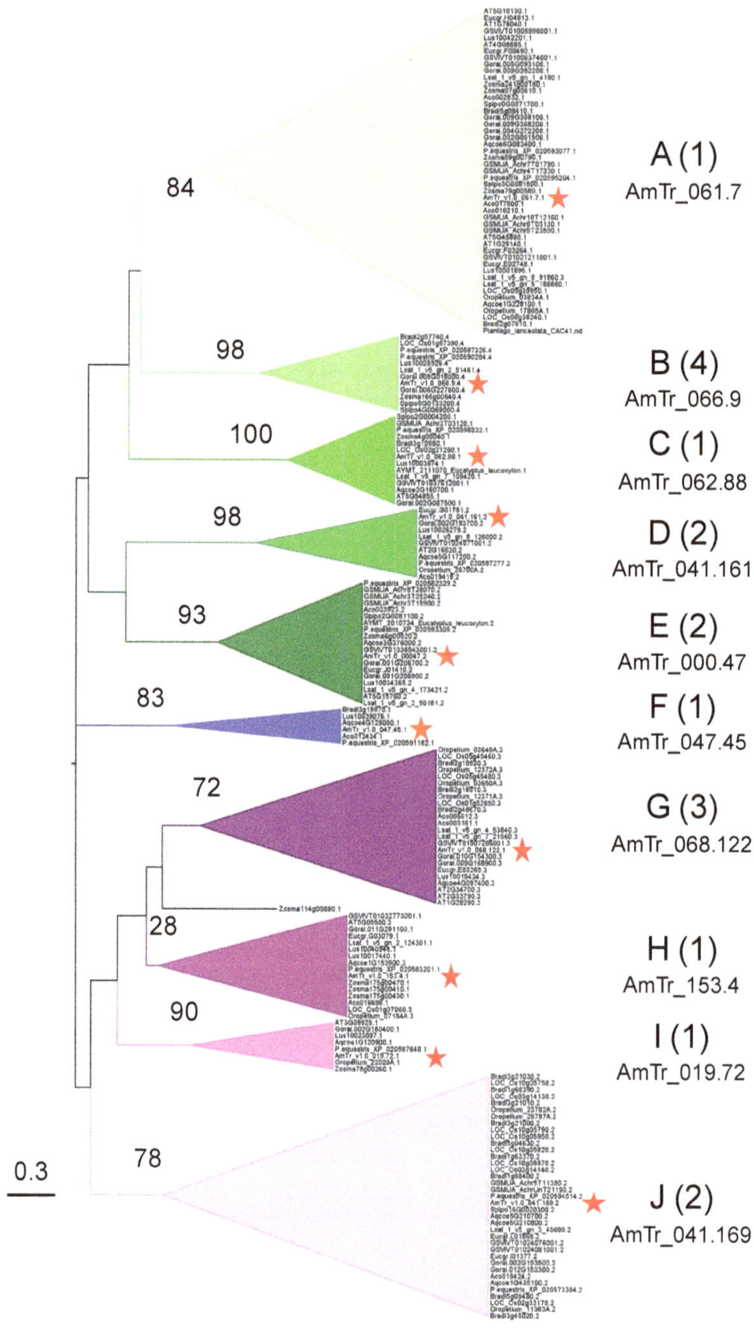

Figure 6. Phylogenetic Tree II. Tree II was built up using 196 PAC domains sequences from *A. trichopoda* to *A. thaliana*. Ten clades (A to I) were defined according to significant bootstrap values (higher than 72, with the exception of clade B). The type of PDPs (e.g., Type 1 is 1, see Figure 2) found in each clade indicated between brackets. The name of the *A. trichopoda* PDP found in each clade is indicated and highlighted with a red star.

2.6. Conserved Amino Acids Motifs Inside Clades

A search for conserved amino acid motifs was done for the PAC domains of each clade of Tree II. The most significant results were found for clades A, B, D, E, G, H, and I (Figure 7). In each clade, the most conserved motifs were detected at the N-terminus of the PAC domain. This was consistent with the definition of the pollen Ole e 1 motif in the Pfam and Prosite databases (PF01190 and PS00925, respectively). However, the consensus defined for the PS00925 domain only exactly fitted with that of clade A PAC domains ([EQT]-G-x-V-Y-C-D-[TNP]-C-R). Furthermore, the most conserved PAC domains were found in the C clade (Figure 8). Their degree of conservation in the green lineage from Lycopodiales to Brassicales is impressive. Finally, the C-terminal W-W domain present in all the proteins belonging to clade D was also very well conserved from the Bryophytes to the Brassicales with common motifs mostly located in its N-terminus half (S10).

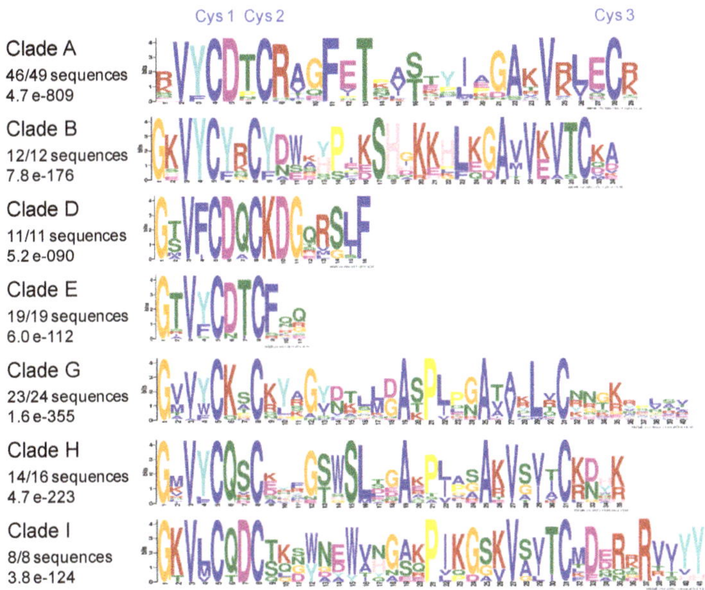

Figure 7. The most conserved motifs of PAC domains inside clades A, B, D, E, G, H, I in PDPs of *A. trichopoda* plant families appeared subsequently. The number of PAC domains in each clade is indicated as well as the score of the conserved motif according to the MEME software.

Figure 8. The most conserved PAC domains from clade C. The comparisons have been made between 18 PAC domain sequences from Lycopodiales to Brassicales.

The combination of sequence conservation with the accessibility of conserved residues on the protein surface shall hint to functional important sites while conserved residues located in the protein core are more likely important for maintaining the fold. Also, conserved residues in the loop regions may have a functional role, although they are less accessible in the static 3D-structural model as loops are often flexible and may move considerably. We, therefore, defined a representative 3D-model for each clade and obtained the solvent accessibility and secondary structure for each residue and aligned this information with the sequence profiles (S12). Indeed, many of the conserved sites are inaccessible to the solvent and located within or close to the β-sheets and, thus, are expected to maintain the fold. Candidates for the functional role are, for example, in clade A a Phe-x-Thr pattern (profile position 11–13); in clade B, a cluster of basic residues at position 18-22; in clade D, the conserved charged residues Lys and Asp at position 9 and 10; or in clade H, the amino acids Lys and Arg at position 35. The reliability of such assumptions depends on the quality of the structural models. We calculated a model quality score with MAESTRO and related the scores of the models to scores of experimentally determined structures (S13). The scores of the models are in the range of the modeling template structure (PDB code 4Z8W), indicating that none of the models should be largely wrong.

The conservation of motifs in PAC domains suggests common biological activities. It is possible to infer that their interactions with cell wall polysaccharides or *O*-glycans assumed from in vitro studies have been conserved and that the distribution of PDPs in the different plant families reflects differences in cell wall polysaccharides. Regarding the W-W C-terminal domain of the clade D PAC domains, its role remains to be unraveled. It is encoded by a distinct exon and could originate from exon shuffling [39].

3. Materials and Methods

3.1. Databases

The sequences used in this study have been retrieved from different databases, such as Orchidstra 2.0 ([40] http://orchidstra2.abrc.sinica.edu.tw/orchidstra2/orchid_blast.php), genome annotation Databases ([41], http://genome.microbedb.jp/blast/blast_search/klebsormidium/genes), Phytozome ([42], https://phytozome.jgi.doe.gov/pz/portal.html), OneKP ([43], https://db.cngb.org/onekp/) (see S1). When necessary, nucleotide sequences have been translated into amino acid sequences using EMBOSS transeq ([44], https://www.ebi.ac.uk/Tools/st/emboss_transeq/).

3.2. Comparisons and Alignment of PAC Domains

The BLAST (https://blast.ncbi.nlm.nih.gov/Blast.cgi) program has been used for sequence comparison. Similarities between PAC domain sequences have been calculated using either Blast2seq (https://blast.ncbi.nlm.nih.gov/Blast.cgi) or needle (http://www.bioinformatics.nl/cgi-bin/emboss/needle). The sub-cellular localization of proteins has been predicted with TargetP-2.0 ([45], http://www.cbs.dtu.dk/services/TargetP/) and the presence of β-sheets and/or α-helices using SABLE ([46], http://sable.cchmc.org/) and NetSurfP ([47], http://www.cbs.dtu.dk/services/NetSurfP/). The selected PAC domains starting at the Gly amino acid located three amino acids upstream of Cys 1 and ending at Cys 6 have been aligned using PROMALS3D ([48], http://prodata.swmed.edu/promals3d/promals3d.php) to take into account the prediction of α-sheets. The phylogeny has been calculated using MEGA7 ([49], https://www.megasoftware.net/) with the maximum likelihood option and 500 bootstraps. The presence of the PROSITE (PS00925, [50], https://prosite.expasy.org/) and Pfam (PF01190, [51], http://pfam.xfam.org/) domains have been checked in the retrieved sequences. Inside clades, conserved motifs have been identified using MEME ([52], http://meme-suite.org/tools/meme) or WebLogo3 ([53], http://weblogo.threeplusone.com/).

3.3. Three-Dimensional Modeling

For a subset of PAC domains, models were generated utilizing MODELLER [54] and I-Tasser [55]. Thereby, disulfide bridges were defined beforehand based on alignments with PDB entry 4Z8W corresponding to the *P. lanceolata* PAC domain [24]. Subsequently, these models were scored with

MAESTRO [56], DOPE [57], and ProSA 2003 [58]. Then the top-scoring models were relaxed with Rosetta [59], and finally, the relaxed models were scored with the same three methods.

We consistently used PAC domains from *A. trichopoda* as representative models for each clade. The relative solvent accessibility of these models was calculated by an adaptation of the Geometry library algorithm [60]. The secondary structure assignment was obtained by DSSP [61,62].

Both MODELLER and I-Tasser depend on template structures. MODELLER is a homology-modeling tool, which assumes significant sequence similarity between target and template structures in order to create a reliable alignment between them. Loops and sidechains are modeled with respect to the target sequence. The overall fold, however, is largely determined by the template structure. I-Tasser is a fold-recognition approach, where sequence similarity between target and template does not play a major role. Moreover, I-Tasser uses structural fragments rather than complete protein (domain) folds, from which the overall fold is built. The final model is not determined by a single template. As such, it should be better applicable for PAC domain sequences with low similarity to the Pla I 1 PAC domain.

4. Conclusions

This study has allowed better defining PDPs by combining amino acid sequences features, secondary structures, and 3D-modeling. This protein family has appeared early during the evolution of the green lineage. It has, however, not been possible to identify with certainty a PAC domain ancestor in the presumed precursor organisms of the green lineage even if the *C. orbicularis* PAC domain appeared as a possible candidate. The association of the PAC domain with Pro-rich sequences seemed to be an ancient event, the most ancient sequence carrying both a PAC domain and a Proline-rich domain being found in Bryophytes, and those carrying both a PAC domain and extensin domains in Psilotales. Despite a great amino acid variability between PAC domains, the tertiary β-barrel structure strengthened by three disulfide bridges has been conserved in all bona fide PAC domains. Finally, the subset of PAC domains belonging to Clade C is intriguing. Their very high level of conservation at the amino acid sequence level suggests that they play critical roles in plant cell walls. Defining the specificity of interaction of the different PAC domains with other cell wall polymers will be one of the next challenges to fully unravel the roles of PDPs in the cell wall architecture.

Supplementary Materials: Supplementary materials can be found at http://www.mdpi.com/1422-0067/21/7/2488/s1 S1 Number of PAC domain and PAC domain-related proteins in different plants from Bryophytes to Brassicales; S2 Amino acid sequences of PAC domain proteins in the green lineage; S3 Some examples of PAC domain-related proteins containing predicted functional domains suggesting intracellular functions; S4 Amino acid sequences of PAC domain-related proteins in ancestors to the green lineage; S5 Amino acid sequences of putative PAC domains with only five Cys residues, more than six Cys residues, or no Gly residue upstream Cys 1; S6 Amino acid sequences of putative PAC domains with six Cys residues, but predicted α-helices; S7 Amino acid sequences of the PAC and W-W domains of Type 2-PDPs including a C-terminal W-W domain; S8 Top-scoring 3D-models of PAC domains and the corresponding scores. Some PAC domain 3D-models; S9 Expanded phylogenetic trees of PAC domains from Bryophytes to *A. trichopoda* (Tree I) and from *A. trichopoda* to Brassicales (Tree II); S10 The conserved W-W domain from PAC domains belonging to clade D from Bryophytes to Brassicales PDPs; S11 *In silico* mutagenesis experiment to test the stability of the 3D-structure of a set of PAC domains mutagenized on one of the six conserved Cys residues; S12 Solvent accessibility and secondary structure for each residue and alignment of this information with the conserved sequence profiles of PAC domains; S13 MAESTRO scores for PAC domain models in relation to MAESTRO scores for experimentally-determined structures taken from the PDB database.

Author Contributions: Conceptualization, E.J. and C.D.; methodology, E.J., C.D., G.G., J.L., P.L., and H.S.C.; validation, J.L., G.G., P.L., and E.J.; investigation, H.N.-K.; data curation, H.N.-K., E.J., H.S.C.; writing—original draft preparation, E.J.; writing—review and editing, all authors.; visualization, G.G., J.L., P.L. and E.J.; supervision, E.J.; project administration, E.J.; funding acquisition, E.J. All authors have read and agreed to the published version of the manuscript.

Funding: The authors are thankful to Université Paul Sabatier-Toulouse III (France) and CNRS for supporting their research work. HNG-K has been granted by the Vietnamese Ministry of Education and Training for his PhD work. This work was also supported by the French Laboratory of Excellence project entitled "TULIP" (ANR-10-LABX-41; ANR-11-IDEX-0002-02). JL is supported by the Austrian Science Fund (FWF, grant P30042).

Conflicts of Interest: The authors declare no conflict of interest.

Abbreviations

PAC	Proline-rich, Arabinogalactan protein, conserved Cysteines
PDP	PAC Domain Protein

References

1. Carpita, N.C.; Gibeaut, D.M. Structural models of primary cell walls in flowering plants, consistency of molecular structure with the physical properties of the walls during growth. *Plant J.* **1993**, *3*, 1–30. [CrossRef] [PubMed]
2. Franková, L.; Fry, S.C. Biochemistry and physiological roles of enzymes that 'cut and paste' plant cell-wall polysaccharides. *J. Exp. Bot.* **2013**, *64*, 3519–3550. [CrossRef] [PubMed]
3. Le Gall, H.; Philippe, F.; Domon, J.-M.; Gillet, F.; Pelloux, J.; Rayon, C. Cell wall metabolism in response to abiotic stress. *Plants* **2015**, *4*, 112–166. [CrossRef] [PubMed]
4. Jamet, E.; Albenne, C.; Boudart, G.; Irshad, M.; Canut, H.; Pont-Lezica, R. Recent advances in plant cell wall proteomics. *Proteomics* **2008**, *8*, 893–908. [CrossRef]
5. Fich, E.A.; Fegerson, N.A.; Rose, J.K.C. The plant polyester cutin: Biosynthesis, structure, and biological roles. *Ann. Rev. Plant Biol.* **2016**, *76*, 207–233. [CrossRef]
6. Francoz, E.; Ranocha, P.; Nguyen-Kim, H.; Jamet, E.; Burlat, V.; Dunand, C. Roles of cell wall peroxidases in plant development. *Phytochemistry* **2015**, *112*, 15–21. [CrossRef]
7. Schaller, A.; Stintzi, A.; Rivas, S.; Serrano, I.; Chichkova, N.V.; Vartapetian, A.B.; Martınez, D.; Guiamet, J.J.; Sueldo, D.J.; van der Hoorn, R.A.L.; et al. From structure to function—A family portrait of plant subtilases. *New Phytol.* **2018**, *218*, 901–915. [CrossRef]
8. Wolf, S.; Hématy, K.; Höfte, H. Growth Control and Cell Wall Signaling in Plants. *Annu. Rev. Plant Boil.* **2012**, *63*, 381–407. [CrossRef]
9. Baldwin, T.C.; van Hengel, A.; Roberts, K. The C-terminal PAC domain of a secreted arabinogalactan protein from carrot defines a family of basic proline-rich proteins. In *Cell and Developmental Biology of Arabinogalactan Proteins*; Kluwer Academic Publishers: New York, NY, USA, 2000; pp. 43–50.
10. Du, H.; Simpson, R.; Clarke, A.E.; Bacic, A. Molecular characterization of a stigma-specific gene encoding an arabinogalactan-protein (AGP) from *Nicotiana alata*. *Plant J.* **1996**, *9*, 313–323. [CrossRef]
11. Hijazi, M.; Roujol, D.; Nguyen-Kim, H.; del Rocio Cisneros Castillo, L.; Saland, E.; Jamet, E.; Albenne, C. Arabinogalactan protein 31 (AGP31), a putative network-forming protein in *Arabidopsis thaliana* cell walls? *Ann. Bot.* **2014**, *114*, 1087–1097. [CrossRef]
12. Baldwin, T.C.; Domingo, C.; Schindler, T.; Seetharaman, G.; Stacey, N.; Roberts, K.J. DcAGP1, a secreted arabinogalactan protein, is related to a family of basic proline-rich proteins. *Plant Mol. Boil.* **2001**, *45*, 421–435. [CrossRef] [PubMed]
13. Van Hengel, A.J.; Roberts, K.J. AtAGP30, an arabinogalactan-protein in the cell walls of the primary root, plays a role in root regeneration and seed germination. *Plant J.* **2003**, *36*, 256–270. [CrossRef]
14. Liu, C.; Mehdy, M.C. A Nonclassical Arabinogalactan Protein Gene Highly Expressed in Vascular Tissues, *AGP31*, is Transcriptionally Repressed by Methyl Jasmonic Acid in Arabidopsis1 [OA]. *Plant Physiol.* **2007**, *145*, 863–874. [CrossRef] [PubMed]
15. Mang, H.G.; Lee, J.-H.; Park, J.-A.; Pyee, J.; Pai, H.-S.; Lee, J.; Kim, W.T. The CaPRP1 gene encoding a putative proline-rich glycoprotein is highly expressed in rapidly elongating early roots and leaves in hot pepper (*Capsicum annuum* L. *cv. Pukang*). *Biochim. Biophys. Acta* **2004**, *1674*, 103–108. [CrossRef] [PubMed]
16. Gong, S.-Y.; Huang, G.-Q.; Sun, X.; Li, P.; Zhao, L.L.; Zhang, D.J.; Li, X.B. GhAGP31, a cotton non-classical arabinogalactan protein, is involved in responsae to cold stress during early seedling development. *Plant Biol.* **2012**, *14*, 447–457. [CrossRef]
17. Twomey, M.C.; Brooks, J.K.; Corey, J.M.; Singh-Cundy, A. Characterization of PhPRP1, a histidine domain arabinogalactan protein from *Petunia hybrida* pistils. *J. Plant Physiol.* **2013**, *170*, 1384–1388. [CrossRef]
18. Hijazi, M.; Durand, J.; Pichereaux, C.; Pont, F.; Jamet, E.; Albenne, C. Characterization of the arabinogalactan protein 31 (AGP31) of *Arabidopsis thaliana*: New advances on the Hyp-*O*-glycosylation of the Pro-rich domain. *J. Biol. Chem.* **2012**, *287*, 9623–9632. [CrossRef]
19. Nguyen-Kim, H. Recherche de la Fonction de Protéines Riches en Hydroxyproline Dans les Parois Végétales. Ph.D. Thesis, Toulouse University, Toulouse, France, 2015.

20. Boron, A.K.; Van Orden, J.; Markakis, M.N.; Mouille, G.; Adriaensen, D.; Verbelen, J.-P.; Höfte, H.; Vissenberg, K. Proline-rich protein-like PRPL1 controls elongation of root hairs in *Arabidopsis thaliana*. *J. Exp. Bot.* **2014**, *65*, 5485–5495. [CrossRef]
21. Hunt, L.; Amsbury, S.; Baillie, A.; Movahedi, M.; Mitchell, A.; Afsharinafar, M.; Swarup, K.; Denyer, T.; Hobbs, J.K.; Swarup, R.; et al. Formation of the Stomatal Outer Cuticular Ledge Requires a Guard Cell Wall Proline-Rich Protein. *Plant Physiol.* **2017**, *174*, 689–699. [CrossRef]
22. van Hengel, A.; Barber, C.; Roberts, K. The expression patterns of arabinogalactan-protein *AtAGP30* and *GLABRA2* reveal a role for abcisic acid in the early stages of root epidermal patterning. *Plant J.* **2004**, *39*, 70–83. [CrossRef]
23. Irshad, M.; Canut, H.; Borderies, G.; Pont-Lezica, R.F.; Jamet, E. A new picture of cell wall protein dynamics in elongating cells of *Arabidopsis thaliana*: Confirmed actors and newcomers. *BMC Plant Boil.* **2008**, *8*, 94. [CrossRef] [PubMed]
24. Stemeseder, T.; Freier, R.; Wildner, S.; Fuchs, J.E.; Briza, P.; Lang, R.; Batanero, E.; Lidholm, J.; Liedl, K.R.; Campo, P.; et al. Crystal structure of Pla l 1 reveals both structural similarity and allergenic divergence within the Ole e 1–like protein family. *J. Allergy Clin. Immunol.* **2016**, *140*, 277–280. [CrossRef] [PubMed]
25. Amborella Genome Project. The Amborella genome and the evolution of flowering plants. *Science* **2013**, *242*, 1241089.
26. Passardi, F.; Longet, D.; Penel, C.; Dunand, C. The class III peroxidase multigenic family in rice and its evolution in land plants. *Phytochemistry* **2004**, *65*, 1879–1893. [CrossRef] [PubMed]
27. Voigt, J.; Frank, R.; Wöstemeyer, J. The chaotrope-soluble glycoprotein GP1 is a constituent of the insoluble glycoprotein framework of the Chlamydomonascell wall. *FEMS Microbiol. Lett.* **2009**, *291*, 209–215. [CrossRef] [PubMed]
28. Voigt, J.; Woestemeyer, J.; Frank, R.; Voigt, J. The Chaotrope-soluble Glycoprotein GP2 is a Precursor of the Insoluble Glycoprotein Framework of the Chlamydomonas Cell Wall. *J. Boil. Chem.* **2007**, *282*, 30381–30392. [CrossRef]
29. Lamport, D.T.A.; Tan, L.; Held, M.; Kieliszewski, M.J. The Role of the Primary Cell Wall in Plant Morphogenesis. *Int. J. Mol. Sci.* **2018**, *19*, 2674. [CrossRef]
30. Barre, A.; Simplicien, M.; Benoist, H.; Van Damme, E.J.M.; Rougé, P. Mannose-Specific Lectins from Marine Algae: Diverse Structural Scaffolds Associated to Common Virucidal and Anti-Cancer Properties. *Mar. Drugs* **2019**, *17*, 440. [CrossRef]
31. Koharudin, L.M.I.; Furey, W.; Gronenborn, A.M. Novel Fold and Carbohydrate Specificity of the Potent Anti-HIV Cyanobacterial Lectin from *Oscillatoria agardhii*. *J. Boil. Chem.* **2010**, *286*, 1588–1597. [CrossRef]
32. Ruhfel, B.R.; A Gitzendanner, M.; Soltis, P.S.; Soltis, D.; Burleigh, J.G. From algae to angiosperms–inferring the phylogeny of green plants (*Viridiplantae*) from 360 plastid genomes. *BMC Evol. Boil.* **2014**, *14*, 23. [CrossRef]
33. Garau, G.; Di Guilmi, A.-M.; Hall, B.G. Structure-based phylogeny of the metallo-lactamases. *Antimicrob. Agents Chemother.* **2005**, *49*, 2778–2784. [CrossRef] [PubMed]
34. Kakarala, K.K.; Jamil, K. Sequence-structure based phylogeny of GRCR classs A rhodopsin receptors. *Mol. Phylogenetics Evol.* **2014**, *74*, 66–96. [CrossRef]
35. Lakshmi, B.; Mishra, M.; Srinivasan, N.; Archunan, G. Structure-Based Phylogenetic Analysis of the Lipocalin Superfamily. *PLoS ONE* **2015**, *10*, e0135507. [CrossRef] [PubMed]
36. Morris, J.; Puttick, M.; Clark, J.; Edwards, D.; Kenrick, P.; Pressel, S.; Wellman, C.; Yang, Z.; Schneider, H.; Donoghue, P. The timescale of early land plant evolution. *Proc. Natl. Acad. Sci. USA* **2018**, *115*, E2274–E2283. [CrossRef] [PubMed]
37. Bell, C.D.; Soltis, U.E.; Soltis, P.S. The age and diversification of the angiosperms re-revisited. *Am. J. Bot.* **2010**, *97*, 1296–1303. [CrossRef] [PubMed]
38. Sarkar, P.; Bosneaga, E.; Auer, M. Plant cell walls throughout evolution: Towards a molecular understanding of their design principles. *J. Exp. Bot.* **2009**, *60*, 3615–3635. [CrossRef]
39. Patthy, L. Modular assembly of genes and the evolution of new functions. *Genetica* **2003**, *118*, 217–231. [CrossRef]
40. Chao, Y.-T.; Yen, S.-H.; Yeh, J.-H.; Chen, W.-C.; Shih, M.-C. Orchidstra 2.0—A Transcriptomics Resource for the Orchid Family. *Plant Cell Physiol.* **2017**, *58*, 9. [CrossRef]

41. Hori, K.; Maruyama, F.; Fujisawa, T.; Togashi, T.; Yamamoto, N.; Seo, M.; Sato, S.; Yamada, T.; Mori, H.; Tajima, N.; et al. *Klebsormidium flaccidum* genome reveals primary factors for plant terrestrial adaptation. *Nat. Commun.* **2014**, *5*, 3978. [CrossRef]
42. Goodstein, D.; Shu, S.; Howson, R.; Neupane, R.; Hayes, R.; Fazo, J.; Mitros, T.; Dirks, W.; Hellsten, U.; Putnam, N.; et al. Phytozome: A comparative platform for green plant genomics. *Nucleic Acids Res.* **2011**, *40*, D1178–D1186. [CrossRef]
43. Carpenter, E.J.; Matasci, N.; Ayyampalayam, S.; Wu, S.; Sun, J.; Yu, J.; Vieira, F.R.J.; Bowler, C.; Dorrell, R.G.; A Gitzendanner, M.; et al. Access to RNA-sequencing data from 1,173 plant species: The 1000 Plant transcriptomes initiative (1KP). *GigaScience* **2019**, *8*, 126. [CrossRef] [PubMed]
44. Madeira, F.; Park, Y.M.; Lee, J.; Buso, N.; Gur, T.; Madhusoodanan, N.; Basutkar, P.; Tivey, A.R.N.; Potter, S.; Finn, R.D.; et al. The EMBL-EBI search and sequence analysis tools APIs in 2019. *Nucleic Acids Res.* **2019**, *47*, W636–W641. [CrossRef] [PubMed]
45. Armenteros, J.J.A.; Salvatore, M.; Emanuelsson, O.; Winther, O.; Von Heijne, G.; Elofsson, A.; Nielsen, H. Detecting sequence signals in targeting peptides using deep learning. *Life Sci. Alliance* **2019**, *2*, e201900429. [CrossRef] [PubMed]
46. Adamczak, R.; Porollo, A.; Meller, J.; Porollo, A. Combining prediction of secondary structure and solvent accessibility in proteins. *Proteins Struct. Funct. Bioinform.* **2005**, *59*, 467–475. [CrossRef] [PubMed]
47. Klausen, M.S.; Jespersen, M.C.; Nielsen, H.; Jensen, K.K.; Jurtz, V.I.; Sønderby, C.K.; Sommer, M.O.A.; Winther, O.; Nielsen, M.; Petersen, B.; et al. NetSurfP-2.0: Improved prediction of protein structural features by integrated deep learning. *Proteins Struct. Funct. Bioinform.* **2019**, *87*, 520–527. [CrossRef] [PubMed]
48. Pei, J.; Kim, B.-H.; Grishin, N.V. PROMALS3D: A tool for multiple protein sequence and structure alignments. *Nucleic Acids Res.* **2008**, *36*, 2295–2300. [CrossRef] [PubMed]
49. Kumar, S.; Stecher, G.; Li, M.; Knyaz, C.; Tamura, K. MEGA X: Molecular Evolutionary Genetics Analysis across Computing Platforms. *Mol. Boil. Evol.* **2018**, *35*, 1547–1549. [CrossRef]
50. Sigrist, C.J.A.; De Castro, E.; Cerutti, L.; Cuche, B.A.; Hulo, N.; Bridge, A.; Bougueleret, L.; Xenarios, I. New and continuing developments at PROSITE. *Nucleic Acids Res.* **2012**, *41*, D344–D347. [CrossRef]
51. El-Gebali, S.; Mistry, J.; Bateman, A.; Eddy, S.R.; Luciani, A.; Potter, S.; Qureshi, M.; Richardson, L.; A Salazar, G.; Smart, A.; et al. The Pfam protein families database in 2019. *Nucleic Acids Res.* **2019**, *47*, D427–D432. [CrossRef]
52. Bailey, T.L.; Bodén, M.; Buske, F.A.; Frith, M.; Grant, C.E.; Clementi, L.; Ren, J.; Li, W.W.; Noble, W.S. MEME SUITE: Tools for motif discovery and searching. *Nucleic Acids Res.* **2009**, *37*, W202–W208. [CrossRef]
53. Crooks, G.E.; Hon, G.; Chandonia, J.-M.; Brenner, S.E. WebLogo: A Sequence Logo Generator. *Genome Res.* **2004**, *14*, 1188–1190. [CrossRef]
54. Šali, A.; Blundell, T.L. Comparative Protein Modelling by Satisfaction of Spatial Restraints. *J. Mol. Boil.* **1993**, *234*, 779–815. [CrossRef]
55. Roy, A.; Kucukural, A.; Zhang, Y. I-TASSER: A unified platform for automated protein structure and function prediction. *Nat. Protoc.* **2010**, *5*, 725–738. [CrossRef]
56. Laimer, J.; Hofer, H.; Fritz, M.; Wegenkittl, S.; Lackner, P. MAESTRO-multi agent stability prediction upon point mutations. *BMC Bioinform.* **2015**, *16*, 116. [CrossRef]
57. Shen, M.-Y.; Sali, A. Statistical potential for assessment and prediction of protein structures. *Protein Sci.* **2006**, *15*, 2507–2524. [CrossRef]
58. Sippl, M.J. Recognition of errors in three-dimensional structures of proteins. *Proteins Struct. Funct. Bioinform.* **1993**, *17*, 355–362. [CrossRef]
59. Tyka, M.D.; Keedy, D.; André, I.; DiMaio, F.; Song, Y.; Richardson, D.C.; Richardson, J.S.; Baker, D. Alternate States of Proteins Revealed by Detailed Energy Landscape Mapping. *J. Mol. Boil.* **2010**, *405*, 607–618. [CrossRef]
60. Voss, N.; Gerstein, M. Calculation of Standard Atomic Volumes for RNA and Comparison with Proteins: RNA is Packed More Tightly. *J. Mol. Boil.* **2005**, *346*, 477–492. [CrossRef]

61. Joosten, R.P.; Beek, T.A.H.T.; Krieger, E.; Hekkelman, M.; Hooft, R.; Schneider, R.; Sander, C.; Vriend, G. A series of PDB related databases for everyday needs. *Nucleic Acids Res.* **2010**, *39*, D411–D419. [CrossRef]
62. Kabsch, W.; Sander, C. Dictionary of protein secondary structure: Pattern recognition of hydrogen-bonded and geometrical features. *Biopolymers* **1983**, *22*, 2577–2637. [CrossRef]

International Journal of
Molecular Sciences

MDPI

Review

Phyllotaxis Turns Over a New Leaf—A New Hypothesis

Derek T. A. Lamport [1,*], Li Tan [2], Michael Held [3] and Marcia J. Kieliszewski [3]

1 School of Life Sciences, University of Sussex, Falmer, Brighton BN1 9QG, UK
2 Complex Carbohydrate Research Center, University of Georgia, Athens, GA 30602, USA; tan@ccrc.uga.edu
3 Department of Chemistry and Biochemistry, Ohio University, Athens, OH 45701, USA; held@ohio.edu (M.H.); kielisze@ohio.edu (M.J.K.)
* Correspondence: derekt.t.a.lamport@googlemail.com; Tel.: +44-1273486430

Received: 1 November 2019; Accepted: 5 February 2020; Published: 9 February 2020

Abstract: Phyllotaxis describes the periodic arrangement of plant organs most conspicuously floral. Oscillators generally underlie periodic phenomena. A hypothetical algorithm generates phyllotaxis regulated by the Hechtian growth oscillator of the stem apical meristem (SAM) protoderm. The oscillator integrates biochemical and mechanical force that regulate morphogenetic gradients of three ionic species, auxin, protons and Ca^{2+}. Hechtian adhesion between cell wall and plasma membrane transduces wall stress that opens Ca^{2+} channels and reorients auxin efflux "PIN" proteins; they control the auxin-activated proton pump that dissociates Ca^{2+} bound by periplasmic arabinogalactan proteins (AGP-Ca^{2+}) hence the source of cytosolic Ca^{2+} waves that activate exocytosis of wall precursors, AGPs and PIN proteins essential for morphogenesis. This novel approach identifies the critical determinants of an algorithm that generates phyllotaxis spiral and Fibonaccian symmetry: these determinants in order of their relative contribution are: (1) size of the apical meristem and the AGP-Ca^{2+} capacitor; (2) proton pump activity; (3) auxin efflux proteins; (4) Ca^{2+} channel activity; (5) Hechtian adhesion that mediates the cell wall stress vector. Arguably, AGPs and the AGP-Ca^{2+} capacitor plays a decisive role in phyllotaxis periodicity and its evolutionary origins.

Keywords: Phyllotaxis; arabinogalactan proteins; auxin; calcium signaling; acid growth; Hechtian oscillator; plasma membrane ion fluxes

1. Introduction

Agnes Arber [1] in "The Natural Philosophy of Plant Form" comprehensively described the development of plant morphology from the ancient philosophers—Plato, Aristotle and Theophrastus—to the more recent Cambridge botanical tradition that extends from William Turner, Nehemiah Grew and "Robin" Hill to the present. William Turner (1508–68), father of English botany. published the first herbal in English (1551) as a Fellow of Pembroke College; Nehemiah Grew (1641–1712), another Pembroke graduate, father of plant anatomy published "The Anatomy of Plants" (1682) depicted in the exquisite stained-glass windows of the college library. Finally, the Hill reaction demonstrated the photolysis of water as the source of atmospheric oxygen and established molecular botany as a new level of scientific enquiry. Arber's historical perspective may help resolve some long-standing problems of plant morphogenesis. Thus, Arber [1] presented "The mechanism of plant morphology" and an insightful approach to the pivotal role of the cell wall and the stress–strain of cell expansion that results in *"form conditioned by pressure"* where *"even a minor [cell wall] alteration may be associated with striking changes in the external form."* In Northcote's laboratory, those ideas catalyzed the first Ph.D. dissertation devoted to the primary cell wall and the discovery of cell wall proteins as a new field of study. These hydroxyproline-rich glycoproteins, especially the arabinogalactan proteins (AGPs), are involved in a hypothetical Hechtian growth oscillator. It involves transduction of the wall stress–strain to the plasma membrane where an auxin-activated proton pump dissociates AGP-Ca^{2+}. Elevated cytosolic Ca^{2+}-activates exocytosis thus regulating plant growth. Discussion of the Hechtian Oscillator vis-a-vis the role of the primary cell wall in plant morphogenesis [2] suggests extrapolating the oscillator to phyllotaxis based on the premise that presence of the oscillator components implies the presence of a functional Hechtian Oscillator. Indeed, recent work suggests the mechanotransduction of stress relocates auxin efflux PIN proteins that generate new protoderm primordia. However, the precise biochemical mechanisms involved in stress transduction and the role of auxin and calcium homeostasis remain to be elucidated. Here, we invoke Hechtian adhesion and AGPs as essential components that lead us to propose a novel biochemical algorithm for floral phyllotaxis and an explanation of its strong tendency towards a periodic series first described by Fibonacci (1170–1240). This approach contrasts with many previous studies with an overwhelming mathematical bias. Indeed, many observations in Nature involve periodicity and the probable underlying oscillations

Oscillatory plant growth, known since Darwin [3], was subsequently confirmed by rapid tip growth of pollen tubes and root hairs [4]. Plant morphogenesis also involves periodicity strikingly displayed by the pattern of leaves and floral organs [5] that often appear as Fibonacci spirals typified by whorls of 3, 5, 8, 13, 21 and 34 petals [6]. Hypothetically, such periodicity depends on an underlying oscillator such as the recently formulated Hechtian growth oscillator [2,7] that involves auxin-driven Ca^{2+} release from arabinogalactan proteins (AGPs) of the cell surface; this hypothesis accounts for the origin of oscillations in molecular detail absent from previous models of tip growth [8]. Here, we extrapolate the Hechtian Oscillator to the challenging problem of phyllotaxis and the generation of primordia in the protoderm, the outermost cell layer of the stem apical meristem SAM. Earlier work emphasized physical factors and a mathematical approach was comprehensively reviewed in [6,9,10]. However, more recent work emphasizes a cell wall stress vector generated by rapid cell expansion in the protoderm that re-orientates auxin efflux PIN proteins of neighboring cells and thus directs auxin transport (and the inferred generation of Ca^{2+} waves) that regulate growth and differentiation (e.g., [11–14]). The present paper complements these and more recent models of [15] but with the notable exception of [16]; none consider a possible role for cell surface AGPs. However, *"Nature keeps some of her secrets longer than others"* [17]. That includes the elusive molecular function of classical AGPs [18–20]. Identified some fifty years ago [21–23], AGPs remained "A Great Puzzle" until the recent demonstration that AGP glycomodules bind Ca^{2+} specifically [24]. They form a cell surface AGP-Ca^{2+} capacitor that involves the interaction of three essential ions auxin, H^+ and Ca^{2+}. These "morphogens" of the Ca^{2+} signal transduction pathway (Figure 1) interact and thus regulate cell expansion and growth.

The pathway begins with the transduction of the cell wall stress vector to the plasma membrane, via AGP57C [25] as the likely molecular basis of Hechtian adhesion between the cell wall and the plasma membrane. Further transmission of a biochemical signal to the cytoplasm involves stretch-activated proton and Ca^{2+} ion fluxes of the plasma membrane generated by the Hechtian growth oscillator [7]. The cytoplasmic response to Ca^{2+} influx presumably involves exocytosis of wall plasticizers and precursors including redirection/reorientation of auxin efflux PIN proteins, eponymously named after their mutant pin-shaped phenotype. These auxin transport proteins channel auxin flow away from slow expansion towards rapid expansion thus generating auxin waves with maxima corresponding to the periodicity of nascent primordia. Turing's classic paper [26] postulated only two morphogens sufficed to generate spiral phyllotactic periodicity. The sections below expand on Turing's original suggestion with recent experimental evidence. Turing's insight was much closer to reality than the "two interacting morphogens" he envisaged.

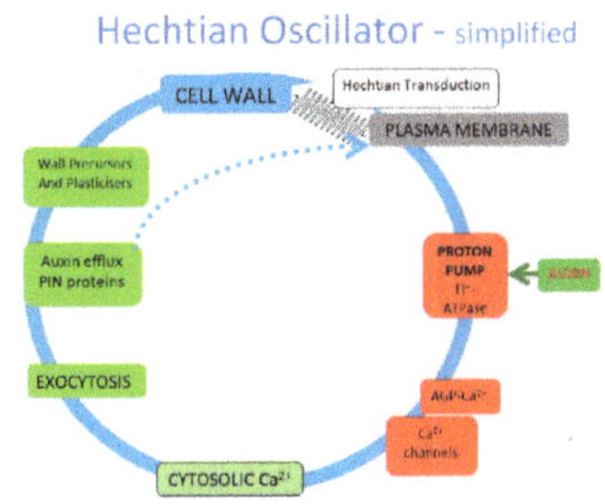

(a)

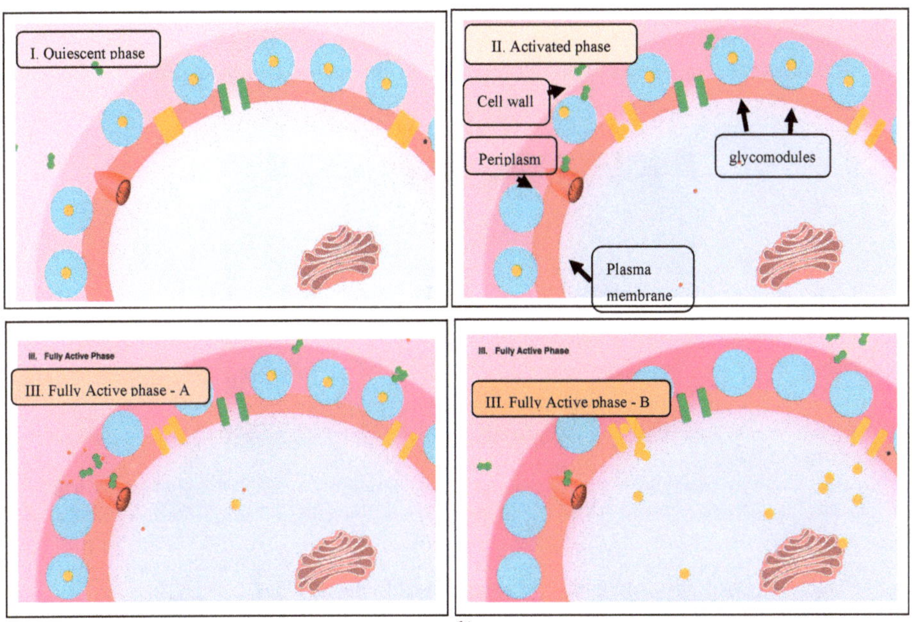

(b)

Figure 1. (**a**) The Hechtian Oscillator ion fluxes regulate growth. Depicts a simplified version of the Hechtian Oscillator in [2]. This figure shows stills from the animation in Supplemental Data. Membrane and ion fluxes are analogous to a molecular "pin-ball machine." KEY: protons: red, Ca^{2+} ions: yellow, auxin: green, stretch-activated Ca^{2+} channels; Ca^{2+} trickle initiates proton pump activity: (**b**) **Phase I. Quiescent: [7s]** Proton pump minimally active; Ca^{2+} channels closed with minimal Ca^{2+} influx. **Phase II. Activation: [6s]** Turgor increases cell expansion and thus wall stress that increases demand for auxin and opens stretch-activated Ca^{2+} channels; Ca^{2+} trickle initiates auxin binding by the proton pump, initiating low-level oscillator activity leading to Phase III. **Phase III. Fully Activated: [12s]** high auxin levels fully activate proton pump. Proton extrusion dissociates periplasmic glycomodule AGP-Ca^{2+}. Entry via Ca^{2+} channels generates cytosolic Ca^{2+} waves that activate: exocytosis of: cell wall precursors, wall plasticizers and redirect auxin efflux "PIN" proteins. **Phase IV: [9s]** Returns to Quiescent state: Stress relaxation closes Ca^{2+} channels. Auxin dissociates from proton pump; cytosolic Ca^{2+} recycles to recharge glycomodules and determine phyllotaxis periodicity as follows.

The ingenuity of Mother Nature exceeds our human imagination by involving three interacting ions, auxin, protons and Ca^{2+} (Figure 1) as the master regulator of plant growth. Although ion accumulation studied for more than 80 years [27] has generally assumed the relative immobility of

Ca^{2+} ionically bound to the cell wall, non-intuitively Ca^{2+} bound by cell surface AGPs now appears to be the major source of dynamic cytosolic Ca^{2+}. Counter-intuitively, the mechanism for the release of dynamic Ca^{2+} from ionically bound AGP-Ca^{2+} is not obvious. However, the paired glucuronic carboxyls of AGP glycomodules explain the remarkable stoichiometric Ca^{2+}-binding properties of periplasmic AGP-Ca^{2+}; its dissociation by an auxin-activated proton pump predicts an essential role of AGPs in Ca^{2+} homeostasis [24].

2. Hechtian Adhesion

The Profound implications of Hecht [28] and many other's observations are becoming clear. Numerous papers emphasize Hechtian strands of plasmolyzed cells but ignore the corollary, strong adhesion between the wall and plasma membrane of turgid cells which until recently has remained a scientific mystery. However, in plasmolyzed pollen tubes [7] and root hair tips [29] (Figure 2), a high density of Hechtian strands correlates rapid tip growth with Hechtian adhesion arguably mediated by AGP57C [25]. This suggests its essential role in transduction of the wall stress vector that initiates oscillations and cytosolic Ca^{2+} waves as hypothesized by the Hechtian Oscillator (Figure 1) [7].

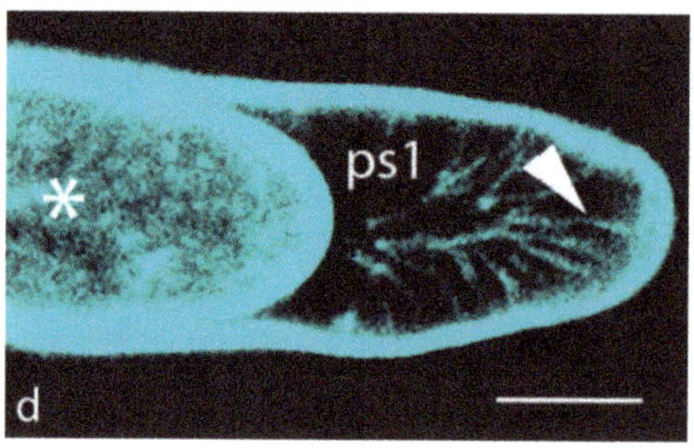

Figure 2. Hechtian strands in root hairs (arrow head) towards the very tip after labeling wheat root hairs with a membrane selective non-permeable fluorescent styryl dye, FM1-43. Reprinted with permission from [29].

3. Is the Hechtian Oscillator Just an Hypothesis? Direct Evidence

The correlation between Hechtian adhesion and tip growth also implies that transduction of the wall stress vector with concomitant activation of the proton pump releases Ca^{2+} from a tip-localized AGP-Ca^{2+} capacitor, hence a source of the tip-focused Ca^{2+} influx. Although initially an inference, direct experimental evidence was described most recently by De Vriese et al. [30]: Tobacco BY-2 cells expressing the bioluminescent Ca^{2+} sensor aequorin responded immediately to addition of the auxin analog 2,4-D, "the luminescent signal rapidly increased and reached a maximum after 90 s." Thus, direct evidence confirms a major prediction of the Hechtian Oscillator hypothesis that connects activation of the proton pump and proton extrusion with rapidly increased cytosolic Ca^{2+} (Figure 1).

The Hechtian Oscillator exemplifies the pollen tube paradigm of rapid tip growth in particular [2,7]. The rapidly growing cell wall transmits its stress–strain status via Hechtian adhesion to the plasma membrane. The role of Hechtian adhesion in stress transduction, inexplicably overlooked for more than a hundred years, also explains how a periplasmic AGP-Ca^{2+}capacitor, as a major component of the oscillator and its auxin-activated proton pump, can regulate plant growth in general. The biochemical physiological and ecological properties of the Hechtian Oscillator also avoid the vagaries of a variable external Ca^{2+} supply; it guarantees immediate access to Ca^{2+} while recycling cytosolic Ca^{2+} replenishes the AGP-Ca^{2+} capacitor. Such efficient use of Ca^{2+} may ensure the survival of calcifuge species in Ca^{2+}-deficient habitats where over-expression of AGPs also observed under salt stress [31] may enhance the ability to scavenge Ca^{2+}. Marine plants such as Zostera (Eelgrass) support that hypothesis. Recent characterization of their AGPs shows an elevated glucuronic acid content suggestive of enhanced Ca^{2+} binding in high salt [32].

4. Auxin Activity Is a Proxy for the Hechtian Oscillator

Heisler et al. [13] concluded that in the shoot apical meristem of Arabidopsis, *"cycles of auxin build-up and depletion accompany, and may direct, different stages of primordium development,"* further confirmed by a more recent study of Arabidopsis embryogenesis [33]. Auxin waves indicate the presence of an auxin-activated proton pump an essential component of the Hechtian Oscillator. Therefore, auxin activity itself can be taken as a proxy for an active Hechtian Oscillator, consistent with the well-known association of auxin with cell expansion.

Thus, H^+ dissociation of periplasmic AGP-Ca^{2+} [24] is the inferred source of cytosolic Ca^{2+} that activates exocytosis in the AGP-rich protoderm. Indeed, ubiquitous distribution of AGPs throughout the Plant Kingdom [34,35] implies an absolute requirement for AGPs. Lethal knockouts of genes encoding pollen AGPs [36] confirm their essential global role. Indeed, AGPs are closely associated with morphogenesis even at the very earliest stages such as microspore embryogenesis [37]. Therefore we hypothesize that a biological oscillator generates oscillatory growth and contributes to primordia periodicity; phyllotaxis is a test case of the Hechtian Oscillator and its general applicability is developed in the following sections.

5. A Molecular Pin-Ball Machine Regulates Ion Fluxes at the Plasma Membrane

Auxin activates plasma membrane H^+-ATPase proton pump by increasing its phosphorylation which increases the rate of proton extrusion [38]. The extent of ATPase phosphorylation [39] exerts fine gain control of the proton pump over a wide range. Hydrolysis of a single ATP molecule fuels the extrusion of about three protons by an H^+-ATPase "turbo-molecular" proton pump. The molecular pathway involves successive glutamate protonation and deprotonation from the cytoplasmic side to the periplasm and cell wall (Figure 3) [39] and initial extremely fast lateral proton diffusion on the plasma membrane surface [40].

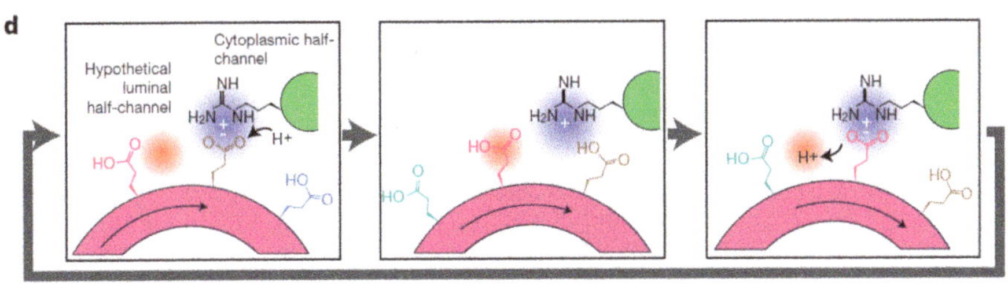

Figure 3. The proton pump pathway. Sequential protonation and deprotonation of the c-ring involve ATP-hydrolysis-driven rotation that causes protonation of a Glu residue at the cytoplasmic half-channel with subsequent deprotonation of a Glu residue at a luminal half-channel. Reprinted with permission from [39].

Proton extrusion and a concomitantly low wall pH associated with cell extension so-called "acid growth" [41] exemplify Lord Rutherford's dictum that *"No experimental result is ever wrong."* A widely accepted (textbook) explanation invokes low pH-dependent wall loosening "enzymes" and expansins all of unknown specificity [42] but ignores the dissociation of AGP-Ca^{2+} that provides an alternative reinterpretation of "acid growth" based on a visual analogy of the plasma membrane depicted as a metaphorical "molecular pin-ball machine" in (Supplemental Data) that regulates three ion fluxes, H^+, Ca^{2+} and auxin (anions at neutral pH, neutral at low pH). When activated by auxin, the proton pump shoots fast protons into the periplasm where they dislodge Ca^{2+} ions from the periplasmic AGP glycomodules; i.e., proton efflux generates Ca^{2+} influx. Thus, free Ca^{2+} ions then enter the cytosol via stretch-activated Ca^{2+}channels. The rapid increase of cytosolic Ca^{2+} [30] activates exocytosis of putative wall plasticizers, namely AGPs and AGP peptides [7] (Figure 1). Regulation of Ca^{2+} homeostasis is thus the major function of the proton pump rather than the regulation of wall pH.

6. Transduction of the Stress Vector through the Protoderm

Primordia are initiated in a "generative annulus" [6]. This thin band of protodermal cells encircles the outermost cell layer of the stem apical meristem (SAM) protoderm where its powerful morphogenetic properties are triggered by the incipient primordia. These most rapidly expanding cells transmit the stress vector to neighboring cells via their anticlinal walls [14,43–45]. Such stress relocates the auxin efflux "PIN" proteins so their polarization in the protoderm results in auxin transport towards incipient primordia. The physical basis of stress transduction depends on Hechtian adhesion while its co-localized auxin efflux "PIN" proteins were shown by immunocytochemistry [45]. Such polarized localization of auxin efflux proteins, e.g., PIN1 in the protoderm [33], suggested a crucial role for auxin transport in the generation of primordia [46]. It was concluded that *"PIN directs auxin to the sites where young primordia are being formed."* Rapid relocation of PIN1 is evidently of huge significance, although the precise mechanism remains obscure. Two components not previously considered essential for primordia formation involve Hechtian adhesion and AGPs. They mediate transduction of the cell wall stress vector, as follows:

7. Stress, PIN Protein Redirection and Auxin Waves

"Symmetries control distribution in space" [47] begs the question: What is the origin of symmetry, and how is it broken? This centuries' old debate gradually developed from "vital force" and the equally unfalsifiable "morphic resonance" to Spemann's organizer, morphogenetic fields, Waddington's "evocators" and Turing's morphogen gradients to current concepts of homeobox genes and a plethora of cognate transcription factors. They illustrate the complexity of animal morphogenesis compared with the sublime sessile simplicity of plants and the view here that auxin gradients control proton and Ca^{2+} fluxes that are predominant regulators of growth and differentiation.

Auxin transport involves diffusion facilitated by auxin efflux, eponymous "PIN" proteins essential to the generation of auxin waves that break the perfect symmetry of the protoderm as follows:

Morphogenesis frequently involves auxin waves [48]. That includes phyllotaxis [5] where a recent theoretical biophysical model involving complex linear wave equations predicts auxin waves that specify the site of new primordia [16]. Those authors noted that: *"The role of auxin transport in phyllotaxis must be universal"* and also inferred that *"electromechanical feedbacks apparently involve the Ca^{2+} and H^+ ions."* Recent direct experimental evidence [7,24] explains how the cell wall stress vector and transcytosis relocate PIN proteins and thus together with AGP-Ca^{2+} generate the auxin and Ca^{2+} waves that initiate primordia formation.

Rapidly expanding cells of the protoderm transmit the stress vector via anticlinal walls towards slower cell expansion (Figure 4). The biophysical basis of stress transduction arguably involves Hechtian adhesion between the wall and plasma membrane as described for the growth of pollen tubes and epidermal cells of root tips [45]. Hechtian adhesion is also evident in the protoderm: for example, Figure 1A of [49] shows a Hechtian strand formation in protodermal areas. Hechtian adhesion is virtually universal and thus also present in actively growing tissue like the protoderm! Furthermore, auxin waves are also evidence of an active Hechtian Oscillator based on Hechtian adhesion.

Transmission of the stress vector relocates PIN auxin efflux proteins to the stressed anticlinal walls of stressed cells [45]. Thus, auxin moves against its concentration gradient towards cells with the highest auxin levels therefore depleting the auxin of less rapidly expanding cells. Channeling auxin towards stress generators, i.e., the most rapidly expanding cells, presumably initiates primordia when auxin reaches a critical threshold level [15]. Attenuation of the stress vector by intervening distal cells depleted of their auxin slows their expansion until a boundary "tipping point" of minimum cell expansion appears (Figure 4) where the stress vector reverses its direction with concomitant reversal of PIN protein polarity [45]. Thus, a new auxin gradient increases towards a new stress vector initiated by cell expansion of a newly formed primordium. Hence, auxin waves appear as peaks that coincide with new primordia separated by auxin troughs; these repeat what is in essence an autocatalytic cycle (Figure 5):

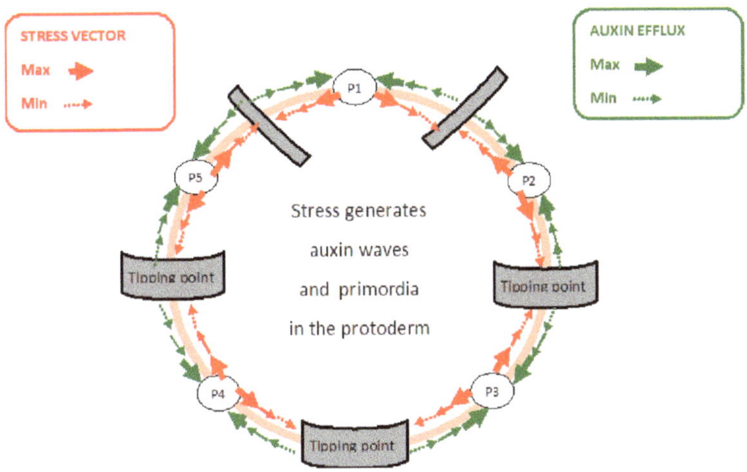

Figure 4. Stress in arabinogalactan proteins (AGP)-rich protoderm generates primordia. This hypothetical scheme illustrates the possible origin of auxin waves in phyllotaxis [6]. Five-fold rotational symmetry predominates as the archetype in dicot floral phyllotaxis. A plausible biochemical algorithm generates auxin waves and new primordia: Rapid cell expansion creates the stress vector (red arrows) that orientates PIN proteins; these channel auxin (green arrows) towards rapidly expanding cells to form incipient primordia and deplete auxin from sites of slow expansion until reaching a "tipping point" (lowest auxin level, slowest cell expansion, minimal stress) where PIN proteins reverse their orientation. Auxin maxima and minima generate regions of rapid expansion at auxin peaks corresponding to incipient primordia P1 to P5 separated by slowest growth at auxin troughs or "tipping points." Precise spacing of growth peaks corresponds to the frequency of auxin waves controlled by three primary determinants, proton pump, auxin flux and AGP-Ca^{2+} capacitor size.

We propose that the generation of successive auxin waves varying in amplitude and frequency depends on the response of the proton pump and cell surface AGP. Indeed, there is an increasingly clear correlation between enhanced AGP expression and tissue morphogenesis.

Membrane-bound PIN proteins recycle rapidly via transcytosis; for example, the relocalization of PIN7 occurs within two minutes after the gravity stimulus [50]. Arguably, the mechanism involves Hechtian adhesion that transmits wall stress directly to the plasma membrane rather than indirect transmission via "statoliths." During rapid tip growth of pollen tubes and root hairs, Hechtian adhesion predominates at the growing tip where wall stress–strain is most apparent and exocytosis is maximal. This correlation suggests that the stress vector relocates Hechtian adhesion sites at a malleable cell wall and this, in turn, directs the exocytosis of wall precursors including auxin efflux PIN proteins (cf. Figure 4). Thus, auxin waves and cytosolic Ca^{2+} [51] generated by transmission of the cell wall stress vector depend on two additional factors, transcytosis and cell wall rheology.

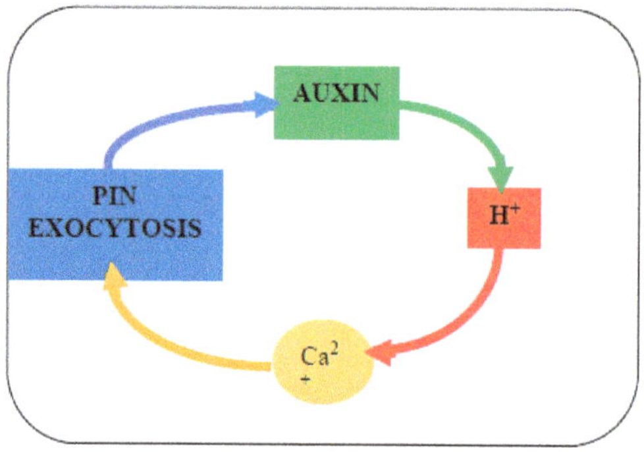

Figure 5. The auxin autocatalytic cycle. While Turing proposed a model based on the interaction between simple diffusion gradients of two morphogens, we suggest the interaction of three ions: PIN proteins boost the uphill diffusion of auxin against the concentration gradient while protons the fastest diffusing ions, dissociate AGP-Ca^{2+} thus enhancing exocytosis of PIN proteins (and Ca^{2+} channels) that propagate the Ca^{2+} message. This is summarized by the canalization theory [12] in which *"small local differences in auxin concentration are amplified by a self-reinforcing accumulation mechanism, resulting in local auxin elevation and auxin depletion in the surrounding tissue."*

8. Cell Wall Rheology

Transmission of the stress vector from rapidly expanding cells of incipient primordia involves plasticity of the anticlinal cell walls. Although Anton Heyn [52] identified wall plasticity as a crucial determinant of cell expansion, even after eighty years the biochemical basis of the Heyn paradigm remains debatable. Despite Heyn's emphasis on plasticity, cleavage of covalent cell wall crosslinks remain the predominant but elusive explanation [53]. Most synthetic plastics depend on plasticizers like phthalates, small molecules that disrupt the alignment of linear polymer chains but do not cleave covalent bonds. Analogous plasticizers of pectins include classical AGPs but their molecular size precludes simple diffusion through a pectic matrix. However, the much smaller diffusible AGP peptides upregulated by auxin [54] are also potential plasticizers; significantly their glucuronic acid content [55] indicates Ca^{2+}-binding capacity similar to the much larger non-diffusible classical AGPs typified by LeAGP1 that transit the wall by extrusion rather than diffusion [31]. Thus small AGP peptides diffusing though the wall can compete for Ca^{2+} crosslinks and thus favor a pectic gel-sol transition with a concomitant increased wall plasticity. Pectin methyl esterases have also been invoked [56]; Cosgrove [57] notes *"localized deesterification of homogalacturonan as a signature event in the auxin-induced patterning of the shoot apical meristem ... this correlation of de-esterified pectin with softer meristem regions is perplexing"* but consistent with electrostatic repulsion of ionized pectic carboxylates their Ca^{2+} depleted and scavenged by AGPs and AGP peptides with a higher affinity for Ca^{2+}. However, Altartouri et al. [58] represent the prevailing view that Ca^{2+} crosslinkage of de-esterified pectin decreases wall plasticity. This implicitly assumes sufficient free Ca^{2+} for pectin crosslinking but ignores AGP-Ca^{2+} homeostasis that determines the availability of both free and bound apoplastic Ca^{2+}.

Fine control of pectin rheology by small diffusible AGP peptides has not previously been considered. Similar reasoning may apply to some monocots where glucuronoxylans largely replace pectin [59]. Finally, we can only agree that: *"Cell expansion thus appears to be intimately linked to these wall sensor pathways in ways we are only beginning to fathom."* [57].

9. A Phyllotaxis Algorithm

"While progress has been made, there are many fascinating challenges in phyllotaxis still open for the curious mind to explore. The story is far from over. While careful experiments are crucial to continued progress, it does not require elaborate experiments for ordinary folk to enjoy the wonderful architectures seen near the

meristems of plants" [6]. Summarizing much recent work: *"Key results stem from the observed facts that phyllotactic patterns are naturally produced by instabilities, connected to both the distribution of the growth hormone auxin and to the local stress–strain fields."* Although those "instabilities" remain undefined, phyllotaxis per se is remarkably stable but with exceptions described by Arber [1] in several species. For example, a completely dimerous flower of Iris on a shoot also bearing a normal trimerous flower, and Potentilla flowers with three, four, five or six petals and concluded that "phyllotaxis depends upon the rhythmic development of primordia at the growing apex." Somewhat ahead of its time, eighty years ago, the insightful observation was made: "it seems reasonable to suppose that these variations are associated with internal chemical oscillations" with a final intuitive leap to "Physico-chemical factors...one such factor has been so universal as to affect the whole of the Plant Kingdom; this is the development of a cell wall encasing each unit of the plant body." "The challenge now is to describe how the stem apical meristem generates phyllotactic patterns de novo" [12]. The historical emphasis on mathematical approaches based on optimal packing shows that Fibonacci patterns can arise naturally in many pattern-forming systems but this is not obviously connected with the biochemical mechanisms involved in patterning. Both approaches achieve optimal packing but in quite different ways. All the components of the Hechtian Oscillator are present. Thus, a dynamic algorithm involving protoderm biochemistry and mechanotransduction is now feasible as a working hypothesis.

Protoderm cell expansion generates new primordia (*N*). For example in a floral phyllotaxis, *N* is a function of the stem apical meristem (SAM) size and the magnitude of major variables that define the symmetry and periodicity of new primordia. To sustain their growth, rapidly expanding cells demand auxin by generating the cell wall stress vector (CW_{sv}) that redirects PIN proteins thus channeling auxin towards these incipient primordia (Figure 4). A resulting auxin gradient then appears as waves in the annulus, a narrow band of morphogenetic cells surrounding the outer protoderm ($SAMP_a$) with auxin maxima and minima corresponding to future primordia and boundary tipping points, respectively. Generally, increasing the magnitude of a variable increases auxin transport towards a primordium hence rapidly depleting distal cells. Reversal of PIN protein orientation then generates a new primordium. Thus, an increased auxin depletion rate increases the number of new primordia. However, they also depend on Ca^{2+} availability determined by the expression of AGPs. We predict that AGPs strongly expressed in the protoderm will increase the periodicity of primordia while weaker expression will decrease it. Thus, to that extent, the algorithm is semi-quantitative and dependent on strong expression of AGPs in the protoderm of Arabidopsis meristems [60], Euphorbia embryonic cultures [61] and during somatic embryogenesis of Arabidopsis [62]. The novel suggestion that AGPs play a decisive role as crucial determinants of phyllotaxis periodicity (Figure 6) depends on a complex function of a hypothetical algorithm derived from the foregoing considerations:

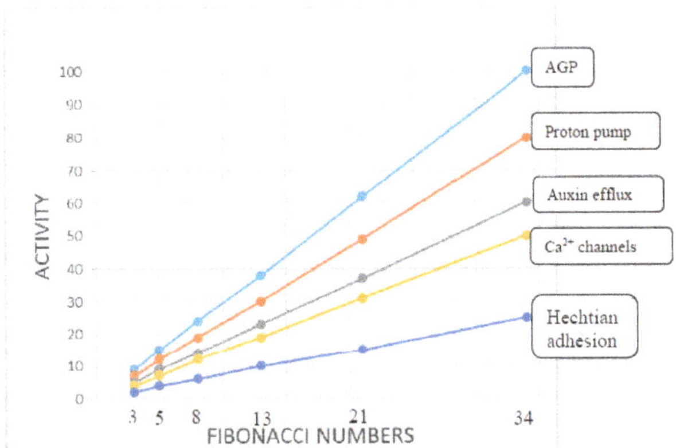

Figure 6. Major variables of the phyllotaxis algorithm.

Stem apical meristem protoderm $SAMP_a$ generates *N* new primordia as a function of an equation comprised of the following variables:

1. AGP-Ca^{2+} capacitor: AGP_c;
2. Stem apical meristem protoderm annulus radius: $SAMP_a$;
3. PP proton pump activity: PP;
4. Auxin efflux activity: A_{efflux} (hence auxin levels: A_{ux});
5. Ca^{2+} channels: C_{ch};
6. Exocytosis is a complex variable regulated by Ca^{2+} influx;
7. Hechtian adhesion: H_{ad};
8. Cell wall stress vector: CW_{sv};

$$\frac{SAMP_a}{AGP^{-1} + (PP \cdot A_{ux})^{-1} + ({C_{ch}}^{-1} + CW_{sv}) + {H_{ad}}^{-1}}$$

The cell wall stress CW_{sv} vector determines auxin, proton and Ca^{2+} ion fluxes involving four phases of the Hechtian Oscillator proton pump that regulate cell expansion.

Phase I. Quiescent: Minimal cell wall stress corresponds to minimal Ca^{2+} influx and minimal activity of the oscillator.

Phase II. Activation: Cell expansion increases wall stress, demand for auxin and opens Ca^{2+} channels; entry of Ca^{2+} initiates auxin binding to the proton pump and initial oscillator activation leading to Phase III.

Phase III. Maximum Activation: Occurs at the high auxin levels supplied by redirected auxin efflux PIN proteins; accelerated proton extrusion dissociates glycomodule AGP-$Ca^{2+}$and supplies the Ca^{2+} channels thus generating cytosolic Ca^{2+}waves that activate exocytosis, notably of wall precursors and plasticizers but also enabling dynamic redirection of PIN proteins. Addition of precursors reinforce the wall and slow its expansion, leading to Phase IV:

Phase IV. Return to quiescent phase: Stress relaxation of reinforced wall closes Ca^{2+} channels.

Cytosolic Ca^{2+} recycles via AGP precursors and Ca^{2+}transporters replenish periplasmic AGP-Ca^{2+}

When attenuation of the stress vector reaches a tipping point of minimal oscillator activity and least rapid cell expansion, distant cells expand more rapidly and now exert a new stress vector in an opposing direction thus generating a new primordium that contributes to phyllotactic symmetry.

$SAMP_a$ stem apical meristem protoderm generates new primordia as a function of the activity of five major variables or determinants shown in the speculative graph taking petal phyllotaxis as an example ranging from 3 to 34 petals. It assumes that the cell wall stress vector CW_{sv} is essentially constant for a given $SAMP_a$.

Protoderm AGP content (activity) is the primary determinant based on the simple hypothesis that size of the AGP-Ca^{2+} capacitor determines the periodicity of primordia initiation (cf. Figure 4) where a large capacitor generates more primordia. The graph correlates the Fibonacci series with AGP expression and activity of other factors particularly the proton pump not previously connected with phyllotaxis. Other determinants illustrate an inferred hierarchy based on the size of their relative input to the Hechtian growth oscillator.

The beguiling simplicity of the above plot inferred largely from biochemical observations is in strong contrast to previous complex formulations based largely on mathematical/geometrical logic.

10. Evolutionary Origin of Angiosperm Phyllotaxis

The evolutionary history of the stem apical meristem from a relatively simple arrangement of apical cells in the Bryophytes and ferns culminates in the morphogenetic protoderm of the angiosperms. Here, we conjecture that hybridization may solve the riddle of Fibonacci phyllotaxis and its evolutionary origins.

Strong AGP expression predicts numerous closely spaced primordia (Figure 4) that generate floral phyllotaxis, arguably determined by the amplitude of the Ca^{2+} signal that depends on the AGP-Ca^{2+} capacitor size, proton flux and Ca^{2+} channel status. Thus, cells of the protoderm with a large AGP-Ca^{2+} capacitor will increase cytosolic Ca^{2+} rapidly and therefore generate numerous primordia within a shorter range than in a protoderm with weaker AGP expression and therefore with more cells between primordia. For example, lower activity of the proton pump and a smaller AGP-Ca^{2+} capacitor increase spacing between primordia. Unlike the animal kingdom, plants have an enormous propensity for polyploidy and hybridization that we suggest provides a simple biochemical explanation for an evolutionary origin of the well-known Fibonacci series of floral organs exemplified by the 3, 5, 8, 13, 21 and 34 petals (Figure 6), e.g., *Ranunculus ficaria* (13), *Erigeron canadensis* (21). Can hybridization between contiguous members of the series generate a Fibonacci sequence? If so, how? Consider a hybrid expressing the sum of AGPs from both parents! If AGPs play a dominant role in defining phyllotaxis, then a hybrid of two-fold and threefold symmetry, with a corresponding increased size of the AGP-Ca^{2+} capacitor, would generate the most common five-fold symmetry, and so on for subsequent members of the series (Figure 6). Thompson [17] viewed the Fibonacci series not as the cause but merely *"a consequence of optimal space filling in systems adding new units at a pole."* Thus, we infer that hybridization generates floral Fibonacci phyllotaxis and accounts for the evolutionary origin of a discrete series rather than a smooth arithmetic progression. This suggestion has the merit of simplicity based on Occam's Razor in contrast to all preceding mathematical conjectures [9] and is supported by the Hechtian Oscillator as a predictive paradigm.

11. D'Arcy Thompson, Alan Turing and Peter Mitchell Revisited

Thompson's classical "Growth and Form" [17] exemplified a purely descriptive mathematical approach that collated a huge corpus of biophysical observations rather than hypothesis-driven experiments. On the other hand, Turing [26] combined a mathematical with a physical-chemical approach. Thus, while a Turing self-replicating machine aptly fits cell replication, Turing postulated that ontogeny based on biological parsimony might involve the diffusion of only two chemical morphogens that would suffice to create morphogenetic gradients. Those ideas preceded the biochemical insights of Mitchell (Mitchell [63], experimentalist par excellence who questioned conventional wisdom and proposed the versatile chemiosmotic proton pump). A universal energy transduction machine couples proton gradients across lipid membranes to generate ATP that energizes all life. In reverse, it consumes ATP and pumps protons. This vectorial chemical system differs fundamentally from conventionally scalar chemical ones as explained by Mitchell [64]: *"It was obviously my hope that the chemiosmotic rationale of vectorial metabolism and biological energy transfer might one day come to be generally accepted, and I have done my best to argue in favour of that state of affairs for more than twenty years ... was it not the great Max Planck who remarked that a new scientific idea does not triumph by convincing its opponents, but rather because its opponents eventually die?"* Although Mitchell's unconventional ideas were initially rejected, they were finally recognized. Their universal applicability has become apparent more recently. In simple photoautotrophs, light-driven proton gradients involve bacteriorhodopsin [65], while in more advanced eukaryotes, an electron transport chain generates mitochondrial proton gradients. Proton pumps and their regulation are thus at the epicenter of plant growth that, stripped to its bare essentials, depend on three morphogen gradients, auxin, protons and Ca^{2+} rather than just two.

However, these gradients do not arise by simple diffusion but are regulated by auxin efflux "PIN" proteins whose discovery began with Rubery and Sheldrake's [66] classic experiments in the laboratory of Northcote [67]. PIN proteins control auxin gradients and auxin levels that activate the proton pump while the cell wall stress vector opens Ca^{2+} channels that generate cytosolic Ca^{2+} gradients. Thus, the Hechtian growth oscillator is an extrapolation of Mitchell's chemiosmosis that unifies physics and chemistry in a minimalist approach to regulating plant growth. Indeed, precursors to life surely involve proton gradients as a basis of prebiotic energy transduction and the universal proton pump of exoplanet life in the habitable zone.

Supplementary Materials: The following are available online at http://www.mdpi.com/1422-0067/21/3/1145/s1.

Author Contributions: D.T.A.L., L.T., M.H. and M.J.K. have made major contributions over many years to the chemistry of hydroxyproline–rich glycoproteins which combined with insightful discussions led directly to this paper.All authors have read and agreed to the published version of the manuscript.

Funding: This research received no external funding.

Acknowledgments: We gratefully acknowledge our home Academic institutions for the past many years of support that has made this and previous work possible. We are indeed greatly indebted to our colleague Ben Coleman for the Supplemental Data animation of the Hechtian Oscillator ion fluxes.

Conflicts of Interest: The authors declare no conflict of interest.

References

1. Arber, A. *The Natural Philosophy of Plant Form*; Cambridge University Press: Cambridge, UK, 1950.
2. Lamport, D.T.A.; Tan, L.; Held, M.A.; Kieliszewksi, M.J. The Role of the Primary Cell Wall in Plant Morphogenesis. *Int. J. Mol. Sci.* **2018**, *19*, 2674. [CrossRef]
3. Darwin, C. *The Power of Movement in Plants*; John Murray: London, UK, 1880.
4. Pierson, E.S.; Li, Y.Q.; Zhang, H.Q.; Willemse, M.T.M.; Linskens, H.F.; Cresti, M. Pulsatory growth of pollen tubes: Investigation of a possible relationship with the periodic distribution of cell wall components. *Acta Bot. Neerl.* **1995**, *44*, 121–128. [CrossRef]
5. Zoulias, N.; Duttke, S.H.C.; Garces, H.; Spencer, V.; Kim, M. The Role of Auxin in the Pattern Formation of the Asteraceae Flower Head (Capitulum). *Plant Physiol.* **2019**, *179*, 391–401. [CrossRef] [PubMed]
6. Pennybacker, M.F.; Shipman, P.D.; Newell, A.C. Phyllotaxis: Some progress, but a story far from over. *Physica D* **2015**, *306*, 48–81. [CrossRef]
7. Lamport, D.T.A.; Tan, L.; Held, M.A.; Kieliszewksi, M.J. Pollen tube growth and guidance: Occam's razor sharpened on a molecular arabinogalactan glycoprotein Rosetta Stone. *New Phytol.* **2018**, *217*, 491–500. [CrossRef] [PubMed]
8. Chebli, Y.; Geitmann, A. Mechanical principles governing pollen tube growth. *Funct. Plant Sci. Biotechnol.* **2007**, *1*, 232–245.
9. Adler, I.; Barabe, D.; Jean, R.V. A history of the study of Phyllotaxis. *Ann. Bot.* **1997**, *80*, 231–244. [CrossRef]
10. Okabe, T. The riddle of phyllotaxis: Exquisite control of divergence angle. *Acta Soc. Bot. Pol.* **2016**, *85*, 3527. [CrossRef]
11. Sachs, T. The Control of the Patterned Differentiation of Vascular Tissues. *Adv. Bot. Res.* **1981**, *9*, 151–262.
12. Reinhardt, D.; Pesce, E.-R.; Stieger, P.; Mandel, T.; Baltensperger, K.; Bennett, M.; Traas, J.; Friml, J. Regulation of phyllotaxis by polar auxin transport. *Nature* **2003**, *426*, 255–260. [CrossRef]
13. Heisler, G.; Ohno, C.; Das, P.; Sieber, P.; Reddy, G.V.; Long, J.A.; Meyerowitz, E.M. Patterns of Auxin Transport and Gene Expression during Primordium Development Revealed by Live Imaging of the *Arabidopsis* Inflorescence Meristem. *Curr. Biol.* **2005**, *15*, 1899–1911. [CrossRef] [PubMed]
14. Heisler, M.G.; Hamant, O.; Krupinski, P.; Uyttewaal, M.; Ohno, C.; Jonsson, H.; Traas, J.; Meyerowitz, E.M. Alignment between PIN1 Polarity and Microtubule Orientation in the Shoot Apical Meristem Reveals a Tight Coupling between Morphogenesis and Auxin Transport. *PLoS Biol.* **2010**, *8*, e1000516. [CrossRef] [PubMed]
15. Bhatia, N.; Heisler, M.G. Self-organising periodicity in development: Organ positioning in plants. *Development* **2018**, *145*. [CrossRef] [PubMed]
16. Abraham-Shrauner, B.; Pickard, B.G. A model for leaf initiation Determination of phyllotaxis by waves in the generative circle. *Plant Signal. Behav.* **2011**, *6*, 1755–1768. [CrossRef] [PubMed]
17. Thompson, D. *On Growth and Form*; Abridged ed.; Cambridge University Press: Cambridge, UK, 1961.
18. Clarke, A.E.; Anderson, R.L.; Stone, B.A. Form and function of arabinogalactans and arabinogalactan-proteins. *Phytochemistry* **1979**, *18*, 521–540. [CrossRef]
19. Gaspar, Y.M.; Johnson, K.L.; McKenna, J.A.; Bacic, A.; Schultz, C.J. The complex structures of arabinogalactan-proteins and the journey towards understanding. *Plant Mol. Biol.* **2001**, *47*, 161–176. [CrossRef]
20. Ellis, M.; Egelund, J.; Schultz, C.J.; Bacic, A. Arabinogalactan-proteins: Key regulators at the cell surface? *Plant Physiol.* **2010**, *153*, 403–419. [CrossRef]

21. Aspinall, G.O.; Malloy, J.A.; Craig, J.W.T. Extracellular polysaccharides from suspension-cultured sycamore cells. *Can. J. Biochem.* **1969**, *47*, 1063–1070. [CrossRef]
22. Lamport, D.T.A. Cell wall metabolism. *Ann. Rev. Plant Physiol.* **1970**, *21*, 235–270. [CrossRef]
23. Jermyn, M.A.; Yeow, Y.M. A class of lectins present in the tissues of seed plants. *Aust. J. Plant Physiol.* **1975**, *2*, 501–531. [CrossRef]
24. Lamport, D.T.A.; Varnai, P. Periplasmic arabinogalactan glycoproteins act as a calcium capacitor that regulates plant growth and development. *New Phytol.* **2013**, *197*, 58–64. [CrossRef] [PubMed]
25. Tan, L.; Eberhard, S.; Pattathil, S.; Warder, C.; Glushka, J.; Yuan, C.; Hao, Z.; Zhu, X.; Avci, U.; Miller, J.S.; et al. An *Arabidopsis* Cell Wall Proteoglycan Consists of Pectin and Arabinoxylan Covalently Linked to an Arabinogalactan Protein. *Plant Cell* **2013**, *25*, 270–287. [CrossRef] [PubMed]
26. Turing, A.M. The chemical basis of morphogenesis. *Philos. Trans. R. Soc. Lond. B* **1952**, *237*, 37–72.
27. Briggs, G.E. The Absorption of Salts by Plant Tissues, considered as Ionic Interchange. *Ann. Bot.* **1933**, *46*, 301–322. [CrossRef]
28. Hecht, K. Studien uber den Vorgang der Plasmophysiologie der Grenzschichten lebender Pflanzenzellen. *Cohns Beitr. Biol. Pflanz.* **1912**, *11*, 137–189.
29. Volgger, M.; Lang, I.; Ovecka, M.; Lichtscheidl, I. Plasmolysis and cell wall deposition in wheat root hairs under osmotic stress. *Protoplasma* **2010**, *243*, 51–62. [CrossRef]
30. De Vriese, K.; Himschoot, E.; Dunser, K.; Nguyen, L.; Drozdzecki, A.; Costa, A.; Nowack, M.K.; Kleine-Vehn, J.; Audenaert, D.; Beeckman, T.; et al. Identification of Novel Inhibitors of Auxin-Induced $Ca2^+$ Signaling via a Plant-Based Chemical Screen. *Plant Physiol.* **2019**, *180*, 480–496. [CrossRef]
31. Lamport, D.T.A.; Kieliszewksi, M.J.; Showalter, A.M. Salt-stress upregulates periplasmic arabinogalactan-proteins: Using salt-stress to analyse AGP function. *New Phytol.* **2006**, *169*, 479–492. [CrossRef]
32. Pfeifer, L.; Classen, B. First structural characterisation of seagrass arabinogalactan-proteins reveals habitat-driven adaptation to marine environment. *Cell Wall Meet.* **2019**, *15*, P003.
33. Moller, B.; Weijers, D. Auxin Control of Embryo Patterning. *Cold Spring Harb. Perspect. Biol.* **2009**, *1*, a001545. [CrossRef]
34. Showalter, A.M. Arabinogalactan-proteins: Structure, expression and function. *Cell. Mol. Life Sci.* **2001**, *58*, 1399–1417. [CrossRef] [PubMed]
35. Palacio-Lopez, K.; Tinaz, B.; Holzinger, A.; Domozych, D.S. Arabinogalactan proteins and the extracellular matrix of Charophytes: A sticky business. *Front. Plant Sci.* **2019**, *10*, 447. [CrossRef] [PubMed]
36. Coimbra, S.; Costa, M.; Jones, B.J.; Mendes, M.A.; Pereira, L.G. Pollen grain development is compromised in *Arabidopsis* agp6 agp11 null mutants. *J. Exp. Bot.* **2009**, *60*, 3133–3142. [CrossRef] [PubMed]
37. Corral-Martinez, P.; Driouich, A.; Segui-Simarro, J.M. Dynamic Changes in Arabinogalactan-Protein, Pectin, Xyloglucan and Xylan Composition of the Cell Wall During Microspore Embryogenesis in Brassica napus. *Front. Plant Sci.* **2019**, *10*, 332. [CrossRef]
38. Koji, T.; Hayashi, K.; Kinoshita, T. Auxin Activates the Plasma Membrane H^+-ATPase by Phosphorylation during Hypocotyl Elongation in *Arabidopsis*. *Plant Physiol.* **2012**, *159*, 632–641.
39. Mazhab-Jafari, M.T.; Rohou, A.; Schmidt, C.; Bueler, S.A.; Benlekbir, S.; Robinson, C.V.; Rubinstein, J.L. Atomic model for the membrane-embedded VO motor of a eukaryotic V-ATPase. *Nature* **2016**, *539*, 118–130. [CrossRef]
40. Amdursky, N.; Lin, Y.; Aho, N.; Groenhof, G. Exploring fast proton transfer events associated with lateral proton diffusion on the surface of membranes. *Proc. Natl. Acad. Sci. USA* **2019**, *116*, 2443–2451. [CrossRef]
41. Rayle, D.L.; Cleland, R.E. The acid growth theory of auxin-induced cell elongation is alive and well. *Plant Physiol.* **1992**, *99*, 1271–1274. [CrossRef]
42. Ramakrishna, P.; Duarte, P.R.; Rance, G.A.; Schubert, A.; Vordermaier, V.; Vu, L.D.; Murphy, E.; Barro, A.V.; Swarup, K.; Moirangthem, K.; et al. EXPANSIN A1-mediated radial swelling of pericycle cells positions anticlinal cell divisions during lateral root initiation. *Proc. Natl. Acad. Sci. USA* **2019**, *116*, 8597–8602. [CrossRef]

43. Hamant, O.; Heisler, M.G.; Jonnson, H.; Krupinski, P.; Uyttewaal, M.; Bokov, P.; Corson, F.; Sahlin, P.; Boudaoud, A.; Meyerowitz, E.M.; et al. Developmental patterning by mechanical signals in *Arabidopsis*. *Science* **2008**, *322*, 1650–1654. [CrossRef]
44. Sampathkumar, A.; Yan, A.; Krupinski, P.; Meyerowitz, E.M. Physical Forces Regulate Plant Development and Morphogenesis. *Curr. Biol.* **2014**, *24*, 475–483. [CrossRef] [PubMed]
45. Feraru, E.; Feraru, M.I.; Kleine-Vehn, J.; Martinie, A.; Mouille, G.; Vanneste, S.; Vernhettes, S.; Runions, J.; Friml, J. PIN Polarity Maintenance by the Cell Wall in *Arabidopsis*. *Curr. Biol.* **2011**, *21*, 338–343. [CrossRef] [PubMed]
46. Vernoux, T.; Besnard, F.; Traas, J. Auxin at the Shoot Apical Meristem. *Cold Spring Harb. Perspect. Biol.* **2010**, *2*, a001487. [CrossRef] [PubMed]
47. Critchlow, K. *The Hidden Geometry of Flowers*; Floris Books: Edinburgh, UK, 2011.
48. Zajaczkowski, S.; Wodzicki, T.J.; Romberger, J.A. Auxin Waves and Plant Morphogenesis. In *Hormonal Regulation of Development II: The Functions of Hormones from the Level of the Cell to the Whole Plant*; Scott, T.K., Ed.; Springer: Berlin/Heidelberg, Germany, 1984; pp. 244–262.
49. Komis, G.; Apostolakos, P.; Galatis, B. Altered patterns of tubulin polymerization in dividing leaf cells of *Chlorophyton comosum* after a hyperosmotic treatment. *New Phytol.* **2001**, *149*, 193–207. [CrossRef]
50. Sato, E.M.; Hijazi, H.; Bennett, M.J.; Vissenjberg, K.; Swarup, R. New insights into root gravitropic signalling. *J. Exp. Bot.* **2015**, *66*, 2155–2165. [CrossRef] [PubMed]
51. Roy, S.; Holdaway-Clarke, T.L.; Hackett, G.R.; Kunkel, J.G.; Lord, E.M.; Hepler, P.K. Uncoupling secretion and tip growth in lily pollen tubes: Evidence for the role of calcium in exocytosis. *Plant J.* **1999**, *19*, 379–386. [CrossRef]
52. Heyn, A.N.J. The physiology of cell elongation. *Bot. Rev.* **1940**, *6*, 515–574. [CrossRef]
53. Dyson, R.J.; Band, L.R.; Jensen, O.E. A model of crosslink kinetics in the expanding plant cell wall: Yield stress and enzyme action. *J. Theor. Biol.* **2012**, *307*, 125–136. [CrossRef]
54. Pacheco-Villalobos, D.; Diaz-Moeeno, M.; van der Schuren, A.; Tamaki, T.; Kang, Y.H.; Gujas, B.; Novak, O.; Jaspert, N.; Li, Z.; Wolf, S.; et al. The Effects of High Steady State Auxin Levels on Root Cell Elongation in Brachypodium. *Plant Cell* **2016**, *28*, 1009–1024. [CrossRef]
55. Tryfona, T.; Liang, H.-C.; Kotake, T.; Kaneko, S.; Marsh, J.; Ichinose, H.; Lovegrove, A.; Tsumaraya, Y.; Shewry, P.R.; Dupree, P. Carbohydrate structural analysis of wheat flour arabinogalactan protein. *Carbohydr. Res.* **2010**, *345*, 2656. [CrossRef]
56. Braybrook, S.A.; Peaucelle, A. Mechano-Chemical Aspects of Organ Formation in *Arabidopsis thaliana*: The Relationship between Auxin and Pectin. *PLoS ONE* **2013**, *8*, e57813. [CrossRef] [PubMed]
57. Cosgrove, D.J. Diffuse Growth of Plant Cell Walls. *Plant Physiol.* **2018**, *176*, 17–27. [CrossRef] [PubMed]
58. Altartouri, B.; Bidhendi, A.; Tani, T.; Suzuki, J.; Conrad, C.; Chebli, Y.; Liu, N.; Karunakaran, C.; Scacelli, G.; Geitmann, A. Pectin chemistry and cellulose crystallinity govern pavement cell morphogenesis in a multi-step mechanism. *Plant Physiol.* **2019**, *181*, 127–141. [CrossRef] [PubMed]
59. Zelaya, V.M.; Fernandez, P.V.; Vega, A.S.; Mantese, A.I.; Federico, A.A.; Ciancia, M. Glucuronoarabinoxylans as major cell walls polymers from young shoots of the woody bamboo *Phyllostachys aurea*. *Carbohydr. Polym.* **2017**, *167*, 240–249. [CrossRef] [PubMed]
60. Abe, M.; Takahashi, T.; Komeda, Y. Cloning and characterization of an LI Layer-specific gene in *Arabidopsis thaliana*. *Plant Cell Physiol.* **1999**, *40*, 571–580. [CrossRef] [PubMed]
61. Saare-Surminski, K.; Preil, W.; Knox, J.P.; Lieberei, R. Arabinogalactan proteins in embryogenic and non-embryogenic callus cultures of Euphorbia pulcherrima. *Physiol. Plant* **2000**, *108*, 180–187. [CrossRef]
62. Potocka, I.; Godel, K.; Dobrowolska, I.; Kurczyńska, E.U. Spatio-temporal localization of selected pectic and arabinogalactan protein epitopes and the ultrastructural characteristics of explant cells that accompany the changes in the cell fate during somatic embryogenesis in *Arabidopsis thaliana*. *Plant Physiol. Biochem.* **2018**, *127*, 573–589. [CrossRef]
63. Mitchell, P. Coupling of phosphorylation to electron and hydrogen transfer by a chemi-osmotic type of mechanism. *Nature* **1961**, *191*, 144–148. [CrossRef]
64. Mitchell, P. David Keilin's Respiratory Chain Concept and Its Chemiosmotic Consequences. *Nobel Lect. Chem.* **1992**, *5*, 295–330.
65. Luckey, M. *Membrane Structural Biology*; Cambridge University Press: Cambridge, UK, 2008.

66. Rubery, P.H.; Sheldrake, A.R. Carrier-mediated Auxin Transport. *Planta* **1974**, *118*, 101–121. [CrossRef]
67. Crompton, D.W.T. Donald Henry Northcote (1921–2004). *Biogr. Mem. Fellows R. Soc.* **2019**, *67*, 359–370. [CrossRef]

International Journal of
Molecular Sciences

MDPI

Concept Paper

The Role of the Primary Cell Wall in Plant Morphogenesis

Derek T. A. Lamport [1,*], Li Tan [2], Michael Held [3] and Marcia J. Kieliszewski [3]

1 School of Life Sciences, University of Sussex, Falmer, Brighton BN1 9QG, UK
2 Complex Carbohydrate Research Center, University of Georgia, Athens, GA 30602, USA; tan@ccrc.uga.edu
3 Department of Chemistry and Biochemistry, Ohio University, Athens, OH 45701, USA; held@ohio.edu (M.H.); kielisze@ohio.edu (M.J.K.)
* Correspondence: lamportd@msu.edu; Tel.: +44-012-7348-6430

Received: 12 June 2018; Accepted: 4 September 2018; Published: 9 September 2018

Abstract: Morphogenesis remains a riddle, wrapped in a mystery, inside an enigma. It remains a formidable problem viewed from many different perspectives of morphology, genetics, and computational modelling. We propose a biochemical reductionist approach that shows how both internal and external physical forces contribute to plant morphogenesis via mechanical stress–strain transduction from the primary cell wall tethered to the plasma membrane by a specific arabinogalactan protein (AGP). The resulting stress vector, with direction defined by Hechtian adhesion sites, has a magnitude of a few piconewtons amplified by a hypothetical Hechtian growth oscillator. This paradigm shift involves stress-activated plasma membrane Ca^{2+} channels and auxin-activated H^+-ATPase. The proton pump dissociates periplasmic AGP-glycomodules that bind Ca^{2+}. Thus, as the immediate source of cytosolic Ca^{2+}, an AGP-Ca^{2+} capacitor directs the vectorial exocytosis of cell wall precursors and auxin efflux (PIN) proteins. In toto, these components comprise the Hechtian oscillator and also the gravisensor. Thus, interdependent auxin and Ca^{2+} morphogen gradients account for the predominance of AGPs. The size and location of a cell surface AGP-Ca^{2+} capacitor is essential to differentiation and explains AGP correlation with all stages of morphogenetic patterning from embryogenesis to root and shoot. Finally, the evolutionary origins of the Hechtian oscillator in the unicellular Chlorophycean algae reflect the ubiquitous role of chemiosmotic proton pumps that preceded DNA at the dawn of life.

Keywords: morphogenesis; cell wall protein; hechtian oscillator; calcium signaling; H^+-ATPase

1. Introduction

"If questions are to be asked about life processes, how can one fail to enquire into what is perhaps the most striking feature of life, morphogenesis? Morphogenesis is the end product of cell differentiation. Cell shape is one aspect of the more general problem of differentiation. Consider the primary cell wall, how it grows, and there is evidence enough for the primary wall playing a decisive role in influencing the shape of a cell. Therefore, the primary cell wall must be studied, but how?" (Lamport 1963) [1].

In 1952, Turing proposed that "morphogens acting together and diffusing through a tissue is adequate to account for morphogenesis." His classic paper [2] described simple reaction–diffusion systems in mathematical detail, no doubt inspired by classic texts such as D'Arcy Thompson's [3] "On Growth and Form" (1917), Wardlaw's "Phylogeny and Morphogenesis" [4] and probably Joseph Needham [5]. However, despite the appealing simplicity of the theologian William of Occam (1287–1347) and his razor "entia non sunt multiplicanda praeter necessitatem", the devil is in the biochemical details where "angels fear to tread". Turing invoked Graham's law of diffusion to infer the existence of morphogen gradients (that includes hormones) and the drift to equilibria analogous to an electronic oscillator, an idea that we recently developed as the Hechtian growth oscillator [6]. Despite Turing's profound insight and subsequent progress, the problem of plant morphogenesis remains a huge challenge far exceeding the original formulation of simple diffusion. Here, we describe the origin of auxin, H^+, and Ca^{2+} gradients and how their interaction creates morphogenetic patterns. We therefore approach the formidable problem of plant morphogenesis from an empirical biochemical perspective based on the dynamic role of AGPs (arabinogalactan glycoproteins) in a novel Hechtian growth oscillator. A cell surface AGP-Ca^{2+} capacitor is a crucial component that depends on the complex carbohydrate chemistry of classical AGPs [7,8] that chelate Ca^{2+} by paired glucuronic carboxyl groups of numerous small glycomodules. Not surprisingly, after their initial discovery as an arabinogalactan polysaccharide [9] and as a hydroxyproline-rich arabinogalactan polysaccharide protein complex [10], it has taken 60 years since the founder event [11] to unravel the structural basis for the proposed central role of AGPs in plant biology as a cell surface Ca^{2+} capacitor [12]. The role of AGPs emerged as an essential component of the Hechtian oscillator when we correlated tip-focussed cytosolic Ca^{2+} and tip-localised AGPs with Hechtian adhesion and rapid tip growth of pollen tubes [6].

Here, we extrapolate these recent results to morphogenesis and propose two new avenues: Firstly, we propose the avenue of cell walls and mechanoperception involving Hechtian transduction as described in detail here but not considered by Turing; secondly, we propose postulated chemical morphogen gradients [2] identified here as auxin and Ca^{2+}. We propose that mechanotransduction generates both gradients, although each has a different biochemical origin; thus, auxin gradients originate from biosynthesis and transport by tissue-specific auxin efflux PIN proteins. On the other hand, cytosolic Ca^{2+} gradients originate from Ca^{2+} stored in a cell surface AGP-Ca^{2+} capacitor [12]. AGPs are essential components of the Hechtian growth oscillator that amplifies the magnitude and direction of stress vectors regulating plant growth [6]; that explains why plants have heavily invested in AGPs [13] and why this investment is so highly diversified [14]. Many AGPs are developmentally regulated and tissue-specific [15–18]. While the significance of a large pool of dynamic Ca^{2+} at the periplasmic cell surface seems inescapable to us, the role of AGPs in Ca^{2+} gradient formation in tissues is less obvious because it depends on a membrane-bound H^+-ATPase that dissociates AGP-Ca^{2+}. That shows the close link between auxin and Ca^{2+} signalling with the primal proton pump. Paired glucuronic carboxyls of AGP glycomodules bind Ca^{2+} tightly [12] (analogous to Ca^{2+} chelation by dicarboxylic acids). This effectively scavenges less tightly bound Ca^{2+} of pectin and also free Ca^{2+} in the transpiration stream. Thus, abundant cell surface AGP-Ca^{2+} acting as a Ca^{2+} sink would enable both cells and tissues to compete for Ca^{2+}, so that Ca^{2+} supply meets demand where it is greatest, such as in meristem primordia. The "choice" of Ca^{2+} as a universal signalling ion can be ascribed to the size (0.9 Å) of its ionic radius; thus, Ca^{2+} sheds water more rapidly and is therefore more reactive than other divalent ions—notably Mg^{2+} [19]—of similar charge but with a smaller ionic radius (0.65 Å) that binds the water of hydration more strongly than Ca^{2+}.

The following five sections of this essay show how morphogenetic patterns originate, involving (1) cell wall mechanotransduction and the developmental sequence, (2) the embryogenesis of the fertilised egg cell, (3) roots, (4) shoots, and finally (5) the evolution of morphogenesis.

Most models of morphogenesis, particularly computer-generated models, generally favour a genetic perspective with an emphasis on signalling cascades. However, they ignore two "unknown unknown" missing pieces of the morphogenetic puzzle, identified here as Hechtian mechanotransduction and the AGP-Ca^{2+} capacitor.

2. Cell Wall Mechanotransduction

Hecht's classical observations of adhesion between wall and plasma membrane in plasmolysed cells [20] and its significance to cell signalling have been overlooked for more than a century with few exceptions [21]. This seems inexplicable and is most likely because the emphasis has been exclusively on signalling molecules rather than a physical connection. Widespread Hechtian adhesion, which is prominent particularly during rapid tip growth, after plasmolysis (Figure 1) suggested the presence of specific tethers that transduce the stress–strain status of the wall to plasma membrane receptors that can respond to piconewton (10^{-12} N) forces (Table 1).

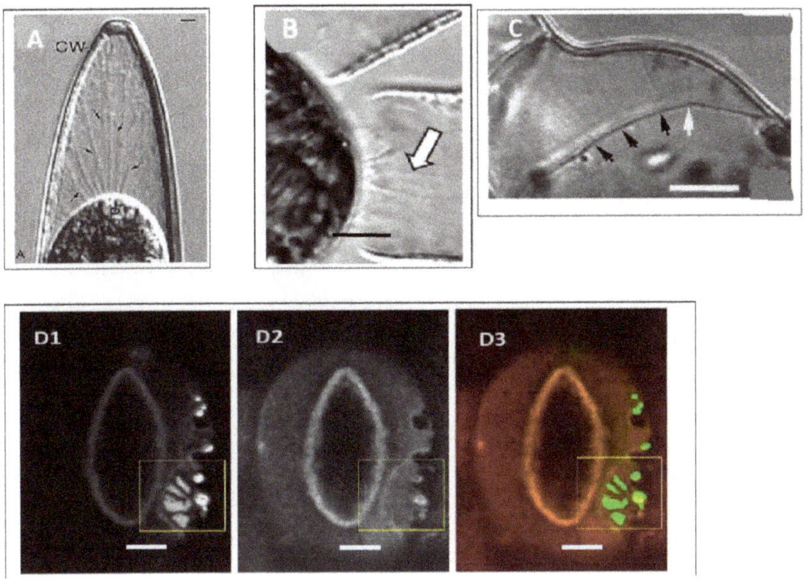

Figure 1. Hechtian strand occurrence: from algae to chloroplast guard cells. (**A**) Closterium desmid in 12% sucrose: reprinted from [22], Scale bar = 2.1 μm; (**B**) *Penium margaritaceum*: reprinted from [23], Scale bar = 2 μm; (**C**). Ginkgo cells plasmolysed in 0.3 M NaCl reprinted from [24], Scale bar is 10 μm; Arrows in (**A–C**) show location of Hechtian strands. (**D1–3**) Guard cell H^+-ATPase and its translocator PATROL (proton ATPase translocation control) in *Arabidopsis*. The plasma membrane ATPase AHA1 and its translocator PATROL1 co-localise at the tips of Hechtian strands in plasmolyzed guard cells transformed with GFP-PATROL1 (**D1**) and RFP–AHA1 (**D2**) merged in (**D3**). Enlarged images of the region enclosed in the yellow square are not shown here. Reprinted from [25], Scale bars = 5 μm.

However, the Hechtian transduction hypothesis creates three dilemmas:

1. How does it distinguish between the huge forces (mega Pascal range) exerted on the membrane by osmotic pressure and the minute piconewton forces that activate Ca^{2+} channels and ATPases?
2. How does it provide a directional signal in response to anisotropic stress? [26]
3. Why does the wall need strong adhesion to the plasma membrane although turgor pressure ensures it? Is the elusive mechanosensor a Boojum or a Snark?

Hechtian adhesion must satisfy several criteria: minimally, it must involve strong, stable and specific interactions with both the cell wall and plasma membrane. Several candidates frequently suggested such as wall-associated kinases and integrins lack hard data and can be eliminated. For example, integrins recognise the RGD motif of the mammalian extracellular matrix but are absent from plants. Other possible candidates include formins and many AGPs anchored securely to the plasma membrane [27–30] by a C-terminal GPI-lipid double tail of long-chain fatty acids that requires a force of ~350 piconewtons to detach from the membrane [31], compared with changes in molecular conformation that are sensitive to far fewer piconewtons [32,33].

At-AGP57C (At3g45230) is covalently linked to pectic RG-I homogalacturonan [34] and is also likely anchored to the plasma membrane by its predicted GPI signal sequence. This identifies a putative Hechtian adhesion site of crucial significance that would enable the instant transmission of the wall stress–strain status and its rate-determining pectin rheology to enable the rapid tip growth of root hairs and pollen tubes.

Wall stress and strain is thus focussed at the molecular level on Ca^{2+} channels and H^+-ATPase; both membrane components are stretch-activated by just a few piconewtons.

Table 1. Piconewton forces crucial to biochemical function and Hechtian adhesion.

System	Action	Adhesion Strength pN	References
Integrin-RGD binding	Binding	~10	[35]
GPI-proteins	Adhesion	103–350 pN	[31]
Protein e.g., talin	Unfolding	5 pN	[32]
Cation channels	Open	4 pN (to open)	[36]

Piezo proteins responsive to mechanical force are well-documented in animal systems [33]. Finally, Hechtian stress–strain transduction and exocytosis comprise the novel Hechtian oscillator [6], an exquisitely designed high gain biological amplifier (Figure 2) that translates piconewton forces [37] into tropisms, with wide ramifications in plant biology.

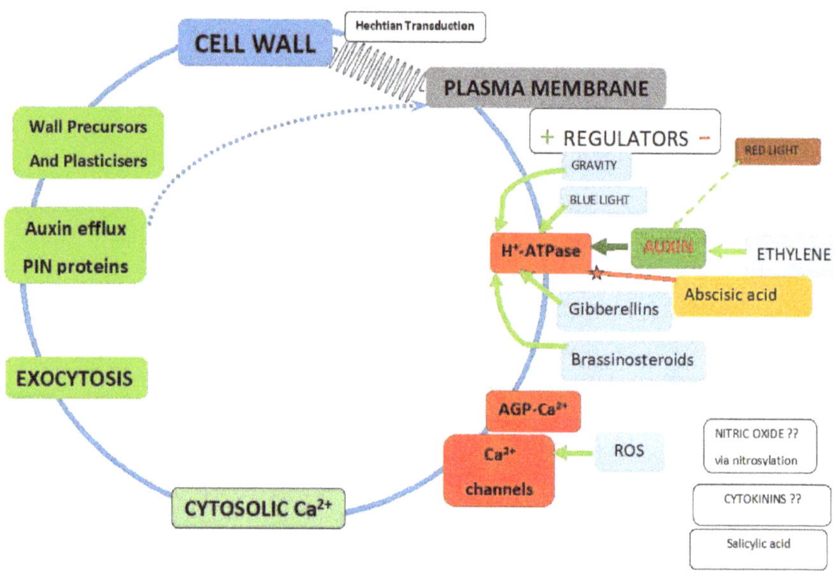

Figure 2. The Hechtian growth oscillator. Regulators of the plasma membrane H^+-ATPase reflect its two major roles: 1. It maintains negative inner membrane potential and enhances anion exit and cation entry; 2. It dissociates AGP-Ca^{2+} as a source of cytosolic Ca^{2+}. Green arrows indicate upregulation or red represent downregulation, with exceptions where the mechanism remains to be elucidated. Auxin activates H^+-ATPase. cf. Figure 3: Plasma membrane H^+-ATPase regulation is central to plant biology [38]; the effects of high steady state auxin levels on root cell elongation in Brachypodium [39]; auxin-driven morphogenetic patterns depend on unidirectional fluxes [40]; the evolution of auxin signaling and PIN proteins [41]; what initiates auxin biosynthesis remains unknown [42]; auxin activates the plasma membrane H^+-ATPase via phosphorylation [38]. Abscisic acid negative regulation: decreases steady-state levels of phosphorylated H^+-ATPase possibly by promoting dephosphorylation via a protein phosphatase [43] and suppresses hypocotyl elongation in Arabidopsis [43]; abscisic acid stress signalling evolved in algal progenitors [44]. Blue light: the blue light photoreceptor pigment phototropin increases cytosolic Ca^{2+} [45]. Brassinosteroids: Increase cytosolic Ca^{2+} via increased auxin

levels [46]. Cytokinins enhance cell division by unknown mechanisms. Ethylene upregulates auxin biosynthesis in the Arabidopsis root apex and inhibits root cell expansion [47]; thus, anthranilate synthase mutants yield ethylene-insensitive root growth phenotypes. Ethylene specifically inhibits the most rapid growth phase of expanding cells—normally the root hair initiation zone but ethylene moves it much closer to the tip. Auxin and ethylene act synergistically to control root elongation, root hair formation, lateral root formation and hypocotyl elongation [48]. Ethylene modulates root elongation through altering auxin transport: Ethylene binds to receptor proteins such as ETR1 and EIN2 controlling transcription factor EIN3 that targets ERF1, the ethylene response factor that regulates diverse genes, most likely tissue-specific. In shoots, auxin moves from the apex to the base [49]. PIN1 mutants decrease auxin transport in inflorescences while PIN2, PIN3, and PIN7 mutants decrease polar auxin transport in roots. In roots, auxin moves rootward via PIN1, PIN3, and PIN7 in the central cylinder and shootward in the outer cell layers via PIN2 which mediates gravitropism. Ethylene-induced auxin synthesis involves the α and β subunits of anthranilate synthase [47]. Fusicoccin: The fungal toxin fusicoccin fills a cavity in the interaction surface between PM H^+-ATPase and 14-3-3 proteins to form a tight bridge that activates the ATPase irreversibly [38]. Gibberellins increase cytosolic Ca^{2+} via a DELLA-independent signalling pathway [50]. Gravity: in root gravitropism, auxin regulates root curvature via apoplastic pH and a Ca^{2+}-dependent signalling pathway [51]; PIN proteins responsible for polar auxin transport and gravitropism reviewed in [42]; in microgravity, cucumber seedling PIN protein distribution is parallel to the minor root axis. However, a 1 g force re-aligns PIN1 to the lower side of the endodermis thus facilitating auxin transport from the upper side of the root to the lower side. Similarly, PIN3 and PIN7 of the gravity sensing columella also re-align to the lower side [52]. Nitric oxide: high apical levels of reactive oxygen species (ROS) in *Arabidopsis* root hairs suggest possible activation of a Ca^{2+} channel that modulates root hair-tip growth [53]. Red light: the elevation of auxin levels is well established as an early event, in response to response to prolonged shade. During an initial triggering phase, phytochrome interacting factors (PIFs) bind to the promoters of auxin synthesis genes and generate a burst of auxin that promotes growth [54]. ROS: NADPH oxidase generates active oxygen species that activate Ca^{2+} channels and regulate cell expansion and root morphogenesis [55].

Although osmotic pressure is equally distributed at the plasma membrane, wall pressure exerts an equal and opposite force. However, a direct covalent connection between pectin and the plasma membrane would enable instant transmission of the wall stress–strain status by fibrous macromolecules.

Two extreme examples, stomata and pollen tubes, demonstrate the efficacy of Hechtian adhesion: the stomatal guard cell wall elasticity transmits stress while the rapid tip growth of a pollen tube involves wall plasticity; when stretched, it transmits strain arguably via AGP57C [34] or its homologous GPI-anchored classical AGPs, supporting a vital role for the Hechtian growth oscillator in morphogenesis as discussed in subsequent sections.

The pollen tube tip has the simplest primary cell wall—almost exclusively pectin and with fast tip growth—so is an ideal system to test hypotheses involving the Hechtian transduction of wall stress–strain. Pectin rheology is deceptively simple and depends largely on the degree of methyl esterification of pectic polygalacturonic acid residues, in contrast to AGP glucuronic acid, which is never esterified. Highly methyl-esterified pectins form gels, while de-esterification followed by Ca^{2+} crosslinking forms much stronger gels. A frequently cited pectic paradigm [56] assumes sufficient available free Ca^{2+} for crosslinking but ignores the regulation of Ca^{2+} in muro by AGPs which, with their demonstrated higher affinity for Ca^{2+}, compete for pectic Ca^{2+}. Such competition effectively strips Ca^{2+} from pectin. The concomitant electrostatic repulsion of negatively charged pectic carboxylates explains the increase in wall plasticity by pectin methylesterase [57]. This effect, not previously considered, may also explain the role of the small diffusible "AGP peptides" [58] that possess two arabinogalactan glycomodules, presumably with paired glucuronic acid residues that bind Ca^{2+}. Significantly, small AGP peptides are upregulated by auxin [39], suggesting another subtle level of control over plasticity of the pectin-rich primary cell walls.

Driven by turgor pressure, cells expand in a direction orthogonal (i.e., at right angles) to the direction of stress that strains the pectic matrix; the concomitant Hechtian transduction of pectic strain initiates Ca^{2+} oscillations and exocytosis (Figure 2). The stress vector also causes the orthogonal orientation of microtubules and cellulose deposition [26] that reinforce the pectic matrix. As noted above, pectin rheology depends on its methyl esterification status regulated by pectin methyl esterase [59], the availability of free Ca^{2+} and borate crosslinking of the RG-II pectic component. Crosslinking of the cellulosic component (notably absent from pollen tube tips) may also be regulated by expansin [60,61].

Such complexity explains why the molecular basis of cell wall plasticity remains recalcitrant since first described [62]. "The difficulty lies not in the new ideas, but in escaping from the old ones" (John Maynard Keynes). Many are Boojums—hypotheses that postulate non-existent entities are not falsifiable and should disappear: for example, the widely accepted effect of low pH on wall loosening by activating enzymic cleavage of load-bearing covalent crosslinks. However, both crosslinks and enzymes remain unidentified. In the absence of an alternative viable unifying hypothesis, the Hechtian oscillator emphasises the role of Ca^{2+} in the exocytosis of wall plasticisers as detailed in [6] rather than covalent bond cleavage. Plant cells share a wall with neighbours and can therefore use Hechtian adhesion to sense which wall is being highly stressed by a rapidly expanding neighbour cell. Such stressed cells respond by the rapid reorientation of auxin efflux PIN proteins to direct auxin efflux towards their rapidly expanding neighbour [26]. Cells also compete for available Ca^{2+} by increasing the size of their cell surface AGP-Ca^{2+} capacitor; this can increase the amplitude of Ca^{2+} oscillations, thereby enhancing exocytosis.

Consider the primary receptors of force transduction. What are they, and how do they propagate tension to the cell interior? Stretch-activated Ca^{2+} channels [63] and auxin-activated H^+-ATPase [64] control Ca^{2+} influx as follows: the plasma membrane ATPase [38,65] is similar to the mitochondrial ATP synthase: a rotary nanomachine fuelled by the proton motive force that generates ATP. However, in reverse, fuelled by cytosolic ATP, it becomes a molecular turbine rotating at up to ~8000 rpm, ejecting proton jets that pump about three protons into the periplasm for each ATP molecule hydrolysed [66].

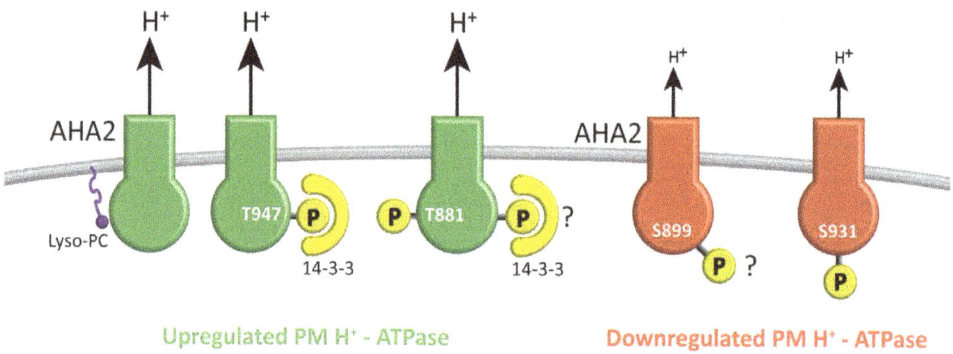

Figure 3. Posttranslational regulation of the PM H^+-ATPase alternates between two states: reprinted from [38]. Red = downregulated pumps hydrolyze ATP with low efficiency, hence the low transport of H^+. Green = upregulated pumps with high ATPase efficiency, and high H^+ transport rates. C-terminal regulatory domains control transition between the two states. The phosphorylation (P) of the C-terminal penultimate threonine residue (Thr-947 in pump AHA2) creates a binding site for a 14-3-3 protein that stabilises the pump. The binding of lysophosphatidylcholine (Lyso-PC) and phosphorylation at Thr-881 also activates the PM H^+-ATPase independently of phosphorylation and 14-3-3 protein binding. It is not known whether phosphorylation at Thr-881 in the C-terminal domain interferes with or depends on 14-3-3 binding (as indicated by the question mark). Phosphorylation at Ser-899 or Ser-931 inactivates the pump. Phosphorylation at Ser-931 blocks the binding of 14-3-3 protein, but is not known for Ser-899. Only well-characterized regulatory events are shown.

These powerful proton jets dissociate the AGP-Ca^{2+} of the cell surface capacitor that supplies the Ca^{2+} channels. The evidence [6] for this is a 100% correlation between tip-localised AGPs and "tip-focussed" cytosolic Ca^{2+} during the rapid tip growth of pollen tubes.

How Ca^{2+} fluxes translate into morphogenesis depends on the local wall plasticity that determines the direction and magnitude of the wall stress vector; its amplification depends on the size of the AGP-Ca^{2+} capacitor (Figure 2) and its obvious abundance at metabolically active sites [67]. Arguably, the stress vector also determines the anisotropic location of Hechtian adhesion sites that activate the local vectorial exocytosis of wall precursors and dynamic reorientation of auxin efflux (PIN) proteins. Mechanical stress causes auxin efflux carrier localisation at the plasma membrane adjacent to the most stressed or strained walls cells and thus directs auxin toward neighbours that are rapidly expanding [68]. Counterintuitively, "the net flow of auxin in shoot tips is up the auxin gradient such that any cell directs its auxin toward neighbouring cells that have a higher auxin concentration" [26]. Auxin flux against the concentration gradient towards regions of active expansion growth, although an apparent contravention of the Second Law of Thermodynamics, is pithily summarised by Mathew's Law: "To him that hath shall be given. And to him that hath not shall be taken away".

Plant cells exploit a "thermodynamic loophole" based on their ability to trap ionised auxin in the cytosol (pH 7.4) where it cannot freely permeate the plasma membrane. However, in the wall at ~pH 5, auxin is protonated and thus freely permeates the plasma membrane of adjacent cells. Thus, auxin wins the Sysiphean uphill struggle.

The regulation of the oscillator at several biochemical levels involves Ca^{2+} channel activity and the plasma membrane H^{+}-ATPase (Figure 2). The proximity of a large or small AGP-Ca^{2+} capacitor and its dissociation by ATPase proton pump thus creates dual gradients of auxin and Ca^{2+}; coincidentally, two interacting morphogens required for the creation of "Turing patterns" [2]. By connecting these gradients, both dependent on the proton motive force, the Hechtian oscillator emerges as a master regulator of plant growth; this paradigm shift demands the reappraisal of the acid growth hypothesis [69] that depends on the widely accepted activation of wall-loosening enzymes. However, alternative explanations for a low wall pH include the PIN-directed auxin transport of protonated auxin and increased plasticity of deprotonated pectin. Finally, the release of Ca^{2+} from AGP-Ca^{2+} by a proton pump [12] is an extrapolation of Mitchell's original chemiosmotic hypothesis [70].

3. Embryogenesis

The above description of stress-dependent auxin and Ca^{2+} gradients shows how polarity can arise at the earliest stages of embryogenesis, particularly with an AGP-Ca^{2+} sink as a source of dynamic cytosolic Ca^{2+}, while tissue tensions determine the orientation of auxin efflux proteins and cellulose deposition. Organogenesis then follows similar biochemical and biomechanical rules. Every budding botanist sees sliced celery stalks curl because internal tissues are under compression and outer tissues under tension. This simple principle applies during embryogenesis as the stress vector changes from the internal meristem (compression) to the outer protoderm (tension). The concomitant rearrangement of auxin efflux proteins in the protoderm under tension results in lateral auxin export from its apex outwards, with consequential slow growth and anisotropy at the torpedo stage that splits the tissue into the two cotyledons [71] (Figure 4).

Angiosperm double fertilisation results in a triploid endosperm and a diploid zygote that elongates and establishes the shoot–root hierarchy with an initial asymmetric division into a small apical cell that becomes the proembryo; the large vacuolated basal cell divides by transverse divisions to produce the suspensor, a short file of cells that act as an "umbilicus" connecting the embryo with maternal tissues. The hypophysis is a remarkable pivotal apical cell of the suspensor, and as the precursor to the embryonic root, it demarcates the crucial zone between shoot and root. Thus, many auxin mutants result in a defective hypophysis and rootless seedlings. Presumably, a simplified morphogenetic programme of hypophysis silences most major shoot programmes and defines the fate of future cells in the root as a "partial shoot" [72] characterised by the root tip meristem, xylem, phloem, pericycle and lateral roots.

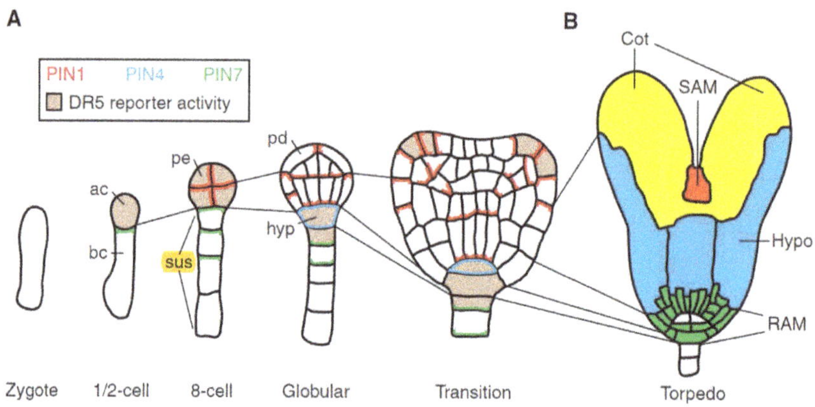

Figure 4. Embryogenesis in Arabidopsis: Cell lineages, PIN protein localization, and auxin response maxima (reprinted from [71]). Auxin response maxima and PIN protein localization follow a regular cell division pattern. Thin lines show lineages between stages: PIN protein localization is shown as follows: red (PIN1), blue (PIN4), and green (PIN7), and DR5 reporter is pink. (**A**) After the division of the zygote, one and two-cell embryos express PIN7 in the basal daughter cell (bc), and the apical cell (ac) expresses the DR5 reporter. After two more cell-divisions, all proembryo (pe) cells express PIN1 and DR5 reporter. Basal suspensor (sus) cells express PIN7 localized on the proembryo side. At the globular stage, central lower cells of the proembryo establish basal PIN1 polarity while PIN1 localizes apically in outer protoderm (pd) cells. At the same time, PIN7 polarity reverses in suspensor cells and PIN4 is activated in the uppermost suspensor cell that now expresses the DR5 reporter and is specified as hypophysis (hyp). During the transition stage, the PIN1 polarity at the flanks of the apical embryo half converges in adjacent cells, accompanied by the appearance of new DR5 maxima. These sites mark the initiation of the cotyledons; (**B**) The torpedo stage shows discrete regions of the embryo: RAM (root apical meristem), green, root apical meristem (white, future quiescent centre); hypo, blue, hypocotyl; Cot, yellow, cotyledons; and SAM (shoot apical meristem), red, the shoot apical meristem.

Plant cells are generally totipotent, and single cultured cells can regenerate an entire plant [73,74] While cell lineage does not necessarily determine cell fate, the aphorism of Hans Driesch that "the fate of a cell is a function of its position" implies that both its lineage [75] and position [76] are pertinent. Indeed, "morphogen" gradients evident as auxin and AGP-Ca^{2+} are highly correlated with the earliest stages of embryogenesis including the polarised haploid egg cell that yields a zygote with polarised AGP distribution increasing at the later two-cell stage and onwards [77]. Auxin biosynthesis evidenced by the DR5 reporter is similarly localized, particularly at the 2-cell, 8-cell and globular stages in *Arabidopsis* [71]. Auxin distribution depends on auxin efflux carriers which are initially inferred [78] and subsequently identified at the genomic level by "PIN" mutants that have phenocopied inhibitors of auxin transport [79]. Unlike the anisotropic protoderm, the early growth of meristematic tissue is isotropic. Variations in the local growth of the meristem arise possibly because cell surface AGPs compete for limited Ca^{2+}; combined with tissue tensions, this leads to boundary regions which are low in Ca^{2+}. These peripheral anisotropic regions, particularly in the ectoderm, result in tge rearrangement of auxin efflux proteins with consequent auxin depletion in boundary zones between the meristem, hence the origin of new primordia and phyllotaxis; i.e. auxin peaks generate a spiral phyllotactic pattern.

Much work over more than 25 years (Table 2) reveals a remarkable family of tissue-specific auxin efflux (PIN) proteins that ensure a regulated supply of auxin during embryogenesis and the ensuing stages of morphogenesis. However, until now, the signals that initiate PIN protein localisation remain unknown; the Hechtian transduction proposed here is the best candidate based on considerable supporting evidence.

Table 2. The PIN Ffamily of auxin efflux proteins.

Auxin Efflux Protein	Major Location or Function	Efflux Orientation	References
PIN 1	Outer protoderm and lower cells of proembryo	Rootward	[71]
PIN 2	Root epidermis and lateral root cap; regulates gravitropism	Shootward	[26] [42]
PIN 3	Gravitropism	Rootward	[26]
PIN 4	Hypophysis		[71]
PIN 5	ER	Cytosol	[42]
PIN 6	ER	Cytosol	[42]
PIN 7	Suspensor	Apical towards embryo Rootward	[80]
PIN 8	ER	cytosol	[42]

The Hechtian growth oscillator generates morphogenetic patterns; arguably, these depend on localised adhesion sites that result in vectorial exocytosis of wall precursors. This includes auxin efflux PIN proteins directed to the sites of the highest strain in expanding cells, [26] thus ensuring a supply of auxin for their continued growth. AGPs specifically localised in the suspensor [81] (Figure 4 in reference [81]) (6 to 9 cells in *Arabidopsis*) with apically polarised PIN4 and PIN7 auxin efflux proteins [80] highlight its active role as an "umbilicus" supplying auxin and nutrients from maternal tissues to the embryo.

Stress also predicts the plane of cell division marked by the preprophase band of microtubules [82] as putative mitosis stabilisers [83] and the appearance of the cell plate templated by extensin self-assembling amphiphiles; [84] hydroxyproline-rich glycoproteins central to plant growth once again cooperate with biomechanical forces. Elegant laser ablation experiments alter the orientation of PIN proteins in the apical meristem [26]. These classic experiments [16,26,85] show that tissue biomechanics create stress vectors that reorientate tissue-specific auxin efflux PIN proteins. We infer that specific Hechtian adhesion is the "missing link" that connects biomechanical stress vectors with auxin action and PIN reorientation: a chemically defined covalent connection between cell wall pectin and a known plasma membrane such as AGP57C likely transduces stress that dissociates AGP-Ca^{2+}, thus increasing cytosolic Ca^{2+} that triggers vectored exocytosis. We infer that Hechtian adhesion and AGPs amplify both the direction and magnitude of stress vectors resulting in growth orientation. Not surprisingly, mutants of Gal31 are lethal [86], possibly by disrupting the 1-6-linked galactose sidechain of the AGP-Ca^{2+} glycomodule [87].

4. Roots

The morphogenesis of root xylem tissues begins with the specific expression of the JIM13 AGP epitope in a single metaxylem initial just above four central cells of the quiescent centre [88]. The how and why have been dominated by the speculative idea of AGPs as signalling molecules per se. Thus, a role for AGPs in Ca^{2+}-signalling was unexpected yet surprisingly close to reality, with AGP-Ca^{2+} belatedly seen here as the major dynamic source of Ca^{2+} regulated by auxin. Significantly, transformed plants that overproduce auxin show excessive xylem and phloem development [89] and bushy phenotypes [90].

Mechanical stress patterns involved in root morphogenesis have yet to be modelled. Nevertheless, a recent comprehensive experimental approach [91] correlated gravitropism with the epidermal location of auxin efflux protein PIN2 which was dramatically co-localised with Hechtian adhesion sites between the plasma membrane and the cell wall (Figure 5).

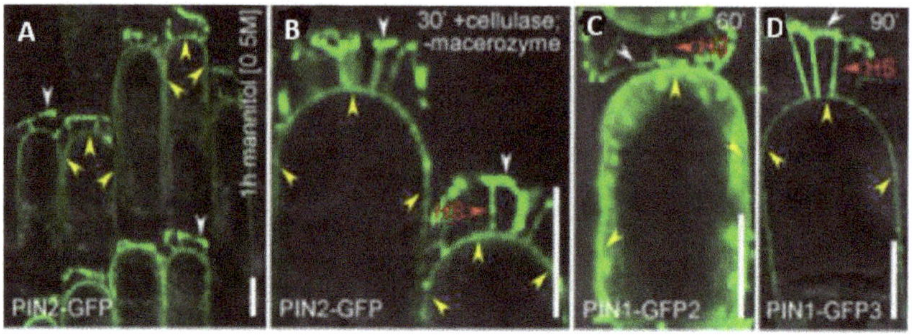

Figure 5. PIN proteins maintain their polarity at the cell wall of plasmolysed cells. Plasmolysis shows that apical (PIN2-GFP and PIN1-GFP-3) and basal (PIN1-GFP-2) proteins remain attached to the cell wall by their polar domains. (**A**–**D**) Note—yellow arrowheads = nonpolar PIN-GFP signal at the plasma membrane. HS = Hechtian strands with PIN-GFP-connections to the wall. White arrowheads = a strong persistent PIN-GFP signal at the cell wall–plasma membrane interface. Scale bars = 10 µm. Reprinted from Figure 4 of [91].

The presence of the cell wall is essential for gravity perception [91]. These observations connect gravitropism with a rapid increase of cytosolic Ca^{2+} [51] and the following conclusion [91]: "Thus, the identified tight link between the cell wall and cell polarity provides the conceptual possibility for regulation of signal fluxes and, ultimately, plant development via signaling from the extracellular matrix." Such evidence suggests that auxin-based mechanoperception and subsequent root morphogenesis involve the Hechtian growth oscillator. This fundamental phenomenon supersedes textbook dogma and hypotheses in highly contentious and confusing fields [49,92–94] that struggle to disentangle cause from effect. As a prime example, consider the challenge to identify the gravity sensor. Although quite unknown, it may involve critical components of the Hechtian oscillator as "the very earliest stages of gravitropic bending depend on auxin-stimulated Ca^{2+} influx" [41,51]. Nevertheless, the Cholodny–Went explanation of auxin redistribution remains unquestioned dogma [95]: "The ability of roots to reorient their growth in response to changes in the gravity vector is dependent on the asymmetric redistribution of auxin at the root tip" [41]. This demands a biochemical explanation.

The root–shoot paradox: root tip gravitropism increases auxin at the lower side that decreases growth while less auxin at the upper side increases growth. However, in shoots, a similar auxin distribution in response to gravity reverses the direction of growth. We resolve this paradox by asking: How does auxin inhibit growth at the lower side of the root? Extensin peroxidase [96] provides the answer: crosslinked extensin decreases the rate of cell extension ([97] cf. [47]) via di-isodityrosine formation [98,99]. Presumably, programmed cells of the root tip respond to high levels of auxin simply by enhancing the enzymic crosslinking of wall proteins. Thus, extensins and AGPs, as the Yin and Yang of cell extension, may illustrate yet again the ingenious parsimony of nature that involves these hydroxyproline-rich glycoproteins as both negative and positive regulators of cell extension.

While the above approach to a complex problem is consistent with Occam's razor, it contrasts with the complexity of recent papers that deal with the dynamic rearrangement of PIN proteins transporting auxin from both root tip and the shoot via xylem and root epidermis (Figure 6).

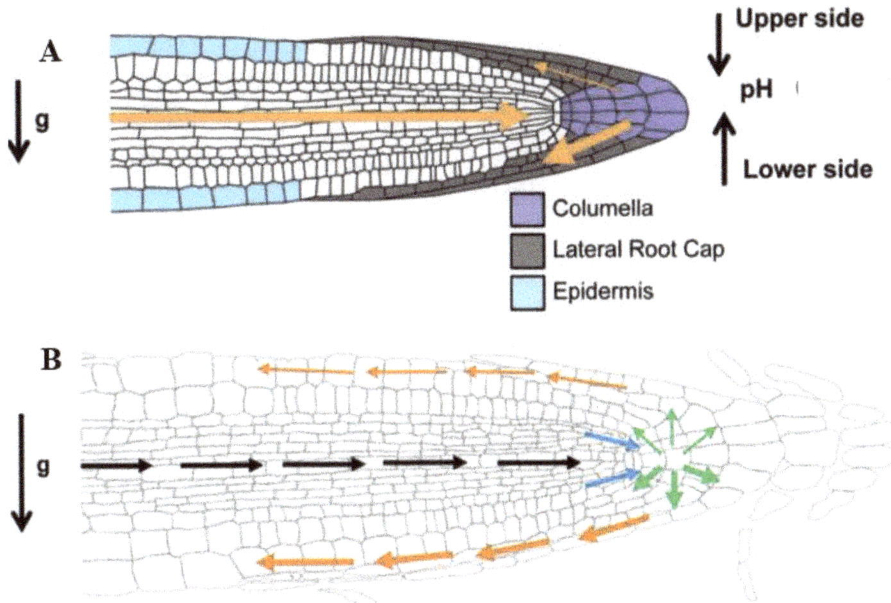

Figure 6. Reprinted with permission from Sato et al. [49]. Root gravitropism in *Arabidopsis*. (**A**) Gravity perception in *Arabidopsis thaliana*. At time point 0, roots grow vertically. After a 90° turn, the following events occur: (1) At 10 s, statoliths are still at the old bottom of the cell. After 3 min, statoliths move towards the new bottom of the cell to be uniformly distributed at 5 min [92]; (2) PIN3 and PIN7 relocalization is achieved 2 min after the gravity stimulus and, in consequence, a lateral auxin gradient is generated between the upper and lower side of the root (thin and thick orange arrows respectively) [100]; (3) the development of differential extracellular pH levels between the upper (acidic) and lower (alkaline) side of the gravistimulated root [51]; (**B**) Gravity signal transduction and transmission via auxin transport and redistribution. Black arrows show that AUX1 and PIN2 channel auxin from the shoot to the root tip. Blue arrows show how PIN4 distributes auxin efflux through the vascular tissue to the columella cells. PIN3 and PIN7 set up the auxin flow (green arrows), with an accumulation on the lower side of the root. PIN2 and AUX1 transport auxin through the lateral root cap to the epidermal cells in the elongation zone (orange arrows) where the actual growth response will occur.

However, no convincing molecular mechanism has been offered for auxin-increased cell extension. Auxin has also not been successfully connected with Ca^{2+} signalling although [49] "auxin somehow activates cytosolic Ca^{2+} waves," the major source being the Hechtian oscillator via auxin-activated H^+-ATPase dissociation of AGP-Ca^{2+} that releases Ca^{2+} to the cytosol via Ca^{2+} channels. How that translates into increased cell extension in the root tip elongation zone remains a conundrum, as discussed above. Although the prevalent statolith hypothesis is now textbook dogma, the precise molecular details of statolith mechanotransduction remain quite unknown. Mutants that lack amyloplast statoliths do not resolve the problem: while they decrease the gravitational response, they do not eliminate it [93]. Finally, both plasma membrane H^+-ATPases [25] (~11 in *Arabidopsis*) and PIN proteins recycle in response to mechanotransduction, thus adding an additional level of complexity.

This raises the pertinent question: What is the biochemical source of signalling Ca^{2+} and its connection to auxin? Regions of high auxin biosynthesis, e.g., meristems, young primordia and reproductive organs, express YUCCA, a flavin mono-oxygenase that decarboxylates indole pyruvate to yield auxin [48]. Arguably, Hechtian adhesion and auxin-activated plasma membrane ATPase dissociate AGP-Ca^{2+} to provide Ca^{2+} that activates exocytosis enabling the gravity vector (9.8 $N{\cdot}kg^{-1}$)

to redistribute auxin by recycling the auxin efflux PIN proteins. As the root tip derives its auxin from transport through central root tissues and also local biosynthesis (Figure 6), the columella is quite possibly the gravity sensor as it also contains the statoliths. However, that view is not consistent with the Hechtian oscillator and the Hechtian transduction of the gravity vector leading to auxin transport and its asymmetric redistribution by PIN1 and PIN3 auxin efflux proteins [26] cf. [41].

The Hechtian oscillator is also evident in mature root tissues such as the pericycle which expresses AGPs [101] and generates adventitious lateral roots and a specific extensin ([102] cf. [75]). Possibly, pericycle progenitor cell AGPs scavenge local Ca^{2+} that enhances exocytosis when stimulated by auxin supplied from root tip biosynthesis [103]. Similarly, root hair positioning is also correlated with AGP-reactive epidermal cells Figure 4A in [101], while defects in root tip auxin biosynthesis result in a shift of root hair emergence towards the shoot.

5. Shoots

Morphogenesis of the shoot involves phyllotaxis of leaves and flowers; a problem of formidable complexity recognised by Egyptian, Greek and Roman scholars since antiquity [104] up to the present. Numerous competing hypotheses largely based on mathematical formulations connect physical forces to growth, but with many biochemical unknowns. Recent significant experimental advances [26,68,85] closely connect the stress vectors of growing tissues with the reorientation of auxin efflux PIN proteins (Figure 7): exocytosis and endocytosis (transcytosis) rapidly redeploy PIN proteins to expanding walls that show the greatest stress or strain. This implies that stress-induced signals from the cell wall promote the accumulation of PIN1 (and thus export of auxin) at the nearest membrane of cells stressed by their rapid expansion as follows.

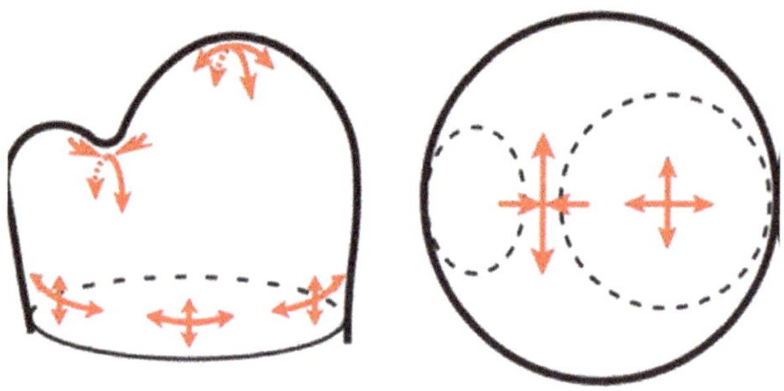

Figure 7. Stress vectors in the shoot apical meristem. The suggested changing meristem stress patterns cause auxin depletion in the boundary region as primordia form [26]. Note that the magnitude and direction of the stress vector are strongly anisotropic (red arrows) in the boundary region between the primordium and shoot apical meristem apex. Reprinted from [85] with permission from AAAS).

Rapid PIN1 redeployment [105] raises the question of how cells perceive the mechanical signals that largely determine PIN1 polarity. In plants, the gap between biophysical stress and biochemical response is generally ignored. However, PIN protein recycling depends on Ca^{2+}-directed vectorial exocytosis, and therefore shares essential components of the Hechtian growth oscillator that translate biophysical force into biochemistry. This connects the oscillator with the generation of new primordia; hypothetically, their regular spatial separation or phyllotaxis results from competing demands for auxin and Ca^{2+} morphogens, channelled to the most rapidly growing cells by PIN proteins and cell surface AGPs that scavenge Ca^{2+}. This results in auxin depletion and slower growth in the boundary or "saddle region" between the apical meristem and primordia. Indeed, competition for auxin and Ca^{2+} may determine the angle of divergence between successive primordia and thus the creation of whorls [106], although the morphogenesis of flowering involving, for example, a protein florigen [107] and development of seeds is beyond the scope of this brief essay.

Auxin initiates phyllotaxis by transport against a concentration gradient to sites of primordia formation; these depend on the polarity of auxin efflux (PIN) proteins maintained by the cell wall [91,108]. However, the apical dominance of the shoot, although generally ascribed to the auxin inhibition of lateral bud growth, may actually depend on the absence of PIN polarity from dormant tissues. Arguably, the removal of the shoot apex restores PIN1 polarisation in the bud [109] via renewed axillary auxin transport that breaks dormancy. A role for the Hechtian growth oscillator in apical dominance seems likely and is consistent with the bushy phenotype of tomato plants over-expressing AGP LeAGP1 [110]. The size of the AGP-Ca^{2+} capacitor could determine the balance between the dormancy and active growth of axillary buds.

6. The Evolution of Morphogenesis

Phylogeny and morphogenesis are inextricably linked [111], and so our brief description can only attempt a broad-brush summary that reconciles old observations with new discoveries. Specifically, these relate wall stress–strain and Hechtian adhesion to the control of cytosolic Ca^{2+} by auxin. Thus, auxin efflux PIN proteins transport auxin to target sites, where the activation of H^+-ATPase releases Ca^{2+} from AGP-Ca^{2+}, triggering exocytosis. This paradigm shift can be traced back to the primordial soup and unifies the chemiosmotic proton motive force with auxin and Ca^{2+} signalling. By analogy with inscriptions on the Rosetta stone, once decoded, these three fundamental transport phenomena translate the same growth message into morphogenesis.

Morphogenesis begins with prebiotic synthesis from CH_4, CO_2, and water to yield HCHO [112], a precursor to simple sugars via the Butlerov (formose) reaction. The subsequent prebiotic generation of self-replicating molecules and self-assembling amphiphiles formed lipid membranes and nanobubble aerosols. Such semipermeable membrane-bounded protocells sequestered and concentrated reactants by chemiosmosis, most likely in Darwin's "warm little pond" (In a letter to J.D. Hooker 1871). Hence, the origin of Peter Mitchell's proton motive force [70], a universal phenomenon that precedes RNA-DNA replication and drives reactions that exploit the chemistry in a watery environment of planets in the "Goldilocks zone". The first simple prokaryotes were chemical machines; their further specialisation and morphological adaption led to unicellular eukaryotes, morphogenetic machines characterised by increasing structural adaptations and complexity, exemplified by the invention of chromosomes as Darlington's "little packets of predestination", internal membrane systems such as ER (endoplasmic reticulum), Golgi, and cell walls to protect the plasma membrane. Indeed, most major evolutionary advances from chemiosmotic energy transduction to cellular organelles appeared very early and remain as essential features of modern eukaryotic cells whose precise origins by endosymbiont capture [113] remain to be elucidated.

From the cell wall viewpoint, the unicellular flagellate *Chlamydomonas* provides the most striking example of evolutionary conservation and diversification, with a large complement of 182 different proteins related to hydroxyproline-rich extensins [114]. The *Chlamydomonas* wall is devoid of cellulose and consists almost entirely of glycoproteins that self-assemble as a crystal lattice [115], a remarkable precursor to their conserved self-assembly role in higher plants [84]. The leap to multicellular cooperation involved two distinct evolutionary lines: one line embedded cells in a hydroxyproline-rich glycoprotein (HRGP) protective matrix typified by volvocine algae; the other Chlorophycean line evolved silicified or calcified protective cell walls (Figure 8) typified by the calcareous plates of Coccolithophores, such as *Emiliana huxleyi* of the cretaceous that built the chalk hills and cliffs of the present era.

Static cell surface $CaCO_3$ coccoliths of Coccolithophores represented by *E. huxleyi* are the probable evolutionary precursor to dynamic cell surface AGP-Ca^{2+}, hence Ca^{2+} signalling. Algae increased the tensile strength of a primary cell wall prototype by co-opting cellulose, thus enabling turgor pressure and the major evolutionary pathway to metaphytes. Indeed, the composition of algal pectin-rich cellulosic walls is remarkably similar to those defined as primary walls in the cambial tissues of higher plants [116]. That raises the question: When did the Hechtian oscillator emerge? Based on the presence of Hechtian strands associated with rapid tip growth [6], we extrapolated Mitchell's original chemiosmotic hypothesis to include Ca^{2+} signalling. This implies incredible evolutionary conservation and an early origin of the Hechtian pscillator in unicellular Chlorophycean algae such as Closterium [22] and Penium [23], as both exhibit tip growth with prominent Hechtian strands (Figure 1).

Figure 8. Chilly Brook chalk cliffs with *E. huxleyi*. Coccolithophores such as *E. huxleyi* (inset) constructed massive soft chalk deposits of calcium carbonate during the cretaceous period (145–65 MYA), now seen here from across the Chilly Brook water meadows dominated by *Ranunculus acris*, adjacent to the River Ouse in the South Downs National Park at Lewes, UK. This view encapsulates the entire evolutionary progression from the simplest unicellular protists such as *E. huxleyi* to advanced dicots such as *R. acris* at the pinnacle of alternation of generations, all dependent on calcium. (Photo: DTAL).

Single cells remaining attached after cytokinesis evolved into simple linear filamentous algae as the first step towards multicellularity and its regulation. Although auxin is a key growth regulator in higher plants, it is evident as a biosynthetic pathway in the simplest unicellular chlorophytes [117]. However, auxin transport also involves efflux PIN proteins, and these first appear in filamentous algae—"living fossils" such as *Spirogyra* [117]—with anchoring rhizoids that differentiate only from terminal cells [118]. Thus, the PIN proteins and polarity of *Spirogyra* rhizoids, as well as stretch-activated Ca^{2+} channels, nicely represent the earliest evolutionary progression from marine via freshwater to land plants (Figure 9).

Further lateral cohesion between adjacent filaments forms a prototype thallus. Other "living fossils" include *Coleochaete* a prototype leafy liverwort (Hepaticae) that retains the fertilised diploid egg cell until "germination". *Coleochaete* cell walls are also significantly enriched in hydroxyproline-rich glycoproteins. Alkaline hydrolysis yields a hydroxyproline glycoside profile (Figure 10) dominated by small hydroxyproline heterooligosaccharides, similar to the profile of *Chlamydomonas*, but surprisingly different from the simple Hyp-arabinoside profile of higher plants.

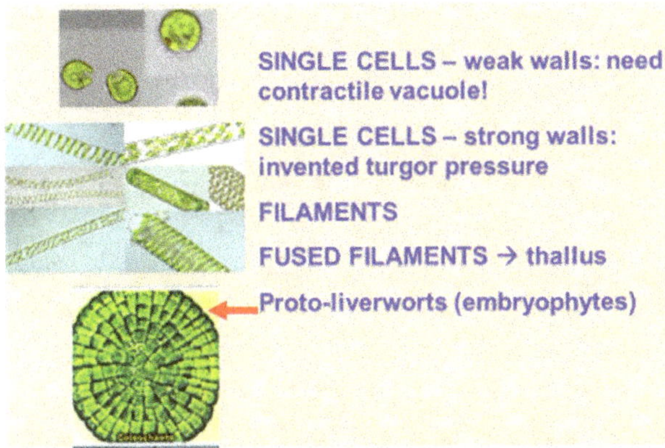

Figure 9. Metaphyte origins. Progression from single cells to filaments that align, forming a flat thallus. The examples here are the biflagellate *Chlamydomonas*, *Spirogyra* and the red arrow points to *Coleochaete scutata*.

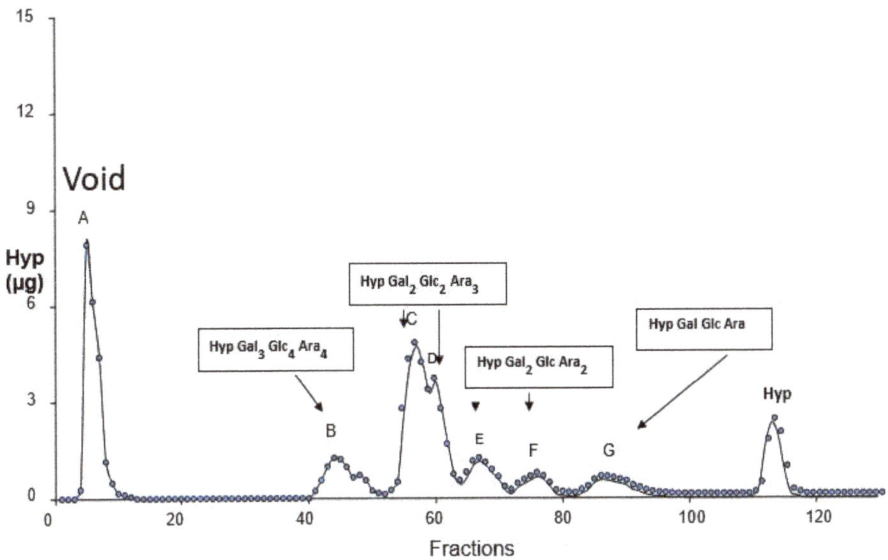

Figure 10. *Coleochaete* hydroxyproline glycosides. Alkaline hydrolysates of *Coleochaete scutata* cell walls fractionated on a cation exchange column gave a Hyp-glycoside profile that appeared similar to the profile of cell walls isolated from higher plants. However, closer inspection and quantitative sugar analyses showed a striking difference; in higher plants, the Hyp-glycosides are simple arabinosides, whereas *Coleochaete* shows heterooligosaccharides of up to seven sugar residues that include galactose, glucose and arabinose, compositions remarkably similar to the profile of cell walls isolated from *Chlamydomonas* [119] but different from other members of the plant kingdom [120]. Presented at the 11th Cell Wall Meeting, Copenhagen [121].

Although more popular notions suggest Chara or Nitella as a land plant precursor [122], that seems unlikely for several reasons; firstly, despite being invoked as a model system, their cell walls lack hydroxyproline-rich cell wall proteins; secondly, the relatively thick walls of Chara are rich in non-methyl esterified pectin; and finally, the large cells are coenocytes but remain locked in an evolutionary cul-de-sac.

Bryophytes initiate the gametophyte–sporophyte "alternation of generations" brilliantly elucidated by Hofmeister [123] (Figure 11).

Figure 11. The moss *Bryum capillare* exemplifies the alternation of generations. *Bryum capillare* with attached diploid sporophytes dependent on the haploid gametophyte generation. (Photo: DTAL).

Further evolutionary development of the attached sporophyte (e.g., with stomata) culminated in its complete dominance in higher plants with their increasingly complex division into root, shoots and leaves. Root-like structures appeared as rhizoids in the Bryophytes, but true roots defined by their gravitropic response, endogenous branching, root hairs, and a protective root cap originate from the suspensor hypophysis during angiosperm embryogenesis. While sharing much in common with true shoots, roots clearly lack many shoot characteristics and have therefore been described as "partial shoots" [72] that evolved as an adaptation to a subterranean life [124].

The ultimate reduction of the gametophyte generation to just a few cells, namely pollen and the ovule as an integumented megasporangium, completes this brief summary based on a novel growth oscillator unifying other growth regulators that control Ca^{2+} release and exocytosis.

7. Postscript

Ion transport dominates this paper, which is greatly influenced (1955–1961) by personal acquaintance (DTAL) with the Cambridge pioneers who included Peter Mitchell (proton pump), David Keilin (cytochrome discovery) and Robert "Robin" Hill Reaction (photolysis of water). The cell wall protein was undoubtedly inspired by Fred Sanger in the adjacent lab (protein structure, 1958 Nobel Prize), and Joseph E. Varner [125] on his sabbatical leave acted as a "midwife to wall protein." H.A. Krebs was one of DTAL's BA examiners. Two years of National Service as a corporal radio instructor at the Royal Air Force Radio School Yatesbury (1953–1955) inspired thoughts about biological oscillators. Finally, the poet Alexander Pope (1688–1744) summarises the effect of the intellectual environment exemplified by Joseph Needham (morphogenesis) and Don Northcote (cell walls) in the Cambridge Department of Biochemistry: "As the twig is bent so is the tree inclined."

Funding: No current funding

Acknowledgments: Dedicated to the memory of Joseph E. Varner, mentor, friend and colleague.

Conflicts of Interest: The authors declare no conflicts of interest.

References

1. Lamport, D.T.A. *The Primary Cell Wall*; University of Cambridge: Cambridge, UK, 1963; pp. 1–181.

2. Turing, A.M. The chemical basis of morphogenesis. *Philos. Trans. R. Soc. Lond. B* **1952**, *237*, 37–72. [CrossRef]
3. Thompson, D.W. *On Growth and Form*; Cambridge University Press: Cambridge, UK, 1917.
4. Wardlaw, C.W. *Phylogeny and Morphogenesis*; Macmillan And Co.: Macmillan, London, UK, 1952.
5. Needham, J. *Biochemistry and Morphogenesis*; Cambridge University Press: Cambridge, UK, 1942.
6. Lamport, D.T.A.; Tan, L.; Held, M.A.; Kieliszewksi, M.J. Pollen tube growth and guidance: Occam's razor sharpened on a molecular arabinogalactan glycoprotein Rosetta Stone. *New Phytol.* **2018**, *217*, 491–500. [CrossRef] [PubMed]
7. Tan, L.; Qiu, F.; Lamport, D.T.A.; Kieliszewski, M.J. Structure of a hydroxyproline-arabinogalactan polysaccharide from repetitive Ala-Hyp expressed in transgenic *Nicotiana tabacum*. *J. Biol. Chem.* **2004**, *279*, 13156–13165. [CrossRef] [PubMed]
8. Tan, L.; Varnai, P.; Lamport, D.T.A.; Yuan, C.; Xu, J.; Qiu, F.; Kieliszewski, M.J. Plant O-Hydroxyproline Arabinogalactans Are Composed of Repeating Trigalactosyl Subunits with Short Bifurcated Side Chains. *J. Biol. Chem.* **2010**, *285*, 24575–24583. [CrossRef] [PubMed]
9. Aspinall, G.O.; Malloy, J.A.; Craig, J.W.T. Extracellular polysaccharides from suspension-cultured sycamore cells. *Can. J. Biochem.* **1969**, *47*, 1063–1070. [CrossRef] [PubMed]
10. Lamport, D.T.A. Cell wall metabolism. *Ann. Rev. Plant Physiol.* **1970**, *21*, 235–270. [CrossRef]
11. Lamport, D.T.A.; Northcote, D.H. Hydroxyproline in primary cell walls of higher plants. *Nature* **1960**, *188*, 665–666. [CrossRef]
12. Lamport, D.T.A.; Varnai, P. Periplasmic arabinogalactan glycoproteins act as a calcium capacitor that regulates plant growth and development. *New Phytol.* **2013**, *197*, 58–64. [CrossRef] [PubMed]
13. Johnson, K.L.; Cassin, A.M.; Lonsdale, A.; Bacic, A.; Doblin, M.; Schultz, C.J. A Motif and Amino Acid Bias Bioinformatics Pipeline to Identify Hydroxyproline-Rich Glycoproteins. *Plant Physiol.* **2017**, *174*, 886–903. [CrossRef] [PubMed]
14. Ma, Y.; Yan, C.; Li, H.; Wu, W.; Liu, Y.; Wang, Y.; Chen, Q.; Ma, H. Bioinformatics Prediction and Evolution Analysis of Arabinogalactan Proteins in the Plant Kingdom. *Front. Plant Sci.* **2017**, *8*, 66. [CrossRef] [PubMed]
15. Pennell, R.I.; Roberts, K. Sexual development in the pea is presaged by altered expression of arabinogalactan protein. *Nature* **1990**, *334*, 547–549. [CrossRef]
16. Pennell, R.I.; Janniche, L.; Scofield, G.N.; Booij, H.; de Vries, S.C.; Roberts, K. Identification of a Transitional Cell State in the Developmental Pathway to Carrot Somatic Embryogenesis. *J. Cell Biol.* **1992**, *119*, 1371–1380. [CrossRef] [PubMed]
17. Showalter, A.M. Arabinogalactan-proteins: Structure, expression and function. *Cell. Mol. Life Sci.* **2001**, *58*, 1399–1417. [CrossRef] [PubMed]
18. Coimbra, S.; Jones, B.; Pereira, L.G. Arabinogalactan proteins (AGPs) related to pollen tube guidance into the embryo sac in Arabidopsis. *Plant Signal. Behav.* **2008**, *3*, 455–456. [CrossRef] [PubMed]
19. Hepler, P.K. The Cytoskeleton and Its Regulation by Calcium and Protons. *Plant Physiol.* **2016**, *170*, 3–22. [CrossRef] [PubMed]
20. Hecht, K. *Studien uber den Vorgang der Plasmolyse*; Beitrage zur Biologie der Pflanzen: Berlin, Germany, 1912; p. 17.
21. Pont-Lezica, R.F.; McNally, J.G.; Pickard, B.G. Wall-to-membrane linkers in onion epidermis: Some hypotheses. *Plant Cell Environ.* **1993**, *16*, 111–123. [CrossRef]
22. Domozych, D.S.; Roberts, R.; Danyow, C.; Flitter, R.; Smith, B. Plasmolysis, Hechtian strand formation, and localized membrane wall adhesions in the desmid *Closterium acerosum* (Chlorophyta). *J. Phycol.* **2003**, *39*, 1194–1206. [CrossRef]
23. Raimundo, S.C.; Sorensen, I.; Tinaz, B.; Ritter, B.; Rose, J.K.C.; Domozych, D.S. Isolation and manipulation of protoplasts from the unicellular green alga *Penium margaritaceum*. *Plant Methods* **2018**, *14*, 18. [CrossRef]
24. Buer, C.S.; Weathers, P.J.; Swartzlander, G.A. Changes in Hechtian strands in cold-hardened cells measured by optical microsurgery. *Plant Physiol.* **2000**, *122*, 1365–1377. [CrossRef] [PubMed]
25. Hashimoto-Sugimoto, M.; Higaki, T.; Yaeno, T.; Nagami, A.; Irie, M.; Fujimi, M.; Miyamoto, M.; Akita, K.; Negi, J.; Shirasu, K.; et al. A Munc13-like protein in Arabidopsis mediates H^+-ATPase translocation that is essential for stomatal responses. *Nat. Commun.* **2013**, *4*, 2215. [CrossRef] [PubMed]
26. Sampathkumar, A.; Yan, A.; Krupinski, P.; Meyerowitz, E.M. Physical Forces Regulate Plant Development and Morphogenesis. *Curr. Biol.* **2014**, *24*, 475–483. [CrossRef] [PubMed]

27. Serpe, M.D.; Nothnagel, E.A. Arabinogalactan-proteins in the multiple domains of the plant cell surface. *Adv. Bot. Res.* **1999**, *30*, 207–289.
28. Oxley, D.; Bacic, A. Structure of the glycosylphosphatidylinositol anchor of an arabinogalactan protein from Pyrus communis suspension-cultured cells. *Proc. Natl. Acad. Sci. USA* **1999**, *96*, 14246–14251. [CrossRef] [PubMed]
29. Eisenhaber, B.; Wildpaner, M.; Schultz, C.J.; Borner, G.H.H.; Dupree, P.; Eisenhaber, F. Glycosylphosphatidylinositol lipid anchoring of plant proteins. Sensitive prediction from sequence- and genome-wide studies for Arabidopsis and rice1. *Plant Physiol.* **2003**, *133*, 1691–1701. [CrossRef] [PubMed]
30. Borner, G.H.H.; Lilley, K.S.; Stevens, T.J.; Dupree, P. Identification of glycosylphosphatidylinositol-anchored proteins in Arabidopsis. A proteomic and genomic analysis. *Plant Physiol.* **2003**, *132*, 568–577. [CrossRef] [PubMed]
31. Cross, B.; Ronzon, F.; Roux, B.; Rieu, J.P. Measurement of the Anchorage Force between GPI-Anchored Alkaline Phosphatase and Supported Membranes by AFM Force Spectroscopy. *Langmuir* **2005**, *21*, 5149–5153. [CrossRef]
32. Dufort, C.C.; Paszek, M.J.; Weaver, V.M. Balancing forces: Architectural control of mechanotransduction. *Nat. Rev. Mol. Cell Biol.* **2011**, *12*, 308–319. [CrossRef] [PubMed]
33. Murthy, S.E.; Dubin, A.E.; Patapoutian, A. Piezos thrive under pressure: Mechanically activated ion channels in health and disease. *Nat. Rev. Mol. Cell Biol.* **2017**, *18*, 771–783. [CrossRef] [PubMed]
34. Tan, L.; Eberhard, S.; Pattathil, S.; Warder, C.; Glushka, J.; Yuan, C.; Hao, Z.; Zhu, X.; Avci, U.; Miller, J.S.; et al. An Arabidopsis Cell Wall Proteoglycan Consists of Pectin and Arabinoxylan Covalently Linked to an Arabinogalactan Protein. *Plant Cell* **2013**, *25*, 270–287. [CrossRef] [PubMed]
35. Moore, S.W.; Roca-Cusachs, P.; Sheetz, M.P. Stretchy Proteins on Stretchy Substrates: The Important Elements of Integrin-Mediated Rigidity Sensing. *Dev. Cell* **2010**, *19*, 194–206. [CrossRef] [PubMed]
36. Sukharev, S.; Sachs, F. Molecular force transduction by ion channels—Diversity and unifying principles. *J. Cell Sci.* **2012**, *125*, 3075–3083. [CrossRef] [PubMed]
37. Yusko, E.C.; Asbury, C.L. Force is a signal that cells cannot ignore. *Mol. Biol. Cell* **2014**, *25*, 3717–3725. [CrossRef] [PubMed]
38. Falhof, J.; Pedersen, J.T.; Fuglsang, J.T.; Palmgren, M. Plasma Membrane H^+-ATPase Regulation in the Center of Plant Physiology. *Mol. Plant* **2016**, *9*, 323–337. [CrossRef] [PubMed]
39. Pacheco-Villalobos, D.; Diaz-Moeeno, M.; van der Schuren, A.; Tamaki, T.; Kang, Y.H.; Gujas, B.; Novak, O.; Jaspert, N.; Li, Z.; Wolf, S.; et al. The Effects of High Steady State Auxin Levels on Root Cell Elongation in Brachypodium. *Plant Cell* **2016**, *28*, 1009–1024. [CrossRef] [PubMed]
40. Cieslak, M.; Runions, A.; Prusinkiewicz, P. Auxin-driven patterning with unidirectional fluxes. *J. Exp. Bot.* **2015**, *66*, 5083–5102. [CrossRef] [PubMed]
41. Leyser, O. Auxin Signaling. *Plant Physiol.* **2018**, *176*, 465–479. [CrossRef] [PubMed]
42. Habets, M.E.; Offringa, R. PIN-driven polar auxin transport in plant developmental plasticity: A key target for environmental and endogenous signals. *New Phytol.* **2014**, *203*, 362–377. [CrossRef] [PubMed]
43. Hayashi, Y.; Takahashi, K.; Inoue, S.; Kinoshita, T. Abscisic Acid Suppresses Hypocotyl Elongation by Dephosphorylating Plasma Membrane H^+-ATPase in *Arabidopsis thaliana*. *Plant Cell Physiol.* **2014**, *55*, 845–853. [CrossRef] [PubMed]
44. De Vries, J.; Curtis, B.A.; Gould, S.B.; Archibald, J.M. Embryophyte stress signaling evolved in the algal progenitors of land plants. *Proc. Natl. Acad. Sci. USA* **2018**, *115*, E3471–E3486. [CrossRef] [PubMed]
45. Folta, K.M.; Lieg, E.G.; Durham, T.; Spalding, E.P. Primary Inhibition of Hypocotyl Growth and Phototropism Depend Differently on Phototropin Mediated Increases in Cytoplasmic Calcium Induced by Blue Light. *Plant Physiol.* **2003**, *133*, 1464–1470. [CrossRef] [PubMed]
46. Zhao, Y.; Qi, Z.; Berkowitz, G.A. Teaching an Old Hormone New Tricks: Cytosolic Ca^{2+} Elevation Involvement in Plant Brassinosteroid Signal Transduction Cascades. *Plant Physiol.* **2013**, *163*, 555–565. [CrossRef] [PubMed]
47. Swarup, R.; Perry, P.; Hagenbeek, D.; Van Der Straeten, D.; Beemster, G.T.S.; Sandberg, G.; Bhalerao, R.; Ljung, K.; Bennett, M. Ethylene Upregulates Auxin Biosynthesis in Arabidopsis Seedlings to Enhance Inhibition of Root Cell Elongation. *Plant Cell* **2007**, *19*, 2186–2196. [CrossRef] [PubMed]
48. Muday, G.K.; Rahman, A.; Binder, B.M. Auxin and ethylene: Collaborators or competitors? *Trends Plant Sci.* **2012**, *17*, 1360–1385. [CrossRef] [PubMed]

49. Sato, E.M.; Hijazi, H.; Bennett, M.J.; Vissenjberg, K.; Swarup, R. New insights into root gravitropic signaling. *J. Exp. Bot.* **2015**, *66*, 2155–2165. [CrossRef] [PubMed]
50. Okada, K.; Ito, T.; Fukazawa, J.; Takahashi, Y. Gibberellin Induces an Increase in Cytosolic Ca^{2+} via a DELLA-Independent Signaling Pathway. *Plant Physiol.* **2017**, *175*, 1536–1542. [CrossRef] [PubMed]
51. Monshausen, G.B.; Miller, N.D.; Murphy, A.S.; Gilroy, J.S. Dynamics of auxin-dependent Ca^{2+} and pH signaling in root growth revealed by integrating high-resolution imaging with automated computer vision-based analysis. *Plant J.* **2011**, *65*, 309–318. [CrossRef] [PubMed]
52. Yamazaki, C.; Fujii, N.; Miyazawa, Y.; Kamada, M.; Kasahara, H.; Osada, I.; Shimazu, T.; Fusejima, Y.; Higashibata, A.; Yamazaki, T.; et al. The gravity-induced re-localization of auxin efflux carrier CsPIN1 in cucumber seedlings: Spaceflight experiments for immunohistochemical microscopy. *Microgravity* **2016**, *2*, 1–7. [CrossRef] [PubMed]
53. Domingos, P.; Prado, A.M.; Wong, A.; Gehring, C.; Feijo, J.A. Nitric Oxide: A Multitasked Signaling Gas in Plants. *Mol. Plant* **2015**, *8*, 506–520. [CrossRef] [PubMed]
54. Pucciariello, O.; Legris, M.; Rojas, C.C.; Iglesias, M.J.; Hernando, C.E.; Dezar, C.; Vazquez, M.; Yanovsky, M.J.; Finlayson, S.A.; Prat, S.; et al. Rewiring of auxin signaling under persistent shade. *Proc. Natl. Acad. Sci. USA* **2018**, *115*, 5612–5617. [CrossRef] [PubMed]
55. Foreman, J.; Demidchik, V.; Bothwell, J.H.F.; Mylona, P.; Miedema, H.; Torres, M.A.; Linstead, P.; Costa, S.; Brownlee, C.; Jones, J.D.G.; et al. Reactive oxygen species produced by NADPH oxidase regulate plant cell growth. *Nature* **2003**, *422*, 442–446. [CrossRef] [PubMed]
56. Proseus, T.E.; Boyer, J.S. Calcium pectate chemistry controls growth rate of *Chara coralline*. *J. Exp. Bot.* **2006**, *57*, 3989–4002. [CrossRef] [PubMed]
57. Levesque-Tremblay, G.; Pelloux, J.; Braybrook, S.A.; Muller, K. Tuning of pectin methylesterification: Consequences for cell wall biomechanics and development. *Planta* **2015**, *242*, 791–811. [CrossRef] [PubMed]
58. Van den Bulck, K.; Swennen, K.; Loosveld, A.M.; Christophe, M.; Brijs, K.; Proost, P.; Van Damme, J.; Campenhout, S.; Mort, A.; Delcour, J.A. Isolation of cereal arabinogalactan-peptides and structural comparison of their carbohydrate and peptide moieties. *J. Cereal Sci.* **2005**, *41*, 59–67. [CrossRef]
59. Braybrook, S.A.; Peaucell, A. Mechano-Chemical Aspects of Organ Formation in Arabidopsis thaliana: The Relationship between Auxin and Pectin. *PLoS ONE* **2013**, *8*, e57813. [CrossRef] [PubMed]
60. Cosgrove, D.J. Catalysts of plant cell wall loosening. *F100Research* **2016**, *5*, 1–13. [CrossRef] [PubMed]
61. Cosgrove, D.J. Diffuse Growth of Plant Cell Walls. *Plant Physiol.* **2018**, *176*, 17–27. [CrossRef] [PubMed]
62. Heyn, A.N.J. The physiology of cell elongation. *Bot. Rev.* **1940**, *6*, 515–574. [CrossRef]
63. Ding, J.P.; Pickard, B.G. Mechanosensory calcium-selective cation channels in epidermal cells. *Plant J.* **1993**, *3*, 83–110. [CrossRef] [PubMed]
64. Takahashi, K.; Hayashi, K.; Kinoshita, T. Auxin Activates the Plasma Membrane H^+-ATPase by Phosphorylation during Hypocotyl Elongation in Arabidopsis. *Plant Physiol.* **2012**, *159*, 632–641. [CrossRef] [PubMed]
65. Mazhab-Jafari, M.T.; Rohou, A.; Schmidt, C.; Bueler, S.A.; Benlekbir, S.; Robinson, C.V.; Rubinstein, J.L. Atomic model for the membrane-embedded VO motor of a eukaryotic V-ATPase. *Nature* **2016**, *539*, 118–130. [CrossRef] [PubMed]
66. Lane, N. *Power, Sex, Suicide: Mitochondria and the Meaning of Life*; Oxford University Press: Cambridge, MA, USA, 2005.
67. Lamport, D.T.A.; Kieliszewksi, M.J.; Showalter, A.M. Salt-stress upregulates periplasmic arabinogalactan-proteins: Using salt-stress to analyse AGP function. *New Phytol.* **2006**, *169*, 479–492. [CrossRef] [PubMed]
68. Heisler, G.; Ohno, C.; Das, P.; Sieber, P.; Reddy, G.V.; Long, J.A.; Meyerowitz, E.M. Patterns of Auxin Transport and Gene Expression during Primordium Development Revealed by Live Imaging of the *Arabidopsis* Inflorescence Meristem. *Curr. Biol.* **2005**, *15*, 1899–1911. [CrossRef] [PubMed]
69. Rayle, D.L.; Cleland, R.E. The acid growth theory of auxin-induced cell elongation is alive and well. *Plant Physiol.* **1992**, *99*, 1271–1274. [CrossRef] [PubMed]
70. Mitchell, P. Coupling of phosphorylation to electron and hydrogen transfer by a chemi-osmotic type of mechanism. *Nature* **1961**, *191*, 144–148. [CrossRef] [PubMed]
71. Moller, B.; Weijers, D. Auxin Control of Embryo Patterning. *Cold Spring Harb. Perspect. Biol.* **2009**, *1*, 1–13. [CrossRef] [PubMed]

72. Arber, A. *The Natural Philosophy of Plant Form*; Cambridge University Press: Cambridge, UK, 1950.
73. Steward, F.C.; Mapes, M.O.; Smith, J. Growth and organised development of cultured cells. I. Growth and division of freely suspended cells. *Am. J. Bot.* **1958**, *45*, 693–703. [CrossRef]
74. Steward, F.C.; Mapes, M.O.; Mears, K. Growth and organised development of cultured cells. II Organisation in cultures grown from freely suspended cells. *Am. J. Bot.* **1958**, *45*, 705–708. [CrossRef]
75. Dolan, L.; Janmaat, K.; Willemsen, V.; Linstead, P.; Poethig, S.; Roberts, K.; Scheres, B. Cellular organisation of the *Arabidopsis thaliana* root. *Development* **1993**, *119*, 71–84. [PubMed]
76. Kidner, C.; Sundaresan, V.; Roberts, K.; Dolan, L. Clonal analysis of the Arabidopsis root confirms that position, not lineage, determines cell fate. *Planta* **2000**, *211*, 191–199. [CrossRef] [PubMed]
77. Qin, Y.; Zhao, J. Localization of arabinogalactan proteins in egg cells, zygotes, and two-celled proembryos and effects of β-D-glucosyl Yariv reagent on egg cell fertilization and zygote division in *Nicotiana tabacum* L. *J. Exp. Bot.* **2006**, *57*, 2061–2074. [CrossRef] [PubMed]
78. Rubery, P.H.; Sheldrake, A.R. Carrier-mediated Auxin Transport. *Planta* **1974**, *118*, 101–121. [CrossRef] [PubMed]
79. Okada, K.; Ueda, J.; Komaki, M.K.; Bell, C.J.; Shimura, Y. Requirement of the auxin polar transport system in early stages of Arabidopsis floral bud formation. *Plant Cell* **1991**, *3*, 677–684. [CrossRef] [PubMed]
80. Peng, X.; Sun, M.X. The suspensor as a model system to study the mechanism of cell fate specification during early embryogenesis. *Plant Reprod.* **2018**, *31*, 59–65. [CrossRef] [PubMed]
81. Pennell, R.I.; Janniche, L.; Kjellbom, P.; Scofield, G.N.; Peart, J.M.; Roberts, K. Developmental regulation of a plasma membrane arabinogalactan protein epitope in oilseed rape flowers. *Plant Cell* **1991**, *3*, 1317–1326. [CrossRef] [PubMed]
82. Pickett-Heaps, J.D.; Northcote, D.H. Organisation of microtubules and endoplasmic reticulum during mitosis and cytokinesis in wheat meristems. *J. Cell Sci.* **1966**, *1*, 109–120. [PubMed]
83. Schaefer, E.; Belcram, K.; Uyttewaal, M.; Duroc, Y.; Goussot, M.; Legland, D.; Laruelle, O.; de Tauzia-Moreau, M.L.; Pastuglia, M.; Bouchez, D. The preprophase band of microtubules controls the robustness of division orientation in plants. *Science* **2017**, *356*, 186–189. [CrossRef] [PubMed]
84. Cannon, M.C.; Terneus, K.; Hall, Q.; Wang, Y.; Wegenhart, B.L.; Chen, L.; Lamport, D.T.A.; Chen, Y.; Kieliszewski, M.J. Self-assembly of the plant cell wall requires an extensin scaffold. *Proc. Natl. Acad. Sci. USA* **2008**, *105*, 2226–2231. [CrossRef] [PubMed]
85. Hamant, O.; Heisler, M.G.; Jonnson, H.; Krupinski, P.; Uyttewaal, M.; Bokov, P.; Corson, F.; Sahlin, P.; Boudaoud, A.; Meyerowitz, E.M.; et al. Developmental patterning by mechanical signals in *Arabidopsis*. *Science* **2008**, *322*, 1650–1654. [CrossRef] [PubMed]
86. Geshi, N.; Johansen, J.N.; Dilokpimol, A.; Rolland, A.; Belcram, K.; Verger, S.; Kotake, T.; Tsumuraya, Y.; Kaneko, S.; Tryfona, T.; et al. A galactosyltransferase acting on arabinogalactan protein glycans is essential for embryo development in Arabidopsis. *Plant J.* **2013**, *76*, 128–137. [CrossRef] [PubMed]
87. Basu, D.; Liang, Y.; Liu, X.; Himmeldirk, K.; Faik, A.; Kieliszewski, M.; Held, M.A.; Showalter, A.M. Functional identification of a hydroxyproline-*O*-galactosyltransferase specific for arabinogalactan protein biosynthesis in *Arabidopsis*. *J. Biol. Chem.* **2013**, *288*, 10132–10143. [CrossRef] [PubMed]
88. Dolan, L.; Linstead, P.; Roberts, K. An AGP epitope distinguishes a central metaxylem initial from other vascular initials in the Arabidopsis root. *Protoplasma* **1995**, *189*, 149–155. [CrossRef]
89. Olatunji, D.; Geelen, D.; Verstraeten, I. Control of Endogenous Auxin Levels in Plant Root Development. *Int. J. Mol. Sci.* **2017**, *18*, 2587. [CrossRef] [PubMed]
90. Sun, W.; Kieliszewski, M.J.; Showalter, A.M. Overexpression of tomato LeAGP-1 arabinogalactan-protein promotes lateral branching and hampers reproductive development. *Plant J.* **2004**, *40*, 70–81. [CrossRef] [PubMed]
91. Feraru, E.; Feraru, M.I.; Kleine-Vehn, J.; Martinie, A.; Mouille, G.; Vanneste, S.; Vernhettes, S.; Runions, J.; Friml, J. PIN Polarity Maintenance by the Cell Wall in Arabidopsis. *Curr. Biol.* **2011**, *21*, 338–343. [CrossRef] [PubMed]
92. Leitz, G.; Kang, B.H.; Schoenwaelder, M.E.A.; Staehelin, L.A. Statolith Sedimentation Kinetics and Force Transduction to the Cortical Endoplasmic Reticulum in Gravity-Sensing Arabidopsis Columella Cells. *Plant Cell* **2009**, *21*, 843–860. [CrossRef] [PubMed]
93. Caspar, T.; Pickard, B.G. Gravitropism in a starchless mutant of Arabidopsis. *Planta* **1989**, *177*, 185–197. [CrossRef] [PubMed]

94. Berut, A.; Chauvet, H.; Legue, V.; Moulia, B.; Pouliquen, V.; Forterre, Y. Gravisensors in plant cells behave like an active granular liquid. *Proc. Natl. Acad. Sci. USA* **2018**, *115*, 5123–5128. [CrossRef] [PubMed]
95. Firn, R.D.; Wagstaff, C.; Digby, J. The use of mutants to probe models of gravitropism. *J. Exp. Bot.* **2000**, *51*, 1323–1340. [CrossRef] [PubMed]
96. Everdeen, D.S.; Kiefer, S.; Willard, J.J.; Muldoon, E.P.; Dey, P.M.; Li, X.B.; Lamport, D.T.A. Enzymic crosslinkage of monomeric extensin precursors *in vitro*. *Plant Physiol.* **1988**, *87*, 616–621. [CrossRef] [PubMed]
97. Cleland, R.E.; Karlsnes, A. A possible role for hydroxyproline-containing proteins in the cessation of cell elongation. *Plant Physiol.* **1967**, *42*, 669–671. [CrossRef] [PubMed]
98. Epstein, L.; Lamport, D.T.A. An intramolecular linkage involving isodityrosine in extension. *Phytochemistry* **1984**, *23*, 1241–1246. [CrossRef]
99. Held, M.A.; Tan, L.; Kamyab, A.; Hare, M.; Shpak, E.; Kieliszewksi, M.J. Di-isodityrosine is the intermolecular cross-link of isodityrosine-rich extensin analogs cross-linked *in vitro*. *J. Biol. Chem.* **2004**, *279*, 55474–55482. [CrossRef] [PubMed]
100. Friml, J.; Wisniewski, J.P.; Benkova, E.; Mengen, K.; Palme, K. Lateral relocation of auxin efflux regulator PIN3 mediates tropism in Arabidopsis. *Nature* **2002**, *415*, 806–809. [CrossRef] [PubMed]
101. Knox, J.P.; Day, S.; Roberts, K. A set of cell surface glycoproteins forms an early marker of cell position but not cell type, in the root apical meristem of *Daucus carota* L. *Development* **1989**, *106*, 47–56.
102. Keller, B.; Lamb, C.J. Specific expression of a novel cell wall hydroxyproline-rich glycoprotein gene in lateral root initiation. *Genes Dev.* **1989**, *3*, 1639–1646. [CrossRef] [PubMed]
103. Ikeda, Y.; Men, S.; Fischer, U.; Stepanova, A.N.; Alonso, J.M.; Ljung, K.; Grebe, M. Local auxin biosynthesis modulates gradient-directed planar polarity in *Arabidopsis*. *Nat. Cell Biol.* **2009**, *11*, 731–739. [CrossRef] [PubMed]
104. Adler, I.; Barabe, D.; Jean, R.V. A History of the Study of Phyllotaxis. *Ann. Bot.* **1997**, *80*, 231–244. [CrossRef]
105. Vernoux, T.; Besnard, F.; Traas, J. Auxin at the Shoot Apical Meristem. *Cold Spring Harb. Perspect. Biol.* **2010**, *2*, a001487. [CrossRef] [PubMed]
106. Coen, E.S.; Meyerowitz, E.M. The war of the whorls: Genetic interactions controlling flower development. *Nature* **1991**, *353*, 31–37. [CrossRef] [PubMed]
107. Taoka, K.; Ohki, I.; Tsuji, H.; Kojima, C.; Shimamoto, K. Structure and function of florigen and the receptor complex. *Trends Plant Sci.* **2013**, *18*, 287–294. [CrossRef] [PubMed]
108. Luschnig, C.; Vert, G. The dynamics of plant plasma membrane proteins: PINs and beyond. *Development* **2014**, *141*, 2924–2938. [CrossRef] [PubMed]
109. Balla, J.; Medvedova, Z.; Kalousek, P.; Matiješèuková, N.; Friml, J.; Reinohl, V.; Prochazka, S. Auxin flow-mediated competition between axillary buds to restore apical dominance. *Sci. Rep.* **2016**, *6*, 35955. [CrossRef] [PubMed]
110. Zhao, Z.D.; Tan, L.; Showalter, A.M.; Lamport, D.T.A.; Kieliszewski, M.J. Tomato LeAGP-1 arabinogalactan-protein purified from transgenic tobacco corroborates the Hyp contiguity hypothesis. *Plant J.* **2002**, *31*, 431–444. [CrossRef] [PubMed]
111. Wardlaw, C.W. Evidence relating to the diffusion reaction theory of morphogenesis. *New Phytol.* **1955**, *54*, 39–48. [CrossRef]
112. Jalbout, A.F.; Leif, A.; Adamowicz, L.; Polt, R.; Apponi, A.J.; Ziurys, L.M. Sugar Synthesis from a Gas-Phase Formose Reaction. *Astrobiology* **2007**, *7*, 433–442. [CrossRef] [PubMed]
113. Margulis, L. *Origin of Eukaryotic Cells*; Yale University Press: New Haven, CT, USA, 1970.
114. Liu, X.; Wolfe, R.; Welch, L.R.; Domozych, D.S.; Popper, Z.A.; Showalter, A.M. Bioinformatic Identification and Analysis of Extensins in the Plant Kingdom. *PLoS ONE* **2016**, *11*, e0150177. [CrossRef] [PubMed]
115. Roberts, K. Crystalline glycoprotein cell walls of algae: Their structure, composition and assembly. *Philos. Trans. R. Soc. Lond. B* **1974**, *268*, 129–146. [CrossRef]
116. Kerr, T.; Bailey, I.W. The cambium and its derivative tissues. X Structure, optical properties and chemical composition of the so-called middle lamella. *J. Arnold Arbor.* **1934**, *15*, 327–349.
117. De Smet, I.; Voss, U.; Lau, S.; Wilson, M.; Shao, N.; Timme, R.E.; Swarup, R.; Kerr, I.; Hodgman, C.; Bock, R.; et al. Unraveling the Evolution of Auxin Signaling. *Plant Physiol.* **2010**, *110*, 168161. [CrossRef] [PubMed]
118. Inoue, N.; Yamada, S.; Nagata, Y.; Shimmen, T. Rhizoid Differentiation in Spirogyra: Position Sensing by Terminal Cells. *Plant Cell Physiol.* **2002**, *43*, 479–483. [CrossRef] [PubMed]

119. Miller, D.H.; Lamport, D.T.A.; Miller, M. Hydroxyproline heterooligosaccharides in *Chlamydomonas*. *Science* **1972**, *176*, 918–920. [CrossRef] [PubMed]
120. Lamport, D.T.A.; Miller, D.H. Hydroxyproline arabinosides in the plant kingdom. *Plant Physiol.* **1971**, *48*, 454–456. [CrossRef] [PubMed]
121. Buglass, S.; Lamport, D.T.A.; Xu, J.; Tan, L.; Kieliszewksi, M.J. Origin of the Land Plants: Is *Coleochaete* their closest living relative? The writing is on the wall. In Proceedings of the 11th Cell Wall Meeting, Copenhagen, Denmark, 12–17 August 2007; p. 130.
122. Delwiche, C.F.; Cooper, E.D. The Evolutionary Origin of a Terrestrial Flora. *Curr. Biol.* **2015**, *2015 25*, R899–R910. [CrossRef]
123. Kaplan, D.R.; Cooke, T.J. The genius of Wilhelm Hofmeister: The origin of causal-analytical research in plant development. *Am. J. Bot.* **1996**, *83*, 1647–1660. [CrossRef]
124. Kenrick, P.; Strullu-Derrien, C. The Origin and Early Evolution of Roots. *Plant Physiol.* **2014**, *166*, 570–580. [CrossRef] [PubMed]
125. Chrispeels, M.J.; Varner, J.E. *Biographical Memoirs*; The National Academy of Sciences: Washington, DC, USA, 1997; pp. 353–371.

International Journal of
Molecular Sciences

MDPI

Review

Feeding the Walls: How Does Nutrient Availability Regulate Cell Wall Composition?

Michael Ogden [1,2], Rainer Hoefgen [2], Ute Roessner [1], Staffan Persson [1] and Ghazanfar Abbas Khan [1,*]

1 School of Biosciences, University of Melbourne, Victoria 3010, Australia; ogdenm@student.unimelb.edu.au (M.O.); u.roessner@unimelb.edu.au (U.R.); staffan.persson@unimelb.edu.au (S.P.)
2 Max Planck Institute of Molecular Plant Physiology, Potsdam-Golm 14476, Germany; hoefgen@mpimp-golm.mpg.de
* Correspondence: ghazanfar.khan@unimelb.edu.au

Received: 15 July 2018; Accepted: 29 August 2018; Published: 10 September 2018

Abstract: Nutrients are critical for plants to grow and develop, and nutrient depletion severely affects crop yield. In order to optimize nutrient acquisition, plants adapt their growth and root architecture. Changes in growth are determined by modifications in the cell walls surrounding every plant cell. The plant cell wall, which is largely composed of complex polysaccharides, is essential for plants to attain their shape and to protect cells against the environment. Within the cell wall, cellulose strands form microfibrils that act as a framework for other wall components, including hemicelluloses, pectins, proteins, and, in some cases, callose, lignin, and suberin. Cell wall composition varies, depending on cell and tissue type. It is governed by synthesis, deposition and remodeling of wall components, and determines the physical and structural properties of the cell wall. How nutrient status affects cell wall synthesis and organization, and thus plant growth and morphology, remains poorly understood. In this review, we aim to summarize and synthesize research on the adaptation of root cell walls in response to nutrient availability and the potential role of cell walls in nutrient sensing.

Keywords: cell wall; nutrients; root system architecture

1. Introduction

Plants are the primary producers on earth, accounting for 80% of all living biomass, and provide us with food, feed, and shelter [1]. Nutrient availability controls how plants grow and thus their ability to produce biomass [2]. The levels of macro- and micro-nutrients in the soil, and the ability of plants to access them, therefore have major implications on agriculture and ecology [3]. To maximize acquisition of nutrients, plants adjust their growth and metabolic processes [4–6]. By deliberately limiting their chlorophyll content and rate of photosynthetic carbon fixation to coordinate their growth with resource availability, plants achieve optimal morphology to survive [3]. Cell division and cell elongation are the major factors that drive growth and morphology in plants. Unlike animal cells, plant cells are surrounded by a protective and supportive structure, referred to as the cell wall, which contributes the bulk of a plant's biomass. Dynamic remodeling of this structure is crucial for plant cells to divide and elongate [7,8]. Because the cell wall dictates cell and tissue morphology, and as nutrient availability drives changes in plant growth, it follows that changes to the cell wall structure should, indirectly or directly, be controlled by the nutrients available to the plant. There is mounting evidence from gene expression data that cell walls are actively regulated in response to nutrient availability [9–13]. However, while it is clear that reprogramming of cell wall genes is essential for plant adaptations to nutrient status, knowledge relating to molecular mechanisms controlling these changes is just starting to emerge. In the agricultural context, plant biomass and nutrient availability are the two most important variables. Research in this field is critically important and has direct links to societal and economic benefits. In this review, we summarize the current understanding of how cell walls are dynamically regulated in response to nutrient availability and how cell wall defects cause changes to nutrient sensitivity. In order to understand the relationship between cell wall and nutrients, we first introduce the basic concepts of cell wall synthesis and structure, along with nutrient transport. Unless otherwise noted, examples presented are from the model plant *Arabidopsis thaliana* (Arabidopsis).

1.1. Cell Walls

Plant cell walls are highly dynamic structures. They are composed primarily of polysaccharides, including cellulose, pectins, hemicelluloses, and callose. In this review, we focus only on the areas of cell wall synthesis and regulation that are directly relevant to nutrient response. There are excellent reviews that cover key topics of cell wall synthesis and regulation [7,14–19].

1.1.1. Cellulose

Cellulose is composed of β-(1→4)-D-glucan chains, which are crystallized into cellulose microfibrils through inter- and intra- molecular hydrogen bonds and Van der Waals forces. These microfibrils are the primary loadbearing polymers of cell walls and act as a framework for tethering and deposition of other wall components [19].

Cellulose is synthesized by CELLULOSE SYNTHASE A (CESA) catalytic subunits, which are organized at the plasma membrane in large multiprotein complexes called cellulose synthase complexes (CSCs). The CSC constitutes a heteromeric arrangement of 18 to 24 CESAs, with CESA1, CESA3, and a CESA6-like protein (CESA2, 5, 6 or 9) being required for primary cell wall synthesis [14,20]. Certain specialized cells, such as xylem tracheary elements, also synthesize a secondary cell wall that is deposited between the primary cell wall and the plasma membrane. During secondary cell wall synthesis, the CSC is comprised of CESA4, CESA7, and CESA8 [21,22]. Most secondary cell walls contain a significantly increased amount of lignins, which are hydrophobic aromatic polymers typically derived from phenylalanine [23].

Apart from their plasma membrane localization, CSCs are also localized at the Golgi apparatus, trans-Golgi network, small CESA compartments, or microtubule-associated cellulose synthase compartments [24]. These latter compartments may be involved in delivery or internalization of the CSCs [14]. During synthesis, newly formed cellulose microfibrils become entangled in cell walls through cross-linking with cell wall polymers. Further synthesis of cellulose pushes the CSCs forward along the plasma membrane [25,26]. CSC speed is therefore often used as a proxy for CESA catalytic activity. The direction of CSC movement, as well as its targeted delivery to the plasma membrane, is guided by cortical microtubules [27]. Several proteins are involved in guiding the CESAs along microtubules; their functions are necessary to maintain cellulose synthesis. These proteins include CELLULOSE SYNTHASE-MICROTUBULE UNCOUPLING (CMU), COMPANIONS OF CELLULOSE SYNTHASE (CCs), and CELLULOSE SYNTHASE INTERACTING PROTEIN 1(CSI1) [28–30]. Of particular note, CCs bind to CSCs and microtubules, and regulate cellulose synthesis under salt stress conditions by re-establishing the microtubule array following salt stress-mediated microtubule depolymerization [29]. KORRIGAN 1 (KOR1), an endo-1,4-β-D-glucanase, also functions in cellulose synthesis by interacting with the CSC at the plasma membrane and during intracellular trafficking [31]. Although the precise function of KOR1 is unknown, *kor1* mutants display reduced cellulose synthesis, and KOR1 is thought to play a role in relieving tensional stress generated during microfibril synthesis, or by releasing microfibrils from the CSC during cessation of cellulose synthesis [32]. COBRA and COBRA-like proteins encode glycosylphosphatidylinositol anchored proteins, and are involved in cellulose synthesis and modify cellulose crystallinity [33]. Finally, the chitinase-like protein homologs, CHITINASE-LIKE PROTEIN 1 (CTL1) and CTL2, bind cellulose and impair CSC activity at the plasma membrane [34–36]. Although the exact function of CTL1 and CTL2 is unclear, CTL1 colocalizes with CESAs during secretion to the apoplast [35,36].

1.1.2. Pectins

Pectins are a diverse family of complex acidic polysaccharides that act as a hydrophilic gel in which other cell wall components are embedded. Pectin composition can vary widely in chain length and branching complexity; however, all pectins contain 1,4-linked α-D-galacturonic acid residues [19]. Pectins are synthesized in the Golgi, and require a minimum of 67 transferases, including glycosyltransferases, acetyltransferases, and methyltransferases, many of which remain unknown or uncharacterized. A large number of transferases may be required due to the many diverse linkages present in pectins [16]. Following synthesis, pectins are packaged into vesicles and trafficked to the plasma membrane for secretion to the cell wall [37]. In the cell wall, cellulose and pectins are closely linked via hydrogen bonds [38]. Pectin deposition and pectin modification play a central role in mediating cell growth [39]. Pectins are typically secreted to the cell wall in a highly methyl-esterified form, and cell wall-localized pectin methylesterases can act upon pectins to cleave methyl ester bonds to remove the methyl groups [40]. This results in the production of free carboxylic groups, which drastically alters the physical properties of pectins, where low levels of methyl-esterified pectins are typically associated with decreased cell wall extensibility and growth inhibition [41].

1.1.3. Hemicelluloses

Similar to pectins, hemicelluloses are synthesized within the Golgi, packaged into vesicles, and secreted to the apoplast [42]. Hemicelluloses are characterized as molecules containing a backbone of β-(1→4)-linked xylose, glucose, or mannose. Mixed linked glucan (MLG) is a special hemicellulose containing a backbone of unbranched (1,3)- and (1,4)-linked b-glucosyl residues and is typically found in grasses, such as cereals [43], which contain distinct differences in cell wall composition compared to dicotyledons, like Arabidopsis [44]. Within the cell wall, hemicelluloses function by crosslinking cellulose microfibrils via hydrogen bonds [42]. Cosgrove (2014) [45] proposed a model, where xyloglucans, a hemicellulose, act as an adhesive layer between cellulose microfibrils to bundle them together at biomechanical hotspots, thus performing an important role in cell wall structure and integrity by maintaining a strong network of interconnected cellulose microfibrils.

1.1.4. Callose

Although plant cells are surrounded by cell walls, they are also symplastically connected through plasmodesmata (PD) (analogous in function to gap junction in animal cells). These are symplastic channels that transverse the cell walls and connect the cytoplasm of adjacent cells [46]. For cell-to-cell transport of signaling molecules to occur through PD, cell wall composition is important, as the major regulation of PD transport is controlled by the deposition or removal of callose (β-1,3-glucan) in the cell walls surrounding PD. Callose deposition causes restriction of the symplastic channels, isolating the cells from each other and interrupting symplastic signaling [46]. Callose is also deposited in the cell plate in dividing cells, pollen tubes, roots, and can be induced by various biotic and abiotic stresses, including wounding [46–48].

1.1.5. Suberin

Suberin is a hydrophobic lipid phenolic polyester comprised of aliphatic, glycerol, and phenolic monomers, with α-ω-dicarboxylic acids and ω-hydroxy acids being the main monomers [49]. Suberin precursors are transported from the endoplasmic reticulum to the plasma membrane for secretion to the apoplast, where suberin polymerization occurs; however, the pathways that mediate these processes remain largely unknown [49]. Suberin localizes to the cell walls of specialized cells, including seed coats and root endodermal cells. Suberin deposition can occur in response to various biotic and abiotic stresses. Most importantly, suberin is deposited to the inner surface of the primary cell wall of root endodermal cells to act as a barrier for diffusion of water and nutrients to the stele [49].

1.1.6. Lignin

Lignin is a complex phenolic polymer typically deposited to secondary cell walls and commonly associated with woody tissues [50]. Lignin is mainly comprised of three monolignols (hydroxycinnamyl alcohols) synthesized from phenylalanine, including ρ-coumaryl, sinapyl, and coniferyl alcohols [23]. Within the cytoplasm, a complex array of reactions is required for monolignol biosynthesis, including deamination, hydroxylation, methylation, and reduction [50]. Monolignols are then transported to the plasma membrane and secreted to the cell wall, where polymerization and crosslinking occurs. Lignification induces unique physical characteristics to cells and tissues by enhancing rigidity through increased cross-linking of cell wall polysaccharides [51]. Lignin deposition is developmentally regulated within specialized cells and tissues, such as the endodermis and xylem [52]. Lignin deposition can also be induced in response to various biotic and abiotic stresses [53]. In the root endodermis, lignin is deposited in maturing cells, specifically to the radial and transverse domains of the cell wall, in a ring-shaped conformation known as the Casparian strip [54]. This acts as a barrier to the apoplastic diffusion of water and nutrients and is important in maintaining nutrient homeostasis. Characteristic hydrophobic properties of lignin are also essential for efficiently transporting water and minerals in water-conducting xylem cells [55].

1.1.7. Structural Proteins and Enzymes

A wide variety of structural proteins and enzymes is required to remodel structures or interactions of cell wall components, leading to wall extensibility and loosening. These proteins include xyloglucan hydrolases, β-1,4-glucanases, peroxidases, extensins, and expansins [56]. Their expression is regulated by a variety of abiotic stresses, including drought and salt stress. Several studies found that these proteins regulate cell wall stiffness through the cross-linking of different polymers. Consequently, ectopic expression of several cell wall modifying proteins leads to abiotic stress tolerance or hyper-susceptibility [57,58].

2. Importance of Cell Walls in Nutrient Transport

In addition to carbon dioxide and water, both of which are required for photosynthetic production of carbohydrates, plant growth depends on nutrient acquisition from the soil. Balanced proportions of a range of macronutrients (nitrogen (N), phosphorus (P), potassium (K), and sulfur (S)), and micronutrients (iron (Fe), copper, molybdenum, selenium, zinc, and others) are essential for optimal growth and crop yield. Plant growth under nutrient-deficient conditions typically results in nutrient-specific phenotypes, including an overall reorganization of root system architecture (RSA), and oftentimes, growth inhibition [59].

Nutrients and water are absorbed at the root-soil interface and are translocated from the soil to the stele via three distinct pathways: An apoplastic pathway, a symplastic pathway, and a coupled transcellular pathway [60]. Water and nutrients freely diffuse through the apoplastic pathway in an unregulated manner. The symplastic pathway is characterized by nutrient movement between cells via PD, with the coupled-transcellular pathway describing nutrient movement in and out of the cells, mediated by activity of influx and efflux carriers (Figure 1) [60]. Cell walls play a key role in regulating nutrient transport between soil and the stele. Casparian strips are highly lignified, hydrophobic cell walls that encompass the radial and transverse domains of root endodermal cells in the zone of maturation. Casparian strips fill the cell wall between each endodermal cell, creating an apoplastic barrier to water and nutrient diffusion. Within the maturation zone, nutrients must therefore move through the symplastic and coupled transcellular pathways to bypass Casparian strips [61]. Following Casparian strip formation, suberin is deposited to the inner surface of the primary cell wall, entirely enveloping each endodermal cell [49,60]. The so-called suberin lamellae initiates along patches of endodermal cells apical to the Casparian strip initiation site, eventually displaying homogenous cell wall distribution across the endodermis at increasing distances from the root tip. This barrier is crucial in limiting the diffusion of water and nutrients across the endodermis, thus protecting the plant from nutrient influx at toxic levels, or nutrient leaching from the stele [62]. Suberin production is tightly regulated by ethylene and abscisic acid (ABA) signaling, and its synthesis and deposition is highly plastic in response to nutrient availability. Moderate deficiency in Mn, Fe, and Zn leads to a delay in suberization, while K- and S-deficiency results in increased suberization [62]. This suggests a nutrient management strategy by the plant that modulates bidirectional movement of nutrients across the endodermis. In agreement with this, targeted degradation of suberin leads to suppression of the mutant phenotype in the Fe uptake mutant *irt*. Moreover, suberin degradation in the endodermis enhances the S deficiency phenotype of S uptake mutants for *SULFATE TRANSPORTER 1* (*sultr1;1sutr1;2*) and leads to K deficiency in plants [62], demonstrating the high plasticity of suberin regulation as an adaptive response to various nutritional cues. The suberization response to nutrient availability is a new field of research that will benefit our understanding of the regulation of nutrient homeostasis within plants.

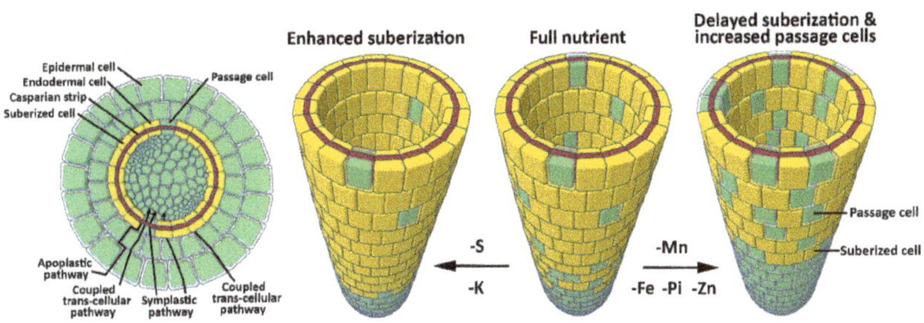

Figure 1. Plasticity in the permeability of endodermal cell walls in response to nutrient imbalance. (**Left** Panel) Cellular schematic cross section of a fully differentiated Arabidopsis root. Suberized endodermal cells (yellow) and the lignified endodermal Casparian strip (red) block nutrient transport via the apoplastic and coupled-trans-cellular pathways. Passage cells are unsuberized endodermal cells that allow symplastic and coupled trans-cellular transport. (**Right** Panel) Magnified view of the endodermal cell layer under varying nutrient availability. K and S deficiency causes an increase in suberization, while deficiency in Fe, Zn, and P causes a delay in suberization and increased number of passage cells. Mn deficiency also delays suberization, but its impact on passage cell occurrence is unknown. This plasticity in root cell wall permeability is an excellent demonstration of how plants modulate nutrient transport in response to varying nutrient concentrations in soil. Fe: iron; K: potassium; Mn: manganese; Pi: phosphate; S: sulfur; Zn: zinc. Schematics were modeled using the online software Tinkercad (www.tinkercad.com).

Within the root zone of continuous suberization, randomly distributed cells, known as passage cells, lack suberin deposition (Figure 1) [61,63,64]. These cells are closely associated with the xylem pole. They lack secondary cell wall deposition and their development is mediated by repression of cytokinin signaling in the root apical meristem (RAM) (Figure 1) [64]. The exact role of passage cells is unclear; however, it is thought that they facilitate water and nutrient transport or cell communication in older parts of the root [61,63,64]. In line with this, expression of the phosphate (Pi) efflux protein family member *PHOSPHATE1* (*PHO1*), which is typically present in pericycle cells, expands into the endodermal passage cells [64]. Furthermore, Pi, Fe, and Zn deficiency simultaneously reduces suberization and increases passage cell occurrence [64]. Dynamic regulation of passage cell occurrence in response to nutritional status and expression of nutrient transporters in these cells supports the hypothesis that passage cells are involved in nutrient transport. Overall, these results suggest that plants tightly regulate nutrient transport across the root by modulating nutrient flux carriers and cell wall permeability (Figure 1).

3. Influence of Nutrient Availability on Cell Wall Composition

3.1. Nitrogen

All plants utilize N in the form of ammonium (NH_4^+) and nitrate (NO_3^-). Various groups have investigated the transcriptomic response of Arabidopsis plants to N availability. These studies have mainly relied upon NO_3^- induction or deprivation, followed by RNA extraction and Affymetrix microarray analyses [9]. Few studies specifically highlight the expression of cell wall related genes; however, it was noted that "glucose catabolic process" is one of the most consistent biological functions associated with NO_3^- response. Genes involved in trichoblast differentiation, which contain many cell wall related genes, were particularly enriched in NO_3^- responsive genes [9]. Additionally, a co-expression analysis of NO_3^- transporters, using the ATTED-II database, revealed that they are co-expressed with many cell wall related genes [65]. However, caution should be exercised in interpreting this data, as a recent meta-analysis of NO_3^--response transcriptomic data revealed a striking inconsistency in the reproducibility between different studies, noting that the common number of differentially expressed genes between any two experiments is only 6.7% [9]. This variation in expression patterns is likely due to redundancy of large gene families, developmental differences, and changes in experimental setups between labs, as gene expression is highly sensitive to environmental conditions and plant age. This is highlighted in plants that were not subjected to changes in NO_3^- availability, but showed gene induction resulting from other stress related to simply moving the plants from one solution to another [66]. Transcriptomic assays would greatly benefit from a standardized experimental setup, with detail paid to the nutrient composition of media, lighting levels, plant handling and transfer between media, and the time at which tissue is harvested following treatment. By adopting a standardized experimental setup, future transcriptomic results may lead to increased reproducibility in gene expression trends.

Under N-limiting conditions, plants adapt their RSA to maximize N acquisition by exploiting a larger soil surface area. This is mostly achieved by the elongation of lateral roots (LRs), whereas high N concentrations result in an inhibition of LR growth, with little effect on primary root (PR) growth [59,67]. In crop plants, N status affects stem mechanical strength and disease resistance. These traits are regulated by cell wall organization and strength, suggesting that cell walls are modulated in response to N status. Expression of genes involved in biosynthesis of both lignin and cellulose was significantly reduced in response to high N in rice [68,69]. In support of this finding, cell wall analysis of roots showed that high N leads to a significant reduction of cellulose and lignin. This reduction in cell wall components is accompanied by reduced stem mechanical strength, increased lodging, and reduced disease resistance [68–70]. Similar to roots, expression of cellulose and lignin synthesis genes is reduced in filling seeds in response to high N. This leads to reduced accumulation of both cellulose and lignin in seed endosperms [71]. In line with this observation, N deficiency leads to an increase in cellulose content in the roots of rice plants [72].

In Arabidopsis, intact cell walls are essential for the adaptation of RSA in response to N availability, as genetic mutants with altered cell wall organization show impaired RSA response to N availability. Cellulose-deficient mutant roots are hypersensitive to high (60 mM) KNO_3 [73]. When grown under high N conditions, *cesa3*, *cesa6*, and *ctl1* mutants show a drastic reduction in PR length, an increase in root hair and LR density, and swollen roots. Further investigation of *ctl1* revealed the same hypersensitive root response to high NaCl, KCl and $CaCl_2$, whereas no response was observed under high PO_4, SO_4, mannitol, or sorbitol, suggesting a hypersensitive response to chloride ions [35]. Hypersensitivity to chloride in conjunction with NO_3^- is not unexpected, as multiple NO_3^- transporters are also known to function in chloride transport [74]. Furthermore, under high KNO_3, *ctl1* and *cesa3* mutants displayed ectopic lignin deposition in the endodermis, whereas *cesa6* and *kor1* exhibited only minor ectopic lignin deposition compared with wild-type controls [73]. If cellulose reduction occurs in Arabidopsis under high N availability (as also observed in rice), this could explain the hypersensitive root phenotype of cellulose deficient mutants when subjected to high N, resulting from an additive effect due to high N-mediated reduction in cellulose in an already cellulose deficient mutant. These results demonstrate the complex response to N availability, inferring a dynamic regulation of cell wall synthesis and remodeling. However, the molecular mechanisms by which N status affects cell wall biosynthesis and organization are unknown.

3.2. Phosphorus

In a recent study of Pi-responsive genes, a detailed transcriptomic analysis using wild-type and mutants for *LOW PHOSPHATE RESPONSE 1* (*LPR1*) and *LPR2*, and *PHOSPHATE DEFICIENCY RESPONSE 2*, which are required for Pi response, highlighted a robust Pi-dependent regulation of cell wall related genes [75]. These genes were sorted into four main groups, which included pectin modification, cell wall relaxation, hemicellulose/cellulose modification, and carbohydrate hydrolytic enzymes; however, it is unknown how these changes in expression impact growth. Additional studies in wild-type plants noted similar changes in expression of cell wall related genes in response to Pi starvation [10,13,76,77]. A transcriptomic analysis for seedlings subjected to Pi-deficiency revealed differential regulation of the largely uncharacterized *CELLULOSE SYNTHASE-LIKE B5* (*CSLB5*) [10]. Further investigation revealed that *cslb5* mutants produce shorter root hairs compared to wild-type under Pi-deficiency, suggesting a role of CSLB5 in Pi response [10]. The exact function of the CSLB family is unknown; however, it is thought that members are involved in the synthesis of β-1,4-linked cell wall polysaccharides [78]. Interestingly, *CSLB5* is expressed in roots, which goes in line with its likely function in regulation of RSA in response to Pi starvation [78].

Under low Pi conditions in Arabidopsis, PR growth is arrested, while growth of secondary roots and root hairs is promoted [59,79]. PR growth arrest is mainly a result of rapid reduction in root cell elongation followed by an arrest of cell division in the stem cell niche. These changes in the RSA are dependent on the external supply of Fe. This suggests an indirect surveillance of Pi status via its interaction with transition metals [80]. During Pi deficiency, LPR1, which is a cell wall localized ferroxidase, modulates Fe deposition in the cell walls of elongating cells and the stem cell niche [48]. Interestingly, in these cells, Fe deposition is promoted by malate secretion to the cell walls. Under low Pi conditions, the transcription factor SENSITIVE TO PROTON TOXICITY 1 (STOP1) targets the malate efflux channel *ALUMINUM-ACTIVATED MALATE TRANSPORTER 1* (*ALMT1*), triggering malate secretion to the cell walls of the root transition zone. Malate deposition promotes Fe^{3+} accumulation via LPR1 activity in the elongating cells [81,82], coinciding with the generation of reactive oxygen species (ROS) and callose deposition [48]. Cell type-specific callose deposition disrupts symplastic communication through plasmodesmata as shown by impaired movement of the transcription factor SHORT ROOT between the stele and the endodermis [48]. Symplastic communication is essential for meristem maintenance, and its disruption is likely the main factor leading to root meristem exhaustion in response to low Pi (Figure 2) [83]; however, the mechanism behind the rapid inhibition of cell elongation, which occurs well before meristem size reduction, is unclear [48,81]. LPR1 mediated Fe accumulation leads to the generation of ROS, which coincides with the initial sites of Fe deposition in the apoplast [48]. Ferroxidase activity of LPR1 leads to Fe^{3+} redox cycling in the cell wall, which creates an overall oxidative environment and ROS generation [48]. ROS may serve as a substrate for the cell wall localized class III peroxidases to catalyze cross-linking between cell wall components (Figure 2) [18], leading to cell wall stiffness and reduced extensibility. Atomic force microscopy revealed an increase in cell wall stiffness in the elongation zone within 30 min following seedling transfer to Pi-deficient media, with the change in stiffness being dependent on the activity of class III peroxidases [81]. Pharmacological inhibition of peroxidases restored root growth in response to low Pi, confirming that increased cell wall rigidity is linked to inhibited root growth under Pi-deficient conditions [81]. Changes in cell wall stiffness need to happen in conjunction with changes in cell turgor in order to coordinate cell expansion [84,85]. This may be achieved through conformational changes in lipids [86], which typically result in changes in cellular trafficking and vacuolar organization [87,88], which are crucial factors for the regulation of intracellular turgor [85].

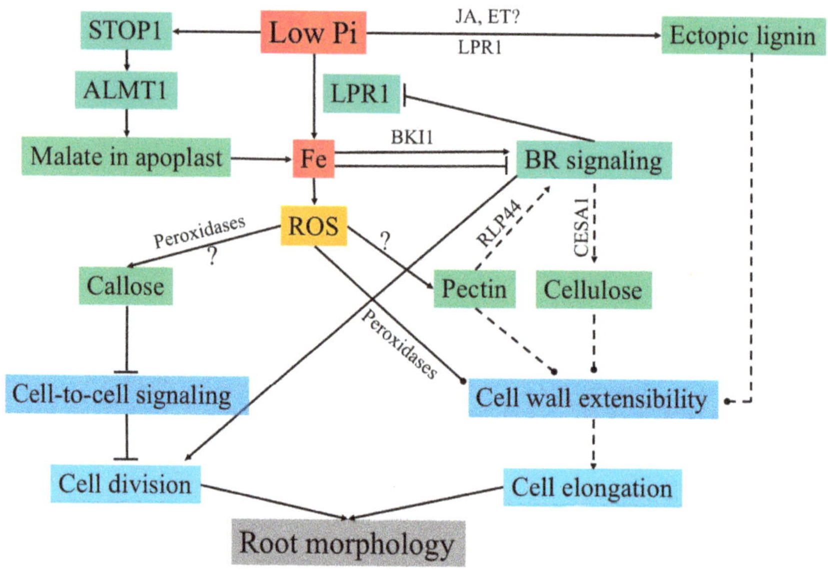

Figure 2. Model of cell wall regulation in response to Pi starvation. Low Pi causes LPR1-dependent Fe accumulation in the apoplast, which is required for the inhibition of primary root growth. Low Pi also leads to activation of the transcription factor STOP1 through an unknown mechanism. STOP1 induces the malate transporter ALMT1 by directly binding to its promoter. Subsequently, ALMT1 secretes malate in the apoplast, which is required for Fe accumulation in the transition zone. Apoplast-localized LPR1 expresses ferroxidase activity and is thought to generate ROS via Fe redox cycling. Cell-type-specific ROS generation is required for callose deposition and impairs symplastic connectivity, leading to inhibition of cell division in the RAM. In a feedback regulation, Fe modulates LPR1 expression via BR signaling in a concentration-dependent manner. Low Fe concentration activates, and high Fe concentration inhibits, BR signaling by translational regulation of BKI1, which is a key inhibitor of BR signaling. BR signaling is known to regulate cellulose synthesis by the transcriptional and post-translational regulation of CESAs. However, it is unknown whether cellulose synthesis is affected in response to low Pi. Additional cell wall modifications include ectopic lignin deposition, possibly via changes in jasmonate and ethylene signaling, peroxidase-dependent cell wall stiffening, and cell-type-specific pectin deposition. Changes in pectin are known to trigger BR signaling via RLP44 to maintain cell wall integrity. Dual control of BR signaling through pectin and Fe may be required to achieve essential changes in cell walls required for growth modulation in response to low Pi, while maintaining cell wall integrity. Cell wall modifications through changes in cellulose, pectin, and lignin, impact cell wall extensibility. These modifications could be responsible for the rapid inhibition of cell expansion in response to low Pi. Solid lines denote known interactions and dashed lines denote inferred interactions. Arrows are positive interactions and terminated lines indicate negative interactions. Round heads suggest interactions responsible for a change in cell wall architecture. STOP1: SENSITIVE TO PROTON TOXICITY 1; ALMT1: ALUMINUM-ACTIVATED MALATE TRANSPORTER 1; ROS: reactive oxygen species; LPR1: LOW PHOSPHATE RESPONSE 1; BR: brassinosteroid.

Varying Fe concentration leads to activation or inhibition of the brassinosteroid (BR) signaling pathway, which in turn regulates the expression of *LPR1* [89]. This feedback loop may be important in mediating distinct root morphologies as an adaptive response to soil microenvironments with varying concentrations of either Pi or Fe [89]. *BRASSINAZOLE RESISTANT 1* (*BZR1*), a key regulator of BR mediated regulation of gene expression [90], represses the expression of LPR1 by directly binding to its promoter [89]. In line with this, BZR1 constitutive mutants (BZR1-D) phenocopy the *lpr1* root phenotype in response to low Pi [91]. Low Fe activates, and high Fe inhibits, BR signaling via translational modulation of its key repressor brassinosteroid kinase inhibitor 1 (BKI1) through an unknown mechanism [89]. Fe concentration-dependent modulation of BR signaling is required for the root growth response in low Pi conditions (Figure 2). Interestingly, BR signaling is known to play a role both in cell proliferation and cell elongation via changes in cell wall composition. Mutants affected in the BR signaling pathway show significant transcriptional changes to various cell wall remodeling enzymes and structural proteins [92]. More importantly, BRASSINOSTEROID INSENSITIVE 2 (BIN2), a key inhibitor of the BR pathway, directly phosphorylates CESA1, leading to a decrease in CESA motility and crystalline cellulose synthesis [93]. BES1 and BZR1, which are the key transcription factors in the BR pathway, positively impact cellulose synthesis by directly binding to *CESA* promoters [94]. Along with the direct regulation of cellulose synthesis, BR is known to influence microtubule rearrangement [95], which is required to guide CSCs during cellulose synthesis. CSI1, CMU, and CCs are the proteins required for the association of CSC with microtubules [28–30]. Investigation of these proteins in context of BR and phosphate starvation response may prove to be an important tool in investigating changes in cellulose synthesis in low Pi stress. Overall, the BR pathway positively impacts cellulose synthesis, and it could be a strong candidate for the regulation of cellulose in response to low Pi.

The cellulose content of rice changes in responses to low Pi [96]; however, it is unknown whether Pi deficiency affects cellulose content in Arabidopsis. Furthermore, rice mutants affected in *CELLULOSE SYNTHASE-LIKE F6* (*CSLF6*) show a constitutively active Pi starvation response and increased Pi transport, even when seedlings are grown on high Pi [97]. CSLF6 is involved in the synthesis of MLG, which typically affects cell wall strength and cell expansion [43]. Interestingly, MLG binds to cellulose in vitro, forming a thick hydrogel at the surface of adsorption [98]. It is suggested that MLG may act as a transient binding surface where other cell wall polysaccharides can attach during periods of cell expansion [43]. Activation of the Pi starvation response in *cslf6* mutants implies an interdependent regulation of cell wall composition and Pi signaling.

Pi deficiency also leads to lignin accumulation in root cell walls; however, lignin deposition was not observed in the *lpr1lpr2* double mutant, suggesting that lignin deposition may be regulated by the same pathway as callose deposition under Pi-deficiency [99]. Lignin deposition is typically induced due to defects in cell wall integrity through the action of jasmonates and ethylene [100,101]. Both jasmonates and ethylene are induced in response to low Pi and possibly mediate the observed ectopic lignin deposition [102,103]. Lignin associates with cellulose and is involved in cell wall strengthening, implicating it in reduced extensibility of cell wall in low Pi conditions [104]. Under Pi-deficiency, the concentration of unesterified pectins significantly increases in cell walls in the elongation zone and the RAM, with a notable increase at the quiescent center and the border between the meristem and the stem cell niche [75]. The sites of increased pectin deposition colocalize to the sites of Fe^{3+} accumulation and callose deposition, indicating that pectins may play a role in cell wall organization in the root elongation inhibition response. Free carboxyl groups in unesterified pectins have a high affinity to bind to Fe^{3+}, Al^{3+}, and Ca^{2+}, and pectins are known to dimerize via Ca^{2+}-pectate crosslinked complexes called egg-boxes [105]. Increase in egg-boxes causes cell wall stiffening and reduced growth [106]. It is tempting to speculate that the so-called egg-box structures can be formed with the Fe^{3+}, leading to an increase in cell wall stiffness. In addition, Fe binding with pectin may promote the release of Pi from Fe-Pi complexes during low Pi stress [107], which would be a fascinating adaptation to maximize Pi absorption. Interestingly, changes in pectin esterification are also known to trigger BR signaling via RECEPTOR-LIKE PROTEIN 44 (RLP44), which is required for the feedback regulation of cell wall modifications and coordination developmental output with cell wall integrity [108]. Dual control of BR signaling via pectin composition and Fe is perhaps required to modulate cell wall composition, driving RSA organization, while maintaining cell wall integrity.

Data summarized here demonstrate that cell wall modulation is essential for root response to low Pi, and that significant progress has been made in dissecting the complex genetic networks involved in this process; however, challenges remain in linking Fe accumulation and ROS generation with the modulation of cell wall chemistry. For example, what molecular mechanisms and pathways are responsible for recruiting callose synthesizing enzymes to the sites of Fe accumulation?

3.3. Other Nutrients

Apart from N and P, little is known about cell wall regulation by other nutrients. A study by Armengaud et al. (2004) [11] revealed that cell wall related genes, including extensins, xyloglucan glucosyltransferases, arabinogalactan proteins, and peroxidases, were enriched in genes that respond to low potassium (K). A meta-analysis of transcriptomic studies revealed that cell wall related genes are enriched in low S responsive genes [12]. As observed in the meta-analysis for genes responding to NO_3^- availability, the reproducibility in consistent gene response was low between the datasets, with only 418 (20%) of the differentially expressed genes shared by a minimum of two experiments. It was found that the gene ontology term "cell wall" had a two-fold enrichment, genes coding apoplast-localized proteins were over-represented, and a predicted glycosyl hydrolase (At3g60140) was significantly up-regulated across all microarray datasets, implicating an S-dependent regulation of cell wall related genes.

The micronutrient boron forms so-called borate bridges by crosslinking the pectin polysaccharide rhamnogalacturonan-II [109]. These crosslinks modify cell wall porosity, mechanical properties, and are critical for plant growth. Consequently, boron deficiency leads to highly de-structured cell walls and severe growth defects [109].

As we understand more of the complex genetic networks in cell wall synthesis, it will be possible to identify more genes that regulate these networks in response to nutrient availability. A combination of in vivo and in vitro methods and systems biology approach [110] revealed a complex transcriptional network regulating secondary cell wall synthesis. Interestingly, this network contains several genes that are transcriptionally regulated in response to N, Fe, and S imbalance, suggesting that secondary cell wall synthesis is regulated by nutrient availability. Investigations revealed that the transcription factor REVOLUTA (REV) has a number of upstream factors that are regulated by Fe deprivation, suggesting that REV plays an important role in regulating secondary cell walls during Fe deficiency. Further experiments demonstrated that REV is a key transcription factor involved in the regulation of lignin biosynthesis in response to Fe deficiency [110]. These results demonstrate that cell wall synthesis is tightly regulated in response to nutritional cues and is underpinned by complex genetic networks.

4. Conclusions and Perspectives

Cell wall synthesis, composition, and remodeling is crucial for the meticulous modulation of RSA in response to nutrient imbalance. For example, Fe-mediated callose deposition in the cell walls of the root meristem and elongation zone is crucial for adapting the RSA in response to P deficiency (although it remains unclear how Fe regulates callose deposition in roots) [46]. Moreover, how other cell wall polymers are modulated in response to P imbalance is unknown. Modulation and reorganization of other cell wall polymers is expected, as a drastic reduction in cell division and cell elongation is observed shortly after exposure to low Pi. Apart from P, imbalances in other nutrients results in RSA reorganization [57], which is likely mediated by regulation of cell wall synthesis and/or architecture. This hypothesis is mainly supported by transcriptomics data due to technical advances, simplicity, and reduced cost of this type of analysis. It will be crucial for future studies to investigate changes in cell wall chemistry and cell wall composition using polymer-specific antibodies and mass spectrometry to understand the role of cell walls in nutrient-specific RSA modulation.

Experiments with rice show that cell wall polymers are reduced in response to excess N, which is mediated through transcriptional regulation of genes involved in cell wall biosynthesis [68,69]. It can be expected that cell walls are regulated in a similar manner in Arabidopsis given the sensitivity of cellulose deficient mutants to high N and Cl. If this is the case, Arabidopsis could prove to be a powerful tool for dissecting the molecular mechanisms of N- and Cl-mediated cell wall regulation.

Investigating the relationship between N availability and cell wall regulation is important in an agricultural context, as high N directly affects important agronomical traits, such as lodging and disease resistance, through cell wall changes.

Researchers typically apply a reductionist approach by modulating the concentration of a single nutrient. However, in the natural environment, plants are typically exposed to changes in various nutrients simultaneously. Emerging evidence highlights the complexity and diversity of growth responses by subjecting plants to multiple nutrient deficiencies simultaneously, where the output is not an additive effect of each individual nutrient deficiency phenotype, but rather a unique phenotype altogether [111]. As soils contain a heterogeneous distribution of nutrients, and in order to understand their complex response pathways, it will be important to study multiple concurrent changes in nutrient concentrations.

Taken as a whole, the recent works referenced here demonstrate that cell wall synthesis, composition, and remodeling are crucial for both the transport and signaling of important nutrients. A robust body of knowledge now exists about how plants control water and nutrient transport across the root by manipulating cell wall permeability. It is clear that Arabidopsis root cell walls are highly plastic in response to the nutritional status of the soil [58]. It can be expected that this plasticity in root permeability helps plants to locally adjust nutrient transport in soil microenvironments with varying nutrient concentrations. Knowledge gained in this field has significant potential for engineering crops with increased nutrient use efficiency by tailoring RSA and endodermis permeability in accordance with soil nutritional status, while also improving growth and disease resistance.

Author Contributions: M.O., S.P., and G.A.K. conceived the topic and wrote the manuscript. R.H. and U.R. proofread and provided valuable comments to improve the manuscript.

Funding: M.O. is supported by the Melbourne-Potsdam PhD Programme. G.A.K. is funded by a Schweizerischer Nationalfonds zur Förderung der Wissenschaftlichen Forschung early postdoc mobility fellowship under the grant number P2LAP3_168408. S.P. is funded by an ARC future fellowship grant (FT160100218) and acknowledges a UoM IRRTF (RNC) grant.

Conflicts of Interest: The authors declare no conflict of interest.

References

1. Bar-On, Y.M.; Phillips, R.; Milo, R. The biomass distribution on Earth. *Proc. Natl. Acad. Sci. USA* **2018**, *115*, 6506–6511. [CrossRef] [PubMed]
2. Hermans, C.; Hammond, J.P.; White, P.J.; Verbruggen, N. How do plants respond to nutrient shortage by biomass allocation? *Trends Plant Sci.* **2006**, *11*, 610–617. [CrossRef] [PubMed]
3. Prinzenberg, A.E.; Barbier, H.; Salt, D.E.; Stich, B.; Reymond, M. Relationships between Growth, Growth Response to Nutrient Supply, and Ion Content Using a Recombinant Inbred Line Population in *Arabidopsis thaliana*. *Plant Physiol.* **2010**, *154*, 1361–1371. [CrossRef] [PubMed]
4. Shahzad, Z.; Amtmann, A. Food for thought: How nutrients regulate root system architecture. *Curr. Opin. Plant Biol.* **2017**, *39*, 80–87. [CrossRef] [PubMed]
5. Jung, J.-Y.; Ried, M.K.; Hothorn, M.; Poirier, Y. Control of plant phosphate homeostasis by inositol pyrophosphates and the SPX domain. *Curr. Opin. Biotechnol.* **2018**, *49*, 156–162. [CrossRef] [PubMed]
6. Bouguyon, E.; Gojon, A.; Nacry, P. Nitrate sensing and signaling in plants. *Semin. Cell Dev. Biol.* **2012**, *23*, 648–654. [CrossRef] [PubMed]
7. Somssich, M.; Khan, G.A.; Persson, S. Cell wall heterogeneity in root development of *Arabidopsis*. *Front. Plant Sci.* **2016**, *7*, 1242. [CrossRef] [PubMed]
8. Cosgrove, D.J. Growth of the plant cell wall. *Nat. Rev. Mol. Cell Biol.* **2005**, *6*, 850–861. [CrossRef] [PubMed]
9. Canales, J.; Moyano, T.C.; Villarroel, E.; Gutiérrez, R.A. Systems analysis of transcriptome data provides new hypotheses about *Arabidopsis* root response to nitrate treatments. *Front. Plant Sci.* **2014**, *5*, 22. [CrossRef] [PubMed]
10. Lin, W.-D.; Liao, Y.-Y.; Yang, T.J.W.; Pan, C.-Y.; Buckhout, T.J.; Schmidt, W. Coexpression-based clustering of *Arabidopsis* root genes predicts functional modules in early phosphate deficiency signaling. *Plant Physiology* **2011**, *155*, 1383–1402. [CrossRef] [PubMed]

11. Armengaud, P. The Potassium-Dependent transcriptome of *Arabidopsis* reveals a prominent role of jasmonic acid in nutrient signaling. *Plant Physiol.* **2004**, *136*, 2556–2576. [CrossRef] [PubMed]
12. Henríquez-Valencia, C.; Arenas-M, A.; Medina, J.; Canales, J. Integrative transcriptomic analysis uncovers novel gene modules that underlie the sulfate response in *Arabidopsis thaliana*. *Front. Plant Sci.* **2018**, *9*, 177. [CrossRef] [PubMed]
13. Wege, S.; Khan, G.A.; Jung, J.-Y.; Vogiatzaki, E.; Pradervand, S.; Aller, I.; Meyer, A.J.; Poirier, Y. The EXS domain of PHO1 participates in the response of shoots to phosphate deficiency via a root-to-shoot signal. *Plant Physiol.* **2016**, *170*, 385–400. [CrossRef] [PubMed]
14. McFarlane, H.E.; Döring, A.; Persson, S. The cell biology of cellulose synthesis. *Annu. Rev. Plant Biol.* **2014**, *65*, 69–94. [CrossRef] [PubMed]
15. Albenne, C.; Canut, H.; Jamet, E. Plant cell wall proteomics: The leadership of Arabidopsis thaliana. *Front. Plant Sci.* **2013**, *4*, 111. [CrossRef] [PubMed]
16. Atmodjo, M.A.; Hao, Z.; Mohnen, D. Evolving views of pectin biosynthesis. *Annu. Rev. Plant Biol.* **2013**, *64*, 747–779. [CrossRef] [PubMed]
17. Scheller, H.V.; Ulvskov, P. Hemicelluloses. *Annu. Rev. Plant Biol.* **2010**, *61*, 263–289. [CrossRef] [PubMed]
18. Wolf, S.; Hématy, K.; Höfte, H. Growth control and cell wall signaling in plants. *Annu. Rev. Plant Biol.* **2012**, *63*, 381–407. [CrossRef] [PubMed]
19. Lampugnani, E.R.; Khan, G.A.; Somssich, M.; Persson, S. Building a plant cell wall at a glance. *J. Cell Sci.* **2018**, *131*. [CrossRef] [PubMed]
20. Persson, S.; Paredcz, A.; Carroll, A.; Palsdottir, H.; Doblin, M.; Poindexter, P.; Khitrov, N.; Auer, M.; Somerville, C.R. Genetic evidence for three unique components in primary cell-wall cellulose synthase complexes in *Arabidopsis*. *Proc. Natl. Acad. Sci. USA* **2007**, *104*, 15566–15571. [CrossRef] [PubMed]
21. Taylor, N.G.; Laurie, S.; Turner, S.R. Multiple cellulose synthase catalytic subunits are required for cellulose synthesis in *Arabidopsis*. *Plant Cell* **2000**, *12*, 2529–2539. [CrossRef] [PubMed]
22. Atanassov, I.I.; Pittman, J.K.; Turner, S.R. Elucidating the mechanisms of assembly and subunit interaction of the cellulose synthase complex of *Arabidopsis* secondary cell walls. *J. Biol. Chem.* **2009**, *284*, 3833–3841. [CrossRef] [PubMed]
23. Vanholme, R.; Demedts, B.; Morreel, K.; Ralph, J.; Boerjan, W. Lignin Biosynthesis and Structure. *Plant Physiol.* **2010**, *153*, 895–905. [CrossRef] [PubMed]
24. Gutierrez, R.; Lindeboom, J.J.; Paredez, A.R.; Emons, A.M.C.; Ehrhardt, D.W. *Arabidopsis* cortical microtubules position cellulose synthase delivery to the plasma membrane and interact with cellulose synthase trafficking compartments. *Nat. Cell Biol.* **2009**, *11*, 797–806. [CrossRef] [PubMed]
25. Diotallevi, F.; Mulder, B. The cellulose synthase complex: A polymerization driven supramolecular motor. *Biophys. J.* **2007**, *92*, 2666–2673. [CrossRef] [PubMed]
26. Morgan, J.L.W.; Strumillo, J.; Zimmer, J. Crystallographic snapshot of cellulose synthesis and membrane translocation. *Nature* **2012**, *493*, 181–186. [CrossRef] [PubMed]
27. Paredez, A.R.; Somerville, C.R.; Ehrhardt, D.W. Visualization of cellulose synthase demonstrates functional association with microtubules. *Science* **2006**, *312*, 1491–1495. [CrossRef] [PubMed]
28. Bringmann, M.; Landrein, B.; Schudoma, C.; Hamant, O.; Hauser, M.-T.; Persson, S. Cracking the elusive alignment hypothesis: The microtubule–cellulose synthase nexus unraveled. *Trends Plant Sci.* **2012**, *17*, 666–674. [CrossRef] [PubMed]
29. Endler, A.; Kesten, C.; Schneider, R.; Zhang, Y.; Ivakov, A.; Froehlich, A.; Funke, N.; Persson, S. A mechanism for sustained cellulose synthesis during salt stress. *Cell* **2015**, *162*, 1353–1364. [CrossRef] [PubMed]
30. Liu, Z.; Schneider, R.; Kesten, C.; Zhang, Y.; Somssich, M.; Zhang, Y.; Fernie, A.R.; Persson, S. Cellulose-microtubule uncoupling proteins prevent lateral displacement of microtubules during cellulose synthesis in *Arabidopsis*. *Dev. Cell* **2016**, *38*, 305–315. [CrossRef] [PubMed]
31. Vain, T.; Crowell, E.F.; Timpano, H.; Biot, E.; Desprez, T.; Mansoori, N.; Trindade, L.M.; Pagant, S.; Robert, S.; Höfte, H.; et al. The cellulase KORRIGAN is part of the cellulose synthase complex. *Plant Physiol.* **2014**, *165*, 1521–1532. [CrossRef] [PubMed]
32. Somerville, C. Cellulose synthesis in higher plants. *Annu. Rev. Cell Dev. Biol.* **2006**, *22*, 53–78. [CrossRef] [PubMed]

33. Liu, L.; Shang-Guan, K.; Zhang, B.; Liu, X.; Yan, M.; Zhang, L.; Shi, Y.; Zhang, M.; Qian, Q.; Li, J.; et al. Brittle Culm1, a COBRA-like protein, functions in cellulose assembly through binding cellulose microfibrils. *PLoS Genet.* **2013**, *9*, e1003704. [CrossRef] [PubMed]
34. Mouille, G.; Robin, S.; Lecomte, M.; Pagant, S.; Höfte, H. Classification and identification of *Arabidopsis* cell wall mutants using Fourier-Transform InfraRed (FT-IR) microspectroscopy. *Plant J.* **2003**, *35*, 393–404. [CrossRef] [PubMed]
35. Hermans, C.; Porco, S.; Vandenbussche, F.; Gille, S.; de Pessemier, J.; van Der Straeten, D.; Verbruggen, N.; Bush, D.R. Dissecting the role of CHITINASE-LIKE1 in nitrate-dependent changes in root architecture. *Plant Physiol.* **2011**, *157*, 1313–1326. [CrossRef] [PubMed]
36. Sánchez-Rodríguez, C.; Bauer, S.; Hématy, K.; Saxe, F.; Ibáñez, A.B.; Vodermaier, V.; Konlechner, C.; Sampathkumar, A.; Rüggeberg, M.; Aichinger, E. vChitinase-like1/pom-pom1 and its homolog CTL2 are glucan-interacting proteins important for cellulose biosynthesis in *Arabidopsis*. *Plant Cell* **2012**, *24*, 589–607. [CrossRef] [PubMed]
37. Kim, S.-J.; Brandizzi, F. The plant secretory pathway: An essential factory for building the plant cell wall. *Plant Cell Physiol.* **2014**, *55*, 687–693. [CrossRef] [PubMed]
38. Wang, T.; Park, Y.B.; Cosgrove, D.J.; Hong, M. Cellulose-pectin spatial contacts are inherent to never-dried *Arabidopsis* primary cell walls: evidence from solid-state nuclear magnetic resonance. *Plant Physiol.* **2015**, *168*, 871–884. [CrossRef] [PubMed]
39. Cosgrove, D.J. Diffuse Growth of Plant Cell Walls. *Plant Physiol.* **2018**, *176*, 16–27. [CrossRef] [PubMed]
40. Wolf, S.; Greiner, S. Growth control by cell wall pectins. *Protoplasma* **2012**, *249*, 169–175. [CrossRef] [PubMed]
41. Derbyshire, P.; McCann, M.C.; Roberts, K. Restricted cell elongation in *Arabidopsis* hypocotyls is associated with a reduced average pectin esterification level. *BMC Plant Biol.* **2007**, *7*, 31. [CrossRef] [PubMed]
42. Pauly, M.; Gille, S.; Liu, L.; Mansoori, N.; de Souza, A.; Schultink, A.; Xiong, G. Hemicellulose biosynthesis. *Planta* **2013**, *238*, 627–642. [CrossRef] [PubMed]
43. Vega-Sánchez, M.E.; Verhertbruggen, Y.; Christensen, U.; Chen, X.; Sharma, V.; Varanasi, P.; Jobling, S.A.; Talbot, M.; White, R.G.; Joo, M. Loss of cellulose synthase-like f6 function affects mixed-linkage glucan deposition, cell wall mechanical properties, and defense responses in vegetative tissues of rice. *Plant Physiol.* **2012**, *159*, 56–69. [CrossRef] [PubMed]
44. Burton, R.A.; Fincher, G.B. Evolution and development of cell walls in cereal grains. *Front. Plant Sci.* **2014**, *5*, 456. [CrossRef] [PubMed]
45. Cosgrove, D.J. Re-constructing our models of cellulose and primary cell wall assembly. *Curr. Opin. Plant Biol.* **2014**, *22*, 122–131. [CrossRef] [PubMed]
46. Zavaliev, R.; Ueki, S.; Epel, B.L.; Citovsky, V. Biology of callose (β-1,3-glucan) turnover at plasmodesmata. *Protoplasma* **2011**, *248*, 117–130. [CrossRef] [PubMed]
47. Chen, X.-Y.; Kim, J.-Y. Callose synthesis in higher plants. *Plant Signal. Behav.* **2009**, *4*, 489–492. [CrossRef] [PubMed]
48. Müller, J.; Toev, T.; Heisters, M.; Teller, J.; Moore, K.L.; Hause, G.; Dinesh, D.C.; Bürstenbinder, K.; Abel, S. Iron-dependent callose deposition adjusts root meristem maintenance to phosphate availability. *Dev. Cell* **2015**, *33*, 216–230. [CrossRef] [PubMed]
49. Vishwanath, S.J.; Delude, C.; Domergue, F.; Rowland, O. Suberin: Biosynthesis, regulation, and polymer assembly of a protective extracellular barrier. *Plant Cell Rep.* **2015**, *34*, 573–586. [CrossRef] [PubMed]
50. Liu, Q.; Luo, L.; Zheng, L. Lignins: Biosynthesis and Biological Functions in Plants. *Int. J. Mol. Sci.* **2018**, *19*, 335. [CrossRef] [PubMed]
51. Nishimura, H.; Kamiya, A.; Nagata, T.; Katahira, M.; Watanabe, T. Direct evidence for α ether linkage between lignin and carbohydrates in wood cell walls. *Sci. Rep.* **2018**, *8*, 6538. [CrossRef] [PubMed]
52. Lee, Y.; Rubio, M.C.; Alassimone, J.; Geldner, N. A mechanism for localized lignin deposition in the endodermis. *Cell* **2013**, *153*, 402–412. [CrossRef] [PubMed]
53. Moura, J.C.M.S.; Bonine, C.A.V.; de Oliveira Fernandes Viana, J.; Dornelas, M.C.; Mazzafera, P. Abiotic and biotic stresses and changes in the lignin content and composition in plants. *J. Integr. Plant Biol.* **2010**, *52*, 360–376. [CrossRef] [PubMed]
54. Naseer, S.; Lee, Y.; Lapierre, C.; Franke, R.; Nawrath, C.; Geldner, N. Casparian strip diffusion barrier in *Arabidopsis* is made of a lignin polymer without suberin. *Proc. Natl. Acad. Sci. USA* **2012**, *109*, 10101–10106. [CrossRef] [PubMed]

55. Boyce, C.K.; Zwieniecki, M.A.; Cody, G.D.; Jacobsen, C.; Wirick, S.; Knoll, A.H.; Holbrook, N.M. Evolution of xylem lignification and hydrogel transport regulation. *Proc. Natl. Acad. Sci. USA* **2004**, *101*, 17555–17558. [CrossRef] [PubMed]
56. Lipchinsky, A. How do expansins control plant growth? A model for cell wall loosening via defect migration in cellulose microfibrils. *Acta Physiol. Plant* **2013**, *35*, 3277–3284. [CrossRef]
57. Tenhaken, R. Cell wall remodeling under abiotic stress. *Front. Plant Sci.* **2014**, *5*, 771. [CrossRef] [PubMed]
58. Wang, T.; McFarlane, H.E.; Persson, S. The impact of abiotic factors on cellulose synthesis. *J. Exp. Bot.* **2016**, *67*, 543–552. [CrossRef] [PubMed]
59. Gruber, B.D.; Giehl, R.F.H.; Friedel, S.; von Wiren, N. Plasticity of the *Arabidopsis* root system under nutrient deficiencies. *Plant Physiol.* **2013**, *163*, 161–179. [CrossRef] [PubMed]
60. Barberon, M. The endodermis as a checkpoint for nutrients. *New Phytol.* **2016**, *213*, 1604–1610. [CrossRef] [PubMed]
61. Geldner, N. The endodermis. *Annu. Rev. Plant Biol.* **2013**, *64*, 531–558. [CrossRef] [PubMed]
62. Barberon, M.; Vermeer, J.E.M.; De Bellis, D.; Wang, P.; Naseer, S.; Andersen, T.G.; Humbel, B.M.; Nawrath, C.; Takano, J.; Salt, D.E. Adaptation of Root Function by Nutrient-Induced Plasticity of Endodermal Differentiation. *Cell* **2016**, *164*, 447–459. [CrossRef] [PubMed]
63. Robbins, N.E.; Trontin, C.; Duan, L.; Dinneny, J.R. Beyond the Barrier: Communication in the Root through the Endodermis. *Plant Physiol.* **2014**, *166*, 551–559. [CrossRef] [PubMed]
64. Andersen, T.G.; Naseer, S.; Ursache, R.; Wybouw, B.; Smet, W.; de Rybel, B.; Vermeer, J.E.M.; Geldner, N. Diffusible repression of cytokinin signalling produces endodermal symmetry and passage cells. *Nature* **2018**, *555*, 529–533. [CrossRef] [PubMed]
65. Landi, S.; Esposito, S. Nitrate uptake affects cell wall synthesis and modeling. *Front. Plant Sci.* **2017**, *8*, 3289. [CrossRef] [PubMed]
66. Menz, J.; Li, Z.; Schulze, W.X.; Ludewig, U. Early nitrogen-deprivation responses in *Arabidopsis* roots reveal distinct differences on transcriptome and (phospho-) proteome levels between nitrate and ammonium nutrition. *Plant J.* **2016**, *88*, 717–734. [CrossRef] [PubMed]
67. Sun, C.-H.; Yu, J.-Q.; Hu, D.-G. Nitrate: A crucial signal during lateral roots development. *Front. Plant Sci.* **2017**, *8*, 485. [CrossRef] [PubMed]
68. Zhang, W.; Wu, L.; Ding, Y.; Yao, X.; Wu, X.; Weng, F.; Li, G.; Liu, Z.; Tang, S.; Ding, C. Nitrogen fertilizer application affects lodging resistance by altering secondary cell wall synthesis in japonica rice (*Oryza sativa*). *J. Plant Res.* **2017**, *130*, 859–871. [CrossRef] [PubMed]
69. Wang, C.; Ruan, R.W.; Yuan, X.H.; Hu, D.; Yang, H.; Li, Y.; Yi, Z.L. Effects of nitrogen fertilizer and planting density on the lignin synthesis in the culm in relation to lodging resistance of buckwheat. *Plant Prod. Sci.* **2015**, *18*, 218–227. [CrossRef]
70. Matsuyama, N. The effect of ample nitrogen fertilizer on cell-wall materials and its significance to rice blast disease. *Jpn. J. Phytopathol.* **1975**, *41*, 56–61. [CrossRef]
71. Midorikawa, K.; Kuroda, M.; Terauchi, K.; Hoshi, M.; Ikenaga, S.; Ishimaru, Y.; Abe, K.; Asakura, T. Additional nitrogen fertilization at heading time of rice down-regulates cellulose synthesis in seed endosperm. *PLoS ONE* **2014**, *9*, e98738. [CrossRef] [PubMed]
72. Zhang, J.; Xu, L.; Wang, F.; Deng, M.; Yi, K. Modulating the root elongation by phosphate/nitrogen starvation in an OsGLU3 dependant way in rice. *Plant Signal. Behav.* **2012**, *7*, 1144–1145. [CrossRef] [PubMed]
73. Hermans, C.; Porco, S.; Verbruggen, N.; Bush, D.R. Chitinase-like protein ctl1 plays a role in altering root system architecture in response to multiple environmental conditions. *Plant Physiol.* **2010**, *152*, 904–917. [CrossRef] [PubMed]
74. Wege, S.; Gilliham, M.; Henderson, S.W. Chloride: Not simply a "cheap osmoticum," but a beneficial plant macronutrient. *J. Exp. Bot.* **2017**, *68*, 3057–3069. [CrossRef] [PubMed]
75. Hoehenwarter, W.; Mönchgesang, S.; Neumann, S.; Majovsky, P.; Abel, S.; Müller, J. Comparative expression profiling reveals a role of the root apoplast in local phosphate response. *BMC Plant Biol.* **2016**, *16*, 106. [CrossRef] [PubMed]
76. Misson, J.; Raghothama, K.G.; Jain, A.; Jouhet, J.; Block, M.A.; Bligny, R.; Ortet, P.; Creff, A.; Somerville, S.; Rolland, N. A genome-wide transcriptional analysis using *Arabidopsis thaliana* Affymetrix gene chips determined plant responses to phosphate deprivation. *Proc. Natl. Acad. Sci. USA* **2005**, *102*, 11934–11939. [CrossRef] [PubMed]

77. Wu, P.; Ma, L.; Hou, X.; Wang, M.; Wu, Y.; Liu, F.; Deng, X.W. Phosphate starvation triggers distinct alterations of genome expression in *Arabidopsis* roots and leaves. *Plant Physiol.* **2003**, *132*, 1260–1271. [CrossRef] [PubMed]
78. Richmond, T.A.; Somerville, C.R. The cellulose synthase superfamily. *Plant Physiol.* **2000**, *124*, 495–498. [CrossRef] [PubMed]
79. Linkohr, B.I.; Williamson, L.C.; Fitter, A.H.; Leyser, H.M.O. Nitrate and phosphate availability and distribution have different effects on root system architecture of *Arabidopsis*. *Plant J.* **2002**, *29*, 751–760. [CrossRef] [PubMed]
80. Ward, J.T.; Lahner, B.; Yakubova, E.; Salt, D.E.; Raghothama, K.G. The Effect of Iron on the Primary Root Elongation of *Arabidopsis* during Phosphate Deficiency. *Plant Physiol.* **2008**, *147*, 1181–1191. [CrossRef] [PubMed]
81. Balzergue, C.; Dartevelle, T.; Godon, C.; Laugier, E.; Meisrimler, C.; Teulon, J.-M.; Creff, A.; Bissler, M.; Brouchoud, C.; Hagège, A. Low phosphate activates STOP1-ALMT1 to rapidly inhibit root cell elongation. *Nat Commun.* **2017**, *8*, 15300. [CrossRef] [PubMed]
82. Mora-Macías, J.; Ojeda-Rivera, J.O.; Gutiérrez-Alanís, D.; Yong-Villalobos, L.; Oropeza-Aburto, A.; Raya-González, J.; Jiménez-Domínguez, G.; Chávez-Calvillo, G.; Rellán-Álvarez, R.; Herrera-Estrella, L. Malate-dependent Fe accumulation is a critical checkpoint in the root developmental response to low phosphate. *Proc. Natl. Acad. Sci. USA* **2017**, *114*, E3563–E3572. [CrossRef] [PubMed]
83. Vatén, A.; Dettmer, J.; Wu, S.; Stierhof, Y.-D.; Miyashima, S.; Yadav, S.R.; Roberts, C.J.; Campilho, A.; Bulone, V.; Lichtenberger, R. Callose biosynthesis regulates symplastic trafficking during root development. *Dev. Cell* **2011**, *21*, 1144–1155.
84. Guerriero, G.; Hausman, J.-F.; Cai, G. No Stress! Relax! Mechanisms Governing Growth and Shape in Plant Cells. *Int. J. Mol. Sci.* **2014**, *15*, 5094–5114. [CrossRef] [PubMed]
85. Karnik, R.; Waghmare, S.; Zhang, B.; Larson, E.; Lefoulon, C.; Gonzalez, W.; Blatt, M.R. Commandeering channel voltage sensors for secretion, cell turgor, and volume control. *Trends Plant Sci.* **2017**, *22*, 81–95. [CrossRef] [PubMed]
86. Nakamura, Y. Phosphate starvation and membrane lipid remodeling in seed plants. *Prog. Lipid Res.* **2013**, *52*, 43–50. [CrossRef] [PubMed]
87. Helms, J.B.; Zurzolo, C. Lipids as targeting signals: lipid rafts and intracellular trafficking. *Traffic* **2004**, *5*, 247–254. [CrossRef] [PubMed]
88. Karunakaran, S.; Sasser, T.; Rajalekshmi, S.; Fratti, R.A. SNAREs, HOPS and regulatory lipids control the dynamics of vacuolar actin during homotypic fusion in S. cerevisiae. *J. Cell Sci.* **2012**, *125*, 1683–1692. [CrossRef] [PubMed]
89. Singh, A.P.; Fridman, Y.; Holland, N.; Ackerman-Lavert, M.; Zananiri, R.; Jaillais, Y.; Henn, A.; Savaldi-Goldstein, S. Interdependent nutrient availability and steroid hormone signals facilitate root growth plasticity. *Dev. Cell* **2018**, *46*, 59–72. [CrossRef] [PubMed]
90. Yin, Y.; Wang, Z.-Y.; Mora-Garcia, S.; Li, J.; Yoshida, S.; Asami, T.; Chory, J. BES1 accumulates in the nucleus in response to brassinosteroids to regulate gene expression and promote stem elongation. *Cell* **2002**, *109*, 181–191. [CrossRef]
91. Singh, A.P.; Fridman, Y.; Friedlander-Shani, L.; Tarkowska, D.; Strnad, M.; Savaldi-Goldstein, S. Activity of the brassinosteroid transcription factors BRASSINAZOLE RESISTANT1 and BRASSINOSTEROID INSENSITIVE1-ETHYL METHANESULFONATE-SUPPRESSOR1/BRASSINAZOLE RESISTANT2 blocks developmental reprogramming in response to low phosphate availability. *Plant Physiol.* **2014**, *166*, 678–688. [CrossRef] [PubMed]
92. Yu, X.; Li, L.; Zola, J.; Aluru, M.; Ye, H.; Foudree, A.; Guo, H.; Anderson, S.; Aluru, S.; Liu, P. A brassinosteroid transcriptional network revealed by genome-wide identification of BESI target genes in *Arabidopsis thaliana*. *Plant J.* **2011**, *65*, 634–646. [CrossRef] [PubMed]
93. Sánchez-Rodríguez, C.; Ketelaar, K.; Schneider, R.; Villalobos, J.A.; Somerville, C.R.; Persson, S.; Wallace, I.S. BRASSINOSTEROID INSENSITIVE2 negatively regulates cellulose synthesis in *Arabidopsis* by phosphorylating cellulose synthase 1. *Proc. Natl. Acad. Sci. USA* **2017**, *114*, 3533–3538. [CrossRef] [PubMed]
94. Xie, L.; Yang, C.; Wang, X. Brassinosteroids can regulate cellulose biosynthesis by controlling the expression of CESA genes in *Arabidopsis*. *J. Exp. Bot.* **2011**, *62*, 4495–4506. [CrossRef] [PubMed]

95. Catterou, M.; Dubois, F.; Schaller, H.; Aubanelle, L.; Vilcot, B.; Sangwan-Norreel, B.S.; Sangwan, R.S. Brassinosteroids, microtubules and cell elongation in *Arabidopsis thaliana*. II. Effects of brassinosteroids on microtubules and cell elongation in the *bul1* mutant. *Planta* **2001**, *212*, 673–683. [CrossRef] [PubMed]
96. Zhang, J.-W.; Xu, L.; Wu, Y.-R.; Chen, X.-A.; Liu, Y.; Zhu, S.-H.; Ding, W.-N.; Wu, P.; Yi, K.-K. OsGLU3, a putative membrane-bound endo-1,4-beta-glucanase, is required for root cell elongation and division in rice (*Oryza sativa* L.). *Mol. Plant* **2012**, *5*, 176–186. [CrossRef] [PubMed]
97. Jin, C.; Fang, C.; Yuan, H.; Wang, S.; Wu, Y.; Liu, X.; Zhang, Y.; Luo, J. Interaction between carbon metabolism and phosphate accumulation is revealed by a mutation of a cellulose synthase-like protein, CSLF6. *J. Exp. Bot.* **2015**, *66*, 2557–2567. [CrossRef] [PubMed]
98. Kiemle, S.N.; Zhang, X.; Esker, A.R.; Toriz, G.; Gatenholm, P.; Cosgrove, D.J. Role of (1,3)(1,4)-β-glucan in cell walls: Interaction with cellulose. *Biomacromolecules* **2014**, *15*, 1727–1736. [CrossRef] [PubMed]
99. Ziegler, J.; Schmidt, S.; Chutia, R.; Müller, J.; Böttcher, C.; Strehmel, N.; Scheel, D.; Abel, S. Non-targeted profiling of semi-polar metabolites in *Arabidopsis* root exudates uncovers a role for coumarin secretion and lignification during the local response to phosphate limitation. *J. Exp. Bot.* **2016**, *67*, 1421–1432. [CrossRef] [PubMed]
100. Ellis, C.; Karafyllidis, I.; Wasternack, C.; Turner, J.G. The *Arabidopsis* mutant *cev1* links cell wall signaling to jasmonate and ethylene responses. *Plant Cell* **2002**, *14*, 1557–1566. [CrossRef] [PubMed]
101. Caño-Delgado, A.; Penfield, S.; Smith, C.; Catley, M.; Bevan, M. Reduced cellulose synthesis invokes lignification and defense responses in *Arabidopsis thaliana*. *Plant J.* **2003**, *34*, 351–362. [CrossRef] [PubMed]
102. Khan, G.A.; Vogiatzaki, E.; Glauser, G.; Poirier, Y. Phosphate deficiency induces the jasmonate pathway and enhances resistance to insect herbivory. *Plant Physiol.* **2016**, *171*, 632–644. [CrossRef] [PubMed]
103. Song, L.; Liu, D. Ethylene and plant responses to phosphate deficiency. *Front. Plant Sci.* **2015**, *6*, 330. [CrossRef] [PubMed]
104. Lodish, H.; Berk, A.; Zipursky, S.L.; Matsudaira, P.; Baltimore, D.; Darnell, J. The dynamic plant cell wall. In *Molecular Cell Biology*, 4th ed.; W. H. Freeman: New York, NY, USA, 2000.
105. Grant, G.T.; Morris, E.R.; Rees, D.A.; Smith, P.J.C.; Thom, D. Biological interactions between polysaccharides and divalent cations: The egg-box model. *FEBS Lett.* **1973**, *32*, 195–198. [CrossRef]
106. Chebli, Y.; Geitmann, A. Cellular growth in plants requires regulation of cell wall biochemistry. *Curr. Opin. Cell Biol.* **2017**, *44*, 28–35. [CrossRef] [PubMed]
107. Abel, S. Phosphate scouting by root tips. *Curr. Opin. Plant Biol.* **2017**, *39*, 168–177. [CrossRef] [PubMed]
108. Wolf, S.; van der Does, D.; Ladwig, F.; Sticht, C.; Kolbeck, A.; Schürholz, A.-K.; Augustin, S.; Keinath, N.; Rausch, T.; Greiner, S. A receptor-like protein mediates the response to pectin modification by activating brassinosteroid signaling. *Proc. Natl. Acad. Sci. USA* **2014**, *111*, 15261–15266. [CrossRef] [PubMed]
109. O'Neill, M.A.; Ishii, T.; Albersheim, P.; Darvill, A.G. Rhamnogalacturonan II: Structure and function of a borate cross-linked cell wall pectic polysaccharide. *Annu. Rev. Plant Biol.* **2004**, *55*, 109–139. [CrossRef] [PubMed]
110. Taylor-Teeples, M.; Lin, L.; de Lucas, M.; Turco, G.; Toal, T.W.; Gaudinier, A.; Young, N.F.; Trabucco, G.M.; Veling, M.T.; Lamothe, R.; et al. An *Arabidopsis* gene regulatory network for secondary cell wall synthesis. *Nature* **2015**, *517*, 571–575. [CrossRef] [PubMed]
111. Kellermeier, F.; Armengaud, P.; Seditas, T.J.; Danku, J.; Salt, D.E.; Amtmann, A. Analysis of the root system architecture of *Arabidopsis* provides a quantitative readout of crosstalk between nutritional signals. *Plant Cell* **2014**, *26*, 1480–1496. [CrossRef] [PubMed]

MDPI
St. Alban-Anlage 66
4052 Basel
Switzerland
Tel. +41 61 683 77 34
Fax +41 61 302 89 18
www.mdpi.com

International Journal of Molecular Sciences Editorial Office
E-mail: ijms@mdpi.com
www.mdpi.com/journal/ijms

MDPI

www.ingramcontent.com/pod-product-compliance
Lightning Source LLC
Chambersburg PA
CBHW041941260926
51850CB00018B/415
9783036575902